VOLUME ONE HUNDRED AND EIGHTEEN

INTERNATIONAL REVIEW OF NEUROBIOLOGY

Neuroimmune Signaling in Drug Actions and Addictions

INTERNATIONAL REVIEW OF NEUROBIOLOGY

VOLUME 118

SERIES EDITORS

R. ADRON HARRIS

Waggoner Center for Alcohol and Drug Addiction Research
The University of Texas at Austin
Austin, Texas, USA

PETER JENNER

Division of Pharmacology and Therapeutics
GKT School of Biomedical Sciences
King's College, London, UK

VOLUME ONE HUNDRED AND EIGHTEEN

INTERNATIONAL REVIEW OF NEUROBIOLOGY

Neuroimmune Signaling in Drug Actions and Addictions

Edited by

CHANGHAI CUI
National Institute on Alcohol Abuse and Alcoholism, National Institutes of Health, Bethesda, Maryland, USA

DAVID SHURTLEFF
National Center for Complementary & Alternative Medicine, National Institutes of Health, Bethesda, Maryland, USA

R. ADRON HARRIS
Waggoner Center for Alcohol and Addiction Research, The University of Texas at Austin, Austin, Texas, USA

ELSEVIER

AMSTERDAM • BOSTON • HEIDELBERG • LONDON
NEW YORK • OXFORD • PARIS • SAN DIEGO
SAN FRANCISCO • SINGAPORE • SYDNEY • TOKYO
Academic Press is an imprint of Elsevier

AP

Academic Press is an imprint of Elsevier
32 Jamestown Road, London NW1 7BY, UK
525 B Street, Suite 1800, San Diego, CA 92101-4495, USA
225 Wyman Street, Waltham, MA 02451, USA
The Boulevard, Langford Lane, Kidlington, Oxford OX5 1GB, UK

First edition 2014

ISBN: 978-0-12-801284-0
ISSN: 0074-7742

For information on all Academic Press publications visit our website at store.elsevier.com

CONTENTS

CONTRIBUTORS

Yuri A. Blednov
Waggoner Center for Alcohol and Addiction Research, The University of Texas at Austin, Austin, Texas, USA

Guy A. Cabral
Department of Microbiology and Immunology, School of Medicine, Virginia Commonwealth University, Richmond, Virginia, USA

Sulie L. Chang
Institute of NeuroImmune Pharmacology, and Department of Biological Sciences, Seton Hall University, South Orange, New Jersey, USA

Lucy G. Chastain
Endocrinology Program, Department of Animal Sciences, Rutgers, The State University of New Jersey, New Brunswick, New Jersey, USA

Kaitlyn P. Connaghan
Institute of NeuroImmune Pharmacology, Seton Hall University, South Orange, New Jersey, USA

Fulton T. Crews
Bowles Center for Alcohol Studies, University of North Carolina at Chapel Hill, Chapel Hill, North Carolina, USA

Changhai Cui
National Institute on Alcohol Abuse and Alcoholism, Bethesda, Maryland, USA

Paul D. Drew
Department of Neurobiology and Developmental Sciences, University of Arkansas for Medical Sciences, Little Rock, Arkansas, USA

Laura B. Ferguson
Waggoner Center for Alcohol and Addiction Research, and The Institute for Neuroscience, The University of Texas at Austin, Austin, Texas, USA

R. Adron Harris
Waggoner Center for Alcohol and Addiction Research, The University of Texas at Austin, Austin, Texas, USA

Kurt F. Hauser
Department of Pharmacology & Toxicology, Institute for Drug and Alcohol Studies, Virginia Commonwealth University, Richmond, Virginia, USA

Keith Heinzerling
Department of Family Medicine, University of California, Los Angeles, Los Angeles, California, USA

Mark R. Hutchinson
Discipline of Physiology, School of Medical Sciences, University of Adelaide, Adelaide, South Australia, Australia

Jonathan Henry W. Jacobsen
Discipline of Pharmacology, School of Medical Sciences, University of Adelaide, Adelaide, South Australia, Australia

Melissa Jamerson
Department of Microbiology and Immunology, School of Medicine, Virginia Commonwealth University, Richmond, Virginia, USA

Aaron Janowsky
Research & Development Service, Portland VA Medical Center; Department of Psychiatry, Oregon Health & Science University, School of Medicine; Methamphetamine Abuse Research Center, and Department of Behavioral Neuroscience, Oregon Health & Science University, Portland, Oregon, USA

Cynthia J.M. Kane
Department of Neurobiology and Developmental Sciences, University of Arkansas for Medical Sciences, Little Rock, Arkansas, USA

Pamela E. Knapp
Department of pharmacology & Toxicology, and Department of Anatomy & Neurobiology, Institute for Drug and Alcohol Studies, Virginia Commonwealth University, Richmond, Virginia, USA

Ming D. Li
Department of Psychiatry and Neurobehavioral Sciences, University of Virginia, Charlottesville, Virginia, USA

Dora Lippai
2nd Department of Internal Medicine, Semmelweis University, Budapest, Hungary

Jennifer M. Loftis
Research & Development Service, Portland VA Medical Center; Department of Psychiatry, Oregon Health & Science University, School of Medicine, and Methamphetamine Abuse Research Center, Oregon Health & Science University, Portland, Oregon, USA

Jody Mayfield
Waggoner Center for Alcohol and Addiction Research, The University of Texas at Austin, Austin, Texas, USA

Olimpia Meucci
Department of Pharmacology and Physiology, and Department of Microbiology and Immunology, Drexel University College of Medicine, Philadelphia, Pennsylvania, USA

Dana Most
Waggoner Center for Alcohol and Addiction Research, and The Institute for Neuroscience, The University of Texas at Austin, Austin, Texas, USA

Bradley Nash
Department of Pharmacology and Physiology, Drexel University College of Medicine, Philadelphia, Pennsylvania, USA

Lara A. Ray
Department of Psychology, University of California, Los Angeles, Los Angeles, California, USA

Gizelle Robinson
Waggoner Center for Alcohol and Addiction Research, The University of Texas at Austin, Austin, Texas, USA

Daniel J.O. Roche
Department of Psychology, University of California, Los Angeles, Los Angeles, California, USA

Dipak K. Sarkar
Endocrinology Program, Department of Animal Sciences, Rutgers, The State University of New Jersey, New Brunswick, New Jersey, USA

Steve Shoptaw
Department of Family Medicine, University of California, Los Angeles, Los Angeles, California, USA

David Shurtleff
National Center for Complementary & Alternative Medicine, Bethesda, Maryland, USA

Gyongyi Szabo
Department of Medicine, University of Massachusetts Medical School, Worcester, Massachusetts, USA

Ryan P. Vetreno
Bowles Center for Alcohol Studies, University of North Carolina at Chapel Hill, Chapel Hill, North Carolina, USA

Linda R. Watkins
Department of Psychology and Centre for Neuroscience, University of Colorado at Boulder, Boulder, Colorado, USA

Yufeng Wei
Institute of NeuroImmune Pharmacology, Seton Hall University, South Orange, New Jersey, USA

CHAPTER ONE

Neuroimmune Mechanisms of Alcohol and Drug Addiction

Changhai Cui*[,1], David Shurtleff[†], R. Adron Harris[‡]
*National Institute on Alcohol Abuse and Alcoholism, Bethesda, Maryland, USA
[†]National Center for Complementary & Alternative Medicine, Bethesda, Maryland, USA
[‡]Waggoner Center for Alcohol and Addiction Research, The University of Texas at Austin, Austin, Texas, USA
[1]Corresponding author: e-mail address: changhai.cui@nih.gov

Contents

Abstract

Alcohol and other drugs of abuse have significant impacts on the neuroimmune system. Studies have demonstrated that drugs of abuse interact with the neuroimmune system and alter neuroimmune gene expression and signaling, which in turn contribute to various aspects of addiction. As the key component of the CNS immune system, neuroimmune factors mediate neuroinflammation and modulate a wide range of brain function including neuronal activity, endocrine function, and CNS development. These neuromodulatory properties of immune factors, together with their essential role in neuroinflammation, provide a new framework to understand neuroimmune mechanisms mediating brain functional and behavioral changes contributing to addiction. This chapter highlights recent advances in understanding neuroimmune changes associated with exposure to alcohol and other drugs of abuse, including opiates, marijuana, methamphetamine, and cocaine. It provides a brief overview on what we know about neuroimmune signaling and its role in drug action and addiction.

International Review of Neurobiology, Volume 118
ISSN 0074-7742
http://dx.doi.org/10.1016/B978-0-12-801284-0.00001-4

1. INTRODUCTION

Alcohol and other drugs of abuse have profound impacts on a variety of neurobiology systems that are related to reward, stress, habit formation, and decision making, which accounts for the reinforcing and addictive properties of these drugs (Koob & Volkow, 2010). The burgeoning field of neuroimmune research has shown that the neuroimmune system modulates a variety of brain function and behaviors related to alcohol and drug addiction. Neuroimmune signaling acts in concert with neurotransmitter and neuropeptide systems that has far-reaching impact on normal brain function and dysfunction involving neurodegenerative diseases, neuropsychiatric disorders, and addiction (Deverman & Patterson, 2009; Frank, Watkins, & Maier, 2011; Haroon, Raison, & Miller, 2012; Mayfield, Ferguson, & Harris, 2013; Rogers, Mastroeni, Leonard, Joyce, & Grover, 2007; Stertz, Magalhaes, & Kapczinski, 2013). In addition to their primary role in mediating neuroinflammation, neuroimmune factors, such as cytokines and chemokines, are critical for a variety of brain functions. Expressed in neurons and glia, these molecules regulate synaptic function, mediate neuron–glia communication (Boulanger, 2009), interact with neuroendocrine and neuropeptide systems, and regulate neurogenesis and CNS development. These findings offer new opportunities and a framework for exploring and understanding the role of the neuroimmune system in addiction.

2. NEUROIMMUNE MODULATION OF SYNAPTIC FUNCTION

Many immune molecules interact with neurotransmitter systems and play essential roles in modulating synaptic function. As a family of G-protein-coupled receptor systems, binding of chemokines to chemokine receptors triggers a cascade of signaling events, which subsequently modulate neurotransmitter release and activities of many receptors and channels. For example, chemokines CCL2 and CXCL-12 regulate the release of several neurotransmitters, including glutamate, GABA, and dopamine (Heinisch & Kirby, 2010; Rostene, Kitabgi, & Parsadaniantz, 2007). The chemokine receptor CCR2 cross-desensitizes GABAa and mu-opioid receptors (Rostene et al., 2007). In addition, activation of CXCR4 by its ligand CXCL-12 triggers the release of glutamate from astrocytes (Cali, Marchaland, Regazzi, & Bezzi, 2008). Studies also demonstrated that proinflammatory cytokines regulate synaptic transmission and plasticity, and contribute to the maintenance of the homeostasis of neuronal networks. For example, TNFα differentially modulates trafficking of AMPA-type glutamate receptors and GABA receptors (Pribiag & Stellwagen, 2013, 2014; Stellwagen & Malenka, 2006). It regulates the synaptic strength by facilitating excitatory synaptic transmission, while decreasing inhibitory synaptic transmission. IL-1β signaling modulates long-term potentiation (Avital et al., 2003; Mori et al., 2014). Type I MHC regulates neural development and activity-dependent synaptic function (Shatz, 2009). In addition, as the key component of the neuroimmune system, microglia dynamically detect the brain environment, even at the resting state, and contribute to postnatal development, neuroplasticity, and circuit function (Kettenmann, Kirchhoff, & Verkhratsky, 2013; Parkhurst et al., 2013; Tremblay et al., 2011; Wake, Moorhouse, Miyamoto, & Nabekura, 2013). It becomes clear that the bidirectional communication between neuron and microglia plays important roles in both normal brain function and neurobiological diseases (Kettenmann et al., 2013; Miyamoto, Wake, Moorhouse, & Nabekura, 2013; Pannell, Szulzewsky, Matyash, Wolf, & Kettenmann, 2014; Schafer, Lehrman, & Stevens, 2013). Thus, the neuroimmune system modulates synaptic functions by presynaptic, postsynaptic, and neural–glial mechanisms. Such actions of the neuroimmune system offer potential neuroimmune mechanisms for brain functional changes associated with alcohol and drug abuse that can alter neuroimmune signaling.

3. NEUROINFLAMMATION

Upon insult by environmental toxins or neuronal damage, microglia release a variety of neuroimmune factors exerting either neuroprotective or neurotoxic effects (Rivest, 2009). At the initial stage of the innate immune response, TNFα and IL1β are the two main cytokines that are produced by microglia. They exert neuroprotective effects by promoting the maturation of oligodendrocytes and increasing the secretion of neurotrophines. However, overactivated microglia releases numerous proinflammatory cytokines, chemokines, and inducible nitric oxide synthase, which synergistically mediate neuroinflammation. To counterbalance neuroinflammation, the brain produces antiinflammatory factors, such as IL-10 and transforming growth factor-b1, to inhibit inflammatory responses. It is important to note that neuroinflammation may not only be provoked by pathological conditions but also can be trigged by increased neuronal activities, such as those associated with noxious stimuli, psychological stress, and epileptic seizures (Xanthos & Sandkuhler, 2014). In addition, neuroimmune factors mediate neurotoxicity through various other mechanisms. Notably, chemokine receptors CXCR4 and CCR5 are important mediators of HIV-associated neurotoxicity (Kaul, Ma, Medders, Desai, & Lipton, 2007).

4. NEUROIMMUNE MOLECULES IN NEURODEVELOPMENT

Neuroimmune molecules are involved in all stages of neurodevelopment. They are expressed in both the developing and adult brain and play important roles in neuro- and gliogenesis, neuronal migration, axonal path finding, and sculpt neurocircuits (Guyon & Nahon, 2007; Paolicelli et al., 2011; Schafer et al., 2012). Dysregulation of CNS immune molecules at the early stage of brain development causes significant behavioral deficits, which are evident by the increased risk of several neurological disorders (Bilbo & Schwarz, 2012; Canetta & Brown, 2012; Garay & McAllister, 2010). Cytokines and chemokines play diverse roles in embryonic brain development and adult neurogenesis. For example, gp130 family cytokines and TNFα regulate neurogenesis, gliogenesis, and neuronal survival in the embryonic brain (Deverman & Patterson, 2009). Chemokine CXCL-12 is considered an indispensable chemoattractant for neuronal migration and axonal path finding in the developing nervous system (Guyon & Nahon, 2007). Disruption of certain chemokine receptors causes the malformation of granule cell layers of the cerebellum, the dentate gyrus, and cortical interneurons (Lu, Grove, & Miller, 2002). In the adult brain, cytokine TNFα and IL-6 inhibit neurogenesis (Monje, Toda, & Palmer, 2003), whereas the constitutive expression of IL-1β is critical for hippocampal neurogenesis. In addition, the chemokine CXCL-12 and its receptor CXCR4 are expressed in the subventricular zone and regulate migration and proliferation of progenitor cells (Tiveron et al., 2006). Evidence suggests that alcohol exposure disrupts cytokine profile during early neuronal differentiation and influences adult neurogenesis, and alters the neuroimmune gene expression in a brain regional-dependent manner (Camarillo, Kumar, Bake, Sohrabji, & Miranda, 2007; Crews & Nixon, 2009; Kane et al., 2014). In addition, adolescent binge drinking leads to persistent upregulation of innate immune signaling in the prefrontal cortex that correlates with adult neurocognitive dysfunction (Crews & Vetreno, 2011). The essential role of neuroimmune molecules in neurodevelopment and adult neurogenesis provide potential mechanisms to understand the effects of alcohol on CNS development.

5. NEUROIMMUNE FACTORS MODULATE NEUROENDOCRINE FUNCTION

The involvement of neuroimmune molecules in the regulation of neuroendocrine function is demonstrated in animal models and by human conditions of stress and depression. Increased levels of cytokines are associated with depression and sickness behavior (Dantzer, O'Connor, Freund, Johnson, & Kelley, 2008; Irwin & Miller, 2007). The underlying molecular and cellular mechanisms of these conditions are believed to be primarily due to the dysregulation of the HPA axis, as well as serotonergic and dopaminergic systems, by neuroimmune factors. A variety of cytokines have potent effects on the HPA axis by regulating the release of neuropeptides and neurohormones, including corticotrophin release factor, ACTH, or cortisol (Hueston & Deak, 2014). Conversely, glucocorticoids induced by chronic stress have a significant impact on the neuroimmune system by regulating expressions of cytokines in the hippocampus, the prefrontal cortex, and the hypothalamus (Munhoz et al., 2006; Sorrells, Caso, Munhoz, & Sapolsky, 2009). In addition, several chemokines are expressed in the paraventricular nucleus of the hypothalamus and regulate the stress-related neuroendocrine responses, such as the release of arginine vasopressin (Callewaere et al., 2006; Callewaere, Banisadr, Rostene, & Parsadaniantz, 2007). One hallmark of alcohol and drug abuse and addiction is the dysregulated HPA axis. The ability of neuroimmune molecules in regulating the HPA axis suggests that neuroimmune molecules may play an integrative role in the close link between stress responses and addiction.

6. NEUROIMMUNE MECHANISM AND ADDICTION

Studies using animal models and postmortem human alcoholic brains suggest that alcohol exposure has a significant impact on the neuroimmune system. Expressions of several immune-related genes are altered in human alcoholic brains and are differentially correlated with the high and low alcohol consuming rodent lines (Flatscher-Bader et al., 2005; Liu et al., 2006; Mulligan et al., 2006). In addition, polymorphisms of genes encoding IL-1β and IL-1, as well as the gene for an antiinflammatory cytokine IL-10, are associated with the susceptibility to alcoholism (Marcos, Pastor, Gonzalez-Sarmiento, & Laso, 2008; Pastor, Laso, Romero, & Gonzalez-Sarmiento, 2005). Furthermore, a study using a mouse binge drinking model revealed a long-lasting increase of the chemokine CCL2 and the cytokine TNFα but a decrease of the anti-inflammatory cytokine IL-10 in the mouse brain (Qin et al., 2008). Together, these studies provide molecular and cellular evidence that ethanol alters the neuroimmune system in the brain. Recent *in vivo* animal studies provide further evidence that neuroimmune modulation contributes to alcohol dependence. Interruption of certain neuroimmune gene expression (Blednov et al., 2005, 2012) or targeted disruption of TLR4 in the central amygdala (Liu et al., 2011) reduced alcohol consumption. In addition, pharmacological suppressions of various neuroimmune signaling pathways reduce alcohol intake in different animal models (Bell et al., 2013; Mayfield et al., 2013). However, it remains largely unclear how neuroimmune alteration may contribute to alcohol dependence. Recent studies begin to shed light on this question. For example, binge and chronic alcohol exposure induce neuroimmune activation through TLRs and HMGB1 (Alfonso-Loeches, Pascual-Lucas, Blanco, Sanchez-Vera, & Guerri, 2010; Crews, Qin, Sheedy, Vetreno, & Zou, 2013); TLR4 and CD14 play an important role in the acute ethanol effects on GABAergic transmission in the central amygdala (Bajo et al., 2014), and cytokines facilitate alcohol withdrawal-induced anxiety via the CRF signaling in the central amygdala (Knapp et al., 2011; Whitman, Knapp, Werner, Crews, & Breese, 2013). In addition, human studies of alcoholics show positive correlations between alcohol craving and serum levels of cytokines and inflammatory endotoxins suggesting that activation of innate immune signaling may increase alcohol craving and consumption (Leclercq et al., 2012; Leclercq, De Saeger, Delzenne, de Timary, & Starkel, 2014). This is consistent with the animal studies discussed above where injection of lipopolysaccharide increased alcohol consumption and deletion of immune-related genes decreased consumption.

Similar to alcohol, opiate drugs interact with the central immune system and glial activation can enhance the rewarding properties of opiates such as morphine (Bland, Hutchinson, Maier, Watkins, & Johnson, 2009). Glia activation results in the release of proinflammatory cytokines and chemokines, which can affect glia-neuronal signaling, modulate neuronal activity (i.e., dopamine release) and behavioral outcomes resulting from opioid exposure. A key proinflammatory cytokine, IL-1β, has been shown to be unregulated following morphine exposure (Raghavendra, Tanga, & DeLeo, 2004) and single nucleotide polymorphisms involved in increased IL-1β production have been associated with risk for opioid dependence in humans (Liu, Hutchinson, White, Somogyi, & Coller, 2009). Toll-like receptors that play an important role in induction of innate immunity have been shown to be important for modulating glia-neuronal signaling and in the reinforcing properties of opiates (Terashvili et al., 2008). Overall, a vast array of research has clearly demonstrated opioid exposure results in a complex pattern, and cascade of neuroimmune changes and interactions with neuronal signaling that modulates opioid-induced reward, dependence, withdrawal, and analgesia.

Cocaine's effects on central immune signaling involve indirect activation of glia-specific mediators and receptors involved in glutamate homeostasis. These changes have been linked to cocaine's rewarding effects (Chiamulera et al., 2001), withdrawal and long-term cocaine-induced incubation of drug reinstatement behavior (Lu et al., 2009). In addition, brain region-specific administration of the chemokine, CXCL-12, and activation of the CXCR4 receptor have been shown to modulate the behavioral effects of cocaine (Trecki & Unterwald, 2009).

Central immune signaling has also been shown to modulate the effects methamphetamine. Pharmacological agents that reduce glial activation, such as ibudilast, have been shown to attenuate methamphetamine-induced relapse (Beardsley, Shelton, Hendrick, & Johnson, 2010). CNS exposure to methamphetamine results in activation of microglial, which has been linked to neurodegeneration through proinflammatory processes (Ehrlich et al., 1998; Gadient & Otten, 1997; McGuire et al., 2001). Evidence from animal models has shown microglia involvement in neurotoxicity associated with methamphetamine exposure, which results in damage to striatal dopaminergic terminals and, overall, a reduction in striatal dopamine (Thomas & Kuhn, 2005). Positron emission tomography imaging data have revealed significant increase in microglial cells in the brains of methamphetamine addicted individuals, further linking methamphetamine exposure to microglia activation and neurotoxicity (Sekine et al., 2008).

7. SUMMARY

Alcohol and other drugs of abuse have significant impacts on the neuroimmune system. Neuroimmune activation may contribute to addiction via a variety of mechanisms. The neuromodulatory properties of immune factors on neuronal activity, endocrine function, and CNS development provide a new framework to understand the role of neuroimmune mechanisms mediating brain functional and behavioral changes contributing to addiction. Importantly, neuroimmune signaling pathways could provide novel drug targets for the treatment of addiction (Coller & Hutchinson,

2012; Mayfield et al., 2013). The chapters included in this volume highlight recent advances in understanding neuroimmune changes associated with the drugs of abuse, including on alcohol, opioids, methamphetamine, marijuana, and cocaine. In addition to providing a critical review on neuroimmune contributions to changes in brain function and behavior associated with addiction, important reviews and perspectives are provided on the alterations of neuroimmune system in FASD, comorbidity of HIV and addictive substances, and neuroimmune therapies for the treatment of addiction. Thus, this volume provides updated and timely information and characterization of neuroimmune mechanisms, at the molecular, cellular, and system, and how the mechanism impacts on behavior contributing to alcohol and drug addiction.

REFERENCES

Alfonso-Loeches, S., Pascual-Lucas, M., Blanco, A. M., Sanchez-Vera, I., & Guerri, C. (2010). Pivotal role of TLR4 receptors in alcohol-induced neuroinflammation and brain damage. *The Journal of Neuroscience*, *30*, 8285–8295.

Avital, A., Goshen, I., Kamsler, A., Segal, M., Iverfeldt, K., Richter-Levin, G., et al. (2003). Impaired interleukin-1 signaling is associated with deficits in hippocampal memory processes and neural plasticity. *Hippocampus*, *13*, 826–834.

Bajo, M., Madamba, S. G., Roberto, M., Blednov, Y. A., Sagi, V. N., Roberts, E., et al. (2014). Innate immune factors modulate ethanol interaction with GABAergic transmission in mouse central amygdala. *Brain, Behavior, and Immunity*, *40*, 191–202.

Beardsley, P. M., Shelton, K. L., Hendrick, E., & Johnson, K. W. (2010). The glial cell modulator and phosphodiesterase inhibitor, AV411 (ibudilast), attenuates prime- and stress-induced methamphetamine relapse. *European Journal of Pharmacology*, *637*, 102–108.

Bell, R. L., Lopez, M. F., Cui, C., Egli, M., Johnson, K. W., Franklin, K. M., et al. (2013). Ibudilast reduces alcohol drinking in multiple animal models of alcohol dependence. *Addiction Biology*. http://dx.doi.org/10.1111/adb.12106. [Epub ahead of print].

Bilbo, S. D., & Schwarz, J. M. (2012). The immune system and developmental programming of brain and behavior. *Frontiers in Neuroendocrinology*, *33*, 267–286.

Bland, S. T., Hutchinson, M. R., Maier, S. F., Watkins, L. R., & Johnson, K. W. (2009). The glial activation inhibitor AV411 reduces morphine-induced nucleus accumbens dopamine release. *Brain, Behavior, and Immunity*, *23*, 492–497.

Blednov, Y. A., Bergeson, S. E., Walker, D., Ferreira, V. M., Kuziel, W. A., & Harris, R. A. (2005). Perturbation of chemokine networks by gene deletion alters the reinforcing actions of ethanol. *Behavioural Brain Research*, *165*, 110–125.

Blednov, Y. A., Ponomarev, I., Geil, C., Bergeson, S., Koob, G. F., & Harris, R. A. (2012). Neuroimmune regulation of alcohol consumption: Behavioral validation of genes obtained from genomic studies. *Addiction Biology*, *17*, 108–120.

Boulanger, L. M. (2009). Immune proteins in brain development and synaptic plasticity. *Neuron*, *64*, 93–109.

Cali, C., Marchaland, J., Regazzi, R., & Bezzi, P. (2008). SDF 1-alpha (CXCL12) triggers glutamate exocytosis from astrocytes on a millisecond time scale: Imaging analysis at the single-vesicle level with TIRF microscopy. *Journal of Neuroimmunology*, *198*, 82–91.

Callewaere, C., Banisadr, G., Desarmenien, M. G., Mechighel, P., Kitabgi, P., Rostene, W. H., et al. (2006). The chemokine SDF-1/CXCL12 modulates the firing pattern of vasopressin neurons and counteracts induced vasopressin release through CXCR4. *Proceedings of the National Academy of Sciences of the United States of America, 103*, 8221–8226.

Callewaere, C., Banisadr, G., Rostene, W., & Parsadaniantz, S. M. (2007). Chemokines and chemokine receptors in the brain: Implication in neuroendocrine regulation. *Journal of Molecular Endocrinology, 38*, 355–363.

Camarillo, C., Kumar, L. S., Bake, S., Sohrabji, F., & Miranda, R. C. (2007). Ethanol regulates angiogenic cytokines during neural development: Evidence from an in vitro model of mitogen-withdrawal-induced cerebral cortical neuroepithelial differentiation. *Alcoholism, Clinical and Experimental Research, 31*, 324–335.

Canetta, S. E., & Brown, A. S. (2012). Prenatal infection, maternal immune activation, and risk for schizophrenia. *Translational Neuroscience, 3*, 320–327.

Chiamulera, C., Epping-Jordan, M. P., Zocchi, A., Marcon, C., Cottiny, C., Tacconi, S., et al. (2001). Reinforcing and locomotor stimulant effects of cocaine are absent in mGluR5 null mutant mice. *Nature Neuroscience, 4*, 873–874.

Coller, J. K., & Hutchinson, M. R. (2012). Implications of central immune signaling caused by drugs of abuse: Mechanisms, mediators and new therapeutic approaches for prediction and treatment of drug dependence. *Pharmacology and Therapeutics, 134*, 219–245.

Crews, F. T., & Nixon, K. (2009). Mechanisms of neurodegeneration and regeneration in alcoholism. *Alcohol and Alcoholism, 44*, 115–127.

Crews, F. T., Qin, L., Sheedy, D., Vetreno, R. P., & Zou, J. (2013). High mobility group box 1/Toll-like receptor danger signaling increases brain neuroimmune activation in alcohol dependence. *Biological Psychiatry, 73*, 602–612.

Crews, F. T., & Vetreno, R. P. (2011). Addiction, adolescence, and innate immune gene induction. *Frontiers in Psychiatry, 2*, 19.

Dantzer, R., O'Connor, J. C., Freund, G. G., Johnson, R. W., & Kelley, K. W. (2008). From inflammation to sickness and depression: When the immune system subjugates the brain. *Nature Reviews. Neuroscience, 9*, 46–56.

Deverman, B. E., & Patterson, P. H. (2009). Cytokines and CNS development. *Neuron, 64*, 61–78.

Ehrlich, L. C., Hu, S., Sheng, W. S., Sutton, R. L., Rockswold, G. L., Peterson, P. K., et al. (1998). Cytokine regulation of human microglial cell IL-8 production. *Journal of Immunology, 160*, 1944–1948.

Flatscher-Bader, T., van der Brug, M., Hwang, J. W., Gochee, P. A., Matsumoto, I., Niwa, S., et al. (2005). Alcohol-responsive genes in the frontal cortex and nucleus accumbens of human alcoholics. *Journal of Neurochemistry, 93*, 359–370.

Frank, M. G., Watkins, L. R., & Maier, S. F. (2011). Stress- and glucocorticoid-induced priming of neuroinflammatory responses: Potential mechanisms of stress-induced vulnerability to drugs of abuse. *Brain, Behavior, and Immunity, 25*(Suppl. 1), S21–S28.

Gadient, R. A., & Otten, U. H. (1997). Interleukin-6 (IL-6)—A molecule with both beneficial and destructive potentials. *Progress in Neurobiology, 52*, 379–390.

Garay, P. A., & McAllister, A. K. (2010). Novel roles for immune molecules in neural development: Implications for neurodevelopmental disorders. *Frontiers in Synaptic Neuroscience, 2*, 136.

Guyon, A., & Nahon, J. L. (2007). Multiple actions of the chemokine stromal cell-derived factor-1alpha on neuronal activity. *Journal of Molecular Endocrinology, 38*, 365–376.

Haroon, E., Raison, C. L., & Miller, A. H. (2012). Psychoneuroimmunology meets neuropsychopharmacology: Translational implications of the impact of inflammation on behavior. *Neuropsychopharmacology, 37*, 137–162.

Heinisch, S., & Kirby, L. G. (2010). SDF-1alpha/CXCL12 enhances GABA and glutamate synaptic activity at serotonin neurons in the rat dorsal raphe nucleus. *Neuropharmacology, 58*, 501–514.

Hueston, C. M., & Deak, T. (2014). The inflamed axis: The interaction between stress, hormones, and the expression of inflammatory-related genes within key structures comprising the hypothalamic–pituitary–adrenal axis. *Physiology and Behavior, 124*, 77–91.
Irwin, M. R., & Miller, A. H. (2007). Depressive disorders and immunity: 20 years of progress and discovery. *Brain, Behavior, and Immunity, 21*, 374–383.
Kane, C. J., Phelan, K. D., Douglas, J. C., Wagoner, G., Johnson, J. W., Xu, J., et al. (2014). Effects of ethanol on immune response in the brain: Region-specific changes in adolescent versus adult mice. *Alcoholism, Clinical and Experimental Research, 38*, 384–391.
Kaul, M., Ma, Q., Medders, K. E., Desai, M. K., & Lipton, S. A. (2007). HIV-1 coreceptors CCR5 and CXCR4 both mediate neuronal cell death but CCR5 paradoxically can also contribute to protection. *Cell Death and Differentiation, 14*, 296–305.
Kettenmann, H., Kirchhoff, F., & Verkhratsky, A. (2013). Microglia: New roles for the synaptic stripper. *Neuron, 77*, 10–18.
Knapp, D. J., Whitman, B. A., Wills, T. A., Angel, R. A., Overstreet, D. H., Criswell, H. E., et al. (2011). Cytokine involvement in stress may depend on corticotrophin releasing factor to sensitize ethanol withdrawal anxiety. *Brain, Behavior, and Immunity, 25*(Suppl. 1), S146–S154.
Koob, G. F., & Volkow, N. D. (2010). Neurocircuitry of addiction. *Neuropsychopharmacology, 35*, 217–238.
Leclercq, S., Cani, P. D., Neyrinck, A. M., Starkel, P., Jamar, F., Mikolajczak, M., et al. (2012). Role of intestinal permeability and inflammation in the biological and behavioral control of alcohol-dependent subjects. *Brain, Behavior, and Immunity, 26*, 911–918.
Leclercq, S., De Saeger, C., Delzenne, N., de Timary, P., & Starkel, P. (2014). Role of inflammatory pathways, blood mononuclear cells, and Gut-derived bacterial products in alcohol dependence. *Biological Psychiatry*, pii: S0006-3223(14) 00097-3. http://dx.doi.org/10.1016/j.biopsych.2014.02.003. [Epub ahead of print].
Liu, L., Hutchinson, M. R., White, J. M., Somogyi, A. A., & Coller, J. K. (2009). Association of IL-1B genetic polymorphisms with an increased risk of opioid and alcohol dependence. *Pharmacogenetics and Genomics, 19*, 869–876.
Liu, J., Lewohl, J. M., Harris, R. A., Iyer, V. R., Dodd, P. R., Randall, P. K., et al. (2006). Patterns of gene expression in the frontal cortex discriminate alcoholic from nonalcoholic individuals. *Neuropsychopharmacology, 31*, 1574–1582.
Liu, J., Yang, A. R., Kelly, T., Puche, A., Esoga, C., June, H. L., Jr., et al. (2011). Binge alcohol drinking is associated with GABAA alpha2-regulated Toll-like receptor 4 (TLR4) expression in the central amygdala. *Proceedings of the National Academy of Sciences of the United States of America, 108*, 4465–4470.
Lu, M., Grove, E. A., & Miller, R. J. (2002). Abnormal development of the hippocampal dentate gyrus in mice lacking the CXCR4 chemokine receptor. *Proceedings of the National Academy of Sciences of the United States of America, 99*, 7090–7095.
Lu, L., Wang, X., Wu, P., Xu, C., Zhao, M., Morales, M., et al. (2009). Role of ventral tegmental area glial cell line-derived neurotrophic factor in incubation of cocaine craving. *Biological Psychiatry, 66*, 137–145.
Marcos, M., Pastor, I., Gonzalez-Sarmiento, R., & Laso, F. J. (2008). Interleukin-10 gene polymorphism is associated with alcoholism but not with alcoholic liver disease. *Alcohol and Alcoholism, 43*, 523–528.
Mayfield, J., Ferguson, L., & Harris, R. A. (2013). Neuroimmune signaling: A key component of alcohol abuse. *Current Opinion in Neurobiology, 23*, 513–520.
McGuire, S. O., Ling, Z. D., Lipton, J. W., Sortwell, C. E., Collier, T. J., & Carvey, P. M. (2001). Tumor necrosis factor alpha is toxic to embryonic mesencephalic dopamine neurons. *Experimental Neurology, 169*, 219–230.
Miyamoto, A., Wake, H., Moorhouse, A. J., & Nabekura, J. (2013). Microglia and synapse interactions: Fine tuning neural circuits and candidate molecules. *Frontiers in Cellular Neuroscience, 7*, 70.

Monje, M. L., Toda, H., & Palmer, T. D. (2003). Inflammatory blockade restores adult hippocampal neurogenesis. *Science, 302*, 1760–1765.
Mori, F., Nistico, R., Mandolesi, G., Piccinin, S., Mango, D., Kusayanagi, H., et al. (2014). Interleukin-1beta promotes long-term potentiation in patients with multiple sclerosis. *Neuromolecular Medicine, 16*, 38–51.
Mulligan, M. K., Ponomarev, I., Hitzemann, R. J., Belknap, J. K., Tabakoff, B., Harris, R. A., et al. (2006). Toward understanding the genetics of alcohol drinking through transcriptome meta-analysis. *Proceedings of the National Academy of Sciences of the United States of America, 103*, 6368–6373.
Munhoz, C. D., Lepsch, L. B., Kawamoto, E. M., Malta, M. B., Lima Lde, S., Avellar, M. C., et al. (2006). Chronic unpredictable stress exacerbates lipopolysaccharide-induced activation of nuclear factor-kappaB in the frontal cortex and hippocampus via glucocorticoid secretion. *The Journal of Neuroscience, 26*, 3813–3820.
Pannell, M., Szulzewsky, F., Matyash, V., Wolf, S. A., & Kettenmann, H. (2014). The subpopulation of microglia sensitive to neurotransmitters/neurohormones is modulated by stimulation with LPS, interferon-gamma, and IL-4. *Glia, 62*, 667–679.
Paolicelli, R. C., Bolasco, G., Pagani, F., Maggi, L., Scianni, M., Panzanelli, P., et al. (2011). Synaptic pruning by microglia is necessary for normal brain development. *Science, 333*, 1456–1458.
Parkhurst, C. N., Yang, G., Ninan, I., Savas, J. N., Yates, J. R., 3rd., Lafaille, J. J., et al. (2013). Microglia promote learning-dependent synapse formation through brain-derived neurotrophic factor. *Cell, 155*, 1596–1609.
Pastor, I. J., Laso, F. J., Romero, A., & Gonzalez-Sarmiento, R. (2005). Interleukin-1 gene cluster polymorphisms and alcoholism in Spanish men. *Alcohol and Alcoholism, 40*, 181–186.
Pribiag, H., & Stellwagen, D. (2013). TNF-alpha downregulates inhibitory neurotransmission through protein phosphatase 1-dependent trafficking of GABA(A) receptors. *The Journal of Neuroscience, 33*, 15879–15893.
Pribiag, H., & Stellwagen, D. (2014). Neuroimmune regulation of homeostatic synaptic plasticity. *Neuropharmacology, 78*, 13–22.
Qin, L., He, J., Hanes, R. N., Pluzarev, O., Hong, J. S., & Crews, F. T. (2008). Increased systemic and brain cytokine production and neuroinflammation by endotoxin following ethanol treatment. *Journal of Neuroinflammation, 5*, 10.
Raghavendra, V., Tanga, F. Y., & DeLeo, J. A. (2004). Attenuation of morphine tolerance, withdrawal-induced hyperalgesia, and associated spinal inflammatory immune responses by propentofylline in rats. *Neuropsychopharmacology, 29*, 327–334.
Rivest, S. (2009). Regulation of innate immune responses in the brain. *Nature Reviews. Immunology, 9*, 429–439.
Rogers, J., Mastroeni, D., Leonard, B., Joyce, J., & Grover, A. (2007). Neuroinflammation in Alzheimer's disease and Parkinson's disease: Are microglia pathogenic in either disorder? *International Review of Neurobiology, 82*, 235–246.
Rostene, W., Kitabgi, P., & Parsadaniantz, S. M. (2007). Chemokines: A new class of neuromodulator? *Nature Reviews. Neuroscience, 8*, 895–903.
Schafer, D. P., Lehrman, E. K., Kautzman, A. G., Koyama, R., Mardinly, A. R., Yamasaki, R., et al. (2012). Microglia sculpt postnatal neural circuits in an activity and complement-dependent manner. *Neuron, 74*, 691–705.
Schafer, D. P., Lehrman, E. K., & Stevens, B. (2013). The "quad-partite" synapse: Microglia-synapse interactions in the developing and mature CNS. *Glia, 61*, 24–36.
Sekine, Y., Ouchi, Y., Sugihara, G., Takei, N., Yoshikawa, E., Nakamura, K., et al. (2008). Methamphetamine causes microglial activation in the brains of human abusers. *The Journal of Neuroscience, 28*, 5756–5761.
Shatz, C. J. (2009). MHC class I: An unexpected role in neuronal plasticity. *Neuron, 64*, 40–45.

Sorrells, S. F., Caso, J. R., Munhoz, C. D., & Sapolsky, R. M. (2009). The stressed CNS: When glucocorticoids aggravate inflammation. *Neuron, 64*, 33–39.

Stellwagen, D., & Malenka, R. C. (2006). Synaptic scaling mediated by glial TNF-alpha. *Nature, 440*, 1054–1059.

Stertz, L., Magalhaes, P. V., & Kapczinski, F. (2013). Is bipolar disorder an inflammatory condition? The relevance of microglial activation. *Current Opinion in Psychiatry, 26*, 19–26.

Terashvili, M., Wu, H. E., Schwasinger, E. T., Hung, K. C., Hong, J. S., & Tseng, L. F. (2008). (+)-Morphine attenuates the (−)-morphine-produced conditioned place preference and the mu-opioid receptor-mediated dopamine increase in the posterior nucleus accumbens of the rat. *European Journal of Pharmacology, 587*, 147–154.

Thomas, D. M., & Kuhn, D. M. (2005). Attenuated microglial activation mediates tolerance to the neurotoxic effects of methamphetamine. *Journal of Neurochemistry, 92*, 790–797.

Tiveron, M. C., Rossel, M., Moepps, B., Zhang, Y. L., Seidenfaden, R., Favor, J., et al. (2006). Molecular interaction between projection neuron precursors and invading interneurons via stromal-derived factor 1 (CXCL12)/CXCR4 signaling in the cortical subventricular zone/intermediate zone. *The Journal of Neuroscience, 26*, 13273–13278.

Trecki, J., & Unterwald, E. M. (2009). Modulation of cocaine-induced activity by intracerebral administration of CXCL12. *Neuroscience, 161*, 13–22.

Tremblay, M. E., Stevens, B., Sierra, A., Wake, H., Bessis, A., & Nimmerjahn, A. (2011). The role of microglia in the healthy brain. *The Journal of Neuroscience, 31*, 16064–16069.

Wake, H., Moorhouse, A. J., Miyamoto, A., & Nabekura, J. (2013). Microglia: Actively surveying and shaping neuronal circuit structure and function. *Trends in Neurosciences, 36*, 209–217.

Whitman, B. A., Knapp, D. J., Werner, D. F., Crews, F. T., & Breese, G. R. (2013). The cytokine mRNA increase induced by withdrawal from chronic ethanol in the sterile environment of brain is mediated by CRF and HMGB1 release. *Alcoholism, Clinical and Experimental Research, 37*, 2086–2097.

Xanthos, D. N., & Sandkuhler, J. (2014). Neurogenic neuroinflammation: Inflammatory CNS reactions in response to neuronal activity. *Nature Reviews. Neuroscience, 15*, 43–53.

CHAPTER TWO

Neuroimmune Pathways in Alcohol Consumption: Evidence from Behavioral and Genetic Studies in Rodents and Humans

Gizelle Robinson*, Dana Most*,†, Laura B. Ferguson*,†, Jody Mayfield*, R. Adron Harris*,[1], Yuri A. Blednov*

*Waggoner Center for Alcohol and Addiction Research, The University of Texas at Austin, Austin, Texas, USA

†The Institute for Neuroscience, The University of Texas at Austin, Austin, Texas, USA

[1]Corresponding author: e-mail address: harris@austin.utexas.edu

Contents

Abstract

Immune or brain proinflammatory signaling has been linked to some of the behavioral effects of alcohol. Immune signaling appears to regulate voluntary ethanol intake in rodent models, and ethanol intake activates the immune system in multiple models. This bidirectional link raises the possibility that consumption increases immune signaling, which in turn further increases consumption in a feed-forward cycle. Data from animal and human studies provide overlapping support for the involvement of immune-related genes and proteins in alcohol action, and combining animal and human data is a promising approach to systematically evaluate and nominate relevant pathways. Based on rodent models, neuroimmune pathways may represent unexplored, nontraditional targets for medication development to reduce alcohol consumption and prevent relapse. Peroxisome proliferator-activated receptor agonists are one class of anti-inflammatory medications that demonstrate antiaddictive properties for alcohol and other drugs of abuse. Expression of immune-related genes is altered in animals

International Review of Neurobiology, Volume 118
ISSN 0074-7742
http://dx.doi.org/10.1016/B978-0-12-801284-0.00002-6

and humans following chronic alcohol exposure, and the regulatory influences of specific mRNAs, microRNAs, and activated cell types are areas of intense study. Ultimately, the use of multiple datasets combined with behavioral validation will be needed to link specific neuroimmune pathways to addiction vulnerability.

1. INTRODUCTION

Effects of excessive alcohol consumption on innate immune signaling have been appreciated for many years, in part because of the critical role of immune dysregulation in alcoholic liver disease (Szabo, Mandrekar, Petrasek, & Catalano, 2011). More recently, the actions of alcohol on neuroimmune function have attracted attention. This area of research likely originated from gene expression profiles of human alcoholic brain, which found an unexpected abundance of changes in immune-related genes (Lewohl et al., 2000). These results were confirmed and extended in subsequent studies of human alcoholics, and changes in neuroimmune gene expression and signaling were also found in rodent models of excessive alcohol consumption. Recent studies led to the surprising finding that neuroimmune signaling is critical for regulation of neuroplasticity and is required for learning and memory (Williamson & Bilbo, 2013; Yirmiya & Goshen, 2011). These findings support the hypothesis that actions of alcohol on neuroimmune function may be important for the development of hallmarks of dependence such as escalation of consumption, craving, tolerance, and withdrawal. This chapter reviews evidence that immune signaling regulates alcohol consumption and that excessive alcohol consumption changes neuroimmune signaling, allowing a positive feedback increase in consumption, craving, and dependence.

2. IMMUNE REGULATION OF ETHANOL CONSUMPTION AND ETHANOL REGULATION OF IMMUNE SIGNALING

Immune signaling has been shown to regulate alcohol consumption in various mouse models and drinking paradigms. Alcohol consumption was studied in mutant mice with deletions of chemokine (*Ccl2, Ccl3*) and chemokine receptor genes (*Ccr2, Ccr5*) (Blednov et al., 2005). Deletion of *Ccr2, Ccl2* (in females), and *Ccl3* reduced ethanol preference and consumption in a two-bottle choice test, and ethanol administration (i.p.) produced a stronger conditioned taste aversion in *Ccr2*-, *Ccl2*-, and *Ccl3*-knock-out mice.

Knock-out mice were used to study six other immune-related genes previously linked to alcohol consumption in a gene expression analysis in mouse brain: *B2m* (beta-2 microglobulin), *Cd14* (cluster of differentiation 14), *Il 1rn* (interleukin-1 receptor antagonist), *Il6* (interleukin 6), *Ctss* (cathepsin S), and *Ctsf* (cathepsin F) (Blednov et al., 2011). Ethanol consumption and preference were reduced in all of these knock-out mice in a 24-h two-bottle choice test with no changes in saccharin or quinine consumption. In addition, ethanol consumption was reduced in *B2m*-, *Il6*-, *Cd14*-, *Il1rn*-, and *Ctss*-knock-out mice in a limited access two-bottle choice test, and in *Il1rn* and *Ctss* knockouts in a limited access test to one bottle of ethanol only. In contrast, a transgenic mouse line overexpressing *Il6* showed increased alcohol preference in females but not in males (Harris & Blednov, 2013).

Manipulations of the Toll-like receptor 4 (TLR4)-myeloid differentiation primary response gene 88 (MyD88) signaling cascade (Fig. 2.1) also alter ethanol responses. Knockout of TLR4 and MyD88 as well as TLR4 antagonism by (+)-naloxone in mice decreased both the duration of loss of righting reflex and motor impairment recovery time induced by ethanol (Wu et al., 2012). C3H/HeJ mice are naturally TLR4 deficient and show decreased operant self-administration of ethanol compared to the control strain (C3H/HeOuJ) (Harris & Blednov, 2013); however, ethanol consumption is not changed in TLR4-knock-out mice (Pascual, Balino, Alfonso-Loeches, Aragon, & Guerri, 2011). In contrast, injection of TLR4 siRNA into the central amygdala of alcohol preferring rats reduced operant self-administration of ethanol (Liu et al., 2011). These targeted disruptions of TLR4 pathways are important for understanding how discrete immune signaling in the brain may impact consumption. TLRs are key regulators of immune activation in the CNS in response to alcohol and have well-established roles in pathogen detection and initiation of innate and adaptive immunity during infection. Future experiments are needed to determine if other TLRs also regulate behavioral responses to alcohol.

Many studies have shown that alcohol increases expression of immune-related mRNA and protein across species (summarized in Table 2.1). Increased expression of the cytokine MCP-1 (monocyte chemoattractant protein 1, also known as CCL2) and microglial markers was observed in postmortem human alcoholic brains (He & Crews, 2008). In rats, binge ethanol (intragastric injection) exposure increases DNA binding of brain nuclear factor kappa-light-chain-enhancer of activated B cells (NF-κB), cyclooxygenase 2 (COX-2) expression, and microglial activation, and ethanol also induces NF-κB binding and production of proinflammatory

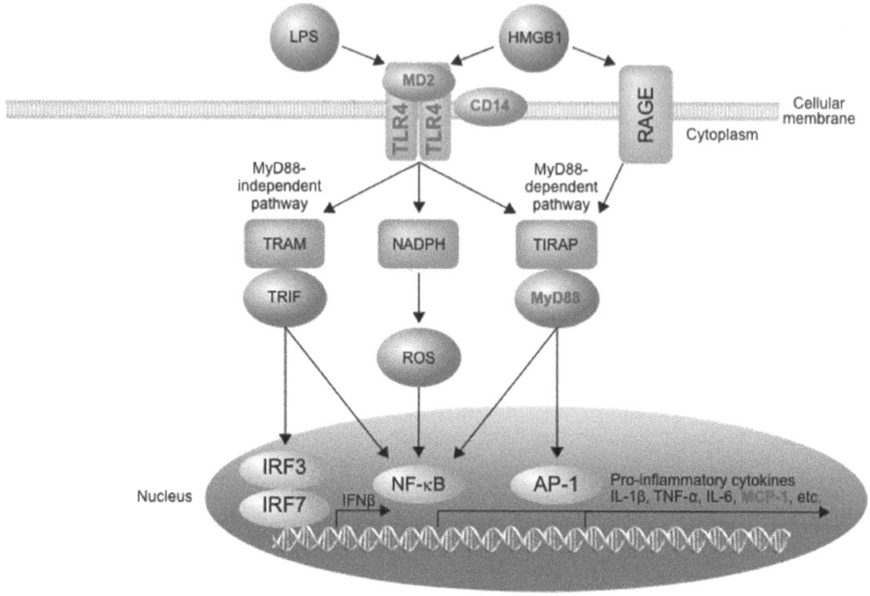

Figure 2.1 TLR4 signaling cascade. TLRs signal as dimers and heterodimers that recruit adaptor proteins such as CD14 and MD2. Depending on the adaptors recruited by the activated TLR, different pathways are triggered, all of which culminate in activation of the proinflammatory transcription factors. One pathway involves MyD88 and TIRAP and activates NF-κB via IκB kinase and also activates AP-1. Another pathway involving NADPH oxidase activates NF-κB via ROS. TRIF and TRAM signaling proteins (via MyD88-independent pathway) also initiate signal cascades, culminating in activation of NF-κB and other proinflammatory transcription factors. RAGE is another transmembrane receptor operating in innate immune cells that is known to respond to HMGB1, and this pathway also induces proinflammatory gene transcription via NF-κB activation. The release of cytokines such as TNF-α, HMGB1, IL-1β, chemokines, proteases, and ROS activate adjacent cells. These cytokines affect the brain and are thought to contribute to the etiology, progression, and persistence of alcohol addiction. Bold red font (bold grey in the print version) indicates a gene that has been manipulated and shown to affect ethanol-related behavior. NF-κB, nuclear factor-κ–light-chain-enhancer of activated B cells; MyD88, myeloid differentiation primary response gene 88; AP-1, activated protein-1; TIRAP, toll-interleukin-1 receptor (TIR) domain containing adaptor protein; ROS, reactive oxygen species; TRIF, TIR-domain-containing adaptor-inducing IFNβ; IRF, interferon regulatory factor; TRAM, TRIF-related adaptor molecule; RAGE, receptor for advanced glycation end products; TNF-α, tumor necrosis factor-α; IL-1β, interleukin-1β; IL-6, interleukin-6; MCP-1, monocyte chemoattractant protein-1.

cytokines in rat brain slice cultures (Crews et al., 2006; Zou & Crews, 2010). Chronic ethanol exposure in mice increases levels of the immune proteins CD14, TNF-α (tumor necrosis factor α), IL-1β (interleukin-1 beta), NF-κB p65, iNOS (inducible nitric oxide synthase), and COX-2 in the cortex, but

Table 2.1 Immune-related expression changes following ethanol exposure

	Increased mRNA	Increased protein	Cell-type activated
Cell cultures			
Macrophages		TNF-α, IL-1β, NO, iNOS, NF-κB p65, COX-2[a]	
Primary microglial cultures		TNF-α, IL-1β, NO iNOS, NF-κB p65, COX-2[b]	Microglial activation[b]
Primary astrocyte cultures		MyD88, iNOS, COX-2[c]	
Brain slices			
Rats	TNF-α, IL-1β, MCP-1/CCL2, iNOS[d]	NF-κB (DNA binding)[d]; HMGB1, IL-1β[e]	
Chronic ethanol*			
Mice	MCP-1/CCL2, IL-1β, TNF-α, TLR4, HMGB1[f]; TLR2, TLR9, RAGE, inflammasome components[g]	CD14, TNF-α, IL-1β, NF-κB p65, iNOS, COX-2[c]; MCP-1/CCL2, phospho-HMGB1, acetyl-HMGB1[g]	Microglia, GFAP + astrocyte[c,g]
Binge ethanol*			
Mice	TLR2, TLR3, TLR4, HMGB1[e]; MCP-1/CCL2, IL-6[h]	TLR2, TLR3, TLR4, HMGB1[e]	GFAP + astrocyte[h]
Rats	COX-2[i]	NF-κB (DNA binding)[i]	Microglial activation[i]
Postmortem brain*			
Humans	RAGE[j]	MCP-1/CCL2[k]; RAGE[j]; IL-1β, TLR2, TLR3, TLR4, HMGB1[e]	Microglia[k]

*Differences may be brain region specific. Abbreviations are defined in the text.
[a]Fernandez-Lizarbe, Pascual, Gascon, Blanco, and Guerri (2008).
[b]Fernandez-Lizarbe, Pascual, and Guerri (2009).
[c]Alfonso-Loeches, Pascual-Lucas, Blanco, Sanchez-Vera, and Guerri (2010).
[d]Zou and Crews (2010).
[e]Crews, Qin, Sheedy, Vetreno, and Zou (2013).
[f]Whitman, Knapp, Werner, Crews, and Breese (2013).
[g]Lippai et al. (2013).
[h]Kane et al. (2014).
[i]Crews et al. (2006).
[j]Vetreno, Qin, and Crews (2013).
[k]He and Crews (2008).

these changes are not observed in TLR4-knock-out mice (Alfonso-Loeches et al., 2010).

Because TLR4 deficiency protects against many alcohol-induced immune changes and normal TLR4 signaling is central to immune function, this pathway is the focus of many alcohol-immune studies. There is even evidence from studies in *Drosophila* showing that ethanol increases expression of genes in the Toll pathway (Kong et al., 2010). Lipopolysaccharide (LPS) is an endotoxin found on Gram-negative bacteria and a ligand for TLR4. In mice, LPS injections increased expression of proinflammatory cytokines (TNF-α, MCP-1, IL-1β, and NF-κB p65) in liver, brain, and serum and increased proliferation of brain microglia (Qin et al., 2008, 2007). Pretreatment with ethanol potentiated LPS-induced increases of TNF-α, MCP-1, and IL-1β (Qin et al., 2008). The cytokine increases observed in mouse brain remain elevated after peripheral levels return to normal, and a single LPS injection can increase alcohol consumption for up to 80 days (Blednov et al., 2011; Qin et al., 2008), suggesting that immune signaling in the brain can be much more persistent compared to peripheral immune activation. It is also noteworthy that chronic alcohol consumption and LPS treatment produce similar changes in mouse brain transcriptomes (Osterndorff-Kahanek, Ponomarev, Blednov, & Harris, 2013).

In the brain, both alcohol and LPS influence GABAergic transmission through immune signaling. Acute ethanol and LPS increased the amplitude of eIPSPs (evoked inhibitory postsynaptic potentials) in the central amygdala of wild-type mouse brain slices, while these increases were attenuated in CD14-knock-out mice (Bajo et al., 2014). Pretreatment with LPS-potentiated ethanol's effect on eIPSP amplitude in wild-type mice and restored the effect in CD14-knock-out mice, while (+)-naloxone blocked the LPS effect and delayed the ethanol effect on eIPSPs in wild-type mice (Bajo et al., 2014). Injection of LPS also decreased dopamine neuron firing in the ventral tegmental area that persisted for at least 1 week after treatment, which may contribute to the increased alcohol consumption seen at this time point (Blednov et al., 2011).

One hypothesis is that chronic alcohol consumption compromises the tight junctions of gut epithelial cells, allowing gut-derived bacterial toxins (e.g., LPS and peptidoglycan) to be released from the gut into the bloodstream. Once in the bloodstream, LPS or peptidoglycan binds to TLR4 receptors on liver Kupffer cells and peripheral blood mononuclear cells to initiate a signaling cascade that culminates in the activation of NF-κB or AP-1 (Fig. 2.1), which increases the transcription of proinflammatory cytokines and toxic intermediates (such as reactive oxygen species and nitric oxide (NO)). Cytokines cross the blood–brain barrier via diffusion or active transport where they promote increased alcohol consumption, which increases the production of proinflammatory cytokines, which in turn increases alcohol consumption in a positive-feedback loop that leads to excessive alcohol consumption. This hypothesis is supported by studies showing an increase in serum levels of LPS, peptidoglycan, and proinflammatory cytokines in human alcoholics compared to control subjects. Moreover, these studies are the first to reveal a correlation between the levels of bacterial products and proinflammatory molecules and alcohol craving and consumption in human alcoholics (Leclercq et al., 2012; Leclercq, De Saeger, Delzenne, de Timary, & Starkel, 2014).

Although it is unlikely that LPS crosses the blood–brain barrier, it is feasible that peripheral TLR4 activation releases endogenous ligands and proinflammatory cytokines that are able to promote immune signaling in the brain (Mayfield, Ferguson, & Harris, 2013). High-mobility group box 1 (HMGB1) is an endogenous ligand for TLR2, TLR4, TLR9, and RAGE (receptor for advanced glycation end products), which is increased following ethanol consumption in mice and is also increased in postmortem alcoholic brains (Crews et al., 2013; Whitman et al., 2013). The role of HMGB1 in alcohol-induced inflammation is discussed in more detail in Chapter 10 by Crews and Vetreno.

In addition to evidence that ethanol consumption increases HMGB1 release and activates TLR signaling, Guerri and colleagues found that ethanol induces the association of TLR4 with TLR2 in lipid rafts, triggering an inflammatory response. They demonstrated that *in vitro* treatment of murine macrophages or astrocytes with ethanol or LPS induces translocation and clustering of TLR4 and its signaling molecules in lipid rafts (Blanco, Perez-Arago, Fernandez-Lizarbe, & Guerri, 2008; Fernandez-Lizarbe et al., 2008). In microglial cultures, ethanol also promotes TLR4/TLR2 association in lipid rafts (Fernandez-Lizarbe, Montesinos, & Guerri, 2013), inducing the production of inflammatory mediators. However, in microglial cultures derived from TLR4-knock-out mice, this inflammatory response and activation of NF-κB are not seen (Fernandez-Lizarbe et al., 2013, 2009). TLR4 activation signals through two distinct pathways: the MyD88-dependent and TRIF-dependent pathways (Fig. 2.1). Ethanol and LPS can promote TLR4 endocytosis through clathrin- and caveolae-dependent pathways, and in cortical astrocytes, TRIF-dependent signaling relies on the clathrin pathway, while disruption of rafts/caveolae inhibits both the MyD88- and TRIF-dependent signaling pathways (Pascual-Lucas, Fernandez-Lizarbe, Montesinos, & Guerri, 2014).

Alcohol treatment in mice has also been shown to activate the NLRP3/ASC inflammasome, leading to IL-1β signaling in the brain (Lippai et al., 2013). Mice-consuming ethanol for 5 weeks showed increased protein levels of MCP-1, TNF-α, IL-1β, acetyl and phosphorylated HMGB1 as well as increased TLR2, TLR4, TLR9, RAGE, and HMGB1 mRNA in the cerebellum (Lippai et al., 2013). Increases in inflammasome components (NLRP1, NLRP3, NLRC4, pannexin-1), the inflammasome adaptor molecule ASC, and the effector proteins procaspase-1 and proIL-1β were found in both the cerebellum and cortex of ethanol-consuming mice compared to controls. After ethanol treatment, TLR4-knock-out mice showed no increase in TNF-α and MCP-1 and reduced levels of IL-β and inflammasome components compared to controls (Lippai et al., 2013). Future experiments will be important to determine how alcohol affects the inflammasome cascade operating in the innate immune system and the resulting actions in brain.

Another important area of study is ethanol-induced neuroimmune signaling in adolescent models and how changes in the adolescent brain could impact consumption as adults. Age-related differences in immune signaling in response to alcohol have been demonstrated in several studies. For example, adult mice given ethanol via gavage for 10 days showed increased mRNA levels of MCP-1/CCL2 in the hippocampus, cerebellum, and cortex and increased IL-6 in the cerebellum, while adolescent mice given the same treatment showed no changes in MCP-1 or IL-6 (Kane et al., 2014). Both adult and adolescent mice showed increased GFAP (glial fibrillary acidic protein) immunostaining, suggesting that astrocytes may play a role in the inflammatory response to ethanol (Kane et al., 2014). A bioinformatics analysis of adult mouse brains showed differential expression of TLR, JAK/STAT (Janus kinase/signal transducers and activators of transcription), MAPK (mitogen-activated protein kinase), T-cell, and chemokine signaling following a drinking in the dark paradigm, while these pathways were not overexpressed in adolescent mice (Agrawal et al., 2014). Also, minocycline, which has anti-inflammatory properties, reduced drinking in adult but not in adolescent mice (Agrawal et al., 2014). However, other studies have reported opposite results. Binge-like ethanol administration increased expression of TLR4, TLR2, TNF-α, and IL-1β mRNA in the prefrontal cortex of adolescent rats, while these changes were not seen in adult rats after the same treatment (Pascual, Pla, Minarro, & Guerri, 2014). These discrepancies might be attributed to differences in drinking paradigms and brain regions. In human postmortem alcoholic brains, there is an increase in protein and mRNA expression of RAGE in the orbitofrontal cortex, and a younger age of drinking onset correlated with increased expression of RAGE, TLR4, and HMGB1 (Vetreno et al., 2013). In rats, RAGE expression declines in the prefrontal cortex with age, and this decline is reduced by adolescent binge alcohol exposure (Vetreno et al., 2013). Adolescent binge ethanol treatment also led to increases in TLR4 and HMGB1 that persisted into early adulthood as well as increased TNF-α, MCP-1, NOX2 (NADPH oxidase), COX-2, and MyD88 mRNA in the young adult frontal cortex (Vetreno et al., 2013). Thus, adolescent drinking may ultimately impact immune signaling in adult brain, producing persistent alterations that could influence the development of alcohol abuse disorders. Changes in inflammatory responses in peripheral cells of teenage binge drinkers and in peripheral cells and brain microglia in a rat model of binge drinking were recently reported (Ward, Lallemand, & de Witte, 2014). Plasma cytokine levels were altered in the adolescent heavy session drinkers, and these peripheral markers point to activation of the innate immune system, which could impact the neuroimmune system, potentially affecting neurogenesis or other developmental processes that are active during this time period.

Overall, there is an ample behavioral evidence to suggest that activation of immune pathways regulates alcohol consumption and other responses in rodent models and that alcohol consumption also activates the immune system in multiple experimental models. We propose an interaction in activated cells where consumption increases neuroimmune signaling, which in turn further increases consumption in a positive-feedback fashion.

3. PEROXISOME PROLIFERATOR-ACTIVATED RECEPTORS: ANTI-INFLAMMATORY ACTION AND ROLE IN ALCOHOL CONSUMPTION

Peroxisome proliferator-activated receptors (PPARs) have an established role in regulating inflammatory responses across many tissues (Clark, 2002). This suggests that PPAR agonists could be promising for treating certain aspects of addiction, where neuroimmune and inflammatory processes play an etiopathogenic role. PPARs are intracellular receptors, belonging to the nuclear hormone receptor family of transcription factors (Issemann & Green, 1990). PPARs are widely distributed, but are most abundant in tissues with a high metabolic rate, e.g., liver, heart, and kidney (Issemann & Green, 1990). There are three isotypes of PPARs coded for by different genes on separate chromosomes: PPARα, PPARγ, and PPARβ/δ, which vary in their tissue distribution and physiological function, and all are present in brain. Most physiological functions of PPARs fall under the umbrella category of lipid homeostasis. In addition to inflammation, PPARs have defined roles in fatty acid metabolism (Aoyama et al., 1998; Schoonjans, Staels, & Auwerx, 1996), insulin sensitization (Moller & Berger, 2003), adipogenesis (Tontonoz, Hu, & Spiegelman, 1994), cell differentiation, and apoptosis (Roberts, James, Woodyatt, Macdonald, & Tugwood, 1998).

Endogenous ligands for PPARs include fatty acids and fatty acid derivatives (Gottlicher, Widmark, Li, & Gustafsson, 1992), such as endocannabinoids, eicosanoids (e.g., prostaglandins and leukotrienes), phosphoinositides, and sphingolipids (Krey et al., 1997; Sun & Bennett, 2007). PPARs can also be activated by synthetic ligands which fall into two major classes: fibrates and thiazolidinediones (TZDs). Fibrates, which include fenofibrate, ciprofibrate, clofibrate, and gemfibrozil, predominantly target PPARα and are used to treat dyslipidemia. TZDs, which include rosiglitazone and pioglitazone, predominately target PPARγ and are used to treat Type 2 diabetes. When activated by a ligand, be it endogenous or synthetic, PPARs translocate to the nucleus where they form a heterodimer with another nuclear hormone receptor, the retinoid X receptor (RXR). The PPAR–RXR complex binds to PPAR response elements in the DNA to regulate the transcription of many target genes. PPARs can also have nongenomic effects through modifying the phosphorylation status of proteins or inhibiting the activities of other transcription factors (the latter is called transrepression).

The ability of PPARs to transrepress other transcription factors, such as NF-κB, NFAT, and AP-1, is thought to be the main mechanism of action for their anti-inflammatory properties. There are three main ways in which ligand-activated PPAR–RXR complexes can negatively regulate the activities of other transcription factors (Fig. 2.2; Daynes & Jones, 2002): they can (1) compete for limited amounts of coactivators, (2) physically associate with transcription factors to inhibit their ability to induce transcription of their target genes, and (3) inhibit protein kinases, e.g., MAPK, from phosphorylating (i.e., activating) other transcription factors.

PPAR agonists have been shown to modulate alcohol consumption. The PPARγ agonists, pioglitazone and rosiglitazone, reduced voluntary alcohol consumption in rats (Stopponi et al., 2013, 2011) and decreased stress-induced (but not cue-induced) relapse and alcohol withdrawal symptoms

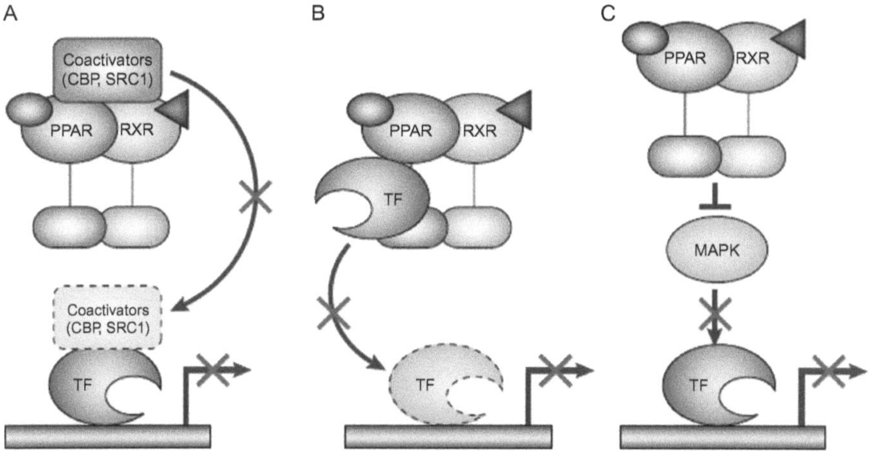

Figure 2.2 Negative regulation of transcription factors by PPARs. The inflammatory and immunomodulatory properties of peroxisome proliferator-activated receptors (PPARs) do not arise primarily through their transcription-factor transactivation abilities, but rather through their ability to antagonize several important signaling cascades. Although different transrepression mechanisms exist, three are mediated by PPARs in cells of the immune system. (A) The first mechanism is the ability of PPARs to compete for limiting amounts of coactivator proteins in a cell, such as cAMP response element binding (CREB)-binding protein (CBP) and steroid receptor coactivator 1 (SRC1), making these coactivators unavailable to other transcription factors (TFs). (B) The second mechanism is known as "cross coupling" or "mutual receptor antagonism" and is facilitated by the ability of PPARs to associate physically with various transcription factors, preventing the transcription factor from binding to its response element and thereby inhibiting its ability to induce gene transcription. (C) Another transrepression mechanism relies on the ability of the PPAR to inhibit activation of a mitogen-activated protein kinase (MAPK). This inhibits the MAPK from phosphorylating and activating downstream transcription factors. RXR, 9-*cis*-retinoic acid receptor. *Figure and legend are from Daynes and Jones (2002).*

in alcohol-dependent rats (Stopponi et al., 2011). PPARα agonists also decrease alcohol consumption. The PPARα agonist gemfibrozil decreased voluntary alcohol consumption in male Sprague-Dawley rats (Barson et al., 2009). One study, however, showed that clofibrate increased voluntary alcohol consumption in male spontaneous hypertensive rats, although a very high ethanol concentration was used (30% ethanol, v/v) (Schlicht, 1987). The PPARα agonist, fenofibrate, and the dual PPARα/γ agonist, tesaglitazar, selectively decreased voluntary alcohol consumption and preference in mice (Blednov et al., submitted; Ferguson et al., 2014). Gene expression profiling revealed that fenofibrate and tesaglitazar changed the transcriptome of mouse amygdala and prefrontal cortex (Ferguson et al.,

2014), two important brain areas for reward and dependence. The list of PPAR agonist-regulated genes was dominated by known neuronal genes, which was unexpected given the importance of glial cells (especially microglia) in regulating neuroimmune and inflammatory signaling in brain (Ferguson et al., 2014). Genes coding for proteins involved in synaptic plasticity and nerve impulse transmission were prominent. In the amygdala, genes involved in GABAergic, dopaminergic, and neuropeptide signaling were regulated in a coordinated manner (coexpressed) by the PPAR agonists, including many already known to alter alcohol consumption based on mutant mouse data (Ferguson et al., 2014).

In addition to alcohol consumption, PPAR agonists effectively reduced intake of other drugs of abuse. PPARα agonists blocked acquisition of nicotine self-administration in naïve rats and nonhuman primates and decreased nicotine intake and cue- and drug-induced relapse in nicotine-dependent rats and nonhuman primates (Mascia et al., 2011; Panlilio et al., 2012). Also, PPAR agonists modulated sensitization to morphine and methamphetamine (but not cocaine) in mice (Fernandez-Espejo, Ramiro-Fuentes, & Rodriguez de Fonseca, 2009; Maeda et al., 2007).

Thus, PPAR agonists represent a class of anti-inflammatory medications that demonstrate antiaddictive properties for several drugs of abuse. Specifically, PPARα- and PPARγ-selective agonists (fibrates and TZDs, respectively) and PPARα/γ dual agonists have been shown to decrease alcohol consumption in rodent models of alcohol dependence. Whether these properties are mediated via their anti-inflammatory actions remains to be proven. Nevertheless, this class of drugs offers a promising new avenue for treating aspects of alcohol dependence. The favorable safety profile and FDA-approval status of some of the PPAR agonists that are effective for reducing alcohol consumption warrant further investigation into their potential as therapeutics for alcohol dependence.

4. ALCOHOL CONSUMPTION AND NEUROIMMUNE-RELATED GENE EXPRESSION

Chronic alcohol consumption and abuse induce long-term changes in brain gene and protein expression, which likely contribute to the neuropathologies associated with alcohol dependence (Kauer & Malenka, 2007; Nestler, 2001). Alcohol-induced transcriptional reprogramming in the brain may account for some of the effects of repeated ethanol exposure (Mayfield, Harris, & Schuckit, 2008; Ron & Messing, 2013). Neuroadaptations induced by alcohol are ultimately controlled by the regulation of many genes expressed within individual neurons or glial cells (Farris & Miles, 2012), causing brain cells to adapt via alterations involving many biological pathways, including growth factors (Ron & Messing, 2013), serine–threonine kinases (Lesscher et al., 2009; Sanna, Simpson, Lutjens, & Koob, 2002), glutathione pathway enzymes, protein translation (Barak et al., 2013; Neasta, Ben Hamida, Yowell, Carnicella, & Ron, 2010), transcription (Edenberg et al., 2008; Okvist et al., 2007), and inflammatory pathways (Blednov et al., 2011; Gorini, Harris, & Mayfield, 2014).

The role of immune-related genes in alcohol dependence is also supported by genetic association studies and transcriptome meta-analysis in peripheral human cells (Boukli et al., 2010), postmortem brains of alcoholics (Liu et al., 2006; Okvist et al., 2007; Ponomarev, Wang, Zhang, Harris, & Mayfield, 2012), drinking models in mice (Gorini, Nunez, & Mayfield, 2013; Gorini, Roberts, & Mayfield, 2013; Nunez et al., 2013; Osterndorff-Kahanek et al., 2013), and vapor models for alcohol dependence in mice (Gorini, Nunez, et al., 2013; Gorini, Roberts, et al., 2013). Many immune-associated genes that are differentially expressed in brains of ethanol-treated mice are also differentially expressed in brains of human alcoholics. Deletion of several immune-related genes that emerged as candidates from the genomic analyses was validated in behavioral studies and found to reduce ethanol consumption in mice, as discussed previously (Blednov et al., 2012).

Upregulation of genes encoding different classes of chaperones has been reported after ethanol treatment (Miles, Diaz, & DeGuzman, 1991; Nunez et al., 2013; Pignataro, Varodayan, Tannenholz, & Harrison, 2009; Varodayan, Pignataro, & Harrison, 2011). Examples include stress-induced chaperones (Hsp 70 and Hsp 90 family members; Nunez et al., 2013), which are implicated in proteostasis (Hutt & Balch, 2010), endotoxin-like effects mediated through TLR4 (Gupta et al., 2013; Lee, Jeong, & Yoo, 2013), and may cocluster with the LPS complex (Triantafilou & Triantafilou, 2002). Alcohol-responsive gene modules from frontal cortex of alcohol-treated mice were also enriched in cytokine pathways and synaptic transmission processes that involve Toll-like and IL-1 receptor signaling (Nunez et al., 2013).

Studies in human alcoholics identified differential expression of genes related to the NF-κB pathway. Genes with NF-κB elements were generally upregulated in alcoholics (Okvist et al., 2007), and *NFKB1* has been associated with alcohol dependence in alcoholics (Edenberg et al., 2008). Furthermore, the NF-κB pathway may undergo a compensatory, adaptive tolerance to alcohol, which could serve to diminish excessive stimulation of this system (Yakovleva, Bazov, Watanabe, Hauser, & Bakalkin, 2011). NF-κB and the signaling pathways that regulate its activity have become a focal point for intense drug discovery efforts (Gupta, Sundaram, Reuter, & Aggarwal, 2010; Karin, Yamamoto, & Wang, 2004) and may also be promising therapeutic targets for alcohol dependence.

A large gene associated study of human alcoholics and healthy controls found a susceptibility locus in neurokinin-1 receptor (NK-1R) that was associated with alcohol dependence (Seneviratne et al., 2009). Furthermore, NK-1R antagonism-altered alcohol consumption in mice (Thorsell, Schank, Singley, Hunt, & Heilig, 2010) and humans (George et al., 2008), and gene therapy in mice using an inhibitory RNA against NK-1R reduced alcohol consumption (Baek et al., 2010). Modulation of the NK-1R system can exacerbate inflammatory immune responses within the CNS, and an NK-1R antagonist attenuates increases in CNS inflammatory cytokine levels and decreases immunosuppressive cytokine production associated with infection (Chauhan, Kluttz, Bost, & Marriott, 2011). NK-1R antagonists are safe clinical drugs that are known to have anti-inflammatory, analgesic, anxiolytic, antidepressant, and antiemetic effects (Rosso, Munoz, & Berger, 2012), and the findings above indicate that these antagonists may warrant further study for the treatment of alcoholism.

5. ALCOHOL CONSUMPTION AND NEUROIMMUNE-RELATED MICRORNAS

microRNAs (miRNAs) comprise a specific class of small noncoding RNAs that bind to complementary sequences on target mRNAs to repress translation and silence gene expression (Ambros, 2001; Lee & Ambros, 2001) and are key regulators of cellular gene expression. They are highly abundant in brain and mediate multiple biological processes, including brain development (Krichevsky, King, Donahue, Khrapko, & Kosik, 2003), synapse formation (Schratt et al., 2006), synaptic plasticity (Cohen, Lee, Chen, Li, & Fields, 2011; Smalheiser & Lugli, 2009), and neuroimmune signaling (Soreq & Wolf, 2011). miRNAs are capable of eliciting targeted actions in innate immunity- and epigenetic-related functions in glial cells (Nunez & Mayfield, 2012) and have been associated with development of the immune system and regulation of multiple immune functions (O'Neill, Sheedy, & McCoy, 2011). Also, TLR signaling can modulate miRNA expression via NF-κB regulation (Taganov, Boldin, Chang, & Baltimore, 2006).

The ability of miRNAs to act as central regulators of multiple cellular pathways may influence the patterns of gene expression produced by alcohol in rodents (Gorini, Nunez, et al., 2013; Pietrzykowski et al., 2008; Tapocik et al., 2013), humans (Lewohl et al., 2011) and human-derived neuroblastoma cells (Yadav et al., 2011). miRNAs are also implicated in the adaptations induced by exposure to other drugs of abuse, including nicotine (Huang & Li, 2009), cocaine (Chandrasekar & Dreyer, 2009, 2011; Im, Hollander, Bali, & Kenny, 2010), and opioids (He, Yang, Kirkmire, & Wang, 2010). Many miRNAs are altered in brains from human alcoholics (Lewohl et al., 2011) and ethanol-treated mice (Nunez et al., 2013), and many are predicted to target neuroimmune processes. Figure 2.3 depicts a model of neuroinflammatory miRNA signaling based on evidence from human alcoholics (Nunez & Mayfield, 2012).

Chemokine receptor 4 (CXCR4) appears to be important in alcohol action and has been reported to interact with the LPS-sensing complex (Triantafilou & Triantafilou, 2002). The CXCR4 gene was significantly over targeted by miRNAs in prefrontal cortex of human alcoholics (Lewohl et al., 2011). Moreover, in mouse frontal cortex, CXCR4 was one of the most highly enriched mRNA pathways following alcohol exposure, and the miRNAs associated with it (let-7g-5p and miR-34c-5p) were highly correlated with the alcohol-responsive miRNAs (Nunez et al., 2013). Examples of alcohol-sensitive, immune-related miRNAs identified in both rodent and human brain are shown in Table 2.2, and some regulate TLR signaling. For example, let-7 family members directly target expression of TLR4 and regulate responsiveness to LPS (Chen, Splinter, O'Hara, & LaRusso, 2007) and form an epigenetic switch with NF-κB and IL-6 that causes persistent inflammation-induced cell transformation (Iliopoulos, Hirsch, & Struhl, 2009). miR-9 may also control the activity of TLR signaling pathway through direct targeting of the *NFKB1* gene in human monocytes and neutrophils (O'Neill et al., 2011). Interestingly, miR-9 also targets the BK channel (large conductance calcium- and voltage-activated potassium channel) and selectively destabilizes splice variants, resulting in downregulation of the alcohol-specific splice variant (Pietrzykowski et al., 2008). This mechanism is proposed to mediate development of cellular tolerance to alcohol. miR-92 and miR-140 are other alcohol-sensitive miRNAs found in both mouse and human brain (Table 2.2), which are associated with immune signaling. Differential expression of miR-92 is characteristic of stages of T lymphocyte development (Sonkoly, Stahle, & Pivarcsi, 2008), and increased plasma levels of miR-92a in patients with traumatic

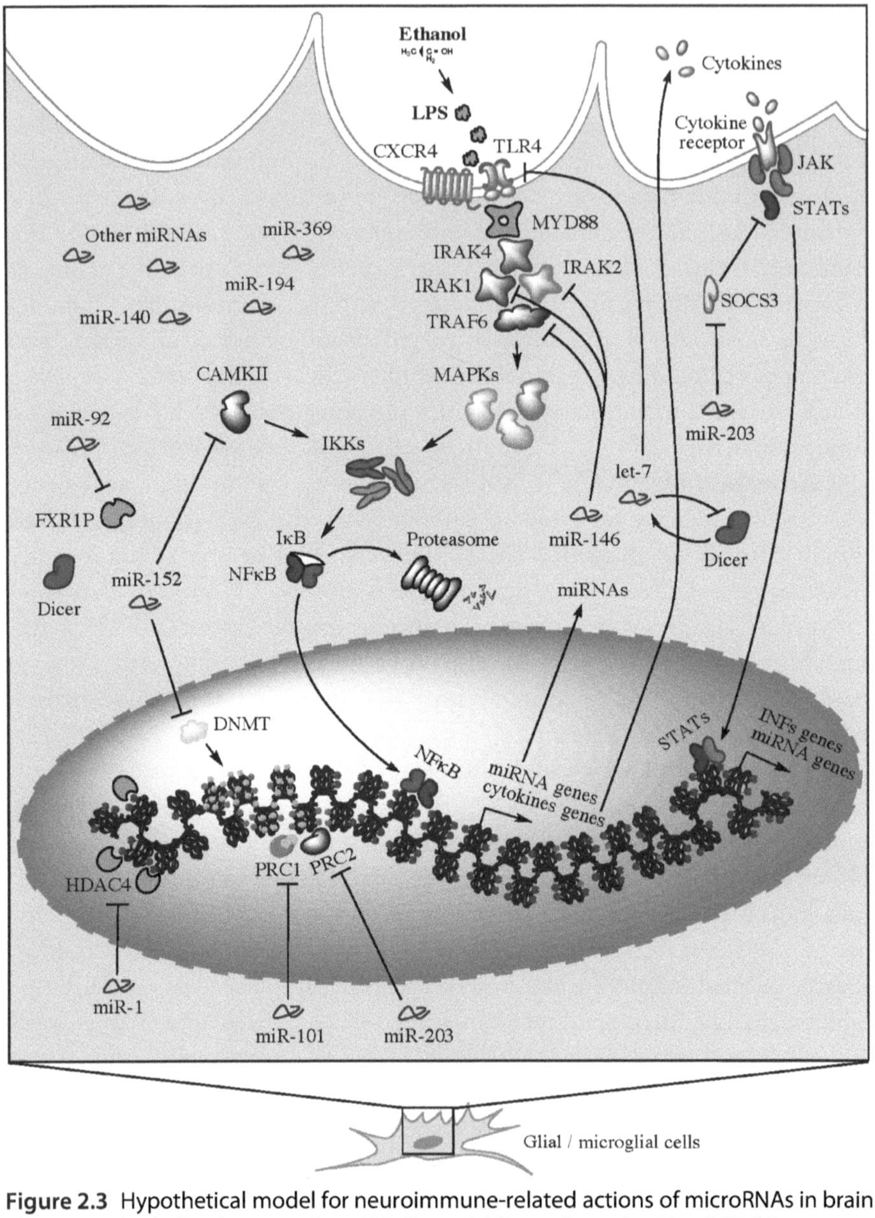

Figure 2.3 Hypothetical model for neuroimmune-related actions of microRNAs in brain of human alcoholics (e.g., in microglia). Bacterial lipopolysaccharides (LPSs) from commensal bacteria may leak into the bloodstream resulting from gut leakiness induced by alcohol consumption. Activation of NF-κB induces the transcription of a variety of proinflammatory and miRNA genes. Newly synthesized proinflammatory factors induce a systemic neuroinflammatory response and a positive-feedback loop in the same activated cell. Subsequent generation of proinflammatory factors results from expression of

brain injury is a good biomarker of the severity of the disease (Redell, Moore, Ward, Hergenroeder, & Dash, 2010). miR-140 is upregulated after LPS treatment in mice (Moschos et al., 2007) and targets the chemokine Cxcl12 (Nicolas et al., 2008). Another member of this chemokine family, Cxcl14, is a predicted target of miR-34b in synaptic preparations from mice treated with ethanol (Most et. al., unpublished results). Ethanol also altered miR-369-3p (miR-369*) in both mouse and human brain (Table 2.2), and miR-369* is known to associate directly with TNF-α mRNA to mediate activation of TNF-α under conditions of arrested growth (Vasudevan & Steitz, 2007). This effect is dependent on the recruitment of the RNA-binding proteins, fragile-X mental retardation-related protein 1 (FXR1) and argonaute 2 (Ago2), and can be reversed when cells are actively proliferating, in which case miR-369* represses TNF-α (Vasudevan, Tong, & Steitz, 2007). Interestingly, miR-369*, TNF-α, FXR1, and Ago2 were all altered in mouse brain synaptoneurosomes from chronic ethanol-treated mice (Most et al., unpublished results).

Synaptoneurosomes (Hollingsworth et al., 1985; Quinlan, Philpot, Huganir, & Bear, 1999) contain pre- and postsynaptic compartments of neurons, astrocytes, oligodendrocytes, and microglia and have been used to study local translation of mRNAs in the synapse (Raab-Graham, Haddick, Jan, & Jan, 2006; Sosanya et al., 2013). Discrete alcohol-mediated changes within the transcriptome can be measured by comparing synaptoneurosome versus total homogenate preparations in the same samples from control and chronic ethanol-treated mice, and miRNAs that may target immune pathways were specifically identified in synaptoneurosomes from ethanol-treated mice (Table 2.2).

---

alternative cytokine receptors in the activated cell, which signal back to the nucleus to induce additional proinflammatory factors (e.g., interferons) and miRNA genes. To avoid overamplification of these signals and excessive inflammation, specific miRNAs (e.g., members of the miR-146, miR-152, and let-7 families) are consequently upregulated, suppressing TLR4/CXCR4 signaling through inhibition of various transducers, such as IL1 receptor-associated kinases (IRAKs), TNF receptor-associated factor 6 (TRAF6), and TLR4/CXCR4 itself. As a compensatory reaction, miRNAs (e.g., miR-203) may also be upregulated to maintain the immune-activated state of the specific cell subtype while promoting a benign, contained inflammatory response. microRNAs that target epigenetic factors are also activated to control and/or fine tune the ongoing remodeling of the cellular epigenome, allowing for long-term homeostatic and cellular adaptations. *Figure and legend are from Nunez and Mayfield (2012).*

Table 2.2 Alcohol-sensitive immune-related microRNAs in rodent and human brain.

microRNAs	Alcohol studies
miR-92	Human brain[a]; mouse brain synaptoneurosomes[b]
miR-140	Human and mouse brain[a,c]; mouse brain synaptoneurosomes[b]
miR-369*	Human brain[a]; mouse brain synaptoneurosomes[b]
let-7	Human and mouse brain[a,c]; mouse brain synaptoneurosomes[b]
miR-9	Mouse and rat brain[b,d]
miR-34	Human and mouse brain[a,c]; mouse brain synaptoneurosomes[b]

[a]Human prefrontal cortex (Lewohl et al., 2011).
[b]Mouse amygdala synaptoneurosomes (unpublished results).
[c]Mouse frontal cortex (Nunez et al., 2013).
[d]Rat striatum (Pietrzykowski et al., 2008).

Furthermore, comparison of ethanol-sensitive mRNAs from synaptoneurosomes (Most et al., 2014) with immune (LPS)-induced mRNAs (Osterndorff-Kahanek et al., 2013) identified 55 mRNAs from mouse brain common to both groups. Astrocyte and microglial transcripts belonging to immune/inflammatory categories (e.g., chemokine- and complement-related transcripts) were upregulated by both treatments, suggesting that some of the ethanol-induced changes in the synaptic transcriptome may be mediated by immune signaling.

Alcohol consumption alters individual or families of miRNAs in human and rodent brain, some of which may regulate the expression of neuroimmune-related mRNAs. The findings in synaptoneurosomes also indicate the importance of the cellular microenvironment in identifying discrete effects of alcohol and suggest that the mouse synaptic transcriptome may be a useful model for predicting expression changes in humans.

6. CONCLUSIONS

The evidence is now quite strong that excessive alcohol consumption not only alters peripheral innate immune signaling but also perturbs neuroimmune signaling in both humans and rodent models of chronic alcohol exposure. The behavioral consequences of these changes in peripheral and central immune signaling are only beginning to emerge, but manipulation of immune genes and administration of LPS to activate innate immune pathways in mice indicate that alcohol consumption is regulated by these signaling systems. In addition, measurement of levels of LPS and proinflammatory molecules in human alcoholics suggests that alcohol craving is correlated with the level of activation of innate immunity (Leclercq et al., 2012, 2014). Overall, there is support for the idea that excessive alcohol consumption in humans and rodents regulates not only peripheral but also central immune signaling, by altering gene expression. These changes in mRNA expression are driven, at least in part, by changes in miRNAs. Some of the diverse changes in immune function appear to be important for regulation of alcohol consumption and craving, indicating a vicious cycle of increased alcohol consumption, leading to increased perturbation of neuroimmune function, which further increases alcohol craving, consumption, and dependence. Despite recent progress, there are many unanswered questions. Most of the approaches (e.g., global gene deletion, measurement of brain gene expression) do not define whether the site of alcohol action is the peripheral immune system (which then affects the brain) or direct action on the brain. Gene expression studies have identified changes in many different mRNAs potentially important for immune signaling, but do not define which are most important. In addition, we have little insight as to how alcohol-mediated changes in neuroimmune signaling might affect circuits important for consumption, reward, or craving. In particular, there are neuroimmune components not only on neurons but also on astrocytes and microglia, and which cell types are critical for the behavioral effects are unknown. Lastly, the recognition of the importance of these signaling systems in alcohol consumption and dependence provides new opportunities for therapeutic interventions (e.g., PPARα- and PPARγ-selective agonists). Indeed, there are many existing and emerging treatments for inflammatory diseases, and interest in the neuroimmune basis of many mental illnesses and repurposing of immune therapies for alcoholism as well as other addictions and mental illnesses is an attractive possibility, but will require better knowledge of the mechanisms of neuroimmune function.

ACKNOWLEDGMENTS

This work was supported by NIAAA grants AA006399, AA012404, and AA013520 (INIA West Consortium).

REFERENCES

Agrawal, R. G., Owen, J. A., Levin, P. S., Hewetson, A., Berman, A. E., Franklin, S. R., et al. (2014). Bioinformatics analyses reveal age-specific neuroimmune modulation as a target for treatment of high ethanol drinking. *Alcoholism, Clinical and Experimental Research*, *38*(2), 428–437. http://dx.doi.org/10.1111/acer.12288.

Alfonso-Loeches, S., Pascual-Lucas, M., Blanco, A. M., Sanchez-Vera, I., & Guerri, C. (2010). Pivotal role of TLR4 receptors in alcohol-induced neuroinflammation and brain damage. *The Journal of Neuroscience, 30*(24), 8285–8295. http://dx.doi.org/10.1523/JNEUROSCI.0976-10.2010.

Ambros, V. (2001). microRNAs: Tiny regulators with great potential. *Cell, 107*(7), 823–826.

Aoyama, T., Peters, J. M., Iritani, N., Nakajima, T., Furihata, K., Hashimoto, T., et al. (1998). Altered constitutive expression of fatty acid-metabolizing enzymes in mice lacking the peroxisome proliferator-activated receptor alpha (PPARalpha). *The Journal of Biological Chemistry, 273*(10), 5678–5684.

Baek, M. N., Jung, K. H., Halder, D., Choi, M. R., Lee, B. H., Lee, B. C., et al. (2010). Artificial microRNA-based neurokinin-1 receptor gene silencing reduces alcohol consumption in mice. *Neuroscience Letters, 475*(3), 124–128. http://dx.doi.org/10.1016/j.neulet.2010.03.051.

Bajo, M., Madamba, S. G., Roberto, M., Blednov, Y. A., Sagi, V. N., Roberts, E., et al. (2014). Innate immune factors modulate ethanol interactions with GABAergic transmission in mouse central amygdala. *Brain, Behavior, and Immunity, 40*, 191–202. http://dx.doi.org/10.1016/j.bbi.2014.03.007.

Barak, S., Liu, F., Ben Hamida, S., Yowell, Q. V., Neasta, J., Kharazia, V., et al. (2013). Disruption of alcohol-related memories by mTORC1 inhibition prevents relapse. *Nature Neuroscience, 16*(8), 1111–1117. http://dx.doi.org/10.1038/nn.3439.

Barson, J. R., Karatayev, O., Chang, G. Q., Johnson, D. F., Bocarsly, M. E., Hoebel, B. G., et al. (2009). Positive relationship between dietary fat, ethanol intake, triglycerides, and hypothalamic peptides: Counteraction by lipid-lowering drugs. *Alcohol, 43*(6), 433–441. http://dx.doi.org/10.1016/j.alcohol.2009.07.003.

Blanco, A. M., Perez-Arago, A., Fernandez-Lizarbe, S., & Guerri, C. (2008). Ethanol mimics ligand-mediated activation and endocytosis of IL-1RI/TLR4 receptors via lipid rafts caveolae in astroglial cells. *Journal of Neurochemistry, 106*(2), 625–639. http://dx.doi.org/10.1111/j.1471-4159.2008.05425.x.

Blednov, Y. A., Benavidez, J. M., Black, M., Goate, A. M., Edenberg, H. J., & Wetherill, L., et al. (submitted). Peroxisome proliferator-activated receptors α and γ are linked with alcohol consumption in mice and withdrawal and dependence in humans. *Alcoholism: Clinical and Experimental Research.*

Blednov, Y. A., Benavidez, J. M., Geil, C., Perra, S., Morikawa, H., & Harris, R. A. (2011). Activation of inflammatory signaling by lipopolysaccharide produces a prolonged increase of voluntary alcohol intake in mice. *Brain, Behavior, and Immunity, 25*(Suppl. 1), S92–S105. http://dx.doi.org/10.1016/j.bbi.2011.01.008.

Blednov, Y. A., Bergeson, S. E., Walker, D., Ferreira, V. M., Kuziel, W. A., & Harris, R. A. (2005). Perturbation of chemokine networks by gene deletion alters the reinforcing actions of ethanol. *Behavioural Brain Research, 165*(1), 110–125. http://dx.doi.org/10.1016/j.bbr.2005.06.026.

Blednov, Y. A., Ponomarev, I., Geil, C., Bergeson, S., Koob, G. F., & Harris, R. A. (2012). Neuroimmune regulation of alcohol consumption: Behavioral validation of genes obtained from genomic studies. *Addiction Biology, 17*(1), 108–120. http://dx.doi.org/10.1111/j.1369-1600.2010.00284.x.

Boukli, N. M., Saiyed, Z. M., Ricaurte, M., Rodriguez, J. W., Rios Olivares, E., Cubano, L. A., et al. (2010). Implications of ER stress, the unfolded protein response, and pro- and anti-apoptotic protein fingerprints in human monocyte-derived dendritic cells treated with alcohol. *Alcoholism, Clinical and Experimental Research, 34*(12), 2081–2088. http://dx.doi.org/10.1111/j.1530-0277.2010.01304.x.

Chandrasekar, V., & Dreyer, J. L. (2009). microRNAs miR-124, let-7d and miR-181a regulate cocaine-induced plasticity. *Molecular and Cellular Neurosciences, 42*(4), 350–362. http://dx.doi.org/10.1016/j.mcn.2009.08.009.

Chandrasekar, V., & Dreyer, J. L. (2011). Regulation of MiR-124, Let-7d, and MiR-181a in the accumbens affects the expression, extinction, and reinstatement of cocaine-induced conditioned place preference. *Neuropsychopharmacology*, *36*(6), 1149–1164. http://dx.doi.org/10.1038/npp.2010.250.

Chauhan, V. S., Kluttz, J. M., Bost, K. L., & Marriott, I. (2011). Prophylactic and therapeutic targeting of the neurokinin-1 receptor limits neuroinflammation in a murine model of pneumococcal meningitis. *Journal of Immunology*, *186*(12), 7255–7263. http://dx.doi.org/10.4049/jimmunol.1100721.

Chen, X. M., Splinter, P. L., O'Hara, S. P., & LaRusso, N. F. (2007). A cellular micro-RNA, let-7i, regulates Toll-like receptor 4 expression and contributes to cholangiocyte immune responses against Cryptosporidium parvum infection. *The Journal of Biological Chemistry*, *282*(39), 28929–28938. http://dx.doi.org/10.1074/jbc.M702633200.

Clark, R. B. (2002). The role of PPARs in inflammation and immunity. *Journal of Leukocyte Biology*, *71*(3), 388–400.

Cohen, J. E., Lee, P. R., Chen, S., Li, W., & Fields, R. D. (2011). MicroRNA regulation of homeostatic synaptic plasticity. *Proceedings of the National Academy of Sciences of the United States of America*, *108*(28), 11650–11655. http://dx.doi.org/10.1073/pnas.1017576108.

Crews, F., Nixon, K., Kim, D., Joseph, J., Shukitt-Hale, B., Qin, L., et al. (2006). BHT blocks NF-kappaB activation and ethanol-induced brain damage. *Alcoholism, Clinical and Experimental Research*, *30*(11), 1938–1949. http://dx.doi.org/10.1111/j.1530-0277.2006.00239.x.

Crews, F. T., Qin, L., Sheedy, D., Vetreno, R. P., & Zou, J. (2013). High mobility group box 1/Toll-like receptor danger signaling increases brain neuroimmune activation in alcohol dependence. *Biological Psychiatry*, *73*(7), 602–612. http://dx.doi.org/10.1016/j.biopsych.2012.09.030.

Daynes, R. A., & Jones, D. C. (2002). Emerging roles of PPARs in inflammation and immunity. *Nature Reviews. Immunology*, *2*(10), 748–759. http://dx.doi.org/10.1038/nri912.

Edenberg, H. J., Xuei, X., Wetherill, L. F., Bierut, L., Bucholz, K., Dick, D. M., et al. (2008). Association of NFKB1, which encodes a subunit of the transcription factor NF-kappaB, with alcohol dependence. *Human Molecular Genetics*, *17*(7), 963–970. http://dx.doi.org/10.1093/hmg/ddm368.

Farris, S. P., & Miles, M. F. (2012). Ethanol modulation of gene networks: Implications for alcoholism. *Neurobiology of Disease*, *45*(1), 115–121. http://dx.doi.org/10.1016/j.nbd.2011.04.013.

Ferguson, L. B., Most, D., Blednov, Y. A., & Harris, R. A. (2014). PPAR agonists regulate brain gene expression: Relationship to their effects on ethanol consumption. *Neuropharmacology*. [Epub ahead of print] doi: 10.1016/j.neuropharm.2014.06.024.

Fernandez-Espejo, E., Ramiro-Fuentes, S., & Rodriguez de Fonseca, F. (2009). The absence of a functional peroxisome proliferator-activated receptor-alpha gene in mice enhances motor sensitizing effects of morphine, but not cocaine. *Neuroscience*, *164*(2), 667–675. http://dx.doi.org/10.1016/j.neuroscience.2009.08.023.

Fernandez-Lizarbe, S., Montesinos, J., & Guerri, C. (2013). Ethanol induces TLR4/TLR2 association, triggering an inflammatory response in microglial cells. *Journal of Neurochemistry*, *126*(2), 261–273. http://dx.doi.org/10.1111/jnc.12276.

Fernandez-Lizarbe, S., Pascual, M., Gascon, M. S., Blanco, A., & Guerri, C. (2008). Lipid rafts regulate ethanol-induced activation of TLR4 signaling in murine macrophages. *Molecular Immunology*, *45*(7), 2007–2016. http://dx.doi.org/10.1016/j.molimm.2007.10.025.

Fernandez-Lizarbe, S., Pascual, M., & Guerri, C. (2009). Critical role of TLR4 response in the activation of microglia induced by ethanol. *Journal of Immunology*, *183*(7), 4733–4744. http://dx.doi.org/10.4049/jimmunol.0803590.

George, D. T., Gilman, J., Hersh, J., Thorsell, A., Herion, D., Geyer, C., et al. (2008). Neurokinin 1 receptor antagonism as a possible therapy for alcoholism. *Science, 319*(5869), 1536–1539. http://dx.doi.org/10.1126/science.1153813.

Gorini, G., Harris, R. A., & Mayfield, R. D. (2014). Proteomic approaches and identification of novel therapeutic targets for alcoholism. *Neuropsychopharmacology, 39*(1), 104–130. http://dx.doi.org/10.1038/npp.2013.182.

Gorini, G., Nunez, Y. O., & Mayfield, R. D. (2013). Integration of miRNA and protein profiling reveals coordinated neuroadaptations in the alcohol-dependent mouse brain. *PLoS One, 8*(12), e82565. http://dx.doi.org/10.1371/journal.pone.0082565.

Gorini, G., Roberts, A. J., & Mayfield, R. D. (2013). Neurobiological signatures of alcohol dependence revealed by protein profiling. *PLoS One, 8*(12), e82656. http://dx.doi.org/10.1371/journal.pone.0082656.

Gottlicher, M., Widmark, E., Li, Q., & Gustafsson, J. A. (1992). Fatty acids activate a chimera of the clofibric acid-activated receptor and the glucocorticoid receptor. *Proceedings of the National Academy of Sciences of the United States of America, 89*(10), 4653–4657.

Gupta, A., Cooper, Z. A., Tulapurkar, M. E., Potla, R., Maity, T., Hasday, J. D., et al. (2013). Toll-like receptor agonists and febrile range hyperthermia synergize to induce heat shock protein 70 expression and extracellular release. *The Journal of Biological Chemistry, 288*(4), 2756–2766. http://dx.doi.org/10.1074/jbc.M112.427336.

Gupta, S. C., Sundaram, C., Reuter, S., & Aggarwal, B. B. (2010). Inhibiting NF-kappaB activation by small molecules as a therapeutic strategy. *Biochimica et Biophysica Acta, 1799*(10–12), 775–787. http://dx.doi.org/10.1016/j.bbagrm.2010.05.004.

Harris, R. A., & Blednov, Y. A. (2013). Neuroimmune genes and alcohol drinking behavior. In C. Cui, L. Grandison, & A. Noronha (Eds.), *Neural-immune interactions in brain function and alcohol related disorders* (pp. 425–440). New York: Springer.

He, J., & Crews, F. T. (2008). Increased MCP-1 and microglia in various regions of the human alcoholic brain. *Experimental Neurology, 210*(2), 349–358. http://dx.doi.org/10.1016/j.expneurol.2007.11.017.

He, Y., Yang, C., Kirkmire, C. M., & Wang, Z. J. (2010). Regulation of opioid tolerance by let-7 family microRNA targeting the mu opioid receptor. *The Journal of Neuroscience, 30*(30), 10251–10258. http://dx.doi.org/10.1523/JNEUROSCI.2419-10.2010.

Hollingsworth, E. B., McNeal, E. T., Burton, J. L., Williams, R. J., Daly, J. W., & Creveling, C. R. (1985). Biochemical characterization of a filtered synaptoneurosome preparation from guinea pig cerebral cortex: Cyclic adenosine 3':5'-monophosphate-generating systems, receptors, and enzymes. *The Journal of Neuroscience, 5*(8), 2240–2253.

Huang, W., & Li, M. D. (2009). Nicotine modulates expression of miR-140*, which targets the 3'-untranslated region of dynamin 1 gene (Dnm1). *International Journal of Neuropsychopharmacology, 12*(4), 537–546. http://dx.doi.org/10.1017/S1461145708009528.

Hutt, D., & Balch, W. E. (2010). Cell biology the proteome in balance. *Science, 329*(5993), 766–767. http://dx.doi.org/10.1126/science.1194160.

Iliopoulos, D., Hirsch, H. A., & Struhl, K. (2009). An epigenetic switch involving NF-kappaB, Lin28, Let-7 MicroRNA, and IL6 links inflammation to cell transformation. *Cell, 139*(4), 693–706. http://dx.doi.org/10.1016/j.cell.2009.10.014.

Im, H. I., Hollander, J. A., Bali, P., & Kenny, P. J. (2010). MeCP2 controls BDNF expression and cocaine intake through homeostatic interactions with microRNA-212. *Nature Neuroscience, 13*(9), 1120–1127. http://dx.doi.org/10.1038/nn.2615.

Issemann, I., & Green, S. (1990). Activation of a member of the steroid hormone receptor superfamily by peroxisome proliferators. *Nature, 347*(6294), 645–650. http://dx.doi.org/10.1038/347645a0.

Kane, C. J., Phelan, K. D., Douglas, J. C., Wagoner, G., Johnson, J. W., Xu, J., et al. (2014). Effects of ethanol on immune response in the brain: Region-specific changes in adolescent versus adult mice. *Alcoholism, Clinical and Experimental Research, 38*(2), 384–391. http://dx.doi.org/10.1111/acer.12244.

Karin, M., Yamamoto, Y., & Wang, Q. M. (2004). The IKK NF-kappa B system: A treasure trove for drug development. *Nature Reviews. Drug Discovery*, *3*(1), 17–26. http://dx.doi.org/10.1038/nrd1279.
Kauer, J. A., & Malenka, R. C. (2007). Synaptic plasticity and addiction. *Nature Reviews. Neuroscience*, *8*(11), 844–858. http://dx.doi.org/10.1038/nrn2234.
Kong, E. C., Allouche, L., Chapot, P. A., Vranizan, K., Moore, M. S., Heberlein, U., et al. (2010). Ethanol-regulated genes that contribute to ethanol sensitivity and rapid tolerance in Drosophila. *Alcoholism, Clinical and Experimental Research*, *34*(2), 302–316. http://dx.doi.org/10.1111/j.1530-0277.2009.01093.x.
Krey, G., Braissant, O., L'Horset, F., Kalkhoven, E., Perroud, M., Parker, M. G., et al. (1997). Fatty acids, eicosanoids, and hypolipidemic agents identified as ligands of peroxisome proliferator-activated receptors by coactivator-dependent receptor ligand assay. *Molecular Endocrinology*, *11*(6), 779–791. http://dx.doi.org/10.1210/mend.11.6.0007.
Krichevsky, A. M., King, K. S., Donahue, C. P., Khrapko, K., & Kosik, K. S. (2003). A microRNA array reveals extensive regulation of microRNAs during brain development. *RNA*, *9*(10), 1274–1281.
Leclercq, S., Cani, P. D., Neyrinck, A. M., Starkel, P., Jamar, F., Mikolajczak, M., et al. (2012). Role of intestinal permeability and inflammation in the biological and behavioral control of alcohol-dependent subjects. *Brain, Behavior, and Immunity*, *26*(6), 911–918. http://dx.doi.org/10.1016/j.bbi.2012.04.001.
Leclercq, S., De Saeger, C., Delzenne, N., de Timary, P., & Starkel, P. (2014). Role of inflammatory pathways, blood mononuclear cells, and gut-derived bacterial products in alcohol dependence. *Biological Psychiatry*. http://dx.doi.org/10.1016/j.biopsych.2014.02.003, (Epub ahead of print).
Lee, R. C., & Ambros, V. (2001). An extensive class of small RNAs in Caenorhabditis elegans. *Science*, *294*(5543), 862–864. http://dx.doi.org/10.1126/science.1065329.
Lee, K. H., Jeong, J., & Yoo, C. G. (2013). Positive feedback regulation of heat shock protein 70 (Hsp70) is mediated through Toll-like receptor 4-PI3K/Akt-glycogen synthase kinase-3beta pathway. *Experimental Cell Research*, *319*(1), 88–95. http://dx.doi.org/10.1016/j.yexcr.2012.09.018.
Lesscher, H. M., Wallace, M. J., Zeng, L., Wang, V., Deitchman, J. K., McMahon, T., et al. (2009). Amygdala protein kinase C epsilon controls alcohol consumption. *Genes, Brain, and Behavior*, *8*(5), 493–499. http://dx.doi.org/10.1111/j.1601-183X.2009.00485.x.
Lewohl, J. M., Nunez, Y. O., Dodd, P. R., Tiwari, G. R., Harris, R. A., & Mayfield, R. D. (2011). Up-regulation of microRNAs in brain of human alcoholics. *Alcoholism, Clinical and Experimental Research*, *35*(11), 1928–1937. http://dx.doi.org/10.1111/j.1530-0277.2011.01544.x.
Lewohl, J. M., Wang, L., Miles, M. F., Zhang, L., Dodd, P. R., & Harris, R. A. (2000). Gene expression in human alcoholism: Microarray analysis of frontal cortex. *Alcoholism, Clinical and Experimental Research*, *24*(12), 1873–1882.
Lippai, D., Bala, S., Petrasek, J., Csak, T., Levin, I., Kurt-Jones, E. A., et al. (2013). Alcohol-induced IL-1beta in the brain is mediated by NLRP3/ASC inflammasome activation that amplifies neuroinflammation. *Journal of Leukocyte Biology*, *94*(1), 171–182. http://dx.doi.org/10.1189/jlb.1212659.
Liu, J., Lewohl, J. M., Harris, R. A., Iyer, V. R., Dodd, P. R., Randall, P. K., et al. (2006). Patterns of gene expression in the frontal cortex discriminate alcoholic from non-alcoholic individuals. *Neuropsychopharmacology*, *31*(7), 1574–1582. http://dx.doi.org/10.1038/sj.npp.1300947.
Liu, J., Yang, A. R., Kelly, T., Puche, A., Esoga, C., June, H. L., Jr., et al. (2011). Binge alcohol drinking is associated with GABAA alpha2-regulated Toll-like receptor 4 (TLR4) expression in the central amygdala. *Proceedings of the National Academy of Sciences of the United States of America*, *108*(11), 4465–4470. http://dx.doi.org/10.1073/pnas.1019020108.

Maeda, T., Kiguchi, N., Fukazawa, Y., Yamamoto, A., Ozaki, M., & Kishioka, S. (2007). Peroxisome proliferator-activated receptor gamma activation relieves expression of behavioral sensitization to methamphetamine in mice. *Neuropsychopharmacology*, *32*(5), 1133–1140. http://dx.doi.org/10.1038/sj.npp.1301213.

Mascia, P., Pistis, M., Justinova, Z., Panlilio, L. V., Luchicchi, A., Lecca, S., et al. (2011). Blockade of nicotine reward and reinstatement by activation of alpha-type peroxisome proliferator-activated receptors. *Biological Psychiatry*, *69*(7), 633–641. http://dx.doi.org/10.1016/j.biopsych.2010.07.009.

Mayfield, J., Ferguson, L., & Harris, R. A. (2013). Neuroimmune signaling: A key component of alcohol abuse. *Current Opinion in Neurobiology*, *23*(4), 513–520. http://dx.doi.org/10.1016/j.conb.2013.01.024.

Mayfield, R. D., Harris, R. A., & Schuckit, M. A. (2008). Genetic factors influencing alcohol dependence. *British Journal of Pharmacology*, *154*(2), 275–287. http://dx.doi.org/10.1038/bjp.2008.88.

Miles, M. F., Diaz, J. E., & DeGuzman, V. S. (1991). Mechanisms of neuronal adaptation to ethanol. Ethanol induces Hsc70 gene transcription in NG108-15 neuroblastoma x glioma cells. *The Journal of Biological Chemistry*, *266*(4), 2409–2414.

Moller, D. E., & Berger, J. P. (2003). Role of PPARs in the regulation of obesity-related insulin sensitivity and inflammation. *International Journal of Obesity and Related Metabolic Disorders*, *27*(Suppl. 3), S17–S21. http://dx.doi.org/10.1038/sj.ijo.0802494.

Moschos, S. A., Williams, A. E., Perry, M. M., Birrell, M. A., Belvisi, M. G., & Lindsay, M. A. (2007). Expression profiling in vivo demonstrates rapid changes in lung microRNA levels following lipopolysaccharide-induced inflammation but not in the anti-inflammatory action of glucocorticoids. *BMC Genomics*, *8*, 240. http://dx.doi.org/10.1186/1471-2164-8-240.

Most, D., Ferguson, L., Blednov, Y., Mayfield, R. D., & Harris, R. A. (2014). The synaptoneurosome transcriptome: A model for profiling the synaptic molecular effects of alcohol. *Pharmacogenomics Journal*. in press.

Neasta, J., Ben Hamida, S., Yowell, Q., Carnicella, S., & Ron, D. (2010). Role for mammalian target of rapamycin complex 1 signaling in neuroadaptations underlying alcohol-related disorders. *Proceedings of the National Academy of Sciences of the United States of America*, *107*(46), 20093–20098. http://dx.doi.org/10.1073/pnas.1005554107.

Nestler, E. J. (2001). Molecular basis of long-term plasticity underlying addiction. *Nature Reviews. Neuroscience*, *2*(2), 119–128. http://dx.doi.org/10.1038/35053570.

Nicolas, F. E., Pais, H., Schwach, F., Lindow, M., Kauppinen, S., Moulton, V., et al. (2008). Experimental identification of microRNA-140 targets by silencing and overexpressing miR-140. *RNA*, *14*(12), 2513–2520. http://dx.doi.org/10.1261/rna.1221108.

Nunez, Y. O., & Mayfield, R. D. (2012). Understanding alcoholism through microRNA signatures in brains of human alcoholics. *Frontiers in Genetics*, *3*, 43. http://dx.doi.org/10.3389/fgene.2012.00043.

Nunez, Y. O., Truitt, J. M., Gorini, G., Ponomareva, O. N., Blednov, Y. A., Harris, R. A., et al. (2013). Positively correlated miRNA-mRNA regulatory networks in mouse frontal cortex during early stages of alcohol dependence. *BMC Genomics*, *14*, 725. http://dx.doi.org/10.1186/1471-2164-14-725.

Okvist, A., Johansson, S., Kuzmin, A., Bazov, I., Merino-Martinez, R., Ponomarev, I., et al. (2007). Neuroadaptations in human chronic alcoholics: Dysregulation of the NF-kappaB system. *PLoS One*, *2*(9), e930. http://dx.doi.org/10.1371/journal.pone.0000930.

O'Neill, L. A., Sheedy, F. J., & McCoy, C. E. (2011). MicroRNAs: The fine-tuners of Toll-like receptor signalling. *Nature Reviews. Immunology*, *11*(3), 163–175. http://dx.doi.org/10.1038/nri2957.

Osterndorff-Kahanek, E., Ponomarev, I., Blednov, Y. A., & Harris, R. A. (2013). Gene expression in brain and liver produced by three different regimens of alcohol consumption in mice: Comparison with immune activation. *PLoS One*, *8*(3), e59870. http://dx.doi.org/10.1371/journal.pone.0059870.

Panlilio, L. V., Justinova, Z., Mascia, P., Pistis, M., Luchicchi, A., Lecca, S., et al. (2012). Novel use of a lipid-lowering fibrate medication to prevent nicotine reward and relapse: Preclinical findings. *Neuropsychopharmacology*, *37*(8), 1838–1847. http://dx.doi.org/10.1038/npp.2012.31.

Pascual, M., Balino, P., Alfonso-Loeches, S., Aragon, C. M., & Guerri, C. (2011). Impact of TLR4 on behavioral and cognitive dysfunctions associated with alcohol-induced neuroinflammatory damage. *Brain, Behavior, and Immunity*, *25*(Suppl. 1), S80–S91. http://dx.doi.org/10.1016/j.bbi.2011.02.012.

Pascual, M., Pla, A., Minarro, J., & Guerri, C. (2014). Neuroimmune activation and myelin changes in adolescent rats exposed to high-dose alcohol and associated cognitive dysfunction: A review with reference to human adolescent drinking. *Alcohol and Alcoholism*, *49*(2), 187–192. http://dx.doi.org/10.1093/alcalc/agt164.

Pascual-Lucas, M., Fernandez-Lizarbe, S., Montesinos, J., & Guerri, C. (2014). LPS or ethanol triggers clathrin- and rafts/caveolae-dependent endocytosis of TLR4 in cortical astrocytes. *Journal of Neurochemistry*, *129*(3), 448–462. http://dx.doi.org/10.1111/jnc.12639.

Pietrzykowski, A. Z., Friesen, R. M., Martin, G. E., Puig, S. I., Nowak, C. L., Wynne, P. M., et al. (2008). Posttranscriptional regulation of BK channel splice variant stability by miR-9 underlies neuroadaptation to alcohol. *Neuron*, *59*(2), 274–287. http://dx.doi.org/10.1016/j.neuron.2008.05.032.

Pignataro, L., Varodayan, F. P., Tannenholz, L. E., & Harrison, N. L. (2009). The regulation of neuronal gene expression by alcohol. *Pharmacology & Therapeutics*, *124*(3), 324–335. http://dx.doi.org/10.1016/j.pharmthera.2009.09.002.

Ponomarev, I., Wang, S., Zhang, L., Harris, R. A., & Mayfield, R. D. (2012). Gene coexpression networks in human brain identify epigenetic modifications in alcohol dependence. *The Journal of Neuroscience*, *32*(5), 1884–1897. http://dx.doi.org/10.1523/JNEUROSCI.3136-11.2012.

Qin, L., He, J., Hanes, R. N., Pluzarev, O., Hong, J. S., & Crews, F. T. (2008). Increased systemic and brain cytokine production and neuroinflammation by endotoxin following ethanol treatment. *Journal of Neuroinflammation*, *5*, 10. http://dx.doi.org/10.1186/1742-2094-5-10.

Qin, L., Wu, X., Block, M. L., Liu, Y., Breese, G. R., Hong, J. S., et al. (2007). Systemic LPS causes chronic neuroinflammation and progressive neurodegeneration. *Glia*, *55*(5), 453–462. http://dx.doi.org/10.1002/glia.20467.

Quinlan, E. M., Philpot, B. D., Huganir, R. L., & Bear, M. F. (1999). Rapid, experience-dependent expression of synaptic NMDA receptors in visual cortex in vivo. *Nature Neuroscience*, *2*(4), 352–357. http://dx.doi.org/10.1038/7263.

Raab-Graham, K. F., Haddick, P. C., Jan, Y. N., & Jan, L. Y. (2006). Activity- and mTOR-dependent suppression of Kv1.1 channel mRNA translation in dendrites. *Science*, *314*(5796), 144–148. http://dx.doi.org/10.1126/science.1131693.

Redell, J. B., Moore, A. N., Ward, N. H., 3rd., Hergenroeder, G. W., & Dash, P. K. (2010). Human traumatic brain injury alters plasma microRNA levels. *Journal of Neurotrauma*, *27*(12), 2147–2156. http://dx.doi.org/10.1089/neu.2010.1481.

Roberts, R. A., James, N. H., Woodyatt, N. J., Macdonald, N., & Tugwood, J. D. (1998). Evidence for the suppression of apoptosis by the peroxisome proliferator activated receptor alpha (PPAR alpha). *Carcinogenesis*, *19*(1), 43–48.

Ron, D., & Messing, R. O. (2013). Signaling pathways mediating alcohol effects. *Current Topics in Behavioral Neurosciences*, *13*, 87–126. http://dx.doi.org/10.1007/7854_2011_161.

Rosso, M., Munoz, M., & Berger, M. (2012). The role of neurokinin-1 receptor in the microenvironment of inflammation and cancer. *Scientific World Journal, 2012*, 381434. http://dx.doi.org/10.1100/2012/381434.
Sanna, P. P., Simpson, C., Lutjens, R., & Koob, G. (2002). ERK regulation in chronic ethanol exposure and withdrawal. *Brain Research, 948*(1–2), 186–191.
Schlicht, I. (1987). Enhancement of voluntary alcohol consumption in rats by clofibrate feeding. *Alcohol, 4*(3), 199–206.
Schoonjans, K., Staels, B., & Auwerx, J. (1996). The peroxisome proliferator activated receptors (PPARS) and their effects on lipid metabolism and adipocyte differentiation. *Biochimica et Biophysica Acta, 1302*(2), 93–109.
Schratt, G. M., Tuebing, F., Nigh, E. A., Kane, C. G., Sabatini, M. E., Kiebler, M., et al. (2006). A brain-specific microRNA regulates dendritic spine development. *Nature, 439*(7074), 283–289. http://dx.doi.org/10.1038/nature04367.
Seneviratne, C., Ait-Daoud, N., Ma, J. Z., Chen, G., Johnson, B. A., & Li, M. D. (2009). Susceptibility locus in neurokinin-1 receptor gene associated with alcohol dependence. *Neuropsychopharmacology, 34*(11), 2442–2449. http://dx.doi.org/10.1038/npp.2009.65.
Smalheiser, N. R., & Lugli, G. (2009). microRNA regulation of synaptic plasticity. *Neuromolecular Medicine, 11*(3), 133–140. http://dx.doi.org/10.1007/s12017-009-8065-2.
Sonkoly, E., Stahle, M., & Pivarcsi, A. (2008). MicroRNAs and immunity: Novel players in the regulation of normal immune function and inflammation. *Seminars in Cancer Biology, 18*(2), 131–140. http://dx.doi.org/10.1016/j.semcancer.2008.01.005.
Soreq, H., & Wolf, Y. (2011). NeurimmiRs: microRNAs in the neuroimmune interface. *Trends in Molecular Medicine, 17*(10), 548–555. http://dx.doi.org/10.1016/j.molmed.2011.06.009.
Sosanya, N. M., Huang, P. P., Cacheaux, L. P., Chen, C. J., Nguyen, K., Perrone-Bizzozero, N. I., et al. (2013). Degradation of high affinity HuD targets releases Kv1.1 mRNA from miR-129 repression by mTORC1. *The Journal of Cell Biology, 202*(1), 53–69. http://dx.doi.org/10.1083/jcb.201212089.
Stopponi, S., de Guglielmo, G., Somaini, L., Cippitelli, A., Cannella, N., Kallupi, M., et al. (2013). Activation of PPARgamma by pioglitazone potentiates the effects of naltrexone on alcohol drinking and relapse in msP rats. *Alcoholism, Clinical and Experimental Research, 37*(8), 1351–1360. http://dx.doi.org/10.1111/acer.12091.
Stopponi, S., Somaini, L., Cippitelli, A., Cannella, N., Braconi, S., Kallupi, M., et al. (2011). Activation of nuclear PPARgamma receptors by the antidiabetic agent pioglitazone suppresses alcohol drinking and relapse to alcohol seeking. *Biological Psychiatry, 69*(7), 642–649. http://dx.doi.org/10.1016/j.biopsych.2010.12.010.
Sun, Y., & Bennett, A. (2007). Cannabinoids: A new group of agonists of PPARs. *PPAR Research, 2007*, 23513. http://dx.doi.org/10.1155/2007/23513.
Szabo, G., Mandrekar, P., Petrasek, J., & Catalano, D. (2011). The unfolding web of innate immune dysregulation in alcoholic liver injury. *Alcoholism, Clinical and Experimental Research, 35*(5), 782–786. http://dx.doi.org/10.1111/j.1530-0277.2010.01398.x.
Taganov, K. D., Boldin, M. P., Chang, K. J., & Baltimore, D. (2006). NF-kappaB-dependent induction of microRNA miR-146, an inhibitor targeted to signaling proteins of innate immune responses. *Proceedings of the National Academy of Sciences of the United States of America, 103*(33), 12481–12486. http://dx.doi.org/10.1073/pnas.0605298103.
Tapocik, J. D., Solomon, M., Flanigan, M., Meinhardt, M., Barbier, E., Schank, J. R., et al. (2013). Coordinated dysregulation of mRNAs and microRNAs in the rat medial prefrontal cortex following a history of alcohol dependence. *The Pharmacogenomics Journal, 13*(3), 286–296. http://dx.doi.org/10.1038/tpj.2012.17.
Thorsell, A., Schank, J. R., Singley, E., Hunt, S. P., & Heilig, M. (2010). Neurokinin-1 receptors (NK1R:s), alcohol consumption, and alcohol reward in mice. *Psychopharmacology, 209*(1), 103–111. http://dx.doi.org/10.1007/s00213-010-1775-1.

Tontonoz, P., Hu, E., & Spiegelman, B. M. (1994). Stimulation of adipogenesis in fibroblasts by PPAR gamma 2, a lipid-activated transcription factor. *Cell*, *79*(7), 1147–1156.

Triantafilou, M., & Triantafilou, K. (2002). Lipopolysaccharide recognition: CD14, TLRs and the LPS-activation cluster. *Trends in Immunology*, *23*(6), 301–304.

Varodayan, F. P., Pignataro, L., & Harrison, N. L. (2011). Alcohol induces synaptotagmin 1 expression in neurons via activation of heat shock factor 1. *Neuroscience*, *193*, 63–71. http://dx.doi.org/10.1016/j.neuroscience.2011.07.035.

Vasudevan, S., & Steitz, J. A. (2007). AU-rich-element-mediated upregulation of translation by FXR1 and Argonaute 2. *Cell*, *128*(6), 1105–1118. http://dx.doi.org/10.1016/j.cell.2007.01.038.

Vasudevan, S., Tong, Y., & Steitz, J. A. (2007). Switching from repression to activation: microRNAs can up-regulate translation. *Science*, *318*(5858), 1931–1934. http://dx.doi.org/10.1126/science.1149460.

Vetreno, R. P., Qin, L., & Crews, F. T. (2013). Increased receptor for advanced glycation end product expression in the human alcoholic prefrontal cortex is linked to adolescent drinking. *Neurobiology of Disease*, *59*, 52–62. http://dx.doi.org/10.1016/j.nbd.2013.07.002.

Ward, R. J., Lallemand, F., & de Witte, P. (2014). Influence of adolescent heavy session drinking on the systemic and brain innate immune system. *Alcohol and Alcoholism*, *49*(2), 193–197. http://dx.doi.org/10.1093/alcalc/agu002.

Whitman, B. A., Knapp, D. J., Werner, D. F., Crews, F. T., & Breese, G. R. (2013). The cytokine mRNA increase induced by withdrawal from chronic ethanol in the sterile environment of brain is mediated by CRF and HMGB1 release. *Alcoholism, Clinical and Experimental Research*, *37*(12), 2086–2097. http://dx.doi.org/10.1111/acer.12189.

Williamson, L. L., & Bilbo, S. D. (2013). Chemokines and the hippocampus: A new perspective on hippocampal plasticity and vulnerability. *Brain, Behavior, and Immunity*, *30*, 186–194. http://dx.doi.org/10.1016/j.bbi.2013.01.077.

Wu, Y., Lousberg, E. L., Moldenhauer, L. M., Hayball, J. D., Coller, J. K., Rice, K. C., et al. (2012). Inhibiting the TLR4-MyD88 signalling cascade by genetic or pharmacological strategies reduces acute alcohol-induced sedation and motor impairment in mice. *British Journal of Pharmacology*, *165*(5), 1319–1329. http://dx.doi.org/10.1111/j.1476-5381.2011.01572.x.

Yadav, S., Pandey, A., Shukla, A., Talwelkar, S. S., Kumar, A., Pant, A. B., et al. (2011). miR-497 and miR-302b regulate ethanol-induced neuronal cell death through BCL2 protein and cyclin D2. *The Journal of Biological Chemistry*, *286*(43), 37347–37357. http://dx.doi.org/10.1074/jbc.M111.235531.

Yakovleva, T., Bazov, I., Watanabe, H., Hauser, K. F., & Bakalkin, G. (2011). Transcriptional control of maladaptive and protective responses in alcoholics: A role of the NF-kappaB system. *Brain, Behavior, and Immunity*, *25*(Suppl. 1), S29–S38. http://dx.doi.org/10.1016/j.bbi.2010.12.019.

Yirmiya, R., & Goshen, I. (2011). Immune modulation of learning, memory, neural plasticity and neurogenesis. *Brain, Behavior, and Immunity*, *25*(2), 181–213. http://dx.doi.org/10.1016/j.bbi.2010.10.015.

Zou, J., & Crews, F. (2010). Induction of innate immune gene expression cascades in brain slice cultures by ethanol: Key role of NF-kappaB and proinflammatory cytokines. *Alcoholism, Clinical and Experimental Research*, *34*(5), 777–789. http://dx.doi.org/10.1111/j.1530-0277.2010.01150.x.

CHAPTER THREE

Fetal Alcohol Spectrum Disorders and Neuroimmune Changes

Paul D. Drew, Cynthia J.M. Kane[1]
Department of Neurobiology and Developmental Sciences, University of Arkansas for Medical Sciences, Little Rock, Arkansas, USA
[1]Corresponding author: e-mail address: kanecynthiaj@uams.edu

Contents

Abstract

The behavioral consequences of fetal alcohol spectrum disorders (FASD) are serious and persist throughout life. The causative mechanisms underlying FASD are poorly understood. However, much has been learned about FASD from human structural and functional studies as well as from animal models, which have provided a greater understanding of the mechanisms underlying FASD. Using animal models of FASD, it has been recently discovered that ethanol induces neuroimmune activation in the

International Review of Neurobiology, Volume 118
ISSN 0074-7742
http://dx.doi.org/10.1016/B978-0-12-801284-0.00003-8

developing brain. The resulting microglial activation, production of proinflammatory molecules, and alteration in expression of developmental genes are postulated to alter neuron survival and function and lead to long-term neuropathological and cognitive defects. It has also been discovered that microglial loss occurs, reducing microglia's ability to protect neurons and contribute to neuronal development. This is important, because emerging evidence demonstrates that microglial depletion during brain development leads to long-term neuropathological and cognitive defects. Interestingly, the behavioral consequences of microglial depletion and neuroimmune activation in the fetal brain are particularly relevant to FASD. This chapter reviews the neuropathological and behavioral abnormalities of FASD and delineates correlates in animal models. This serves as a foundation to discuss the role of the neuroimmune system in normal brain development, the consequences of microglial depletion and neuroinflammation, the evidence of ethanol induction of neuroinflammatory processes in animal models of FASD, and the development of anti-inflammatory therapies as a new strategy for prevention or treatment of FASD. Together, this knowledge provides a framework for discussion and further investigation of the role of neuroimmune processes in FASD.

1. OVERVIEW OF FETAL ALCOHOL SPECTRUM DISORDERS

Maternal consumption of alcohol during pregnancy can lead to a spectrum of defects in their offspring. The range of disorders induced by gestational alcohol exposure is designated as fetal alcohol spectrum disorders (FASD) and includes the severe disorder of fetal alcohol syndrome (FAS). FASD is associated with monumental personal, societal, and economic impacts. In the United States, a staggering 12% of pregnant women consume alcohol despite extensive public health warnings (Floyd, Weber, Denny, & O'Connor, 2009). The consequence is an incidence of FASD of at least 1 in 100 births and an incidence of FAS of 2–7 in 1000 births (May et al., 2009; Sampson et al., 1997). The direct economic cost of FASD in the United States was recently estimated at $4 billion per year (Lupton, Burd, & Harwood, 2004). Thus, FASD is a major public health problem.

FASD represents a range of mild to severe effects on the brain and is the primary cause of mental retardation (Abel & Sokol, 1986; American Academy of Pediatrics Committee on Substance Abuse and Committee on Children with Disabilities, 2000). Both structural and functional defects are produced in the developing brain by fetal alcohol exposure (Riley & McGee, 2005). The CNS pathology and cognitive and behavioral impairments commonly persist throughout life (Riley, Infante, & Warren, 2011; Streissguth, Landesman-Dwyer, Martin, & Smith, 1980). Brain structural defects in individuals with FASD are commonly identified in the corpus callosum, cerebellum, cerebral cortex, hippocampus, amygdala, thalamus, and basal ganglia (Astley et al., 2009; Norman, Crocker, Mattson, & Riley, 2009). These defects range from diminution in the size of the brain region to microstructural pathology at the level of loss of neurons and glial cells, ectopic locations of neurons and glia, or defects in neural connectivity. Clinical imaging studies of individuals with FASD indicate that the extent of brain damage correlates with the extent of cognitive deficits as well as with FAS facial dysmorphology (Astley et al., 2009; Coles & Li, 2011; Lebel, Roussotte, & Sowell, 2011; O'Hare et al., 2005; Suttie et al., 2013).

2. NORMAL BRAIN DEVELOPMENT

The fetal brain develops throughout the entire period of pregnancy and is the most vulnerable organ to alcohol pathology. The nervous system develops from neuroectoderm, which forms the neural tube (see Squire et al., 2013). Neuroepithelial cells in the wall of the neural tube proliferate to form neuroblasts. Glioblasts are produced by neuroepithelial proliferation to produce astrocytes and oligodendrocytes. Microglial cells are produced from nonneuroectodermal, mesenchymal tissue as described below (Saijo & Glass, 2011). The cranial end of the neural tube forms the prosencephalon (forebrain), mesencephalon (midbrain), and rhombencephalon (hindbrain). The cerebral cortex, hippocampus, and cerebellum develop by the proliferation of neuroepithelial cells in the innermost layers of the neural tube. Cerebral cortical and hippocampal neurons are generated from progenitor cells in the ventricular and subventricular zones of the telencephalon by proliferation, differentiation, and migration. Cerebellar Purkinje neurons are generated by the proliferation of neuroepithelial cells in the ventricular zone and migration into the developing cerebellar cortex. Cerebellar granule cells are generated by the proliferation of neuroepithelial cells in the rhombencephalon and migration to form the granular layer. Migrating neurons use radial glial cells, chemoattractant and chemorepulsive molecules, cytokines and chemokines, neurotrophins, and neurotransmitters as guides to their correct position. As described below, studies in animal models reveal that alcohol exposure leads to neuron and glial death and disruption of normal neurogenesis, differentiation, and migration of neurons.

3. FASD NEUROPATHOLOGY IN HUMANS

Because of advanced imaging capabilities and earlier and more reliable diagnosis of FASD, detailed analyses of structural anomalies in the brain of individuals with FASD are advancing. Microencephaly is the most commonly reported structural brain defect. But, magnetic resonance imaging (MRI) has revealed more specific volume reductions in the corpus callosum, cerebral cortex, cerebellum, and subcortical structures including the hippocampus, basal ganglia, amygdala, and thalamus (Astley et al., 2009; Coles & Li, 2011; Lebel et al., 2011; Norman et al., 2009; Riley et al., 2011; Riley, McGee, & Sowell, 2004). Additional studies, as described below, have uncovered both gross and microstructural brain defects that correspond to functional deficits in individuals with FASD.

The corpus callosum is the principal white matter track in the brain and is very sensitive to fetal alcohol exposure. It exhibits gross changes with partial to complete agenesis and changes in volume, length, and thickness in individuals with FASD (Yang et al., 2012). At a higher level of resolution, diffusion tensor imaging (DTI) reveals microstructural abnormalities throughout the region (Wozniak et al., 2009). There is a striking correspondence between gross and microstructural pathologies in the corpus callosum and FASD disease severity, facial dysmorphology, and functional deficits (Sowell et al., 2001; Wozniak & Muetzel, 2011). For example, reduction in the size of the corpus callosum and displacement correspond to decreased executive function, learning and memory, and impaired motor skills (Kodituwakku, Kalberg, & May, 2001; Roebuck-Spencer, Mattson, Marion, Brown, & Riley, 2004; Sowell et al., 2001).

The cerebral cortex is also highly vulnerable to fetal alcohol exposure. At the gross level, aberrant formation of the gyri and sulci occurs (De Guio et al., 2013). MRI reveals that the frontal cortex atrophies (Leigland, Ford, Lerch, & Kroenke, 2013; Sowell et al., 2008; Yang et al., 2012; Zhou et al., 2011). Structural and functional imaging also indicates deficits in the frontal, parietal, and temporal lobes (Archibald et al., 2001; Bhatara et al., 2002; Lebel et al., 2012; Riley et al., 2004). There is a reduction in the volume and altered morphology of both the cortical gray matter and white matter (Bjorkquist, Fryer, Reiss, Mattson, & Riley, 2010; Nardelli, Lebel, Rasmussen, Andrew, & Beaulieu, 2011; Riley et al., 2004). Further, there are changes throughout the subcortical region with particular losses in the hippocampus, thalamus, and globus pallidus (Riley et al., 2004; Wozniak et al., 2009). Correlation to disease severity is strong for structural and functional defects in the cerebral cortex (Norman et al., 2009; Yang et al., 2012).

Gross and fine structural defects are also produced in the cerebellum by fetal alcohol exposure. The volume of the entire cerebellum is smaller, and the reduced volume and displacement of the anterior vermis and posterior inferior regions are particularly striking (Archibald et al., 2001; Autti-Ramo et al., 2002; O'Hare et al., 2005; Sowell et al., 1996). Defects in cerebellar development have been often documented in individuals with moderate to severe FASD and correlate with disease severity (O'Hare et al., 2005; Wozniak et al., 2009). For example, DTI studies of the cerebellum have revealed defects in the cerebellar white matter tracts that correspond to behavioral deficits (Spottiswoode et al., 2011).

There is an emerging wealth of information defining functional connectivity defects within and between brain regions in individuals with FASD (Norman et al., 2013; Wozniak et al., 2013). There are increases or decreases in regional brain activity that correlate to deficits in cognitive function including executive function, learning and memory, mathematical processing, attention, and inhibition (Coles & Li, 2011). For example, there is differential activation of circuitry in regions of the cerebral cortex, corpus callosum, striatum, and cerebellum during cognitive tasks (see, e.g., Coles & Li, 2011; Diwadkar et al., 2013; Lebel, Rasmussen, Wyper, Andrew, & Beaulieu, 2010; Norman et al., 2013; Wozniak et al., 2011, 2013. These studies are particularly enlightening, because the degree of difference in brain activity correlates to the degree of cognitive impairment and disease severity.

4. FASD NEUROPATHOLOGY IN ANIMAL MODELS

Human studies confirm that findings in animal models accurately reproduce major aspects of the neuropathology and behavioral dysfunction seen in FASD. Treatment of rodents and nonhuman primates with alcohol (ethanol) has proved to be highly informative for understanding the neuroanatomical and cellular events that occur as a consequence of fetal alcohol exposure. For comparison to human gestation, it is important to align the temporal stages of brain development with the corresponding stage in human brain development. For example, rodent birth represents human midgestation for the development of several brain regions, whereas birth represents the beginning of the human third trimester for the development of other brain regions (Clancy, Darlington, & Finlay, 2001; Clancy, Finlay, Darlington, & Anand, 2007; Dobbing & Sands, 1979). Understanding parallels between human and rodent brain development is particularly important because so much of what we know about FASD neuropathology and its causes have been identified first in rodent models of these disorders.

Imaging studies of mice exposed to gestational ethanol have revealed significant parallels with human FASD including facial dysmorphology and defects in brain development. There is a highly specific pattern of regional brain malformation depending on the temporal period of exposure to ethanol (O'Leary-Moore, Parnell, Lipinski, & Sulik, 2011). For example, exposure on mouse gestational day 7 produces defects in the cerebral cortex, hippocampus, and basal ganglia in addition to abnormalities in fiber tracts (Godin et al., 2010). However, exposure on gestational day 8 produces neuronal death in the hindbrain and cranial neural crest cells, cranial nerve defects with the loss of ganglia, and abnormalities in fiber tracts (Parnell et al., 2009). The volume of the cerebral cortex, cerebellum, and hippocampus is reduced, and hypoplasia or agenesis of the corpus callosum is also present. Other exposure periods produce different unique patterns of structural and functional brain defects as demonstrated in numerous studies (e.g., O'Leary-Moore et al., 2011).

Investigation of nonhuman primates as models of FASD has provided important insights. These studies provide strong evidence that the cerebral cortex is vulnerable to ethanol-induced neuropathology. Offspring of Macaque monkeys exposed to ethanol during early pregnancy experienced significant neuronal loss in the cortex, brainstem, and subcortical structures (Farber, Creeley, & Olney, 2010; Miller, 2006, 2007; Mooney & Miller, 2001, 2009). Studies in rodents also demonstrate loss of cortical neurons in offspring following ethanol exposure during gestation or in the postnatal period (Ikonomidou et al., 2000; Miller & Potempa, 1990; Wilson, Peterson, Basavaraj, & Saito, 2011). Cortical volume, thickness, and surface area are reduced by ethanol exposure throughout gestation in rodents (Leigland et al., 2013). Proper neuron migration into the cerebral cortex is also disrupted by ethanol (Gressens, Lammens, Picard, & Evrard, 1992; Miller, 1993) apparently through inhibition of radial glia development and ethanol-induced changes in the expression of molecules central to migration (Aronne, Guadagnoli, Fontanet, Evrard, & Brusco, 2011; Miller & Robertson, 1993). Rodent models of fetal ethanol exposure further demonstrate neuroanatomical, synaptic, and electrophysiological defects in neocortical circuitry that correspond to deficits in behavior (El Shawa, Abbott, & Huffman, 2013; Inomata, Nasu, & Tanaka, 1987; Slawecki, Thomas, Riley, & Ehlers, 2004; Whitcher & Klintsova, 2008; Wilson et al., 2011).

The developing hippocampus of the rodent is vulnerable to ethanol exposure. The size of the individual neuron populations in all regions of the hippocampus is reduced by ethanol exposure during gestation or in the postnatal period (Gil-Mohapel, Boehme, Kainer, & Christie, 2010). This includes significant loss of pyramidal neurons in CA1 and CA3 and granule neurons in the dentate gyrus (Ieraci & Herrera, 2007; Ikonomidou et al., 2000; Klintsova et al., 2007; Livy, Miller, Maier, & West, 2003; Tran & Kelly, 2003; Uban et al., 2010). Developmental ethanol exposure also produces neuroanatomical and electrophysiological defects in hippocampal circuitry (Diaz et al., 2014; Everett, Licon-Munoz, & Valenzuela, 2012; Lindquist, Sokoloff, Milner, & Steinmetz, 2013; Sakata-Haga et al., 2003). Defects in the hippocampus correlate with behavioral deficits, particularly with learning and memory (Banuelos et al., 2012; El Shawa et al., 2013; Hamilton et al., 2010; Idrus, McGough, Riley, & Thomas, 2014; Kelly, Pierce, & West, 1987; Thomas, Idrus, Monk, & Dominguez, 2010; Wagner, Zhou, & Goodlett, 2014; West, Kelly, & Pierce, 1987; Zink et al., 2011).

The developing cerebellum is vulnerable to ethanol exposure. Administration of ethanol to Macaque monkeys demonstrates significant cerebellar Purkinje neuron loss in their offspring (Bonthius et al., 1996; Farber et al., 2010). In rodents, cerebellar volume is reduced by midgestational ethanol exposure (Parnell et al., 2013). The cerebellum is also vulnerable to ethanol in the postnatal period where there is volume loss as well as marked loss of Purkinje neurons, granule neurons, and deep cerebellar neurons (Goodlett & Eilers, 1997; Green, Tran, Steinmetz, & Goodlett, 2002; Hamre & West, 1993; Napper & West, 1995; Pierce, Kane, Serbus, & Light, 1997; Pierce, Serbus, & Light, 1993; Pierce, Williams, & Light, 1999). Functional development of neurons is inhibited as evidenced by impaired migration, stunted development of dendrites, reduction in the number of synapses, and impaired electrophysiological activity (Allam et al., 2013; Jiang, Kumada, Cameron, & Komuro, 2008; Kane et al., 2011; Servais et al., 2007; Smith & Davies, 1990; Valenzuela, Lindquist, & Zamudio-Bulcock, 2010). These defects contribute to the deficits in motor coordination and classical conditioning commonly observed in rodent models of FASD (Brown, Calizo, & Stanton, 2008; Goodlett, Thomas, & West, 1991; Idrus, McGough, Riley, & Thomas, 2011; Klintsova et al., 1998; Lewis, Wellmann, & Barron, 2007; Murawski, Jablonski, Brown, & Stanton, 2013; Wagner, Klintsova, Greenough, & Goodlett, 2013).

5. BEHAVIORAL CONSEQUENCES IN HUMANS WITH FASD

The negative cognitive and behavioral outcomes associated with FASD are far ranging. This reflects the extended vulnerability of the developing brain to ethanol throughout gestation. The cognitive consequences of

FASD include deficits in executive function, deficits in learning and memory, impaired information processing, deficits in vigilance, delayed reaction time, restricted mathematical ability, poor speech and language skills, and impaired visual-spatial ability (Guerri, Bazinet, & Riley, 2009; Lebel et al., 2010; Mattson & Riley, 1998; Mattson & Roebuck, 2002; Mattson et al., 2013; Santhanam, Li, Hu, Lynch, & Coles, 2009). Impairment of executive function is particularly common in FASD, leading to problems with planning, set shifting, fluency, response inhibition, and working memory (Connor, Sampson, Bookstein, Barr, & Streissguth, 2000; Kodituwakku et al., 2001). Psychiatric disorders include abuse and addiction to alcohol and other drugs, internalizing and externalizing behavioral disorders, mood disturbances, and major depressive disorder (Barr et al., 2006; Fryer, McGee, Matt, Riley, & Mattson, 2007; Olson, Feldman, Streissguth, Sampson, & Bookstein, 1998; Streissguth et al., 2004; Ware et al., 2014). It is becoming increasingly clear due to functional imaging studies that these cognitive deficits are closely linked to ethanol-induced neuropathology in the cerebral cortex, hippocampus, cerebellum, and corpus callosum, as well as in other structures. Neuropathology in the cerebellum and basal ganglia is evident as delayed motor development and ataxia, including difficulty with balance, gait, and fine motor skills (Adnams et al., 2001; Connor, Sampson, Streissguth, Bookstein, & Barr, 2006; Mattson & Riley, 1998; Roebuck, Simmons, Mattson, & Riley, 1998; Roebuck-Spencer et al., 2004).

6. BEHAVIORAL CONSEQUENCES IN RODENT MODELS OF FASD

Human FASD behavioral outcomes have been modeled in rodents. Cerebral cortical, hippocampal, and cerebellar functions have been assessed with cognitive tests of learning and memory, motivation, anxiety, and locomotion, among others. Gestational exposure produces altered social behavior, impaired learning and memory, delayed development of spontaneous alternation behavior, increased anxiety, and poor motor coordination (El Shawa et al., 2013; Hamilton et al., 2010; Thomas et al., 2010). Postnatal exposure produces impaired learning and memory, altered delay discounting, and increased hyperactivity (Banuelos et al., 2012; Idrus et al., 2014; Kelly et al., 1987; Wagner et al., 2014; West et al., 1987; Zink et al., 2011). Ethanol-induced cerebellar pathology underlies deficits in classical eyeblink conditioning tasks (Brown et al., 2008; Murawski et al., 2013; Wagner et al., 2013). The cerebellum is centrally involved in motor function and ethanol impairs balance and coordination in rodents exposed to ethanol during early postnatal development (Goodlett et al., 1991; Idrus et al., 2011; Klintsova et al., 1998; Lewis et al., 2007).

The correspondence between ethanol-induced neuroanatomical and behavioral outcomes in humans and that in animals allows the use of animal models to elucidate links between structural and functional neuropathology and cognitive and motor deficits. Thus, a foundation for further investigation of the cellular and molecular mechanisms underlying FASD is established. This is important because, as there is no treatment that targets the causes underlying FASD and the long-term consequences can be so severe, there is a pressing need to develop therapeutic interventions to ameliorate the spectrum of deficits associated with the disorders.

7. OVERVIEW: ALCOHOL EFFECTS ON IMMUNE RESPONSE IN THE BRAIN

It is well established that alcohol has potent effects on immune activity in the CNS throughout the life span (Drew & Kane, 2013). Ethanol effects on immune response have begun to be evaluated in animal models of maternal consumption, resulting in FASD, models of binge drinking common to adolescents, and chronic alcohol abuse common in adults and in aged populations. Alcohol abuse at each of these stages can result in neuroinflammation, which is believed to contribute to ethanol-induced neurodegeneration. This chapter focuses primarily on the effects of ethanol on immune activation in the developing CNS.

The effects of ethanol on the peripheral immune system have been extensively investigated. Therefore, ethanol has complex effects on immune activity that depend on a variety of factors including age, gender, pattern of ethanol exposure (acute, binge, or chronic administration), the timing of evaluation following ethanol exposure, and the cells or tissues under investigation (Goral, Karavitis, & Kovacs, 2008). These studies demonstrate that ethanol can induce or alternatively suppress immune activation in the periphery, depending on the specific experimental paradigm. An example of the dichotomous effects of alcohol on peripheral immune responses that occurs in human alcoholics who are generally immunosuppressed and yet frequently exhibit elevated serum levels of proinflammatory cytokines (Goral et al., 2008). Relative to the periphery, the effects of ethanol on the CNS is understudied (Blanco & Guerri, 2007; Crews & Nixon, 2009; Crews et al., 2006).

A series of seminal studies demonstrated that ethanol increases the expression of proinflammatory cytokines and chemokines, as well as molecules such as nitric oxide and Cox-2 in the CNS (Blanco, Pascual, Valles, & Guerri, 2004; Crews et al., 2006; Davis & Syapin, 2004; Knapp & Crews, 1999; Ward et al., 1996; Zou & Crews, 2006, 2010). These studies further determined that transcription factors including NF-κB and CREB are activated by ethanol (Blanco et al., 2004; Crews et al., 2006; Davis & Syapin, 2004; Ward et al., 1996; Zou & Crews, 2006, 2010), suggesting that these molecules may play critical roles in ethanol induction of proinflammatory molecules, oxidative stress, and ethanol-induced pathology in the CNS. Ethanol has been demonstrated to alter the function and viability of CNS glia including microglia and astrocytes, as well as neurons. However, much work is required to better understand the effects of ethanol on neuroinflammation and neurodegeneration.

8. MICROGLIA

8.1. Overview

Together with astrocytes and oligodendrocytes, microglia comprise the glial cells of the CNS. While oligodendrocytes play a critical role in forming myelin, astrocytes and microglia maintain the health of neurons through the production of neurotrophic factors and the removal of potentially neurotoxic molecules from the CNS. Microglia also maintain the CNS by the removal of debris and are involved in synaptic pruning critical in the formation of functional synapses. However, in response to CNS insult, microglia can become activated, resulting in the production of inflammatory molecules that can contribute to neurodegenerative disorders.

8.2. Role in the CNS development

Distinct from other glia and neurons that are derived from neuroectoderm, microglia are believed to be of hematopoietic origin (Saijo & Glass, 2011). Microglia are generated in the primitive yolk sac and migrate to the CNS early during fetal development prior to formation of the blood–brain barrier. As myeloid lineage cells, microglia are most closely related to macrophages. Microglia and macrophages both express molecules such as CD11b and CD14 (Kettenmann, Hanisch, Noda, & Verkhratsky, 2011). Like macrophages, the differentiation of microglia is dependent of the transcription factor PU.1, as well as colony stimulating factor 1 and its receptor. The fact that

PU.1-knock-out mice are devoid of both microglia and macrophages supports a common myeloid lineage for these cells (McKercher et al., 1996).

Controversy has existed regarding whether macrophages are able to migrate into the mature CNS and form microglia, thus serving as a source to replenish the microglial cell population. Bone marrow chimera studies suggested that this is the case (Eglitis & Mezey, 1997; Hickey, 1991; Hickey & Kimura, 1988; Priller et al., 2001). However, these studies are complicated by the fact that the irradiation used in these protocols compromised the blood–brain barrier. More recent parabiosis studies that involve direct connection of the circulation of the donor and recipient animals demonstrate that few donor cells enter the CNS in the absence of irradiation (Ajami, Bennett, Krieger, Tetzlaff, & Rossi, 2007; Mildner et al., 2007) and macrophages that do enter do not appear to differentiate into microglia (Ajami, Bennett, Krieger, McNagny, & Rossi, 2011). In addition, microglia appear to be long-lived cells that can proliferate in response to pathological conditions in the CNS (Fellner, Jellinger, Wenning, & Stefanova, 2011; Glass, Saijo, Winner, Marchetto, & Gage, 2010; Lawson, Perry, & Gordon, 1992; Reitz, Brayne, & Mayeux, 2011), suggesting that they are replenished from stores of microglial precursors in the CNS when needed.

8.3. Function in the healthy CNS

Microglia comprise approximately 5–20% of the cells in the CNS. The relative abundance of microglia varies regionally as does the morphology of these cells (Lawson, Perry, Dri, & Gordon, 1990). This may reflect regional differences in the function and phenotype of microglia. Microglia are more abundant in gray matter than in white matter (Rivest, 2009). In the healthy mature CNS, microglia generally exhibit a ramified appearance characterized by a small soma, limited cytoplasm, and a series of thin, highly branched processes (Fig. 3.1A). Although these branched processes extend significantly away from the soma, it is believed that the territory occupied by individual microglia does not overlap with adjacent microglia (Ransohoff & Perry, 2009). Microglia contribute to homeostasis of the healthy brain by producing growth factors critical to the survival of neurons as well as protecting neurons by removing potentially neurotoxic molecules from the parenchyma. Microglia play a critical surveillance role in the CNS. They are motile and phagocytose cellular debris including that derived from apoptotic cells (Sierra et al., 2010). Removal of cellular debris in this manner

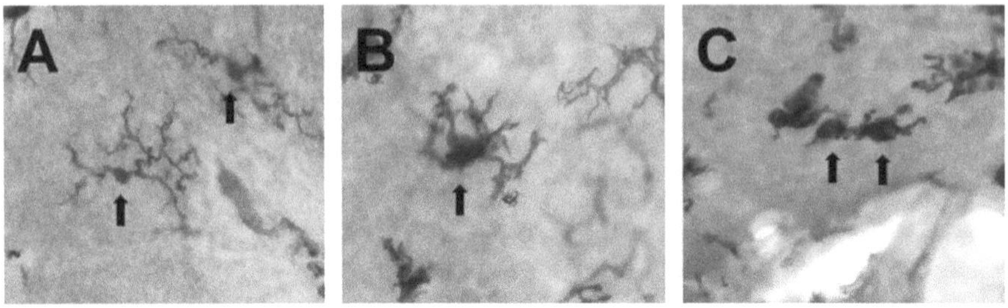

Figure 3.1 Microglial morphology exhibits a stepwise modification as the process of activation changes the functional phenotype of the cell. Microglial cells in the normal, unperturbed CNS exhibit a ramified morphology with a small soma, limited cytoplasm, and long, thin, highly branched processes (A). With activation, the cells hypertrophy and the processes retract, becoming broader and stunted with less branching (B). Highly activated cells progress toward an ameboid, hypertrophied form with few or no processes (C). These images are from the postnatal mouse cerebellum and are representative of the described morphological phenotypes of microglia. The microglial cells in these images were visualized with immunohistochemistry against the ionized calcium-binding adapter molecule (Iba-1), classically used to identify CNS microglia. Arrows indicate microglial cells with the described morphologies.

occurs in the absence of microglial activation and inflammation, thus maintaining homeostasis. Microglia also play a critical role in the formation and remodeling of synapses. Multiphoton microscopy studies indicate that microglia are dynamic *in vivo*, constantly sampling the microenvironment by extending and retracting processes, particularly at the synapse (Davalos et al., 2005; Nimmerjahn, Kirchhoff, & Helmchen, 2005). Microglia surveillance is particularly important in synaptic plasticity during development through activity-dependent synaptic pruning (Tremblay, Lowery, & Majewska, 2010). The role of microglia in synaptic plasticity is also supported by the documented physical association of these cells with developing and mature synapses (Dalmau, Finsen, Zimmer, Gonzalez, & Castellano, 1998; Fiske & Brunjes, 2000; Perry, Hume, & Gordon, 1985). Microglia also associate with dendritic spines, which further support a role for microglia in modulating the structure and function of synapses (Tremblay et al., 2010; Wake, Moorhouse, Jinno, Kohsaka, & Nabekura, 2009). Interestingly, microglia-derived products are critical to synapse development and function. This includes molecules classically defined as proinflammatory including major histocompatibility (MHC) class I, MHC I-binding receptors, complement proteins, and cytokines including TNF-α and IL-6 (Boulanger, 2009). The chemokine receptor CX3CR1 is expressed specifically on microglia and is critical in microglia–neuron interactions through association with the chemokine CX3CL1/fractalkine present on neurons. Interestingly, CX3CR1-deficient mice exhibit reduced microglial numbers during development, and these mice exhibit deficits in synapse formation and plasticity in the hippocampus (Paolicelli et al., 2011). This further supports a role of microglia in the modulation of neuronal circuitry in the developing brain.

8.4. Immune response

Traditionally, the CNS has been considered to be an immune-privileged site. This is due to the absence of lymphatic drainage from the CNS, the observation that tissue grafted into the CNS is relatively protected from immune-mediated destruction, and the presence of a blood–brain barrier that limits the movement of immune cells from the periphery. However, it is now known that peripheral immune cells are capable of entering the CNS parenchyma. In addition, resident CNS cells including microglia and to a lesser extent astrocytes provide immune functions (Carson, Doose, Melchior, Schmid, & Ploix, 2006). As noted earlier, microglia provide homeostatic functions in the healthy CNS. Upon insult or aging, microglia can become activated, changing morphology from a highly ramified cell with a small soma to a hypertrophied cell with broad, stunted processes, and then to an ameboid cell (Fig. 3.1). When activated, they can produce a variety of proinflammatory molecules including cytokines, chemokines, reactive oxygen species, and nitric oxide. These molecules can be protective, for example, in ridding the CNS of infectious agents. In this regard, microglia are the primary cell type that performs innate immune functions in the CNS. However, chronically activated microglia and these same proinflammatory molecules can be toxic to parenchymal cells including neurons. Activated microglia are more aggressively phagocytic. They also express increased levels of MHC class I and II, which are central to antigen presentation during immune responses, and are, thus, capable of serving as antigen-presenting cells. In this manner, microglia are capable of presenting antigen to T cells and mediating adaptive immune responses (Ransohoff & Perry, 2009; Saijo & Glass, 2011).

Microglial activation is triggered following the interaction of Toll-like receptors (TLRs) present on the surface of these cells with conserved motifs associated with pathogens. In addition, microglia can also be activated in response to nonpathogenic insults through TLR engagement by endogenous danger signals including heat shock protein 70 and high mobility group box protein (HMGB) 1. TLR activation results in the activation of signal transduction pathways and changes in the functional phenotype of microglia including increased cell proliferation, increased production of proinflammatory molecules, increased antigen presentation, and increased phagocytic activity. Microglial activation occurs in a stepwise and not in an all or none manner (Carson et al., 2007; Colton, 2009; Hanisch & Kettenmann, 2007; Perry, Cunningham, & Holmes, 2007; Schwartz, Butovsky, Bruck, & Hanisch, 2006; Streit, Walter, & Pennell, 1999). Microglia can also undergo distinct forms of activation characterized by differential protein expression patterns and function in a manner similar to that first described as classical or alternative activation of macrophages (Ransohoff & Perry, 2009). Following the removal of inflammatory stimuli, microglia can revert back toward a quiescent form with a ramified morphological phenotype. However, studies indicate that these microglia are not fully deactivated and are primed for subsequent activation (Hanisch & Kettenmann, 2007).

9. POTENTIAL LONG-TERM CONSEQUENCES OF IMMUNE ACTIVATION IN THE DEVELOPING CNS

As discussed previously, proinflammatory molecules including cytokines and chemokines play an important role in ridding the brain of pathogens, but expression of these molecules is low in the mature brain in the absence of inflammatory insults. In contrast, in the developing brain, proinflammatory molecules are expressed at significant levels in the absence of inflammatory insults. This reflects the fact that these molecules classically recognized for their inflammatory activity have other important roles in CNS development that are independent of an immune response. For example, expression of the cytokines IL-1β and TNF-α is developmentally regulated, and these molecules are believed to play important roles in CNS development including modulation of neural cell migration, proliferation, differentiation, and death (Merrill, 1992). The cytokine IL-6 is believed to serve as a neurotrophic factor and to contribute to vascular development in the CNS (Fee et al., 2000; Gadient & Otten, 1994). MHC class I molecules have been demonstrated to mediate activity-dependent synapse formation in the developing visual system (Corriveau, Huh, & Shatz, 1998). Complement, a traditional immune molecule central to antigen presentation, is necessary for microglial sculpting of early postnatal neural circuits (Schafer et al., 2012). Clearly, altered expression of traditional immune system molecules that have important nonimmune roles in the developing brain will have significant, detrimental consequences for brain development.

Recent studies have shown that the induction of an inflammatory response in the developing rodent brain can have dramatic, lasting effects as evidenced by cognitive deficits and behavioral disorders in adulthood (Bilbo, Smith, & Schwarz, 2012; Schwarz & Bilbo, 2012). Immune activation during early development can also alter the immune response later in life (Bilbo et al., 2012; Schwarz & Bilbo, 2012). Furthermore, even transient reduction in the number of microglia can have dramatic impact on the developing brain. For example, genetic deletion of the molecule CX3CR1, which is expressed by microglia, was used to generate a transient loss of microglia in the early postnatal mouse. Transient loss of microglia resulted in deficits in synaptic pruning, leading to weak synaptic transmission and behavioral deficits in adulthood (Zhan et al., 2014). In addition, mice with deletion of CX3CR1 demonstrate impaired hippocampal synaptic plasticity and cognitive function (Rogers et al., 2011). Collectively, these studies indicate that immune activation in the developing brain has dramatic long-term effects on synapse formation and plasticity, cognition, development of behavioral disorders, and altered immune responses. Future studies are needed to determine if ethanol induction of immune activity in the developing brain contributes to the long-term sequelae associated with FASD.

10. LINK BETWEEN ETHANOL AND IMMUNE RESPONSES

10.1. Human studies

The effects of alcohol on peripheral immune responses in humans are well established. For example, serum from alcoholics has been demonstrated to exhibit increased levels of cytokines including IL-1β, IL-6, IL-12, and TNF-α (Achur, Freeman, & Vrana, 2010). Increased cytokine expression in the serum of alcoholics is believed to result in part from release of these molecules into the circulation from the alcohol-damaged liver. In addition, peripheral blood mononuclear cells derived from alcoholics express increased levels of proinflammatory cytokines (Laso, Vaquero, Almeida, Marcos, & Orfao, 2007). Even in healthy individuals, excessive acute or binge alcohol and associated hangover is marked by increased cytokine expression in peripheral blood (Kim et al., 2003). A vicious cycle of alcohol abuse develops since proinflammatory cytokines increase alcohol craving behavior and result in increased consumption (Kiefer, Jahn, Schick, & Wiedemann, 2002).

Relatively little is known regarding the effects of alcohol on the immune response in the human brain. Postmortem evaluation of adult alcoholic brains indicates increased expression of mRNAs encoding immune molecules in the prefrontal cortex of alcoholics relative to nonalcoholics (Liu et al., 2006). The expression of the transcription factor NF-κB, which activates the expression of a wide variety of neuroimmune genes, is also increased in alcoholic brain (Okvist et al., 2007; Yakovleva, Bazov, Watanabe, Hauser, & Bakalkin, 2011). NF-κB also modulates synaptic plasticity suggesting a mechanism by which neuroimmune activation may impair cognition and memory in alcoholics. Contribution of neuroimmune activation to alcohol neuropathology is further suggested through the activity of the proinflammatory chemokine CCL2/MCP-1. CCL2 expression is increased in the hippocampus, amygdala, substantia nigra, and ventral tegmentum in alcoholics (He & Crews, 2008). Studies in transgenic mice overexpressing CCL2 in the CNS reveal that CCL2 alters synaptic transmission in the hippocampus (Nelson, Hao, Manos, Ransohoff, & Gruol, 2011) suggesting another mechanism by which neuroimmune activation may impair cognition and memory. These studies suggest that neuroimmune activity and proinflammatory molecules may alter synaptic plasticity in alcoholics.

Relatively little is known concerning the effects of alcohol on immune response in the developing human brain. It has been demonstrated that increased expression of proinflammatory cytokines including IL-1β, IL-6, and TNF-α is detected shortly after birth in the blood of neonates and mothers, when the mother chronically consumed alcohol during pregnancy (Ahluwalia et al., 2000). It was not possible in these studies to determine if these cytokines were of neonatal or maternal origin, but it is clear that the neonate is exposed to these proinflammatory molecules. Studies are needed to determine if alcohol alters the expression of proinflammatory molecules in the developing brain.

10.2. Animal studies

Animal models of FASD have played a critical role in defining the mechanisms by which ethanol damages the developing CNS. These models involve prenatal or neonatal exposure to ethanol. Prenatal exposure to ethanol is toxic to neurons *in vivo* including cerebral cortical, hippocampal, and cerebellar neurons as described above. Ethanol is also toxic to cultured primary neurons. For example, conditioned medium from ethanol-treated microglia was toxic to primary basal hippocampal neurons. Furthermore, TNF-α neurtralizing antibodies blocked ethanol-induced neuronal death

(Boyadjieva & Sarkar, 2010). This suggests that ethanol induces the production of TNF-α and possibly other proinflammatory molecules by activated microglia, which contributes to ethanol-induced neuronal cell death.

The neonatal rodent brain is also highly susceptible to the toxic effects of ethanol with neuropathology in the cerebral cortex, cerebellum, corpus callosum, hippocampus, and other regions as detailed above. The relevance of the neonatal rodent is that it is developmentally similar to the brain during the second half to third trimester of pregnancy (Clancy et al., 2001; Dobbing & Sands, 1979). We have demonstrated that ethanol exposure of 3- to 5-day-old neonatal mice results in a significant loss of cerebellar Purkinje neurons (Kane et al., 2011) as has been well documented in a variety of FASD models (Dikranian, Qin, Labruyere, Nemmers, & Olney, 2005; Hamre & West, 1993; Pierce et al., 1999). In addition to neurons, we demonstrated that microglia are highly susceptible to ethanol-induced death. Strikingly, the surviving microglia in ethanol-treated animals exhibit an altered morphology with enlarged soma and shorter, broader, less branched processes characteristic of activated microglia, as opposed to the ramified appearance of normal microglia in vehicle-treated control animals. Our studies in this FASD model further demonstrated that agonists of the anti-inflammatory nuclear receptor PPAR-γ block microglial activation and protect neurons and microglia against ethanol cytotoxicity, suggesting a possible link between ethanol-induced microglial activation and loss of neurons in the developing brain. As discussed previously, microglia perform a variety of functions in the developing brain and their loss or activation during the period of brain development produces persistent neuropathological and negative behavioral consequences. Together, these observations suggest that loss of microglia and microglial activation in response to ethanol in the developing brain may contribute to the long-term behavioral and structural anomalies associated with FASD.

Neonatal exposure to ethanol also increases the expression of cytokines in the brain, which can persist following ethanol withdrawal (Tiwari & Chopra, 2011). In these studies, neonatal rats were exposed to ethanol on postnatal days 7–9, and cytokine expression was evaluated on postnatal day 28. IL-1β, TGF-β, and TNF-α expressions were elevated in the cerebral cortex and hippocampus. The transcription factor NF-κB, which activates the expression of a number of genes that encode proinflammatory molecules, was increased following ethanol exposure in these brain regions, as were markers of oxidative stress including catalase, superoxide dismutase, nitric oxide, glutathione, and lipid peroxidation. Importantly, treatment with the antioxidant resveratrol suppressed the production of oxidative stress in the developing CNS and this resulted in improved cognition in adult animals. Other studies demonstrated that a single dose of ethanol administered at postnatal day 7 resulted in lipid peroxidation and expression of the apoptotic marker caspase-3 in the developing cerebellum (Kumar, Singh, Lavoie, Dipette, & Singh, 2011). In another study, a single dose of ethanol administered at postnatal day 7 resulted in altered morphology and increased Iba-1 staining in microglia, suggesting microglial activation (Saito et al., 2010). Collectively, these studies suggest that microglial activation, neuroinflammation, and oxidative stress contribute to ethanol-induced degeneration in the developing CNS.

As discussed previously, microglia play critical roles in CNS development and activation of immune responses in the developing brain can result in long-term consequences in the mature brain (Schwarz & Bilbo, 2012). These studies suggest that ethanol-induced immune activation in the developing CNS may produce similar long-term consequences. In this regard, prenatal exposure of rats to ethanol resulted in an altered neuroimmune response to trauma in adults (DeVito & Stone, 2001). In these studies, rats exposed to ethanol prenatally were inflicted with a stab wound to the brain as adults in order to produce a neuroimmune response. The expression of the neuroimmune activation markers GFAP, ICAM-1, and TNF-α was reduced in animals exposed to ethanol relative to vehicle-treated controls, indicating that developmental ethanol exposure alters the neuroimmune response in adulthood. In contrast, expression of ED1, which is commonly associated with microglial activation, was increased by trauma in animals exposed to ethanol prenatally, indicating that the alteration of neuroimmune responses in adulthood due to developmental ethanol exposure is complex. Future studies are needed to determine how fetal exposure to ethanol alters immune response and long-term function of the mature brain. These studies will have important implications concerning the contribution of neuroinflammation in the developing brain to the long-term consequences of FASD.

10.3. Signaling pathways

Innate immunity provides the first line of defense against CNS insult. Resident microglia and astrocytes are the principal cells that mediate innate immune responses. These cells react to insults such as pathogens or endogenous danger signals through TLRs present on their surface. Recent studies have begun to investigate the role of TLRs in modulating ethanol activation of neuroimmune responses. Studies suggest that ethanol may cause release of HMGB1 protein, which can serve as an endogenous danger signal to activate TLR signaling and generate neuroimmune activation (Vetreno, Qin, & Crews, 2013; Whitman, Knapp, Werner, Crews, & Breese, 2013; Zou & Crews, 2014). Ethanol can activate the TLR-dependent transcription factors AP1 and NF-κB in astrocytes (Blanco, Valles, Pascual, & Guerri, 2005). Ethanol can also increase the expression of downstream TLR-dependent genes including Cox-2 and iNOS via activation of p38 MAP kinase, ERK1/2, and SAPK/JNK signaling pathways (Blanco et al., 2005). Ethanol was demonstrated to cause IL-1R and TLR4 to move into lipid rafts in astrocytes (Blanco, Perez-Arago, Fernandez-Lizarbe, & Guerri, 2008). IL-1 and TLR4 signaling is important in these processes as demonstrated by the fact that neutralizing antibodies to IL-1R and TLR4 can suppress ethanol induction of these proinflammatory molecules and signaling pathways (Blanco et al., 2005). TLR4 was also shown to be critical in ethanol induction of Cox-2 and Src phosphorylation in astrocytes (Floreani et al., 2010). These observations suggest that ethanol induction of immune responses in the CNS occurs through the movement of these receptors to lipid rafts, resulting in the activation of signaling pathways that ultimately lead to the production of proinflammatory molecules.

The role of TLR4 in modulating ethanol effects on immune responses in the CNS is further supported by studies utilizing TLR4-deficient mice. Ethanol induced the expression of Cox-2 and iNOS and activated p38 MAP kinase, ERK, and JNK signaling pathways in microglia from wild-type but not TLR4-deficient mice. Conditioned medium from microglia treated with ethanol from wild-type but not TLR4-deficient mice caused apoptosis of cortical neurons. These studies indicate that ethanol effects on microglia result in the production of molecules that are indirectly toxic to neurons in a TLR4-dependent manner (Fernandez-Lizarbe, Pascual, & Guerri, 2009). In other studies, ethanol was demonstrated to increase the expression of CD11b by microglia and GFAP by astrocytes in the frontal cortex of wild-type but not TLR4-deficient mice. Furthermore, these studies indicated that ethanol increased the expression of IL-6, TNF-α, NF-κB, and the apoptotic marker caspase-3 in a TLR4-dependent manner (Alfonso-Loeches, Pascual-Lucas, Blanco, Sanchez-Vera, & Guerri, 2010). Collectively, these studies demonstrate that TLR4 plays a critical role in ethanol-induced neuroinflammation and neurodegeneration.

Proinflammatory molecules are known to increase alcohol consumption in rodents (Blednov et al., 2005, 2012). Thus, ethanol induction of neuroinflammation may create a vicious cycle, which supports additional alcohol consumption, abuse, and addiction. As noted previously, TLR4-deficient mice exhibit decreased ethanol-induced neuroinflammation. In addition, these mice consume less alcohol than their wild-type counterparts and exhibit less alcohol-mediated anxiety and cognitive impairment than control animals (Pascual, Balino, Alfonso-Loeches, Aragon, & Guerri, 2011). These studies also indicate that alcohol decreases histone acetylation in the brain in a TLR4-dependent manner. This suggests that ethanol triggers TLR4-dependent epigenetic changes in chromatin configuration, which are linked to downstream effects on neuroinflammation and ultimately to ethanol-induced alterations in cognition and behavior. Further studies indicate that TLR4-deficient mice exhibit less sedation in response to ethanol than wild-type control animals (Wu et al., 2012). This altered behavior was not due to altered alcohol pharmacodynamics, which was similar in wild-type and TLR4-deficient mice. Furthermore, these studies demonstrated that TLR4 siRNA infused into the amygdala inhibited binge drinking behavior in alcohol-preferring rats (Wu et al., 2012). Based on our understanding of the role of GABA in alcohol consumption, it may be significant that $GABA_A$ $\alpha 2$ plays a crucial role in TLR4-dependent changes in alcohol consumption (Liu et al., 2011). These studies collectively support a role for TLR4 in modulating ethanol-induced neuroinflammation, alcohol consumption, and alcohol-mediated changes in cognition and behavior.

The studies outlined above clearly demonstrate a role for TLR4 in ethanol-induced neuroinflammation and neurodegeneration. Ligand binding to TLR4 can result in the activation of the MyD88-dependent or alternatively the MyD88-independent signaling pathway. In the MyD88-dependent signaling pathway, TLR4 physically interacts with MyD88, which results in the activation of NF-κB. The MyD88-independent pathway involves interaction of TLR4 with the adaptor protein TRIF. This leads to the activation of NF-κB and IRF-3, both transcription factors that activate the expression of genes encoding proinflammatory molecules. Future studies utilizing MyD88- and TRIF-deficient mice are required to determine if these TLR4-dependent effects occur through MyD88-dependent or -independent mechanisms. In this regard, studies indicate that MyD88-deficient mice exhibit altered ethanol-induced sedation and motor impairment (Wu et al., 2012); this supports a role for MyD88-dependent signaling in modulating ethanol consumption. However, additional studies indicate that ethanol increases the activation of NF-κB and IRF-3 in microglia (Fernandez-Lizarbe et al., 2009), which suggests that MyD88-independent signaling may also mediate effects of ethanol.

As mentioned above, IL-1 signaling is believed to play a significant role in ethanol-induced neuroinflammation, where IL-1β expression is increased by ethanol (Alfonso-Loeches et al., 2010; Crews, Qin, Sheedy, Vetreno, & Zou, 2013; Lippai et al., 2013; Tiwari & Chopra, 2011). Recent studies demonstrate that in addition to IL-1β, caspase-1 activity is induced by ethanol in the adult brain (Lippai et al., 2013). Both IL-1β and caspase-1 are produced as precursor molecules, which must be processed to form mature, active IL-1β and caspase-1. Caspase-1 functions to process pro-IL-1β to mature IL-1β. Inflammasomes are protein complexes which function as a part of the innate immune system and respond to pathogens as well as to endogenous danger signals including those resulting from the toxic effects of ethanol. Caspase-1-activating inflammasomes are the best-characterized inflammasomes and function to process pro-caspase-1 to mature caspase-1. The caspase-1-activating NLRP3 inflammasome is composed of a number of proteins including NLRP3 and the adaptor protein apoptosis-associated speck-like protein (ASC). Ethanol increases caspase-1 and IL-1β expression in adult brain of wild-type but not NLRP3- or ASC-deficient mice (Lippai et al., 2013). These studies suggest that caspase-1-activating inflammasomes may modulate IL-1β-dependent and ethanol-induced neuroinflammation and neurodegeneration. However, future studies are needed to determine the mechanisms by which ethanol-induced IL-1β signaling modulates neuroinflammation and neurodegeneration in the developing and mature brain.

11. THERAPIES

11.1. Early diagnosis of FASD

Early diagnosis is believed to be essential to maximize the potential for therapies to limit the devastating effects of fetal ethanol exposure. Thus, extensive effort has been put forth to identify biomarkers of gestational ethanol exposure (Joya et al., 2012; Memo, Gnoato, Caminiti, Pichini, & Tarani, 2013). Clinical identification of characteristic facial dysmorphologies is valuable, but is most applicable for identification of FASD in children (Jones et al., 2009, 2010). Identification of FASD in younger infants and children is progressing through the use of sophisticated imaging techniques (Coles & Li, 2011; Lebel et al., 2011; Norman et al., 2009). In fact, it is becoming possible to identify fetal ethanol exposure in newborns. Analysis of non-oxidative metabolites of ethanol including fatty acid ethyl esters, ethyl glucuronide, and ethyl sulfate in meconium allows identification of fetal ethanol exposure in newborns (Joya et al., 2012; Memo et al., 2013). It has recently been demonstrated that fetal ethanol exposure can be identified effectively and at low cost using newborn screening for phosphatidylethanol in dried blood spots (Bakhireva et al., 2014). In addition, advanced MRI, ultrasound imaging, and functional analysis show promise for prenatal diagnosis of FASD (Bookstein et al., 2005; Hepper, Dornan, & Lynch, 2012; Kfir et al., 2009; Memo et al., 2013). Mouse models of FASD are forecasting even earlier detection of fetal ethanol exposure (O'Leary-Moore et al., 2011; Sudheendran, Bake, Miranda, & Larin, 2013). Together, these findings suggest that it will be realistic to diagnose FASD in neonates and infants in the hospital or clinic in the foreseeable future, opening the possibility that early temporal windows will be available for intervention with therapeutics. It is hoped that early diagnosis of FASD will lead to the treatment of FASD infants in addition, or alternatively, to pregnant women with therapeutics (Kodituwakku & Kodituwakku, 2011; Roach & Anderson, 2008), including those exhibiting anti-inflammatory properties.

11.2. Development of novel anti-inflammatory therapies for FASD

As discussed previously, our work and that of others indicate that ethanol exposure in neonatal rodents results in potent neuroinflammation. Since neuroinflammation is believed to contribute to ethanol-induced neurodegeneration, this suggests that anti-inflammatory agents may be effective in the treatment of FASD. Our discussion will focus on the potential of the following anti-inflammatory agents—pioglitazone, minocycline, and naltrexone—for the treatment of FASD. However, it is recognized that other anti-inflammatory agents could also proved to be effective in the treatment of FASD.

11.2.1 Pioglitazone

Pioglitazone is an FDA-approved pharmaceutical agonist for PPAR-γ, a member of a nuclear receptor family of proteins that modulate neuroinflammatory responses and other neurodegenerative responses. We (Diab et al., 2002; Drew & Chavis, 2001; Storer, Xu, Chavis, & Drew, 2005a, 2005b) and others (Bernardo, Levi, & Minghetti, 2000; Kielian, Bearden, Baldwin, & Esen, 2004; Petrova, Akama, & Van Eldik, 1999) demonstrated that PPAR-γ agonists, including pioglitazone, inhibit activation of microglia. PPAR-γ agonists also inhibit the production of proinflammatory cytokines and chemokines by microglia (Cunard et al., 2002; Lovett-Racke et al., 2004; Xu, Chavis, Racke, & Drew, 2006; Xu & Drew, 2007; Xu, Storer, Chavis, Racke, & Drew, 2005). We and others also demonstrated that PPAR-γ agonists are protective in animal models of many CNS disorders characterized by neuroinflammation and neurodegeneration (Diab et al., 2004; Heneka, Landreth, & Feinstein, 2001; Mandrekar-Colucci, Sauerbeck, Popovich, & McTigue, 2013; Niino et al., 2001). Most recently, we demonstrated that PPAR-γ agonists provide protection for neurons and microglia from ethanol-induced cell death and blocked ethanol-induced activation of microglia in a mouse model of FASD (Kane et al., 2011). We further have demonstrated that pioglitazone suppresses ethanol increases in proinflammatory cytokines and chemokines in neonatal mice P.D. Drew, J.W. Johnson, J.C. Douglas, K.D. Phelan, & C.J.M. Kane (Unpublished observations). Taken together, these studies support the possibility that PPAR-γ agonists may protect microglia and neurons against ethanol neurotoxicity and prevent ethanol induction of inflammatory processes in FASD. Interestingly, PPAR-γ agonists have been demonstrated to suppress alcohol consumption and relapse to alcohol-seeking behavior (Stopponi et al., 2011), which may be related to their anti-inflammatory activity.

11.2.2 Minocycline

Minocycline is a tetracycline derivative antibiotic used in the treatment of bacterial infections. However, it is now clear that minocycline also possesses potent anti-inflammatory activity. In fact, minocycline has been demonstrated to strongly suppress microglial activation in a variety of neuroinflammatory and neurodegenerative disorders (reviewed in Garrido-Mesa, Zarzuelo, & Galvez, 2013). However, the effects of minocycline on ethanol-induced neuroinflammation have only just begun to be evaluated and have not been evaluated in models of FASD. Minocycline was shown to block ethanol-induced microglial activation in an 8-week old male mice treated intragastrically with 5 g/kg ethanol for 10 days (Qin & Crews, 2012). Interestingly, like PPAR-γ agonists, minocycline and the related antibiotic doxycycline reduced alcohol consumption in rodents (Agrawal, Hewetson, George, Syapin, & Bergeson, 2011; McIver, Muccigrosso, & Haydon, 2012; Wu et al., 2011). Minocycline also increased the motor impairing effects of ethanol, a phenomenon associated with reduced alcohol consumption (McIver et al., 2012; Wu et al., 2011). Further, in our investigations, we discovered that ethanol increases neuroinflammation in adult, but not adolescent mice (Kane et al., 2014). This is interesting in light of the observation that minocycline decreases alcohol consumption in adult but not adolescent mice supporting the concept that minocycline suppresses alcohol consumption by suppressing neuroinflammation (Agrawal et al., 2014).

11.2.3 Naltrexone

Naltrexone is one of the few approved treatments for alcoholism. It is an opioid antagonist with suppressive effects on alcohol consumption that have been thought to be mediated primarily through opioid receptors (Maisel, Blodgett, Wilbourne, Humphreys, & Finney, 2013). However, naltrexone (and its stereoisomer naloxone) also functions as a TLR4 antagonist (Hutchinson et al., 2008), and TLR4 plays a critical role in ethanol-induced neuroinflammation and neurodegeneration as outlined above (Alfonso-Loeches et al., 2010; Blanco et al., 2005, 2008; Fernandez-Lizarbe et al., 2009; Lippai et al., 2013; Pascual et al., 2011; Wu et al., 2012). It is notable then, that in addition to suppressing alcohol consumption, naltrexone can block ethanol-induced microglial activation in adult mice (Qin & Crews, 2012). However, the effects of naltrexone on microglial activation and neuroinflammation in response to ethanol in models of FASD have not been evaluated. The PPAR-γ agonist pioglitazone was demonstrated to potentiate the suppressive effects on naltrexone on alcohol consumption (Stopponi et al., 2013). This suggests that although PPAR-γ agonists and naltrexone are both anti-inflammatory, these agents may act through distinct pathways. Collectively, these studies suggest that a combination of antiinflammatories such as pioglitazone, minocycline, and naltrexone may act cooperatively to suppress ethanol-induced neuroinflammation and neurodegeneration.

12. CONCLUSIONS

FASD is associated with neuropathology and functional deficits that persist throughout life. Animal models are providing not only a foundation for understanding the causes and consequences of FASD but also new knowledge that is leading to understanding the cellular and molecular mechanisms at play, when the fetal brain is exposed to alcohol. Investigation of alcohol's effect on microglia and neuronal–microglial interactions has led to a model of the novel contributions of microglia to the neuropathology and functional deficits of FASD (Fig. 3.2). This is based on the recent findings in models of FASD that alcohol causes loss of not only neurons but also microglia in the developing brain, depleting the population of protective microglia. Further, alcohol also induces activation of microglia, changing their phenotype to a proinflammatory form that can lead to neuron death. These observations are particularly important given emerging evidence that the loss of microglia in the developing brain is associated with loss of synapses, loss of neuronal plasticity, reduced neurogenesis, cognitive

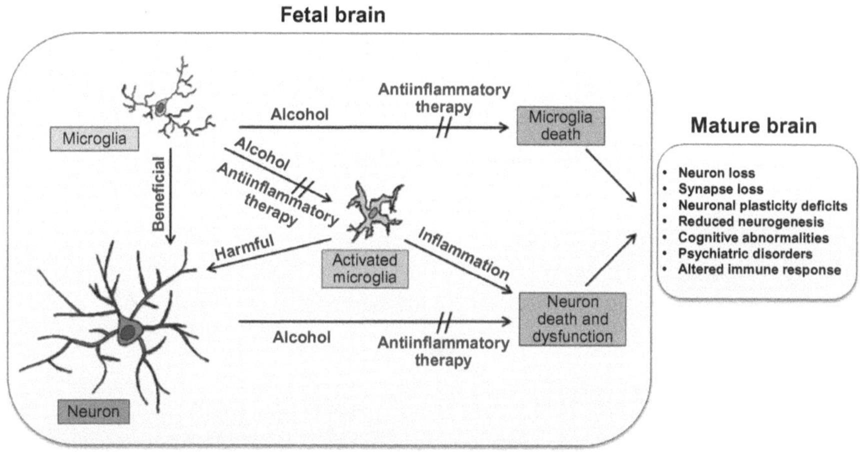

Figure 3.2 Together, direct toxic effects of alcohol on neurons and postulated indirect effects of alcohol mediated through effects on microglia lead to significant neuronal cell death and dysfunction in the developing brain. Microglia in the unperturbed fetal brain serve beneficial, protective roles for neurons. Alcohol depletes the microglial population in the developing brain through microglia cell loss. In addition, alcohol exposure induces microglial activation and expression of neuroinflammatory molecules in the developing brain. These effects of alcohol on microglia reduce the beneficial activity of microglia exerted during normal development and induce proinflammatory activity of microglia that leads to neurodegenerative processes. The direct consequences of microglial loss and neuroinflammation in the developing brain are now known to generate long-term neuroanatomical and behavioral consequences in the adult brain. These include persistent loss of neurons, loss of synapses, deficits in neuronal plasticity, cognitive abnormalities, psychiatric disorders, and altered reactivity of the immune system. These are consistent with some of the long-term consequences in individuals with FASD. Further, based on our findings, it is suggested that treatment with anti-inflammatory pharmaceuticals will prevent alcohol-induced microglial and neuronal cells' loss, microglial activation, and neuroinflammation, leading to better long-term outcomes in FASD.

abnormalities, behavioral deficits, and altered immune response in adulthood. These same abnormalities are produced by microglial activation and inflammation in the developing brain. This knowledge elucidates potentially critical consequences of microglial loss, microglial activation, and inflammatory processes during brain development. Thus, alcohol's effects on microglia during fetal development may contribute to the neuropathological and behavioral abnormalities associated with FASD. The age at which FASD can be diagnosed in the clinic is progressively declining, making therapeutic intervention in infants a realistic possibility in the foreseeable future. Based on these observations, therapeutic intervention in the ethanol-induced microglial activation and neuroinflammation with anti-inflammatory pharmaceuticals is suggested to be of potential long-term benefit to infants born with FASD. Clearly, this area of research is yielding new understanding of cellular and molecular mechanisms underlying FASD and holds promise for even greater knowledge.

ACKNOWLEDGMENTS

This work was supported by NIH Grants AA18834, AA18839, and AA19108.

REFERENCES

Abel, E. L., & Sokol, R. J. (1986). Fetal alcohol syndrome is now leading cause of mental retardation. *Lancet*, *2*(8517), 1222.

Achur, R. N., Freeman, W. M., & Vrana, K. E. (2010). Circulating cytokines as biomarkers of alcohol abuse and alcoholism. *Journal of Neuroimmune Pharmacology*, *5*(1), 83–91.

Adnams, C. M., Kodituwakku, P. W., Hay, A., Molteno, C. D., Viljoen, D., & May, P. A. (2001). Patterns of cognitive-motor development in children with fetal alcohol syndrome from a community in South Africa. *Alcoholism, Clinical and Experimental Research*, *25*(4), 557–562.

Agrawal, R. G., Hewetson, A., George, C. M., Syapin, P. J., & Bergeson, S. E. (2011). Minocycline reduces ethanol drinking. *Brain, Behavior, and Immunity*, *25*(Suppl. 1), S165–S169.

Agrawal, R. G., Owen, J. A., Levin, P. S., Hewetson, A., Berman, A. E., Franklin, S. R., et al. (2014). Bioinformatics analyses reveal age-specific neuroimmune modulation as a target for treatment of high ethanol drinking. *Alcoholism, Clinical and Experimental Research*, *38*(2), 428–437.

Ahluwalia, B., Wesley, B., Adeyiga, O., Smith, D. M., Da-Silva, A., & Rajguru, S. (2000). Alcohol modulates cytokine secretion and synthesis in human fetus: An in vivo and in vitro study. *Alcohol*, *21*(3), 207–213.

Ajami, B., Bennett, J. L., Krieger, C., McNagny, K. M., & Rossi, F. M. (2011). Infiltrating monocytes trigger EAE progression, but do not contribute to the resident microglia pool. *Nature Neuroscience*, *14*(9), 1142–1149.

Ajami, B., Bennett, J. L., Krieger, C., Tetzlaff, W., & Rossi, F. M. (2007). Local self-renewal can sustain CNS microglia maintenance and function throughout adult life. *Nature Neuroscience*, *10*(12), 1538–1543.

Alfonso-Loeches, S., Pascual-Lucas, M., Blanco, A. M., Sanchez-Vera, I., & Guerri, C. (2010). Pivotal role of TLR4 receptors in alcohol-induced neuroinflammation and brain damage. *The Journal of Neuroscience*, *30*(24), 8285–8295.

Allam, A., Abdul-Hamjid, M., Allam, G., Al-hroob, A., Ibraheem, G., & Alsubaie, M. (2013). Perinatal ethyl alcohol effects on the development of cerebellar cortex in albino rat. *African Journal of Pharmacy and Pharmacology*, 7, 1293–1301.

American Academy of Pediatrics Committee on Substance Abuse and Committee on Children with Disabilities. (2000). Fetal alcohol syndrome and alcohol-related neurodevelopmental disorders. *Pediatrics*, *106*(2 Pt. 1), 358–361.

Archibald, S. L., Fennema-Notestine, C., Gamst, A., Riley, E. P., Mattson, S. N., & Jernigan, T. L. (2001). Brain dysmorphology in individuals with severe prenatal alcohol exposure. *Developmental Medicine and Child Neurology*, *43*(3), 148–154.

Aronne, M. P., Guadagnoli, T., Fontanet, P., Evrard, S. G., & Brusco, A. (2011). Effects of prenatal ethanol exposure on rat brain radial glia and neuroblast migration. *Experimental Neurology*, *229*(2), 364–371.

Astley, S. J., Aylward, E. H., Olson, H. C., Kerns, K., Brooks, A., Coggins, T. E., et al. (2009). Magnetic resonance imaging outcomes from a comprehensive magnetic resonance study of children with fetal alcohol spectrum disorders. *Alcoholism, Clinical and Experimental Research*, *33*(10), 1671–1689.

Autti-Ramo, I., Autti, T., Korkman, M., Kettunen, S., Salonen, O., & Valanne, L. (2002). MRI findings in children with school problems who had been exposed prenatally to alcohol. *Developmental Medicine and Child Neurology*, *44*(2), 98–106.

Bakhireva, L. N., Leeman, L., Savich, R. D., Cano, S., Gutierrez, H., Savage, D. D., et al. (2014). The validity of phosphatidylethanol in dried blood spots of newborns for the identification of prenatal alcohol exposure. *Alcoholism, Clinical and Experimental Research*, *38*(4), 1078–1085.

Banuelos, C., Gilbert, R. J., Montgomery, K. S., Fincher, A. S., Wang, H., Frye, G. D., et al. (2012). Altered spatial learning and delay discounting in a rat model of human third trimester binge ethanol exposure. *Behavioural Pharmacology*, *23*(1), 54–65.

Barr, H. M., Bookstein, F. L., O'Malley, K. D., Connor, P. D., Huggins, J. E., & Streissguth, A. P. (2006). Binge drinking during pregnancy as a predictor of psychiatric disorders on the structured clinical interview for DSM-IV in young adult offspring. *The American Journal of Psychiatry*, *163*(6), 1061–1065.

Bernardo, A., Levi, G., & Minghetti, L. (2000). Role of the peroxisome proliferator-activated receptor-gamma (PPAR-gamma) and its natural ligand 15-deoxy-delta12, 14-prostaglandin J_2 in the regulation of microglial functions. *The European Journal of Neuroscience*, *12*(7), 2215–2223.

Bhatara, V. S., Lovrein, F., Kirkeby, J., Swayze, V., II, Unruh, E., & Johnson, V. (2002). Brain function in fetal alcohol syndrome assessed by single photon emission computed tomography. *South Dakota Journal of Medicine*, *55*(2), 59–62.

Bilbo, S. D., Smith, S. H., & Schwarz, J. M. (2012). A lifespan approach to neuroinflammatory and cognitive disorders: A critical role for glia. *Journal of Neuroimmune Pharmacology*, *7*(1), 24–41.

Bjorkquist, O. A., Fryer, S. L., Reiss, A. L., Mattson, S. N., & Riley, E. P. (2010). Cingulate gyrus morphology in children and adolescents with fetal alcohol spectrum disorders. *Psychiatry Research*, *181*(2), 101–107.

Blanco, A. M., & Guerri, C. (2007). Ethanol intake enhances inflammatory mediators in brain: Role of glial cells and TLR4/IL-1RI receptors. *Frontiers in Bioscience*, *12*, 2616–2630.

Blanco, A. M., Pascual, M., Valles, S. L., & Guerri, C. (2004). Ethanol-induced iNOS and COX-2 expression in cultured astrocytes via NF-kappa B. *Neuroreport*, *15*(4), 681–685.

Blanco, A. M., Perez-Arago, A., Fernandez-Lizarbe, S., & Guerri, C. (2008). Ethanol mimics ligand-mediated activation and endocytosis of IL-1RI/TLR4 receptors via lipid rafts caveolae in astroglial cells. *Journal of Neurochemistry*, *106*(2), 625–639.

Blanco, A. M., Valles, S. L., Pascual, M., & Guerri, C. (2005). Involvement of TLR4/type I IL-1 receptor signaling in the induction of inflammatory mediators and cell death induced by ethanol in cultured astrocytes. *Journal of Immunology*, *175*(10), 6893–6899.

Blednov, Y. A., Bergeson, S. E., Walker, D., Ferreira, V. M., Kuziel, W. A., & Harris, R. A. (2005). Perturbation of chemokine networks by gene deletion alters the reinforcing actions of ethanol. *Behavioural Brain Research*, *165*(1), 110–125.

Blednov, Y. A., Ponomarev, I., Geil, C., Bergeson, S., Koob, G. F., & Harris, R. A. (2012). Neuroimmune regulation of alcohol consumption: Behavioral validation of genes obtained from genomic studies. *Addiction Biology*, *17*(1), 108–120.

Bonthius, D. J., Bonthius, N. E., Napper, R. M., Astley, S. J., Clarren, S. K., & West, J. R. (1996). Purkinje cell deficits in nonhuman primates following weekly exposure to ethanol during gestation. *Teratology*, *53*(4), 230–236.

Bookstein, F. L., Connor, P. D., Covell, K. D., Barr, H. M., Gleason, C. A., Sze, R. W., et al. (2005). Preliminary evidence that prenatal alcohol damage may be visible in averaged ultrasound images of the neonatal human corpus callosum. *Alcohol*, *36*(3), 151–160.

Boulanger, L. M. (2009). Immune proteins in brain development and synaptic plasticity. *Neuron*, *64*(1), 93–109.

Boyadjieva, N. I., & Sarkar, D. K. (2010). Role of microglia in ethanol's apoptotic action on hypothalamic neuronal cells in primary cultures. *Alcoholism, Clinical and Experimental Research*, *34*(11), 1835–1842.

Brown, K. L., Calizo, L. H., & Stanton, M. E. (2008). Dose-dependent deficits in dual interstimulus interval classical eyeblink conditioning tasks following neonatal binge alcohol exposure in rats. *Alcoholism, Clinical and Experimental Research*, *32*(2), 277–293.

Carson, M. J., Bilousova, T. V., Puntambekar, S. S., Melchior, B., Doose, J. M., & Ethell, I. M. (2007). A rose by any other name? The potential consequences of microglial heterogeneity during CNS health and disease. *Neurotherapeutics*, *4*(4), 571–579.

Carson, M. J., Doose, J. M., Melchior, B., Schmid, C. D., & Ploix, C. C. (2006). CNS immune privilege: Hiding in plain sight. *Immunological Reviews*, *213*, 48–65.

Clancy, B., Darlington, R. B., & Finlay, B. L. (2001). Translating developmental time across mammalian species. *Neuroscience*, *105*(1), 7–17.

Clancy, B., Finlay, B. L., Darlington, R. B., & Anand, K. J. (2007). Extrapolating brain development from experimental species to humans. *Neurotoxicology*, *28*(5), 931–937.

Coles, C. D., & Li, Z. (2011). Functional neuroimaging in the examination of effects of prenatal alcohol exposure. *Neuropsychology Review*, *21*(2), 119–132.

Colton, C. A. (2009). Heterogeneity of microglial activation in the innate immune response in the brain. *Journal of Neuroimmune Pharmacology*, *4*(4), 399–418.

Connor, P. D., Sampson, P. D., Bookstein, F. L., Barr, H. M., & Streissguth, A. P. (2000). Direct and indirect effects of prenatal alcohol damage on executive function. *Developmental Neuropsychology*, *18*(3), 331–354.

Connor, P. D., Sampson, P. D., Streissguth, A. P., Bookstein, F. L., & Barr, H. M. (2006). Effects of prenatal alcohol exposure on fine motor coordination and balance: A study of two adult samples. *Neuropsychologia*, *44*(5), 744–751.

Corriveau, R. A., Huh, G. S., & Shatz, C. J. (1998). Regulation of class I MHC gene expression in the developing and mature CNS by neural activity. *Neuron*, *21*(3), 505–520.

Crews, F. T., Bechara, R., Brown, L. A., Guidot, D. M., Mandrekar, P., Oak, S., et al. (2006). Cytokines and alcohol. *Alcoholism, Clinical and Experimental Research*, *30*(4), 720–730.

Crews, F. T., & Nixon, K. (2009). Mechanisms of neurodegeneration and regeneration in alcoholism. *Alcohol and Alcoholism*, *44*(2), 115–127.

Crews, F. T., Qin, L., Sheedy, D., Vetreno, R. P., & Zou, J. (2013). High mobility group box 1/Toll-like receptor danger signaling increases brain neuroimmune activation in alcohol dependence. *Biological Psychiatry*, *73*(7), 602–612.

Cunard, R., Ricote, M., DiCampli, D., Archer, D. C., Kahn, D. A., Glass, C. K., et al. (2002). Regulation of cytokine expression by ligands of peroxisome proliferator activated receptors. *Journal of Immunology*, *168*(6), 2795–2802.

Dalmau, I., Finsen, B., Zimmer, J., Gonzalez, B., & Castellano, B. (1998). Development of microglia in the postnatal rat hippocampus. *Hippocampus*, *8*(5), 458–474.

Davalos, D., Grutzendler, J., Yang, G., Kim, J. V., Zuo, Y., Jung, S., et al. (2005). ATP mediates rapid microglial response to local brain injury in vivo. *Nature Neuroscience*, *8*(6), 752–758.

Davis, R. L., & Syapin, P. J. (2004). Ethanol increases nuclear factor-kappa B activity in human astroglial cells. *Neuroscience Letters, 371*(2–3), 128–132.

De Guio, F., Mangin, J. F., Riviere, D., Perrot, M., Molteno, C. D., Jacobson, S. W., et al. (2013). A study of cortical morphology in children with fetal alcohol spectrum disorders. *Human Brain Mapping, 35*(5), 2285–2296.

DeVito, W. J., & Stone, S. (2001). Prenatal exposure to ethanol alters the neuroimmune response to a central nervous system wound in the adult rat. *Alcohol, 25*(1), 39–47.

Diab, A., Deng, C., Smith, J. D., Hussain, R. Z., Phanavanh, B., Lovett-Racke, A. E., et al. (2002). Peroxisome proliferator-activated receptor-gamma agonist 15-deoxy-delta (12,14)-prostaglandin J_2 ameliorates experimental autoimmune encephalomyelitis. *Journal of Immunology, 168*(5), 2508–2515.

Diab, A., Hussain, R. Z., Lovett-Racke, A. E., Chavis, J. A., Drew, P. D., & Racke, M. K. (2004). Ligands for the peroxisome proliferator-activated receptor-gamma and the retinoid X receptor exert additive anti-inflammatory effects on experimental autoimmune encephalomyelitis. *Journal of Neuroimmunology, 148*(1–2), 116–126.

Diaz, M. R., Vollmer, C. C., Zamudio-Bulcock, P. A., Vollmer, W., Blomquist, S. L., Morton, R. A., et al. (2014). Repeated intermittent alcohol exposure during the third trimester-equivalent increases expression of the GABA(A) receptor delta subunit in cerebellar granule neurons and delays motor development in rats. *Neuropharmacology, 79*, 262–274.

Dikranian, K., Qin, Y. Q., Labruyere, J., Nemmers, B., & Olney, J. W. (2005). Ethanol-induced neuroapoptosis in the developing rodent cerebellum and related brain stem structures. *Brain Research. Developmental Brain Research, 155*(1), 1–13.

Diwadkar, V. A., Meintjes, E. M., Goradia, D., Dodge, N. C., Warton, C., Molteno, C. D., et al. (2013). Differences in cortico-striatal-cerebellar activation during working memory in syndromal and nonsyndromal children with prenatal alcohol exposure. *Human Brain Mapping, 34*(8), 1931–1945.

Dobbing, J., & Sands, J. (1979). Comparative aspects of the brain growth spurt. *Early Human Development, 3*(1), 79–83.

Drew, P. D., & Chavis, J. A. (2001). The cyclopentone prostaglandin 15-deoxy-delta(12,14) prostaglandin J_2 represses nitric oxide, TNF-alpha, and IL-12 production by microglial cells. *Journal of Neuroimmunology, 115*(1–2), 28–35.

Drew, P. D., & Kane, C. J. M. (2013). Neuroimmune mechanisms of glia and their interplay with alcohol exposure across the lifespan. In C. Cui, L. Grandison, & A. Noronha (Eds.), *Neural–immune interactions in brain function and alcohol related disorders* (pp. 359–386). New York, NY: Springer.

Eglitis, M. A., & Mezey, E. (1997). Hematopoietic cells differentiate into both microglia and macroglia in the brains of adult mice. *Proceedings of the National Academy of Sciences of the United States of America, 94*(8), 4080–4085.

El Shawa, H., Abbott, C. W., III, & Huffman, K. J. (2013). Prenatal ethanol exposure disrupts intraneocortical circuitry, cortical gene expression, and behavior in a mouse model of FASD. *The Journal of Neuroscience, 33*(48), 18893–18905.

Everett, J. C., Licon-Munoz, Y., & Valenzuela, C. F. (2012). Effects of third trimester-equivalent ethanol exposure on Cl^- co-transporter expression, network activity, and GABAergic transmission in the CA3 hippocampal region of neonatal rats. *Alcohol, 46*(6), 595–601.

Farber, N. B., Creeley, C. E., & Olney, J. W. (2010). Alcohol-induced neuroapoptosis in the fetal macaque brain. *Neurobiology of Disease, 40*(1), 200–206.

Fee, D., Grzybicki, D., Dobbs, M., Ihyer, S., Clotfelter, J., Macvilay, S., et al. (2000). Interleukin 6 promotes vasculogenesis of murine brain microvessel endothelial cells. *Cytokine, 12*(6), 655–665.

Fellner, L., Jellinger, K. A., Wenning, G. K., & Stefanova, N. (2011). Glial dysfunction in the pathogenesis of alpha-synucleinopathies: Emerging concepts. *Acta Neuropathologica, 121*(6), 675–693.

Fernandez-Lizarbe, S., Pascual, M., & Guerri, C. (2009). Critical role of TLR4 response in the activation of microglia induced by ethanol. *Journal of Immunology, 183*(7), 4733–4744.

Fiske, B. K., & Brunjes, P. C. (2000). Microglial activation in the developing rat olfactory bulb. *Neuroscience, 96*(4), 807–815.

Floreani, N. A., Rump, T. J., Abdul Muneer, P. M., Alikunju, S., Morsey, B. M., Brodie, M. R., et al. (2010). Alcohol-induced interactive phosphorylation of SRC and Toll-like receptor regulates the secretion of inflammatory mediators by human astrocytes. *Journal of Neuroimmune Pharmacology, 5*(4), 533–545.

Floyd, R. L., Weber, M. K., Denny, C., & O'Connor, M. J. (2009). Prevention of fetal alcohol spectrum disorders. *Developmental Disabilities Research Reviews, 15*(3), 193–199.

Fryer, S. L., McGee, C. L., Matt, G. E., Riley, E. P., & Mattson, S. N. (2007). Evaluation of psychopathological conditions in children with heavy prenatal alcohol exposure. *Pediatrics, 119*(3), e733–e741.

Gadient, R. A., & Otten, U. (1994). Expression of interleukin-6 (IL-6) and interleukin-6 receptor (IL-6R) mRNAs in rat brain during postnatal development. *Brain Research, 637*(1–2), 10–14.

Garrido-Mesa, N., Zarzuelo, A., & Galvez, J. (2013). Minocycline: Far beyond an antibiotic. *British Journal of Pharmacology, 169*(2), 337–352.

Gil-Mohapel, J., Boehme, F., Kainer, L., & Christie, B. R. (2010). Hippocampal cell loss and neurogenesis after fetal alcohol exposure: Insights from different rodent models. *Brain Research Reviews, 64*(2), 283–303.

Glass, C. K., Saijo, K., Winner, B., Marchetto, M. C., & Gage, F. H. (2010). Mechanisms underlying inflammation in neurodegeneration. *Cell, 140*(6), 918–934.

Godin, E. A., O'Leary-Moore, S. K., Khan, A. A., Parnell, S. E., Ament, J. J., Dehart, D. B., et al. (2010). Magnetic resonance microscopy defines ethanol-induced brain abnormalities in prenatal mice: Effects of acute insult on gestational day 7. *Alcoholism, Clinical and Experimental Research, 34*(1), 98–111.

Goodlett, C. R., & Eilers, A. T. (1997). Alcohol-induced Purkinje cell loss with a single binge exposure in neonatal rats: A stereological study of temporal windows of vulnerability. *Alcoholism, Clinical and Experimental Research, 21*(4), 738–744.

Goodlett, C. R., Thomas, J. D., & West, J. R. (1991). Long-term deficits in cerebellar growth and rotarod performance of rats following "binge-like" alcohol exposure during the neonatal brain growth spurt. *Neurotoxicology and Teratology, 13*(1), 69–74.

Goral, J., Karavitis, J., & Kovacs, E. J. (2008). Exposure-dependent effects of ethanol on the innate immune system. *Alcohol, 42*(4), 237–247.

Green, J. T., Tran, T., Steinmetz, J. E., & Goodlett, C. R. (2002). Neonatal ethanol produces cerebellar deep nuclear cell loss and correlated disruption of eyeblink conditioning in adult rats. *Brain Research, 956*(2), 302–311.

Gressens, P., Lammens, M., Picard, J. J., & Evrard, P. (1992). Ethanol-induced disturbances of gliogenesis and neuronogenesis in the developing murine brain: An in vitro and in vivo immunohistochemical and ultrastructural study. *Alcohol and Alcoholism, 27*(3), 219–226.

Guerri, C., Bazinet, A., & Riley, E. P. (2009). Foetal alcohol spectrum disorders and alterations in brain and behaviour. *Alcohol and Alcoholism, 44*(2), 108–114.

Hamilton, D. A., Akers, K. G., Rice, J. P., Johnson, T. E., Candelaria-Cook, F. T., Maes, L. I., et al. (2010). Prenatal exposure to moderate levels of ethanol alters social behavior in adult rats: Relationship to structural plasticity and immediate early gene expression in frontal cortex. *Behavioural Brain Research, 207*(2), 290–304.

Hamre, K. M., & West, J. R. (1993). The effects of the timing of ethanol exposure during the brain growth spurt on the number of cerebellar Purkinje and granule cell nuclear profiles. *Alcoholism, Clinical and Experimental Research, 17*(3), 610–622.

Hanisch, U. K., & Kettenmann, H. (2007). Microglia: Active sensor and versatile effector cells in the normal and pathologic brain. *Nature Neuroscience, 10*(11), 1387–1394.

He, J., & Crews, F. T. (2008). Increased MCP-1 and microglia in various regions of the human alcoholic brain. *Experimental Neurology, 210*(2), 349–358.

Heneka, M. T., Landreth, G. E., & Feinstein, D. L. (2001). Role for peroxisome proliferator-activated receptor-gamma in Alzheimer's disease. *Annals of Neurology, 49*(2), 276.

Hepper, P. G., Dornan, J. C., & Lynch, C. (2012). Fetal brain function in response to maternal alcohol consumption: Early evidence of damage. *Alcoholism, Clinical and Experimental Research, 36*(12), 2168–2175.

Hickey, W. F. (1991). Migration of hematogenous cells through the blood–brain barrier and the initiation of CNS inflammation. *Brain Pathology, 1*(2), 97–105.

Hickey, W. F., & Kimura, H. (1988). Perivascular microglial cells of the CNS are bone marrow-derived and present antigen in vivo. *Science, 239*(4837), 290–292.

Hutchinson, M. R., Zhang, Y., Brown, K., Coats, B. D., Shridhar, M., Sholar, P. W., et al. (2008). Non-stereoselective reversal of neuropathic pain by naloxone and naltrexone: Involvement of toll-like receptor 4 (TLR4). *The European Journal of Neuroscience, 28*(1), 20–29.

Idrus, N. M., McGough, N. N., Riley, E. P., & Thomas, J. D. (2011). Administration of memantine during ethanol withdrawal in neonatal rats: Effects on long-term ethanol-induced motor incoordination and cerebellar Purkinje cell loss. *Alcoholism, Clinical and Experimental Research, 35*(2), 355–364.

Idrus, N. M., McGough, N. N., Riley, E. P., & Thomas, J. D. (2014). Administration of memantine during withdrawal mitigates overactivity and spatial learning impairments associated with neonatal alcohol exposure in rats. *Alcoholism, Clinical and Experimental Research, 38*(2), 529–537.

Ieraci, A., & Herrera, D. G. (2007). Single alcohol exposure in early life damages hippocampal stem/progenitor cells and reduces adult neurogenesis. *Neurobiology of Disease, 26*(3), 597–605.

Ikonomidou, C., Bittigau, P., Ishimaru, M. J., Wozniak, D. F., Koch, C., Genz, K., et al. (2000). Ethanol-induced apoptotic neurodegeneration and fetal alcohol syndrome. *Science, 287*(5455), 1056–1060.

Inomata, K., Nasu, F., & Tanaka, H. (1987). Decreased density of synaptic formation in the frontal cortex of neonatal rats exposed to ethanol in utero. *International Journal of Developmental Neuroscience, 5*(5–6), 455–460.

Jiang, Y., Kumada, T., Cameron, D. B., & Komuro, H. (2008). Cerebellar granule cell migration and the effects of alcohol. *Developmental Neuroscience, 30*(1–3), 7–23.

Jones, K. L., Hoyme, H. E., Robinson, L. K., del Campo, M., Manning, M. A., Bakhireva, L. N., et al. (2009). Developmental pathogenesis of short palpebral fissure length in children with fetal alcohol syndrome. *Birth Defects Research, Part A: Clinical and Molecular Teratology, 85*(8), 695–699.

Jones, K. L., Hoyme, H. E., Robinson, L. K., Del Campo, M., Manning, M. A., Prewitt, L. M., et al. (2010). Fetal alcohol spectrum disorders: Extending the range of structural defects. *American Journal of Medical Genetics Part A, 152A*(11), 2731–2735.

Joya, X., Friguls, B., Ortigosa, S., Papaseit, E., Martinez, S. E., Manich, A., et al. (2012). Determination of maternal-fetal biomarkers of prenatal exposure to ethanol: A review. *Journal of Pharmaceutical and Biomedical Analysis, 69*, 209–222.

Kane, C. J. M., Phelan, K. D., Douglas, J. C., Wagoner, G., Johnson, J. W., Xu, J., et al. (2014). Effects of ethanol on immune response in the brain: Region-specific changes in adolescent versus adult mice. *Alcoholism, Clinical and Experimental Research, 38*(2), 384–391.

Kane, C. J. M., Phelan, K. D., Han, L., Smith, R. R., Xie, J., Douglas, J. C., et al. (2011). Protection of neurons and microglia against ethanol in a mouse model of fetal alcohol spectrum disorders by peroxisome proliferator-activated receptor-gamma agonists. *Brain, Behavior, and Immunity, 25*(Suppl. 1), S137–S145.

Kelly, S. J., Pierce, D. R., & West, J. R. (1987). Microencephaly and hyperactivity in adult rats can be induced by neonatal exposure to high blood alcohol concentrations. *Experimental Neurology, 96*(3), 580–593.

Kettenmann, H., Hanisch, U. K., Noda, M., & Verkhratsky, A. (2011). Physiology of microglia. *Physiological Reviews, 91*(2), 461–553.

Kfir, M., Yevtushok, L., Onishchenko, S., Wertelecki, W., Bakhireva, L., Chambers, C. D., et al. (2009). Can prenatal ultrasound detect the effects of in-utero alcohol exposure? A pilot study. *Ultrasound in Obstetrics & Gynecology, 33*(6), 683–689.

Kiefer, F., Jahn, H., Schick, M., & Wiedemann, K. (2002). Alcohol intake, tumour necrosis factor-alpha, leptin and craving: Factors of a possibly vicious circle? *Alcohol and Alcoholism, 37*(4), 401–404.

Kielian, T., Bearden, E. D., Baldwin, A. C., & Esen, N. (2004). IL-1 and TNF-alpha play a pivotal role in the host immune response in a mouse model of staphylococcus aureus-induced experimental brain abscess. *Journal of Neuropathology and Experimental Neurology, 63*, 381–396.

Kim, D. J., Kim, W., Yoon, S. J., Choi, B. M., Kim, J. S., Go, H. J., et al. (2003). Effects of alcohol hangover on cytokine production in healthy subjects. *Alcohol, 31*(3), 167–170.

Klintsova, A. Y., Cowell, R. M., Swain, R. A., Napper, R. M., Goodlett, C. R., & Greenough, W. T. (1998). Therapeutic effects of complex motor training on motor performance deficits induced by neonatal binge-like alcohol exposure in rats. I. Behavioral results. *Brain Research, 800*(1), 48–61.

Klintsova, A. Y., Helfer, J. L., Calizo, L. H., Dong, W. K., Goodlett, C. R., & Greenough, W. T. (2007). Persistent impairment of hippocampal neurogenesis in young adult rats following early postnatal alcohol exposure. *Alcoholism, Clinical and Experimental Research, 31*(12), 2073–2082.

Knapp, D. J., & Crews, F. T. (1999). Induction of cyclooxygenase-2 in brain during acute and chronic ethanol treatment and ethanol withdrawal. *Alcoholism, Clinical and Experimental Research, 23*(4), 633–643.

Kodituwakku, P. W., Kalberg, W., & May, P. A. (2001). The effects of prenatal alcohol exposure on executive functioning. *Alcohol Research & Health, 25*(3), 192–198.

Kodituwakku, P. W., & Kodituwakku, E. L. (2011). From research to practice: An integrative framework for the development of interventions for children with fetal alcohol spectrum disorders. *Neuropsychology Review, 21*(2), 204–223.

Kumar, A., Singh, C. K., Lavoie, H. A., Dipette, D. J., & Singh, U. S. (2011). Resveratrol restores NRF2 level and prevents ethanol-induced toxic effects in the cerebellum of a rodent model of fetal alcohol spectrum disorders. *Molecular Pharmacology, 80*(3), 446–457.

Laso, F. J., Vaquero, J. M., Almeida, J., Marcos, M., & Orfao, A. (2007). Production of inflammatory cytokines by peripheral blood monocytes in chronic alcoholism: Relationship with ethanol intake and liver disease. *Cytometry. Part B, Clinical Cytometry, 72*(5), 408–415.

Lawson, L. J., Perry, V. H., Dri, P., & Gordon, S. (1990). Heterogeneity in the distribution and morphology of microglia in the normal adult mouse brain. *Neuroscience, 39*(1), 151–170.

Lawson, L. J., Perry, V. H., & Gordon, S. (1992). Turnover of resident microglia in the normal adult mouse brain. *Neuroscience, 48*(2), 405–415.

Lebel, C., Mattson, S. N., Riley, E. P., Jones, K. L., Adnams, C. M., May, P. A., et al. (2012). A longitudinal study of the long-term consequences of drinking during pregnancy: Heavy in utero alcohol exposure disrupts the normal processes of brain development. *The Journal of Neuroscience, 32*(44), 15243–15251.

Lebel, C., Rasmussen, C., Wyper, K., Andrew, G., & Beaulieu, C. (2010). Brain microstructure is related to math ability in children with fetal alcohol spectrum disorder. *Alcoholism, Clinical and Experimental Research, 34*(2), 354–363.

Lebel, C., Roussotte, F., & Sowell, E. R. (2011). Imaging the impact of prenatal alcohol exposure on the structure of the developing human brain. *Neuropsychology Review, 21*(2), 102–118.

Leigland, L. A., Ford, M. M., Lerch, J. P., & Kroenke, C. D. (2013). The influence of fetal ethanol exposure on subsequent development of the cerebral cortex as revealed by magnetic resonance imaging. *Alcoholism, Clinical and Experimental Research, 37*(6), 924–932.

Lewis, B., Wellmann, K. A., & Barron, S. (2007). Agmatine reduces balance deficits in a rat model of third trimester binge-like ethanol exposure. *Pharmacology, Biochemistry, and Behavior, 88*(1), 114–121.

Lindquist, D. H., Sokoloff, G., Milner, E., & Steinmetz, J. E. (2013). Neonatal ethanol exposure results in dose-dependent impairments in the acquisition and timing of the conditioned eyeblink response and altered cerebellar interpositus nucleus and hippocampal CA1 unit activity in adult rats. *Alcohol, 47*(6), 447–457.

Lippai, D., Bala, S., Petrasek, J., Csak, T., Levin, I., Kurt-Jones, E. A., et al. (2013). Alcohol-induced IL-1beta in the brain is mediated by NLRP3/ASC inflammasome activation that amplifies neuroinflammation. *Journal of Leukocyte Biology, 94*(1), 171–182.

Liu, J., Lewohl, J. M., Harris, R. A., Iyer, V. R., Dodd, P. R., Randall, P. K., et al. (2006). Patterns of gene expression in the frontal cortex discriminate alcoholic from non-alcoholic individuals. *Neuropsychopharmacology, 31*(7), 1574–1582.

Liu, J., Yang, A. R., Kelly, T., Puche, A., Esoga, C., June, H. L., Jr., et al. (2011). Binge alcohol drinking is associated with GABAa alpha2-regulated toll-like receptor 4 (TLR4) expression in the central amygdala. *Proceedings of the National Academy of Sciences of the United States of America, 108*(11), 4465–4470.

Livy, D. J., Miller, E. K., Maier, S. E., & West, J. R. (2003). Fetal alcohol exposure and temporal vulnerability: Effects of binge-like alcohol exposure on the developing rat hippocampus. *Neurotoxicology and Teratology, 25*(4), 447–458.

Lovett-Racke, A. E., Hussain, R. Z., Northrop, S., Choy, J., Rocchini, A., Matthes, L., et al. (2004). Peroxisome proliferator-activated receptor alpha agonists as therapy for autoimmune disease. *Journal of Immunology, 172*(9), 5790–5798.

Lupton, C., Burd, L., & Harwood, R. (2004). Cost of fetal alcohol spectrum disorders. *American Journal of Medical Genetics Part C: Seminars in Medical Genetics, 127C*(1), 42–50.

Maisel, N. C., Blodgett, J. C., Wilbourne, P. L., Humphreys, K., & Finney, J. W. (2013). Meta-analysis of naltrexone and acamprosate for treating alcohol use disorders: When are these medications most helpful? *Addiction, 108*(2), 275–293.

Mandrekar-Colucci, S., Sauerbeck, A., Popovich, P. G., & McTigue, D. M. (2013). PPAR agonists as therapeutics for CNS trauma and neurological diseases. *ASN Neuro, 5*(5), e00129.

Mattson, S. N., & Riley, E. P. (1998). A review of the neurobehavioral deficits in children with fetal alcohol syndrome or prenatal exposure to alcohol. *Alcoholism, Clinical and Experimental Research, 22*(2), 279–294.

Mattson, S. N., & Roebuck, T. M. (2002). Acquisition and retention of verbal and nonverbal information in children with heavy prenatal alcohol exposure. *Alcoholism, Clinical and Experimental Research, 26*(6), 875–882.

Mattson, S. N., Roesch, S. C., Glass, L., Deweese, B. N., Coles, C. D., Kable, J. A., et al. (2013). Further development of a neurobehavioral profile of fetal alcohol spectrum disorders. *Alcoholism, Clinical and Experimental Research, 37*(3), 517–528.

May, P. A., Gossage, J. P., Kalberg, W. O., Robinson, L. K., Buckley, D., Manning, M., et al. (2009). Prevalence and epidemiologic characteristics of FASD from various research methods with an emphasis on recent in-school studies. *Developmental Disabilities Research Reviews, 15*(3), 176–192.

McIver, S. R., Muccigrosso, M. M., & Haydon, P. G. (2012). The effect of doxycycline on alcohol consumption and sensitivity: Consideration for inducible transgenic mouse models. *Experimental Biology and Medicine (Maywood, N.J.), 237*(10), 1129–1133.

McKercher, S. R., Torbett, B. E., Anderson, K. L., Henkel, G. W., Vestal, D. J., Baribault, H., et al. (1996). Targeted disruption of the PU.1 gene results in multiple hematopoietic abnormalities. *The EMBO Journal, 15*(20), 5647–5658.

Memo, L., Gnoato, E., Caminiti, S., Pichini, S., & Tarani, L. (2013). Fetal alcohol spectrum disorders and fetal alcohol syndrome: The state of the art and new diagnostic tools. *Early Human Development, 89*(Suppl. 1), S40–S43.

Merrill, J. E. (1992). Tumor necrosis factor alpha, interleukin 1 and related cytokines in brain development: Normal and pathological. *Developmental Neuroscience, 14*(1), 1–10.

Mildner, A., Schmidt, H., Nitsche, M., Merkler, D., Hanisch, U. K., Mack, M., et al. (2007). Microglia in the adult brain arise from LY-6CHICCR2 + monocytes only under defined host conditions. *Nature Neuroscience, 10*(12), 1544–1553.

Miller, M. W. (1993). Migration of cortical neurons is altered by gestational exposure to ethanol. *Alcoholism, Clinical and Experimental Research, 17*(2), 304–314.

Miller, M. W. (2006). Effect of prenatal exposure to ethanol on glutamate and GABA immunoreactivity in macaque somatosensory and motor cortices: Critical timing of exposure. *Neuroscience, 138*(1), 97–107.

Miller, M. W. (2007). Exposure to ethanol during gastrulation alters somatosensory-motor cortices and the underlying white matter in the macaque. *Cerebral Cortex, 17*(12), 2961–2971.

Miller, M. W., & Potempa, G. (1990). Numbers of neurons and glia in mature rat somatosensory cortex: Effects of prenatal exposure to ethanol. *The Journal of Comparative Neurology, 293*(1), 92–102.

Miller, M. W., & Robertson, S. (1993). Prenatal exposure to ethanol alters the postnatal development and transformation of radial glia to astrocytes in the cortex. *The Journal of Comparative Neurology, 337*(2), 253–266.

Mooney, S. M., & Miller, M. W. (2001). Episodic exposure to ethanol during development differentially affects brainstem nuclei in the macaque. *Journal of Neurocytology, 30*(12), 973–982.

Mooney, S. M., & Miller, M. W. (2009). Vulnerability of macaque cranial nerve neurons to ethanol is time- and site-dependent. *Alcohol, 43*(4), 323–331.

Murawski, N. J., Jablonski, S. A., Brown, K. L., & Stanton, M. E. (2013). Effects of neonatal alcohol dose and exposure window on long delay and trace eyeblink conditioning in juvenile rats. *Behavioural Brain Research, 236*(1), 307–318.

Napper, R. M., & West, J. R. (1995). Permanent neuronal cell loss in the cerebellum of rats exposed to continuous low blood alcohol levels during the brain growth spurt: A stereological investigation. *The Journal of Comparative Neurology, 362*(2), 283–292.

Nardelli, A., Lebel, C., Rasmussen, C., Andrew, G., & Beaulieu, C. (2011). Extensive deep gray matter volume reductions in children and adolescents with fetal alcohol spectrum disorders. *Alcoholism, Clinical and Experimental Research, 35*(8), 1404–1417.

Nelson, T. E., Hao, C., Manos, J., Ransohoff, R. M., & Gruol, D. L. (2011). Altered hippocampal synaptic transmission in transgenic mice with astrocyte-targeted enhanced CCL2 expression. *Brain, Behavior, and Immunity, 25*(Suppl. 1), S106–S119.

Niino, M., Iwabuchi, K., Kikuchi, S., Ato, M., Morohashi, T., Ogata, A., et al. (2001). Amelioration of experimental autoimmune encephalomyelitis in C57BL/6 mice by an agonist of peroxisome proliferator-activated receptor-gamma. *Journal of Neuroimmunology*, *116*(1), 40–48.

Nimmerjahn, A., Kirchhoff, F., & Helmchen, F. (2005). Resting microglial cells are highly dynamic surveillants of brain parenchyma in vivo. *Science*, *308*(5726), 1314–1318.

Norman, A. L., Crocker, N., Mattson, S. N., & Riley, E. P. (2009). Neuroimaging and fetal alcohol spectrum disorders. *Developmental Disabilities Research Reviews*, *15*(3), 209–217.

Norman, A. L., O'Brien, J. W., Spadoni, A. D., Tapert, S. F., Jones, K. L., Riley, E. P., et al. (2013). A functional magnetic resonance imaging study of spatial working memory in children with prenatal alcohol exposure: Contribution of familial history of alcohol use disorders. *Alcoholism, Clinical and Experimental Research*, *37*(1), 132–140.

O'Hare, E. D., Kan, E., Yoshii, J., Mattson, S. N., Riley, E. P., Thompson, P. M., et al. (2005). Mapping cerebellar vermal morphology and cognitive correlates in prenatal alcohol exposure. *Neuroreport*, *16*(12), 1285–1290.

Okvist, A., Johansson, S., Kuzmin, A., Bazov, I., Merino-Martinez, R., Ponomarev, I., et al. (2007). Neuroadaptations in human chronic alcoholics: Dysregulation of the NF-kappaB system. *PLoS One*, *2*(9), e930.

O'Leary-Moore, S. K., Parnell, S. E., Lipinski, R. J., & Sulik, K. K. (2011). Magnetic resonance-based imaging in animal models of fetal alcohol spectrum disorder. *Neuropsychology Review*, *21*(2), 167–185.

Olson, H. C., Feldman, J. J., Streissguth, A. P., Sampson, P. D., & Bookstein, F. L. (1998). Neuropsychological deficits in adolescents with fetal alcohol syndrome: Clinical findings. *Alcoholism, Clinical and Experimental Research*, *22*(9), 1998–2012.

Paolicelli, R. C., Bolasco, G., Pagani, F., Maggi, L., Scianni, M., Panzanelli, P., et al. (2011). Synaptic pruning by microglia is necessary for normal brain development. *Science*, *333*(6048), 1456–1458.

Parnell, S. E., Holloway, H. T., O'Leary-Moore, S. K., Dehart, D. B., Paniaqua, B., Oguz, I., et al. (2013). Magnetic resonance microscopy-based analyses of the neuroanatomical effects of gestational day 9 ethanol exposure in mice. *Neurotoxicology and Teratology*, *39*, 77–83.

Parnell, S. E., O'Leary-Moore, S. K., Godin, E. A., Dehart, D. B., Johnson, B. W., Allan Johnson, G., et al. (2009). Magnetic resonance microscopy defines ethanol-induced brain abnormalities in prenatal mice: Effects of acute insult on gestational day 8. *Alcoholism, Clinical and Experimental Research*, *33*(6), 1001–1011.

Pascual, M., Balino, P., Alfonso-Loeches, S., Aragon, C. M., & Guerri, C. (2011). Impact of TLR4 on behavioral and cognitive dysfunctions associated with alcohol-induced neuroinflammatory damage. *Brain, Behavior, and Immunity*, *25*(Suppl. 1), S80–S91.

Perry, V. H., Cunningham, C., & Holmes, C. (2007). Systemic infections and inflammation affect chronic neurodegeneration. *Nature Reviews Immunology*, *7*(2), 161–167.

Perry, V. H., Hume, D. A., & Gordon, S. (1985). Immunohistochemical localization of macrophages and microglia in the adult and developing mouse brain. *Neuroscience*, *15*(2), 313–326.

Petrova, T. V., Akama, K. T., & Van Eldik, L. J. (1999). Cyclopentenone prostaglandins suppress activation of microglia: Down-regulation of inducible nitric-oxide synthase by 15-deoxy-delta12,14-prostaglandin J_2. *Proceedings of the National Academy of Sciences of the United States of America*, *96*(8), 4668–4673.

Pierce, D. R., Kane, C. J. M., Serbus, D. C., & Light, K. E. (1997). Microencephaly and selective decreases in cerebellar Purkinje cell numbers following combined exposure to ethanol and methadone during rat brain development. *Developmental Neuroscience*, *19*(5), 438–445.

Pierce, D. R., Serbus, D. C., & Light, K. E. (1993). Intragastric intubation of alcohol during postnatal development of rats results in selective cell loss in the cerebellum. *Alcoholism, Clinical and Experimental Research, 17*(6), 1275–1280.

Pierce, D. R., Williams, D. K., & Light, K. E. (1999). Purkinje cell vulnerability to developmental ethanol exposure in the rat cerebellum. *Alcoholism, Clinical and Experimental Research, 23*(10), 1650–1659.

Priller, J., Flugel, A., Wehner, T., Boentert, M., Haas, C. A., Prinz, M., et al. (2001). Targeting gene-modified hematopoietic cells to the central nervous system: Use of green fluorescent protein uncovers microglial engraftment. *Nature Medicine*, 7(12), 1356–1361.

Qin, L., & Crews, F. T. (2012). Chronic ethanol increases systemic TLR3 agonist-induced neuroinflammation and neurodegeneration. *Journal of Neuroinflammation, 9*, 130.

Ransohoff, R. M., & Perry, V. H. (2009). Microglial physiology: Unique stimuli, specialized responses. *Annual Review of Immunology, 27*, 119–145.

Reitz, C., Brayne, C., & Mayeux, R. (2011). Epidemiology of Alzheimer disease. *Nature Reviews Neurology*, 7(3), 137–152.

Riley, E. P., Infante, M. A., & Warren, K. R. (2011). Fetal alcohol spectrum disorders: An overview. *Neuropsychology Review, 21*(2), 73–80.

Riley, E. P., & McGee, C. L. (2005). Fetal alcohol spectrum disorders: An overview with emphasis on changes in brain and behavior. *Experimental Biology and Medicine (Maywood, N.J.), 230*(6), 357–365.

Riley, E. P., McGee, C. L., & Sowell, E. R. (2004). Teratogenic effects of alcohol: A decade of brain imaging. *American Journal of Medical Genetics Part C: Seminars in Medical Genetics, 127C*(1), 35–41.

Rivest, S. (2009). Regulation of innate immune responses in the brain. *Nature Reviews Immunology, 9*(6), 429–439.

Roach, D., & Anderson, S. M. (2008). Interagency coordinating committee on fetal alcohol syndrome work group on women, drinking, and pregnancy. In D. Roach, & S. M. Anderson (Eds.), *Preventing alcohol tobacco, and other substance-exposed pregnancies: A community affair* (pp. 1–62). Rockville, MD: NIAAA, NIH, American Legacy Foundation.

Roebuck, T. M., Simmons, R. W., Mattson, S. N., & Riley, E. P. (1998). Prenatal exposure to alcohol affects the ability to maintain postural balance. *Alcoholism, Clinical and Experimental Research, 22*(1), 252–258.

Roebuck-Spencer, T. M., Mattson, S. N., Marion, S. D., Brown, W. S., & Riley, E. P. (2004). Bimanual coordination in alcohol-exposed children: Role of the corpus callosum. *Journal of the International Neuropsychological Society, 10*(4), 536–548.

Rogers, J. T., Morganti, J. M., Bachstetter, A. D., Hudson, C. E., Peters, M. M., Grimmig, B. A., et al. (2011). CX3CR1 deficiency leads to impairment of hippocampal cognitive function and synaptic plasticity. *The Journal of Neuroscience, 31*(45), 16241–16250.

Saijo, K., & Glass, C. K. (2011). Microglial cell origin and phenotypes in health and disease. *Nature Reviews Immunology, 11*(11), 775–787.

Saito, M., Chakraborty, G., Mao, R. F., Paik, S. M., Vadasz, C., & Saito, M. (2010). Tau phosphorylation and cleavage in ethanol-induced neurodegeneration in the developing mouse brain. *Neurochemical Research, 35*(4), 651–659.

Sakata-Haga, H., Sawada, K., Ohta, K., Cui, C., Hisano, S., & Fukui, Y. (2003). Adverse effects of maternal ethanol consumption on development of dorsal hippocampus in rat offspring. *Acta Neuropathologica, 105*(1), 30–36.

Sampson, P. D., Streissguth, A. P., Bookstein, F. L., Little, R. E., Clarren, S. K., Dehaene, P., et al. (1997). Incidence of fetal alcohol syndrome and prevalence of alcohol-related neurodevelopmental disorder. *Teratology, 56*(5), 317–326.

Santhanam, P., Li, Z., Hu, X., Lynch, M. E., & Coles, C. D. (2009). Effects of prenatal alcohol exposure on brain activation during an arithmetic task: An fMRI study. *Alcoholism, Clinical and Experimental Research, 33*(11), 1901–1908.

Schafer, D. P., Lehrman, E. K., Kautzman, A. G., Koyama, R., Mardinly, A. R., Yamasaki, R., et al. (2012). Microglia sculpt postnatal neural circuits in an activity and complement-dependent manner. *Neuron, 74*(4), 691–705.

Schwartz, M., Butovsky, O., Bruck, W., & Hanisch, U. K. (2006). Microglial phenotype: Is the commitment reversible? *Trends in Neurosciences, 29*(2), 68–74.

Schwarz, J. M., & Bilbo, S. D. (2012). The immune system and the developing brain. In M. M. McCarthy (Ed.), *Colloquium series on the developing brain* (pp. 1–118). San Rafael, CA: Morgan & Claypool Life Sciences.

Servais, L., Hourez, R., Bearzatto, B., Gall, D., Schiffmann, S. N., & Cheron, G. (2007). Purkinje cell dysfunction and alteration of long-term synaptic plasticity in fetal alcohol syndrome. *Proceedings of the National Academy of Sciences of the United States of America, 104*(23), 9858–9863.

Sierra, A., Encinas, J. M., Deudero, J. J., Chancey, J. H., Enikolopov, G., Overstreet-Wadiche, L. S., et al. (2010). Microglia shape adult hippocampal neurogenesis through apoptosis-coupled phagocytosis. *Cell Stem Cell, 7*(4), 483–495.

Slawecki, C. J., Thomas, J. D., Riley, E. P., & Ehlers, C. L. (2004). Neurophysiologic consequences of neonatal ethanol exposure in the rat. *Alcohol, 34*(2–3), 187–196.

Smith, D. E., & Davies, D. L. (1990). Effect of perinatal administration of ethanol on the CA1 pyramidal cell of the hippocampus and Purkinje cell of the cerebellum: An ultrastructural survey. *Journal of Neurocytology, 19*(5), 708–717.

Sowell, E. R., Jernigan, T. L., Mattson, S. N., Riley, E. P., Sobel, D. F., & Jones, K. L. (1996). Abnormal development of the cerebellar vermis in children prenatally exposed to alcohol: Size reduction in lobules I–V. *Alcoholism, Clinical and Experimental Research, 20*(1), 31–34.

Sowell, E. R., Mattson, S. N., Kan, E., Thompson, P. M., Riley, E. P., & Toga, A. W. (2008). Abnormal cortical thickness and brain-behavior correlation patterns in individuals with heavy prenatal alcohol exposure. *Cerebral Cortex, 18*(1), 136–144.

Sowell, E. R., Mattson, S. N., Thompson, P. M., Jernigan, T. L., Riley, E. P., & Toga, A. W. (2001). Mapping callosal morphology and cognitive correlates: Effects of heavy prenatal alcohol exposure. *Neurology, 57*(2), 235–244.

Spottiswoode, B. S., Meintjes, E. M., Anderson, A. W., Molteno, C. D., Stanton, M. E., Dodge, N. C., et al. (2011). Diffusion tensor imaging of the cerebellum and eyeblink conditioning in fetal alcohol spectrum disorder. *Alcoholism, Clinical and Experimental Research, 35*(12), 2174–2183.

Squire, L. R., Berg, D., Bloom, F. E., du Lac, S., Ghosh, A., & Spitzer, N. C. (2013). *Fundamental neuroscience*. Oxford, UK: Elsevier, Inc, pp. 1–1426.

Stopponi, S., de Guglielmo, G., Somaini, L., Cippitelli, A., Cannella, N., Kallupi, M., et al. (2013). Activation of PPARgamma by pioglitazone potentiates the effects of naltrexone on alcohol drinking and relapse in MSP rats. *Alcoholism, Clinical and Experimental Research, 37*(8), 1351–1360.

Stopponi, S., Somaini, L., Cippitelli, A., Cannella, N., Braconi, S., Kallupi, M., et al. (2011). Activation of nuclear PPARgamma receptors by the antidiabetic agent pioglitazone suppresses alcohol drinking and relapse to alcohol seeking. *Biological Psychiatry, 69*(7), 642–649.

Storer, P. D., Xu, J., Chavis, J. A., & Drew, P. D. (2005a). Peroxisome proliferator-activated receptor-gamma agonists inhibit the activation of microglia and astrocytes: Implications for multiple sclerosis. *Journal of Neuroimmunology, 161*(1–2), 113–122.

Storer, P. D., Xu, J., Chavis, J. A., & Drew, P. D. (2005b). Cyclopentenone prostaglandins PGA2 and 15-deoxy-delta12,14 PGJ_2 suppress activation of murine microglia and astrocytes: Implications for multiple sclerosis. *Journal of Neuroscience Research*, *80*(1), 66–74.
Streissguth, A. P., Bookstein, F. L., Barr, H. M., Sampson, P. D., O'Malley, K., & Young, J. K. (2004). Risk factors for adverse life outcomes in fetal alcohol syndrome and fetal alcohol effects. *Journal of Developmental and Behavioral Pediatrics*, *25*(4), 228–238.
Streissguth, A. P., Landesman-Dwyer, S., Martin, J. C., & Smith, D. W. (1980). Teratogenic effects of alcohol in humans and laboratory animals. *Science*, *209*(4454), 353–361.
Streit, W. J., Walter, S. A., & Pennell, N. A. (1999). Reactive microgliosis. *Progress in Neurobiology*, *57*(6), 563–581.
Sudheendran, N., Bake, S., Miranda, R. C., & Larin, K. V. (2013). Comparative assessments of the effects of alcohol exposure on fetal brain development using optical coherence tomography and ultrasound imaging. *Journal of Biomedical Optics*, *18*(2), 20506.
Suttie, M., Foroud, T., Wetherill, L., Jacobson, J. L., Molteno, C. D., Meintjes, E. M., et al. (2013). Facial dysmorphism across the fetal alcohol spectrum. *Pediatrics*, *131*(3), e779–e788.
Thomas, J. D., Idrus, N. M., Monk, B. R., & Dominguez, H. D. (2010). Prenatal choline supplementation mitigates behavioral alterations associated with prenatal alcohol exposure in rats. *Birth Defects Research, Part A: Clinical and Molecular Teratology*, *88*(10), 827–837.
Tiwari, V., & Chopra, K. (2011). Resveratrol prevents alcohol-induced cognitive deficits and brain damage by blocking inflammatory signaling and cell death cascade in neonatal rat brain. *Journal of Neurochemistry*, *117*(4), 678–690.
Tran, T. D., & Kelly, S. J. (2003). Critical periods for ethanol-induced cell loss in the hippocampal formation. *Neurotoxicology and Teratology*, *25*(5), 519–528.
Tremblay, M. E., Lowery, R. L., & Majewska, A. K. (2010). Microglial interactions with synapses are modulated by visual experience. *PLoS Biology*, *8*(11), e1000527.
Uban, K. A., Sliwowska, J. H., Lieblich, S., Ellis, L. A., Yu, W. K., Weinberg, J., et al. (2010). Prenatal alcohol exposure reduces the proportion of newly produced neurons and glia in the dentate gyrus of the hippocampus in female rats. *Hormones and Behavior*, *58*(5), 835–843.
Valenzuela, C. F., Lindquist, B., & Zamudio-Bulcock, P. A. (2010). A review of synaptic plasticity at Purkinje neurons with a focus on ethanol-induced cerebellar dysfunction. *International Review of Neurobiology*, *91*, 339–372.
Vetreno, R. P., Qin, L., & Crews, F. T. (2013). Increased receptor for advanced glycation end product expression in the human alcoholic prefrontal cortex is linked to adolescent drinking. *Neurobiology of Disease*, *59*, 52–62.
Wagner, J. L., Klintsova, A. Y., Greenough, W. T., & Goodlett, C. R. (2013). Rehabilitation training using complex motor learning rescues deficits in eyeblink classical conditioning in female rats induced by binge-like neonatal alcohol exposure. *Alcoholism, Clinical and Experimental Research*, *37*(9), 1561–1570.
Wagner, J. L., Zhou, F. C., & Goodlett, C. R. (2014). Effects of one- and three-day binge alcohol exposure in neonatal C57BL/6 mice on spatial learning and memory in adolescence and adulthood. *Alcohol*, *48*(2), 99–111.
Wake, H., Moorhouse, A. J., Jinno, S., Kohsaka, S., & Nabekura, J. (2009). Resting microglia directly monitor the functional state of synapses in vivo and determine the fate of ischemic terminals. *The Journal of Neuroscience*, *29*(13), 3974–3980.
Ward, R. J., Zhang, Y., Crichton, R. R., Piret, B., Piette, J., & de Witte, P. (1996). Identification of the nuclear transcription factor NFkappaB in rat after in vivo ethanol administration. *FEBS Letters*, *389*(2), 119–122.
Ware, A. L., Glass, L., Crocker, N., Deweese, B. N., Coles, C. D., Kable, J. A., et al. (2014). Effects of prenatal alcohol exposure and attention-deficit/hyperactivity disorder on adaptive functioning. *Alcoholism, Clinical and Experimental Research*, *38*(5), 1439–1447.

West, J. R., Kelly, S. J., & Pierce, D. R. (1987). Severity of alcohol-induced deficits in rats during the third trimester equivalent is determined by the pattern of exposure. *Alcohol and Alcoholism. Supplement*, *1*, 461–465.

Whitcher, L. T., & Klintsova, A. Y. (2008). Postnatal binge-like alcohol exposure reduces spine density without affecting dendritic morphology in rat MPFC. *Synapse*, *62*(8), 566–573.

Whitman, B. A., Knapp, D. J., Werner, D. F., Crews, F. T., & Breese, G. R. (2013). The cytokine mRNA increase induced by withdrawal from chronic ethanol in the sterile environment of brain is mediated by CRF and HMGB1 release. *Alcoholism, Clinical and Experimental Research*, *37*(12), 2086–2097.

Wilson, D. A., Peterson, J., Basavaraj, B. S., & Saito, M. (2011). Local and regional network function in behaviorally relevant cortical circuits of adult mice following postnatal alcohol exposure. *Alcoholism, Clinical and Experimental Research*, *35*(11), 1974–1984.

Wozniak, J. R., Mueller, B. A., Bell, C. J., Muetzel, R. L., Hoecker, H. L., Boys, C. J., et al. (2013). Global functional connectivity abnormalities in children with fetal alcohol spectrum disorders. *Alcoholism, Clinical and Experimental Research*, *37*(5), 748–756.

Wozniak, J. R., Mueller, B. A., Muetzel, R. L., Bell, C. J., Hoecker, H. L., Nelson, M. L., et al. (2011). Inter-hemispheric functional connectivity disruption in children with prenatal alcohol exposure. *Alcoholism, Clinical and Experimental Research*, *35*(5), 849–861.

Wozniak, J. R., & Muetzel, R. L. (2011). What does diffusion tensor imaging reveal about the brain and cognition in fetal alcohol spectrum disorders? *Neuropsychology Review*, *21*(2), 133–147.

Wozniak, J. R., Muetzel, R. L., Mueller, B. A., McGee, C. L., Freerks, M. A., Ward, E. E., et al. (2009). Microstructural corpus callosum anomalies in children with prenatal alcohol exposure: An extension of previous diffusion tensor imaging findings. *Alcoholism, Clinical and Experimental Research*, *33*(10), 1825–1835.

Wu, Y., Lousberg, E. L., Moldenhauer, L. M., Hayball, J. D., Coller, J. K., Rice, K. C., et al. (2012). Inhibiting the TLR4-MYD88 signalling cascade by genetic or pharmacological strategies reduces acute alcohol-induced sedation and motor impairment in mice. *British Journal of Pharmacology*, *165*(5), 1319–1329.

Wu, Y., Lousberg, E. L., Moldenhauer, L. M., Hayball, J. D., Robertson, S. A., Coller, J. K., et al. (2011). Attenuation of microglial and IL-1 signaling protects mice from acute alcohol-induced sedation and/or motor impairment. *Brain, Behavior, and Immunity*, *25*(Suppl. 1), S155–S164.

Xu, J., Chavis, J. A., Racke, M. K., & Drew, P. D. (2006). Peroxisome proliferator-activated receptor-alpha and retinoid X receptor agonists inhibit inflammatory responses of astrocytes. *Journal of Neuroimmunology*, *176*(1–2), 95–105.

Xu, J., & Drew, P. D. (2007). Peroxisome proliferator-activated receptor-gamma agonists suppress the production of IL-12 family cytokines by activated glia. *Journal of Immunology*, *178*(3), 1904–1913.

Xu, J., Storer, P. D., Chavis, J. A., Racke, M. K., & Drew, P. D. (2005). Agonists for the peroxisome proliferator-activated receptor-alpha and the retinoid X receptor inhibit inflammatory responses of microglia. *Journal of Neuroscience Research*, *81*(3), 403–411.

Yakovleva, T., Bazov, I., Watanabe, H., Hauser, K. F., & Bakalkin, G. (2011). Transcriptional control of maladaptive and protective responses in alcoholics: A role of the NF-kappaB system. *Brain, Behavior, and Immunity*, *25*(Suppl. 1), S29–S38.

Yang, Y., Phillips, O. R., Kan, E., Sulik, K. K., Mattson, S. N., Riley, E. P., et al. (2012). Callosal thickness reductions relate to facial dysmorphology in fetal alcohol spectrum disorders. *Alcoholism, Clinical and Experimental Research*, *36*(5), 798–806.

Zhan, Y., Paolicelli, R. C., Sforazzini, F., Weinhard, L., Bolasco, G., Pagani, F., et al. (2014). Deficient neuron-microglia signaling results in impaired functional brain connectivity and social behavior. *Nature Neuroscience*, *17*(3), 400–406.

Zhou, D., Lebel, C., Lepage, C., Rasmussen, C., Evans, A., Wyper, K., et al. (2011). Developmental cortical thinning in fetal alcohol spectrum disorders. *NeuroImage*, *58*(1), 16–25.
Zink, M., Ferbert, T., Frank, S. T., Seufert, P., Gebicke-Haerter, P. J., & Spanagel, R. (2011). Perinatal exposure to alcohol disturbs spatial learning and glutamate transmission-related gene expression in the adult hippocampus. *The European Journal of Neuroscience*, *34*(3), 457–468.
Zou, J. Y., & Crews, F. T. (2006). CREB and NF-kappaB transcription factors regulate sensitivity to excitotoxic and oxidative stress induced neuronal cell death. *Cellular and Molecular Neurobiology*, *26*(4–6), 385–405.
Zou, J. Y., & Crews, F. T. (2010). Induction of innate immune gene expression cascades in brain slice cultures by ethanol: Key role of NF-kappaB and proinflammatory cytokines. *Alcoholism, Clinical and Experimental Research*, *34*(5), 777–789.
Zou, J. Y., & Crews, F. T. (2014). Release of neuronal HMGB1 by ethanol through decreased HDAC activity activates brain neuroimmune signaling. *PLoS One*, *9*(2), e87915.

CHAPTER FOUR

Role of Microglia in Regulation of Ethanol Neurotoxic Action

Lucy G. Chastain, Dipak K. Sarkar[1]
Endocrinology Program, Department of Animal Sciences, Rutgers, The State University of New Jersey, New Brunswick, New Jersey, USA
[1]Corresponding author: e-mail address: sarkar@aesop.rutgers.edu

Contents

Abstract

Exposure to alcohol, during development or adulthood, may result in damage to the nervous system, which underlies neurological and cognitive disruptions observed in patients with alcohol-related disorders, including fetal alcohol spectrum disorders (FASDs) and alcohol-use disorders (AUDs). Both clinical and preclinical evidence suggest microglia, the immune cells of the central nervous system, play a key role in modulating alcohol-induced neurotoxicity. Particularly, microglia are implicated in alcohol-induced neuroinflammation and in alcohol-induced increases in oxidative stress, which can lead to neuronal apoptosis. Recent studies also suggest a regenerative role for microglia in reestablishing homeostasis after alcohol exposure. These studies are summarized and reviewed in this chapter with emphasis on relevance to FASD and AUD.

International Review of Neurobiology, Volume 118
ISSN 0074-7742
http://dx.doi.org/10.1016/B978-0-12-801284-0.00004-X

ABBREVIATIONS

AUD alcohol-use disorder
BEP beta-endorphin
CNS central nervous system
DAMPs damage-associated molecular patterns
FAS fetal alcohol syndrome
FASD fetal alcohol spectrum disorder
GSH glutathione
GSH-Px GSH peroxidase
HAM human alveolar macrophage
HMGB1 high-mobility group box 1
Iba1 ionized calcium-binding adaptor molecule 1
IFN-γ interferon γ
IL interleukin
LPS lipopolysaccharide
MCP-1 monocyte chemoattractant protein-1
MHC major histocompatibility complex
MIP-2 macrophage inflammatory protein 2
NF-κB nuclear factor kappa-light-chain-enhancer of activated B cells
O_2^- superoxide
ROS reactive oxygen species
SOD superoxide dismutase
TGF-β transforming growth factor-β
TLR toll-like receptor
TNF-α tumor necrosis factor-α

1. INTRODUCTION

According to the World Health Organization alcohol exposure is one of the highest risk factors for a multitude of health problems, including brain damage (WHO, 2011). Both alcohol exposure during development and during adulthood have potentially toxic effects on the peripheral and central nervous system (CNS). Prenatal exposure to alcohol has potentially neurotoxic effects to the fetus in several brain regions, including cerebellum, cortex, hippocampus, and hypothalamus, which underlie the development of a range of developmental, cognitive, and neurobehavioral symptoms termed as fetal alcohol spectrum disorders (FASDs) (Riley, Infante, & Warren, 2011). Consumption of alcohol during adolescence and adulthood, especially heavy "binge" drinking or chronic drinking may result in development of an alcohol-use disorder (AUD). AUDs are associated with impaired cognition and neurotoxicity; both binge drinking and chronic alcohol use are linked to decreased frontal lobe volume, and alcoholism is correlated with neuronal loss in the cerebellum, cortex, and hypothalamus (Alfonso-Loeches & Guerri, 2011). In addition, chronic alcoholism can result in peripheral neuropathy, causing chronic pain and disability (Chopra & Tiwari, 2012). There are currently no treatments for alcohol-associated damage to the nervous system, necessitating a better understanding of the underlying mechanisms of alcohol-induced neurotoxicity.

Both preclinical and clinical studies implicate microglia in the neurotoxic effects of alcohol and in alcohol-related health disorders. Microglia are the major immune cells of the CNS and are first responders to tissue injury and infection. In addition, microglia can also be activated by environmental toxins and drugs, such as alcohol. When activated, microglia respond to possible threats with increased release of inflammatory factors and reactive oxygen species (ROS), allowing for removal of pathogens and dead or dying neural tissue. Under normal conditions, when the threat is resolved, microglia will become inactivated and the inflammation will be resolved, but under pathological conditions, such as harmful alcohol exposure, microglia may become chronically activated and become neurotoxic (Takeuchi, 2010). In addition, a few recent studies suggest microglia may have potentially beneficial effects when stimulated by alcohol under certain conditions, possibly countering alcohol-induced neurotoxicity. In this chapter, we review recent studies on the role of microglia in ethanol-induced neurotoxicity, focusing on both evidence in cellular mechanisms and clinical studies.

2. MICROGLIAL CELL FUNCTIONS IN THE CNS

Microglia are the primary immune cells of the CNS and comprise about 10% of the cells in the brain (Lawson, Perry, Dri, & Gordon, 1990; Mittelbronn, Dietz, Schluesener, & Meyermann, 2001). Microglia are derived from myeloid origin early on in embryogenesis and migrate to the nervous system (Ginhoux, Lim, Hoeffel, Low, & Huber, 2013; Harry, 2013). Specifically, microglia originate from a large family of immune cells known as tissue-resident macrophages, derived from yolk sac during development (Cronk & Kipnis, 2013; Schulz et al., 2012). After migration of microglia from the yolk sac to the brain, microglia undergo stages of proliferation and migration throughout prenatal and postnatal development through adolescence (Bilbo, 2013; Bilbo, Smith, & Schwarz, 2012; Harry, 2013). While the initial origin of microglia is from macrophage precursors in the yolk sac, later during development and possibly during pathological conditions, microglia may also originate from blood-borne monocytes (Ginhoux et al., 2013).

2.1. Microglial response to pathogens and tissue damage

Microglia play several functional roles, including classical immune cell functions. Microglia play an essential role in the CNS immune system, acting as first responders to infection and tissue injury. Microglial response to viruses and bacteria has been well characterized *in vitro* and *in vivo* by injection of polyinosinic-polycytidylic acid (poly I:C), which is structurally similar to double-stranded RNA and is present in some viruses, and lipopolysaccharide (LPS), an endotoxin found on the cell membrane of Gram negative bacteria and potent immune activator, while microglial response to tissue damage has been modeled using targeted nerve damage in animal models (Jonas et al., 2012; Kreutzberg, 1996). In their quiescent, "resting state" microglia exist as sentinels, monitoring the extracellular environment for possible threats. In this state, microglia show a "ramified" morphology with multiple thin, branched processes extending from the soma. Pathogens, neuronal damage, and inflammation stimulate microglia, inducing stereotyped morphological changes. Specifically, microglia undergo a graded decrease in branched processes, an enlargement of the soma, and, in the fully activated state, achieve an amoeboid-like appearance allowing for phagocytosis of pathogens or cellular debris (Kreutzberg, 1996). While some studies utilize a binary approach to characterizing microglial activation (i.e., resting or activated), other studies distinguish between partial activation and full activation, and others distinguish several levels of microglial activation (Jonas et al., 2012; Raivich, Bohatschek, et al., 1999). Different levels of microglial activation are associated with increased expression of specific receptors and cytokines. In their activated states, in order to recognize pathogens and cellular debris, microglia increase expression of pattern recognition receptors, such as toll-like receptors (TLRs) and scavenger receptors (Takeuchi, 2010). In addition, activated microglia increase expression of major histocompatibility complex (MHC) molecules and can also serve as antigen-presenting cells, thus, playing a role in both innate and adaptive immune system function (Town, Nikolic, & Tan, 2005). In their activated states, microglia are also capable of proliferating, specifically through the action of colony-stimulating factors (CSFs) and show increased motility, migrating to different brain areas allowing for a large-scale, region-specific immune response to local threats in the CNS (Badoer, 2010; Deverman & Patterson, 2009).

2.2. Microglia, inflammation, and cell death

As discussed above, microglia respond to regional inflammation by morphological activation, proliferation, and migration to the inflamed tissue. Microglia can detect released inflammatory factors via TLRs and scavenger receptors. In addition, activated microglia can also secrete pro- or anti-inflammatory factors themselves, potentially magnifying or inhibiting the inflammatory response to pathogens or cell death (Takeuchi, 2010). At higher levels of activation, microglia secrete proinflammatory factors including cytokines, such as tumor necrosis factor-α (TNF-α), interferon γ (IFN-γ), interleukins (ILs), and transforming growth factor-β (TGF-β), as well as ROS, excitatory amino acids such as glutamate, nucleic acids, and proteases (Raivich, Jones, et al., 1999; Takeuchi, 2010). Ideally, these secreted inflammatory factors can be beneficial in that they function to regulate microglial removal of pathogens or apoptotic cells and minimize further damage. Specifically, through their secretion of inflammatory factors, activated microglia can facilitate apoptosis, a process of programmed cell death, of neurons (Saijo & Glass, 2011). However, excessive secretion of proinflammatory factors by activated microglia can play a role in pathological conditions involving neurodegeneration (Takeuchi, 2010), including alcohol-related disorders (Crews, Zou, & Qin, 2011), as will be discussed in this review.

In addition to secretion of proinflammatory factors, at lower levels of activation, microglia can secrete anti-inflammatory factors to halt the inflammatory response, including ILs such as IL-3, IL-6, and IL-10 and growth factors like brain-derived neurotrophic factor (BDNF) and nerve growth factor (NGF) (Badoer, 2010; Block & Hong, 2005; Raivich, Jones, et al., 1999). Indeed, microglia are often characterized as having two distinct activated macrophage-like phenotypes: (1) the M1 phenotype, associated with secretion of proinflammatory factors and neurotoxic activity and (2) the M2 phenotype, which is associated with secretion of anti-inflammatory factors (Bilbo, 2013; Marshall et al., 2013; Polazzi & Monti, 2010). These distinct phenotypes demonstrate a complex role for microglia in regulating inflammation and determining cell death or survival.

2.3. Microglial modulation of neuronal activity and development

In addition to traditional immune system functions, it is becoming increasingly clear that microglia also play a broader homeostatic role in modulating synaptic structure and function in the CNS (Tremblay et al., 2011). In resting state, microglia are closely associated with neuronal synapses and show slight motility, extending and retracting their processes in response to small changes in their microenvironment (Nimmerjahn, Kirchhoff, & Helmchen, 2005; Tremblay et al., 2011). Recent studies show microglia play an important role in pruning excess synapses during postnatal development (Paolicelli et al., 2011), a role which is mediated through the receptor for chemokine (C-X3-C motif) ligand 1 (CX3CL1 or fractaline), as mice lacking this receptor (CX3CR1) show aberrant synapse formation and disrupted functional neural connectivity (Zhan et al., 2014). A similar process might also take place in the adult brain (Tremblay, Lowery, & Majewska, 2010). In the mature brain, microglia also regulate hippocampal neurogenesis, by influencing differentiation of neural progenitor cells and clearing apoptotic progenitor cells in the dentate gyrus (Avidan, Kipnis, Butovsky, Caspi, & Schwartz, 2004; Sierra et al., 2010). Finally, microglia express various neurotransmitter receptors, allowing for quick response to neuronal activity, and the possibility of microglial modulation neuronal activity (Pocock & Kettenmann, 2007; Tremblay et al., 2011). These recent studies demonstrate the broad role that microglia play in modulating neural activity and the development of neural networks. Future studies will likely increase the depth and complexity of these findings.

3. MICROGLIAL MECHANISMS OF THE NEUROTOXIC EFFECTS OF ALCOHOL

The neurotoxic effects of alcohol have been well studied *in vitro* and in animal models. Several mechanisms for alcohol-induced cell death have been identified, two of which include (1) the release of proinflammatory factors and (2) release of ROS. Studies have implicated microglia in mediating the toxic effects of alcohol via both of these mechanisms. Alcohol has been observed to induce activation of microglia *in vitro* (Boyadjieva & Sarkar, 2010, 2013a, 2013b) and in some animal models (Alfonso-Loeches & Guerri, 2011; Alfonso-Loeches et al., 2010; Qin & Crews, 2012a; Zhao et al., 2013). In addition, clinical studies have found increased microglial activation in the *postmortem* brains of alcoholics (Crews et al., 2011; He & Crews, 2008). Notably, alcohol-induced microglial activation occurs via stimulation of microglial toll-like receptor 4 (TLR4) (Fernandez-Lizarbe, Pascual, & Guerri, 2009). While many studies suggest microglia facilitates the neurotoxic effects of alcohol, others studies suggest microglia may possibly have an ameliorative function in restoring homeostasis after alcohol exposure. These studies are summarized below.

3.1. Alcohol-induced microglial activation, inflammation, and neurotoxicity

Given their role in facilitating inflammation, it is not surprising that alcohol-activated microglia have been implicated in alcohol-induced inflammatory pathways. In rats, intermittent and chronic alcohol exposure have been shown to activate microglia, while concomitantly increasing expression of proinflammatory cytokines and neuronal cell death, providing indirect evidence for the role of microglia in alcohol-induced neuroinflammation and neurotoxicity (Alfonso-Loeches & Guerri, 2011; Zhao et al., 2013). Alcohol can activate microglia via stimulation of TLRs and via neuronal damage and subsequent release of damage-associated molecular patterns (DAMPs), resulting in reactive microgliosis (Alfonso-Loeches & Guerri, 2011). Specifically, microglia express TLR4 receptors and microglial TLR4 receptors appear to be necessary in alcohol-induced activation of microglia and subsequent microglial production of inflammatory mediators and neuronal apoptosis (Fernandez-Lizarbe, Montesinos, & Guerri, 2013; Fernandez-Lizarbe et al., 2009). In an *in vitro* study from our lab (Boyadjieva & Sarkar, 2010), the presence of microglial-conditioned media enhanced ethanol-induced apoptosis of cultured fetal hypothalamic neurons. Interestingly, immunoneutralization of TNF-α abolished neuronal cell death induced by microglial-conditioned media, suggesting that microglial production of TNF-α plays a key role in ethanol-induced neurotoxicity in developing neurons (Fig. 4.1). The mechanism by which alcohol induces neuronal cell death may involve upregulation of glial nuclear factor kappa-light-chain-enhancer of activated B cells (NF-κB) expression, which stimulates release of TNF-α, resulting in neuronal apoptosis (Crews & Nixon, 2009; Guadagno, Xu, Karajgikar, Brown, & Cregan, 2013; Fig. 4.1). Stimulation of the transcription factor activator protein 1 (AP-1) and release of IL-1β, IL-6, and TGF-β1 also may contribute to alcohol-induced neuronal apoptosis (Alfonso-Loeches & Guerri, 2011; Chen, Kuhn, Chaturvedi, Boyadjieva, & Sarkar, 2006).

3.2. Alcohol-induced microglial production of ROS and neurotoxicity

Another mechanism by which alcohol induces neurotoxicity is by stimulation of ROS in the brain (Alfonso-Loeches & Guerri, 2011). ROS are reactive oxidative molecules, such as superoxide (O_2-), hydrogen peroxide (H_2O_2), and nitric oxide (NO) that are capable of breaking down cell membranes and inducing cell death. Increases in ROS are created by alcohol

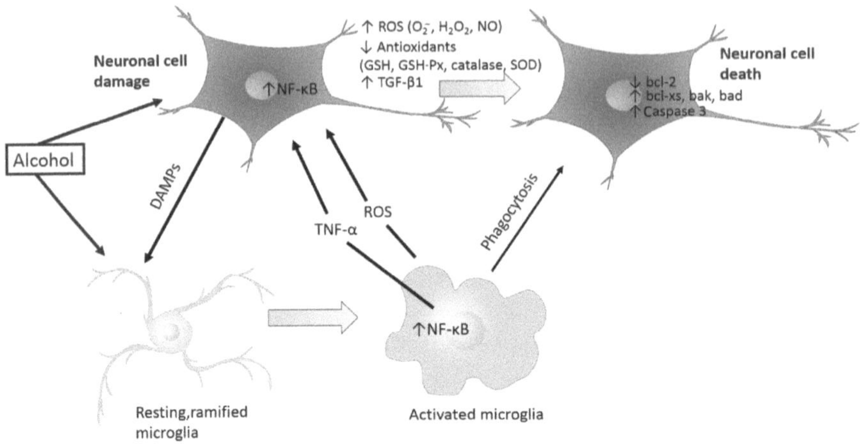

Figure 4.1 Mechanisms by which microglia facilitate alcohol-induced apoptosis of developing neurons. Alcohol damages developing neurons, resulting in increased levels of ROS, decreased levels of antioxidants, and increased stimulation of TGF-β1 apoptotic signaling. Alcohol activates microglia directly and via DAMPs resulting in increased microglial release of TNF-α and ROS, exacerbating neuroinflammation and oxidative stress. This results in TGF-β1 mediated increases in proapoptotic proteins bak, bad, and bcl-xs, lowered levels of the antiapoptotic protein bcl-2, increased production of apoptotic enzyme caspase 3, increased neuronal cell death, and increased phagocytosis of dead neurons by microglia.

exposure as a natural by-product of alcohol metabolism and by increasing cellular respiration, thus, creating an increase in oxidative stress and neuronal cell death (Guerri, Montoliu, & Renau-Piqueras, 1994; Montoliu et al., 1995). In addition, alcohol exposure decreases endogenous antioxidant levels, thus, decreasing natural inhibition of ROS and increasing oxidative stress (Henderson, Devi, Perez, & Schenker, 1995).

In addition to releasing cytokines, stimulated microglia can also secrete ROS, contributing to neurotoxicity (Takeuchi, 2010). Several studies implicate microglia in alcohol-induced production of ROS and resulting neurotoxicity. Qin and Crews (2012a) found mice exposed to chronic alcohol showed increased expression of nicotinamide adenine dinucleotide phosphate-oxidase (NADPH) oxidase (an enzyme that produces ROS), O_2-, microglial activation, and cell death in cortical and hippocampal brain regions. In addition, inhibition of NADPH oxidase during alcohol administration decreased O_2-, microglial activation, and cell death, directly linking production of ROS to alcohol-induced microglial activation and neurotoxicity. In agreement with these *in vivo* findings, *in vitro* studies by our lab

(Boyadjieva & Sarkar, 2013a) showed that ethanol-activated microglial-conditioned media enhanced oxidative stress in cultured fetal hypothalamic neuronal cells and increased apoptotic cell death. Additionally, it was shown that ethanol induces oxidative stress in neurons by increasing the cellular production of O_2^-, ROS, and nitrite, while decreasing the level of antioxidant glutathione (GSH) and the cellular activity of antioxidative enzymes (glutathione peroxidase, GSH-Px; catalase; superoxide dismutase, SOD). Furthermore, synthetic SOD/catalase mimetic (EUK-134) and a water-soluble analog of vitamin E (Trolox), two well-known antioxidants, protected developing hypothalamic neurons from oxidative stress and cellular apoptosis caused by ethanol-treated microglia medium. Therefore, ethanol-activated microglia may produce ROS and nitrite to decrease cellular activity of antioxidative enzymes leading to induction of apoptotic cell death (Fig. 4.1).

In sum, alcohol-induced activation of microglia results in microglial release of proinflammatory factors, specifically TNF-α and ROS, which augment neurotoxicity. Specifically, alcohol-induced release of these microglial proinflammatory factors appears to exacerbate neuronal oxidative stress and neuronal release of proinflammatory cytokines, triggering apoptotic cell death pathways (Fig. 4.1). One mechanism by which alcohol appears to induce neuronal apoptosis specifically is via increased neuronal release of TGF-β1. It has been shown that alcohol-increased TGF-β1 levels in neuronal cells is accompanied by increased expression levels of E2F1 (whose overexpression sensitizes cells to apoptosis), reduced expression of cyclin D1 and cyclin-dependent kinase-4 (key regulators of cell cycle progression), elevated levels of mitochondrial proapoptotic proteins bak, bad, and bcl-xs, lowered levels of the antiapoptotic protein bcl 2, increased production of apoptotic enzyme caspase 3, and increased neuronal cell death (Chen et al., 2006; Kuhn & Sarkar, 2008). It is interesting to note that in a transformed cell culture model, inhibition of NF-κB or ROS abrogates TGF-β1 stimulation of cell functions (Tobar, Villar, & Santibanez, 2010). Hence, it could be postulated that the ROS–NF-κB–TGF-β1 signaling cascade is a possible mechanism by which alcohol induces the apoptotic process in neurons, a process that is modulated by microglia (Fig. 4.1).

4. PROTECTIVE EFFECTS OF MICROGLIA

The aforementioned studies suggest alcohol administration induces microglial activation and release of proinflammatory cytokines and ROS, which contributes to alcohol-induced neurotoxicity. However, a few recent studies suggest alcohol may elicit an anti-inflammatory response from microglia under some conditions. In one study, binge-like exposure to alcohol results in partial (not full) activation of microglia, increased release of IL-10 (an anti-inflammatory factor), and no change in release of proinflammatory factors IL-6 and TNF-α (further discussed in Section 5) (Marshall et al., 2013). The authors of this study concluded that under some conditions, alcohol may stimulate microglia to have beneficial or homeostatic effects, rather than contribute to neurodegeneration. In line with this study, a recent study compared the proteomic expression profiles of cultured rat microglia exposed to LPS with the profiles elicited by acute ethanol (Bell-Temin et al., 2013). The authors found LPS and ethanol elicited distinct expression profiles in microglia: LPS-elicited upregulation of classic proinflammatory cytokines (e.g., TNF-α and IL-6), while ethanol did not elicit upregulation of these factors, leading the authors to conclude that the activation profile elicited by ethanol in microglia might be distinct from the classic M1 proinflammatory phenotype elicited by immune challenges like LPS.

In addition, alcohol administration also has distinct effects on neurogenesis, effects that may be mediated by microglia. Alcohol exposure can result in decreased hippocampal neurogenesis, an effect that may underlie alcohol-related neurodegeneration (Crews et al., 2006; Morris, Eaves, Smith, & Nixon, 2010). However, alcohol exposure followed by a period of abstinence results in increased hippocampal neurogenesis, a process which may serve a regenerative purpose. Interestingly, this process is preceded by microglial proliferation, raising the possibility that microglia may facilitate some regenerative mechanisms in recovery from alcohol exposure (McClain et al., 2011; Nixon, Kim, Potts, He, & Crews, 2008).

5. ALCOHOL-RELATED DISEASE, NEUROTOXICITY, AND MICROGLIA

In many pathological states, microglia may become dysfunctional or chronically activated, as is the case with neurodegenerative diseases, such as Alzheimer's and Parkinson's disease (Block & Hong, 2007). While not classified as traditional neurodegenerative disorders, alcohol-related disorders, such as AUDs and FASD do involve neuronal cell death and degenerative processes. Evidence for the role of microglia in these disorders from clinical and preclinical studies is reviewed below.

5.1. Adult alcohol-use disorders

AUDs are characterized by recurrent or excessive consumption of alcohol that may result in persistent social or interpersonal problems or in physical dependence or alcoholism (APA, 2013). AUDs may result in neuronal death in the central and/or peripheral nervous systems. Imaging and *postmortem* studies of the brains of alcoholics have consistently found decreased frontal lobe volume, and atrophy in cortical and subcortical regions, due mainly to decreased amounts of white matter (Alfonso-Loeches & Guerri, 2011; Crews & Nixon, 2009). Increased duration and amount of alcohol consumption also correlates with the degree of atrophy in alcoholics. In addition, some *postmortem* studies have found a decrease in neurons in the cortex, hypothalamus, and cerebellum of alcoholics. Brain damage in these patients correlates with cognitive impairment such as deficits in learning, memory, and perceptual motor skills, and confers a risk of dementia (Sullivan & Pfefferbaum, 2005). Finally, AUDs can result in peripheral neuropathy, which is characterized by severe, chronic pain in the limbs (Chopra & Tiwari, 2012).

Several studies implicate microglia in neurotoxicity associated with AUDs. In addition to showing neural damage, some studies of *postmortem* brains of alcoholics show increased microglial activation alongside increased levels of proinflammatory cytokines and chemokines, leading to the theory that neural damage observed in AUDs may be due to chronic overstimulation of microglia by alcohol consumption (Fig. 4.2; Crews et al., 2011; He & Crews, 2008). This is observed in both the CNS and peripheral nervous system in peripheral alcoholic neuropathy (Chopra & Tiwari, 2012).

Models relevant to AUD in rats and mice also suggest long-term alcohol abuse results in microglial activation, although the length of exposure may determine the extent of microglial activation, neuroinflammation, or neurotoxicity (Table 4.1; Fig. 4.2). Rodent models for AUD-related neuronal damage include exposure to binge-like alcohol exposure administered over a couple of days and chronic or intermittent exposure administered over several weeks or months. Chronic alcohol treatment (2–3 weeks) results in an increase in fully activated, phagocytic microglia in the hippocampus and cortex accompanied by increased neuronal cell death in these regions and release of proinflammatory factors (Qin & Crews, 2012a, 2012b; Ward et al., 2009; Zhao et al., 2013). In one study, mice exposed to 5 months of chronic alcohol showed increased markers for activated microglia

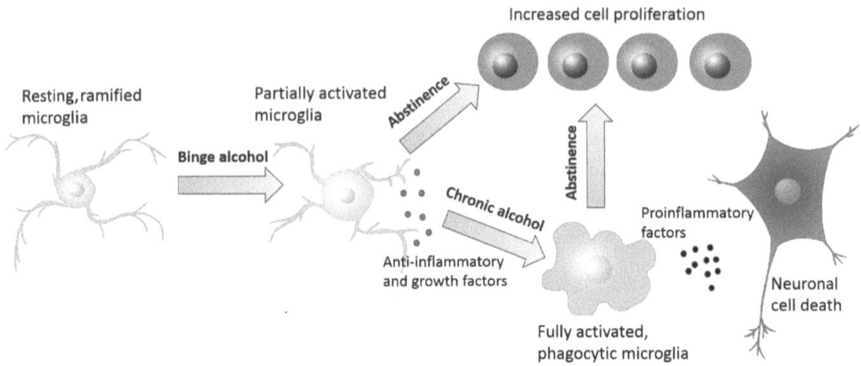

Figure 4.2 The effect of binge or chronic alcohol exposure on microglia. Binge alcohol exposure partially activates microglia and may induce increased release of anti-inflammatory or growth factors. Chronic alcohol fully activates microglia, increasing release of proinflammatory cytokines and ROS, possibly resulting in neurotoxicity. Abstinence from alcohol results in increased microglial proliferation and a possible return to homeostasis.

(CD11b) as well as increased inflammatory cytokines and cell death markers in cortex (Alfonso-Loeches et al., 2010). In addition, in a rat model using 12 months of alcohol exposure, alcohol-exposed rats showed a decrease in cholinergic neurons, an increase in cortical microglial markers, increased levels of proinflammatory chemokine monocyte chemoattractant protein-1 (MCP-1), and memory impairment (Ehrlich, Pirchl, & Humpel, 2012). These studies coincide with the aforementioned clinical studies of chronic alcoholics, providing further evidence that microglia play a role in neuroinflammation, neurotoxicity, and cognitive impairment in AUDs.

In vivo binge-like alcohol exposure in rodents also results in neurotoxicity (Crews, Braun, Hoplight, Switzer, & Knapp, 2000; Obernier, Bouldin, & Crews, 2002), but, in contrast to chronic exposure to alcohol, microglia show a distinct response (Table 4.1; Fig. 4.2). Specifically, binge-like alcohol models show partial microglial activation and proliferation, but in these studies, microglial activation was not associated with alcohol-induced release of proinflammatory factors (Marshall et al., 2013; McClain et al., 2011; Nixon et al., 2008). As mentioned previously, in one binge model of alcohol utilizing a 4-day exposure to alcohol, only partial activation of microglia was noted, accompanied by an increase in growth factors and anti-inflammatory cytokines, and no change in proinflammatory cytokines (Marshall et al., 2013). In addition, in binge-like alcohol models, alcohol-induced hippocampal neurogenesis is preceded by increased proliferation of microglia,

Table 4.1 Summary of studies investigating the effects of alcohol exposure on microglial activation and proliferation, inflammation, and neurodegeneration utilizing *in vivo* animal models

Exposure	Animal	Brain region	Microglia markers	Microglia morphology	Inflammatory factors	Neuronal cell death	References
Chronic/ intermittent (3 weeks)	Rat 6 weeks	Hipp	↑ OX-6/ MHC-II	Large soma, thick processes	↑ Glutamate	–	Ward et al. (2009)
Chronic (10 days)	Mouse 8 weeks	Hipp, cortex	↑ Iba1	Large soma, irregular shape	↑ NF-κB, NADPH oxidase, O_2^-	↑ Cell death	Qin and Crews (2012a)
Chronic (10 days)	Mouse 8 weeks	Hipp, cortex,	↑ Iba1	–	↑ IL-6, TNF-α, MCP1, TLR3, HMGB1, O_2^-, NADPH oxidase	↑ Cell death	Qin and Crews (2012b)
Chronic/ intermittent (25 days)	Rat Adult	Hipp, cortex	↑ CD11b ↑ ED-1/ CD68	Large soma, few processes	↑ IL-1β, TNF-α	↑ Cell death, dendritic degeneration	Zhao et al. (2013)
Chronic (5 months)	Mouse adult	cortex	↑ CD11b	–	↑ IL-1β, TNF-α, IL-6, NF-κB	↑ caspase-3 activation	Alfonso-Loeches et al., 2010
Chronic (12 months)	Rat Adult	Cortex, nuc. bas. Meynert	↑ Iba1 N.C.: OX-42	–	↑ MCP-1 N.C.: IL-1β, MIP-2, TNF-α	↓ Cholinergic cell count	Ehrlich et al. (2012)
Binge (4 days)	Rat Adult	Hipp, cortex	N.C ED-1/ CD68	–	–	–	Nixon et al. (2008)

Continued

Table 4.1 Summary of studies investigating the effects of alcohol exposure on microglial activation and proliferation, inflammation, and neurodegeneration utilizing *in vivo* animal models—cont'd

Exposure	Animal	Brain region	Microglia markers	Microglia morphology	Inflammatory factors	Neuronal cell death	References
Binge (4 days)	Rat Juvenile	Hipp	N.C. MHC-II, ED-1/ CD-68	Large soma, thick processes	N.C.; TNF-α	–	McClain et al. (2011)
Binge (4 days)	Rat Adult	Hipp, cortex	↑ OX-42, Iba1	–	↑ IL-10, TGF-β N.C.: IL-6, TNF-α	–	Marshall et al. (2013)
Neonatal (PND 7)	Mouse Neonate	Cortex, corpus callosum	Iba1	Large soma, thick processes, or amoeboid	–	↑ Phagocytosis of cells and processes	Saito et al. (2010)
Neonatal (PND 3–5)	Mouse Neonate	Cerebell	↓ Isolectin B4	Short, thick processes	–	↓ Cell count	Kane et al. (2011)
Prenatal	Sheep Fetal	Cortical white matter	↑ HAM56	–	–	–	Watari et al. (2006)

↑, Increase in gene or protein; ↓, decrease in gene or protein; N.C., no change in gene or protein; –, not measured; PND, postnatal day; hipp, hippocampus; cerebell, cerebellum; nuc. bas. Meynert, nucleus basalis of Meynert.

which may suggest microglia has a role in homeostatic regenerative mechanisms during abstinence after binge alcohol exposure. Interestingly, in one study utilizing a more chronic alcohol exposure (25 days of intermittent alcohol treatments), an increase in microglial proliferation was noted at 7–14 days of abstinence after alcohol, suggesting microglia might also play a regenerative role during abstinence even after chronic alcohol exposure (Zhao et al., 2013). It is likely that different amounts and time courses of alcohol exposure may yield different levels of microglial activation, which may result in either release of proinflammatory factors and neurotoxicity or, conversely, release in anti-inflammatory factors and homeostatic outcomes. Particularly, microglia may play a regenerative role during abstinence after binge drinking and perhaps chronic drinking, but with long-term, chronic exposure to alcohol, microglia may become chronically activated and contribute to neuroinflammation and neurotoxicity (Fig. 4.2).

It has been proposed that, instead of being a "side effect" of the excessive amounts of alcohol consumed, neural damage due to alcohol consumption might actually underlie some of the mechanisms of developing an alcohol addiction or an AUD (Crews & Boettiger, 2009). Particularly, the development of addiction is thought to be due in part to lack of inhibition of the subcortical mesolimbic reward system, by the frontal cortex (Koob & Le Moal, 1997). Cell death in regions such as the prefrontal cortex due to excessive alcohol use may lead to lack of inhibition in subcortical reward areas such as the striatum, which may result in lack of behavioral inhibition and increased motivation to drink. Repeated stimulation of the innate immune system during chronic or heavy alcohol consumption might also facilitate this process and lead to decreased inhibition of the mesolimbic reward system, and, thus, increased drinking (Crews et al., 2011).

As microglia are implicated in alcohol-induced neurotoxicity, they might also be implicated in the processes that lead to alcohol addiction or an AUD. Recent studies utilizing rodent models support a role for microglia in voluntary alcohol drinking and preference. In a quantitative trait loci analysis of six strains of mice known to differ in voluntary alcohol drinking behavior, an increase in expression of the gene encoding β-2-microglobulin (*β2m*), an NF-κB target gene involved in microglial MHC immune signaling, was found in the brains of alcohol-preferring mice (Mulligan et al., 2006). In addition, knockout of the *β2m* gene in mice decreased voluntary alcohol consumption and preference (Blednov et al., 2012). Finally, treatment with minocycline, an antibiotic and selective inhibitor of microglia, reduced voluntary alcohol consumption in adult mice (Agrawal et al.,

2014). These studies suggest microglia might mediate alcohol preference and might contribute to the development of AUDs.

5.2. Fetal alcohol spectrum disorders

The CDC reports approximately 7.6% of U.S. mothers drink alcohol during pregnancy and at least 1.4% engage in binge drinking throughout pregnancy (May & Gossage, 2001). Fetal exposure to alcohol may result in FASD, a range of developmental, cognitive, and neurobehavioral symptoms, which vary in presentation and severity depending on the magnitude and timing of alcohol exposure during pregnancy (Alfonso-Loeches & Guerri, 2011; Hellemans et al., 2010). On the more extreme end of the spectrum is fetal alcohol syndrome (FAS), which is characterized by pre- and postnatal growth retardation, craniofacial abnormalities, and CNS dysfunction, including intellectual impairment and developmental delays (Jones, 2011). Less severe manifestations of FASD include partial FAS, alcohol-related birth defects, and alcohol-related neurobehavioral disorder. The prevalence of FASD is about 1 out of every 100 live births in the United States, making it one of the most common developmental disorders (May & Gossage, 2001).

Underlying the intellectual impairment and neurobehavioral disorder observed in FASD is alcohol-related neural damage in several brain regions, including the corpus callosum, cerebellum, cortex, hippocampus, and hypothalamus (Alfonso-Loeches & Guerri, 2011; Hannigan, O'leary-Moore, & Berman, 2007; Mattson & Riley, 1998). Exposure to alcohol earlier in development (embryonic stage) is associated with neural tube defects, abnormal cell migration, and corpus callosum malformation, while exposure later in pregnancy (fetal stage) is associated with glial abnormalities, increased apoptotic death, altered neural connections, and cerebellar abnormalities (reviewed in Alfonso-Loeches & Guerri, 2011).

These effects are paralleled in rodent models with prenatal exposure in rodents being equivalent to embryonic and early fetal exposure in humans and neonatal exposure in rodents being equivalent to third trimester exposure in humans. Indeed ethanol exposure prenatally and postnatally in rodents results in apoptotic death in the hippocampus, hypothalamus, and cerebellum (Brocardo, Gil-Mohapel, & Christie, 2011; Gil-Mohapel, Boehme, Kainer, & Christie, 2010; Ikonomidou et al., 2000; Kane et al., 2011; Sarkar, Kuhn, Marano, Chen, & Boyadjieva, 2007). As in the adult brain, developmental alcohol-induced neurotoxicity appears to occur through similar mechanisms, including production of proinflammatory cytokines and ROS (see Fig. 4.1; Alfonso-Loeches & Guerri, 2011; Boyadjieva & Sarkar, 2010, 2013a; Brocardo et al., 2011).

In vitro and *in vivo* models suggest microglia may mediate ethanol-induced neurotoxicity during development (Fig. 4.1). As mentioned previously, studies from our lab utilizing cultured microglia from fetal rat cortex and cultured neurons from fetal rat hypothalamus found that ethanol treatment activated fetal microglia, and activated microglia augmented ethanol-induced neurotoxicity via release of ROS and proinflammatory cytokines, particularly TNF-α (Boyadjieva & Sarkar, 2010, 2013a, 2013b).These studies coincide with *in vivo* work in our lab showing exposure to alcohol neonatally in rodents, induces cell death in beta-endorphin (BEP)-producing neurons in the hypothalamus, and suggests microglia may mediate this BEP neurodegeneration (Sarkar et al., 2007). As BEP neurons in the hypothalamus play an important role in regulating endocrine function and response to stress, it is possible that BEP neuronal loss in patients with FASD may underlie stress hyperresponsiveness observed in these patients, and that microglia may mediate this dysregulation (Sarkar, 2012). Other *in vivo* studies (see Table 4.1), also provide indirect evidence that microglia may mediate alcohol-induced neurotoxicity during early development by showing alcohol-induced increases in microglial activation and phagocytic activity alongside neuronal cell death, although follow-up studies directly testing the role of microglia in developmental alcohol-induced neurotoxicity are warranted (Kane et al., 2011; Saito et al., 2010; Watari, Born, & Gleason, 2006).

5.3. Microglia priming hypothesis

Microglia are highly sensitive cells and are capable of being sensitized to a stimulus with multiple presentations (Perry & Teeling, 2013). This is particularly relevant to understanding the role of microglia in both AUDs involving chronic or intermittent alcohol use and to FASD. In a mouse model of chronic alcohol exposure, prior chronic exposure of mice to ethanol increased microglial inflammatory response to a subsequent immune challenge by a systemic TLR3 agonist (Qin & Crews, 2012b). Similarly, in another study, prenatal alcohol exposure in rats exacerbated the severity of adjuvant-induced inflammation in an animal model of rheumatoid arthritis (Zhang et al., 2012). These and other studies have led to the hypothesis that either chronic alcohol exposure during adulthood or developmental alcohol exposure prenatally may sensitize the immune system to subsequent insult and lead to excess neuroinflammation, neurotoxicity, and the development of alcohol-related disorders such as adult AUD or FASD. Given the role of microglia in ethanol-induced neurotoxicity and the susceptibility of microglia sensitization by repeated stimuli, it is hypothesized that microglial priming by alcohol exposure may be a key factor in neurotoxicity observed in AUDs and FASD. The mechanism for this microglial priming is unknown, but it has been proposed that alcohol-induced chronic activation of microglia, alcohol-induced upregulation of microglial receptors mediating inflammation, or failure of microglia to inactivate after alcohol exposure may be responsible (Bilbo, 2013, Takeuchi, 2010). Future studies might test these theories directly.

6. SUMMARY AND FUTURE DIRECTIONS

The studies summarized and reviewed in this chapter strongly suggest a role for microglia in alcohol-induced neurotoxicity. *In vitro* and *in vivo* studies implicate microglia in mediating neuronal cell death via excess production of proinflammatory cytokines and ROS. In addition, a role for microglia in the symptoms and development of AUDs and FASD is suggested. Specifically, excessive or developmental stimulation of microglia by alcohol may lead to chronic microglial activation and microglia-mediated neuronal damage and may also prime microglia to be supersensitive to future stimulation, leading to immune system dysregulation and chronic degeneration in these disorders. Future experiments might utilize pharmacological or genetic inhibition of microglia to test this hypothesis more directly.

While chronic and developmental alcohol exposure appear to be associated with microglial-mediated neurotoxicity, acute and binge-like exposure to alcohol may have a distinct effect on microglia, characterized by release of anti-inflammatory cytokines and growth factors, which may have a beneficial or homeostatic effect. Future studies might further characterize this response in microglia. Finally, given the role of microglia in synaptogenesis and neurogenesis during development, future studies might explore whether developmental or chronic alcohol perturbs microglial regulation of these functions, and whether this may contribute to the development of AUDs or FASDs. Progress in the characterization of the role of microglia in alcohol-induced neurotoxicity may elucidate novel targets for the treatment of AUDs and FASDs.

ACKNOWLEDGMENTS

This work is partly supported by a National Institute of Health Grant R37AA08757 and R01AA016695. The authors would like to acknowledge Aaron DeLaRosa for his assistance in generating the figures for this paper.

REFERENCES

Agrawal, R. G., Owen, J. A., Levin, P. S., Hewetson, A., Berman, A. E., Franklin, S. R., et al. (2014). Bioinformatics analyses reveal age-specific neuroimmune modulation as a target for treatment of high ethanol drinking. *Alcoholism, Clinical and Experimental Research, 38*, 428–437.

Alfonso-Loeches, S., & Guerri, C. (2011). Molecular and behavioral aspects of the actions of alcohol on the adult and developing brain. *Critical Reviews in Clinical Laboratory Sciences, 48*, 19–47.

Alfonso-Loeches, S., Pascual-Lucas, M., Blanco, A. M., Sanchez-Vera, I., & Guerri, C. (2010). Pivotal role of TLR4 receptors in alcohol-induced neuroinflammation and brain damage. *The Journal of Neuroscience, 30*, 8285–8295.

APA (Ed.). (2013). *Diagnostic and statistical manual of mental disorders* (5th ed.). Arlington, VA: American Psychiatric Association.

Avidan, H., Kipnis, J., Butovsky, O., Caspi, R. R., & Schwartz, M. (2004). Vaccination with autoantigen protects against aggregated beta-amyloid and glutamate toxicity by controlling microglia: Effect of CD4+CD25+ T cells. *European Journal of Immunology, 34*, 3434–3445.

Badoer, E. (2010). Microglia: Activation in acute and chronic inflammatory states and in response to cardiovascular dysfunction. *The International Journal of Biochemistry & Cell Biology, 42*, 1580–1585.

Bell-Temin, H., Zhang, P., Chaput, D., King, M. A., You, M., Liu, B., et al. (2013). Quantitative proteomic characterization of ethanol-responsive pathways in rat microglial cells. *Journal of Proteome Research, 12*, 2067–2077.

Bilbo, S. D. (2013). Frank A. Beach award: Programming of neuroendocrine function by early-life experience: A critical role for the immune system. *Hormones and Behavior, 63*, 684–691.

Bilbo, S. D., Smith, S. H., & Schwarz, J. M. (2012). A lifespan approach to neuroinflammatory and cognitive disorders: A critical role for glia. *Journal of Neuroimmune Pharmacology, 7*, 24–41.

Blednov, Y. A., Ponomarev, I., Geil, C., Bergeson, S., Koob, G. F., & Harris, R. A. (2012). Neuroimmune regulation of alcohol consumption: Behavioral validation of genes obtained from genomic studies. *Addiction Biology, 17*, 108–120.

Block, M. L., & Hong, J. S. (2005). Microglia and inflammation-mediated neurodegeneration: Multiple triggers with a common mechanism. *Progress in Neurobiology, 76*, 77–98.

Block, M. L., & Hong, J. S. (2007). Chronic microglial activation and progressive dopaminergic neurotoxicity. *Biochemical Society Transactions, 35*, 1127–1132.

Boyadjieva, N. I., & Sarkar, D. K. (2010). Role of microglia in ethanol's apoptotic action on hypothalamic neuronal cells in primary cultures. *Alcoholism, Clinical and Experimental Research, 34*, 1835–1842.

Boyadjieva, N. I., & Sarkar, D. K. (2013a). Cyclic adenosine monophosphate and brain-derived neurotrophic factor decreased oxidative stress and apoptosis in developing hypothalamic neuronal cells: Role of microglia. *Alcoholism, Clinical and Experimental Research, 37*, 1370–1379.

Boyadjieva, N. I., & Sarkar, D. K. (2013b). Microglia play a role in ethanol-induced oxidative stress and apoptosis in developing hypothalamic neurons. *Alcoholism, Clinical and Experimental Research, 37*, 252–262.

Brocardo, P. S., Gil-Mohapel, J., & Christie, B. R. (2011). The role of oxidative stress in fetal alcohol spectrum disorders. *Brain Research Reviews, 67*, 209–225.

Chen, C. P., Kuhn, P., Chaturvedi, K., Boyadjieva, N., & Sarkar, D. K. (2006). Ethanol induces apoptotic death of developing beta-endorphin neurons via suppression of cyclic adenosine monophosphate production and activation of transforming growth factor-beta1-linked apoptotic signaling. *Molecular Pharmacology, 69*, 706–717.

Chopra, K., & Tiwari, V. (2012). Alcoholic neuropathy: Possible mechanisms and future treatment possibilities. *British Journal of Clinical Pharmacology, 73*, 348–362.

Crews, F. T., Bechara, R., Brown, L. A., Guidot, D. M., Mandrekar, P., Oak, S., et al. (2006). Cytokines and alcohol. *Alcoholism, Clinical and Experimental Research, 30*, 720–730.

Crews, F. T., & Boettiger, C. A. (2009). Impulsivity, frontal lobes and risk for addiction. *Pharmacology, Biochemistry, and Behavior, 93*, 237–247.

Crews, F. T., Braun, C. J., Hoplight, B., Switzer, R. C., 3rd., & Knapp, D. J. (2000). Binge ethanol consumption causes differential brain damage in young adolescent rats compared with adult rats. *Alcoholism, Clinical and Experimental Research, 24*, 1712–1723.

Crews, F. T., & Nixon, K. (2009). Mechanisms of neurodegeneration and regeneration in alcoholism. *Alcohol and Alcoholism, 44*, 115–127.

Crews, F. T., Zou, J., & Qin, L. (2011). Induction of innate immune genes in brain create the neurobiology of addiction. *Brain, Behavior, and Immunity, 25*(Suppl. 1), S4–S12.

Cronk, J. C., & Kipnis, J. (2013). Microglia—The brain's busy bees. *F1000Prime Reports, 5*, 53.

Deverman, B. E., & Patterson, P. H. (2009). Cytokines and CNS development. *Neuron, 64*, 61–78.

Ehrlich, D., Pirchl, M., & Humpel, C. (2012). Effects of long-term moderate ethanol and cholesterol on cognition, cholinergic neurons, inflammation, and vascular impairment in rats. *Neuroscience, 205*, 154–166.

Fernandez-Lizarbe, S., Montesinos, J., & Guerri, C. (2013). Ethanol induces TLR4/TLR2 association, triggering an inflammatory response in microglial cells. *Journal of Neurochemistry, 126*, 261–273.

Fernandez-Lizarbe, S., Pascual, M., & Guerri, C. (2009). Critical role of TLR4 response in the activation of microglia induced by ethanol. *Journal of Immunology, 183*, 4733–4744.

Gil-Mohapel, J., Boehme, F., Kainer, L., & Christie, B. R. (2010). Hippocampal cell loss and neurogenesis after fetal alcohol exposure: Insights from different rodent models. *Brain Research Reviews, 64*, 283–303.

Ginhoux, F., Lim, S., Hoeffel, G., Low, D., & Huber, T. (2013). Origin and differentiation of microglia. *Frontiers in Cellular Neuroscience, 7*, 45.

Guadagno, J., Xu, X., Karajgikar, M., Brown, A., & Cregan, S. P. (2013). Microglia-derived TNFalpha induces apoptosis in neural precursor cells via transcriptional activation of the Bcl-2 family member Puma. *Cell Death and Disease, 4*, e538.

Guerri, C., Montoliu, C., & Renau-Piqueras, J. (1994). Involvement of free radical mechanism in the toxic effects of alcohol: Implications for fetal alcohol syndrome. *Advances in Experimental Medicine and Biology, 366*, 291–305.

Hannigan, J. H., O'leary-Moore, S. K., & Berman, R. F. (2007). Postnatal environmental or experiential amelioration of neurobehavioral effects of perinatal alcohol exposure in rats. *Neuroscience and Biobehavioral Reviews, 31*, 202–211.

Harry, G. J. (2013). Microglia during development and aging. *Pharmacology and Therapeutics, 139*, 313–326.

He, J., & Crews, F. T. (2008). Increased MCP-1 and microglia in various regions of the human alcoholic brain. *Experimental Neurology*, *210*(2), 349–358. http://dx.doi.org/10.1016/j.expneurol.2007.11.017.

Hellemans, K. G., Sliwowska, J. H., Verma, P., & Weinberg, J. (2010). Prenatal alcohol exposure: Fetal programming and later life vulnerability to stress, depression and anxiety disorders. *Neuroscience and Biobehavioral Reviews*, *34*, 791–807.

Henderson, G. I., Devi, B. G., Perez, A., & Schenker, S. (1995). In utero ethanol exposure elicits oxidative stress in the rat fetus. *Alcoholism, Clinical and Experimental Research*, *19*, 714–720.

Ikonomidou, C., Bittigau, P., Ishimaru, M. J., Wozniak, D. F., Koch, C., Genz, K., et al. (2000). Ethanol-induced apoptotic neurodegeneration and fetal alcohol syndrome. *Science*, *287*, 1056–1060.

Jonas, R. A., Yuan, T. F., Liang, Y. X., Jonas, J. B., Tay, D. K., & Ellis-Behnke, R. G. (2012). The spider effect: Morphological and orienting classification of microglia in response to stimuli in vivo. *PLoS One*, *7*, e30763.

Jones, K. L. (2011). The effects of alcohol on fetal development. *Birth Defects Research. Part C, Embryo Today*, *93*, 3–11.

Kane, C. J., Phelan, K. D., Han, L., Smith, R. R., Xie, J., Douglas, J. C., et al. (2011). Protection of neurons and microglia against ethanol in a mouse model of fetal alcohol spectrum disorders by peroxisome proliferator-activated receptor-gamma agonists. *Brain, Behavior, and Immunity*, *25*(Suppl. 1), S137–S145.

Koob, G. F., & Le Moal, M. (1997). Drug abuse: Hedonic homeostatic dysregulation. *Science*, *278*, 52–58.

Kreutzberg, G. W. (1996). Microglia: A sensor for pathological events in the CNS. *Trends in Neurosciences*, *19*, 312–318.

Kuhn, P., & Sarkar, D. K. (2008). Ethanol induces apoptotic death of beta-endorphin neurons in the rat hypothalamus by a TGF-beta 1-dependent mechanism. *Alcoholism, Clinical and Experimental Research*, *32*, 706–714.

Lawson, L. J., Perry, V. H., Dri, P., & Gordon, S. (1990). Heterogeneity in the distribution and morphology of microglia in the normal adult mouse brain. *Neuroscience*, *39*, 151–170.

Marshall, S., McClain, J., McKelso, M., Hopkins, D., Pauly, J., & Nixon, K. (2013). Microglial activation is not equivalent to neuroinflammation in alcohol-induced neurodegeneration: The importance of microglia phenotype. *Neurobiology of Disease*, *54*, 239–251.

Mattson, S. N., & Riley, E. P. (1998). A review of the neurobehavioral deficits in children with fetal alcohol syndrome or prenatal exposure to alcohol. *Alcoholism, Clinical and Experimental Research*, *22*, 279–294.

May, P. A., & Gossage, J. P. (2001). Estimating the prevalence of fetal alcohol syndrome. A summary. *Alcohol Research & Health*, *25*, 159–167.

McClain, J. A., Morris, S. A., Deeny, M. A., Marshall, S. A., Hayes, D. M., Kiser, Z. M., et al. (2011). Adolescent binge alcohol exposure induces long-lasting partial activation of microglia. *Brain, Behavior, and Immunity*, *25*(Suppl. 1), S120–S128.

Mittelbronn, M., Dietz, K., Schluesener, H. J., & Meyermann, R. (2001). Local distribution of microglia in the normal adult human central nervous system differs by up to one order of magnitude. *Acta Neuropathologica*, *101*, 249–255.

Montoliu, C., Sancho-Tello, M., Azorin, I., Burgal, M., Valles, S., Renau-Piqueras, J., et al. (1995). Ethanol increases cytochrome P4502E1 and induces oxidative stress in astrocytes. *Journal of Neurochemistry*, *65*, 2561–2570.

Morris, S. A., Eaves, D. W., Smith, A. R., & Nixon, K. (2010). Alcohol inhibition of neurogenesis: A mechanism of hippocampal neurodegeneration in an adolescent alcohol abuse model. *Hippocampus*, *20*, 596–607.

Mulligan, M. K., Ponomarev, I., Hitzemann, R. J., Belknap, J. K., Tabakoff, B., Harris, R. A., et al. (2006). Toward understanding the genetics of alcohol drinking through transcriptome meta-analysis. *Proceedings of the National Academy of Sciences of the United States of America, 103*, 6368–6373.

Nimmerjahn, A., Kirchhoff, F., & Helmchen, F. (2005). Resting microglial cells are highly dynamic surveillants of brain parenchyma in vivo. *Science, 308*, 1314–1318.

Nixon, K., Kim, D. H., Potts, E. N., He, J., & Crews, F. T. (2008). Distinct cell proliferation events during abstinence after alcohol dependence: Microglia proliferation precedes neurogenesis. *Neurobiology of Disease, 31*, 218–229.

Obernier, J. A., Bouldin, T. W., & Crews, F. T. (2002). Binge ethanol exposure in adult rats causes necrotic cell death. *Alcoholism, Clinical and Experimental Research, 26*, 547–557.

Paolicelli, R. C., Bolasco, G., Pagani, F., Maggi, L., Scianni, M., Panzanelli, P., et al. (2011). Synaptic pruning by microglia is necessary for normal brain development. *Science, 333*, 1456–1458.

Perry, V. H., & Teeling, J. (2013). Microglia and macrophages of the central nervous system: The contribution of microglia priming and systemic inflammation to chronic neurodegeneration. *Seminars in Immunopathology, 35*, 601–612.

Pocock, J. M., & Kettenmann, H. (2007). Neurotransmitter receptors on microglia. *Trends in Neurosciences, 30*, 527–535.

Polazzi, E., & Monti, B. (2010). Microglia and neuroprotection: From in vitro studies to therapeutic applications. *Progress in Neurobiology, 92*, 293–315.

Qin, L., & Crews, F. T. (2012a). NADPH oxidase and reactive oxygen species contribute to alcohol-induced microglial activation and neurodegeneration. *Journal of Neuroinflammation, 9*, 5.

Qin, L., & Crews, F. T. (2012b). Chronic ethanol increases systemic TLR3 agonist-induced neuroinflammation and neurodegeneration. *Journal of Neuroinflammation, 9*, 130.

Raivich, G., Bohatschek, M., Kloss, C. U., Werner, A., Jones, L. L., & Kreutzberg, G. W. (1999). Neuroglial activation repertoire in the injured brain: Graded response, molecular mechanisms and cues to physiological function. *Brain Research. Brain Research Reviews, 30*, 77–105.

Raivich, G., Jones, L. L., Werner, A., Bluthmann, H., Doetschmann, T., & Kreutzberg, G. W. (1999). Molecular signals for glial activation: Pro- and anti-inflammatory cytokines in the injured brain. *Acta Neurochirurgica. Supplement, 73*, 21–30.

Riley, E. P., Infante, M. A., & Warren, K. R. (2011). Fetal alcohol spectrum disorders: An overview. *Neuropsychology Review, 21*, 73–80.

Saijo, K., & Glass, C. K. (2011). Microglial cell origin and phenotypes in health and disease. *Nature Reviews. Immunology, 11*, 775–787.

Saito, M., Chakraborty, G., Mao, R. F., Paik, S. M., Vadasz, C., & Saito, M. (2010). Tau phosphorylation and cleavage in ethanol-induced neurodegeneration in the developing mouse brain. *Neurochemical Research, 35*, 651–659.

Sarkar, D. K. (2012). Circadian genes, the stress axis, and alcoholism. *Alcohol Research, 34*, 362–366.

Sarkar, D. K., Kuhn, P., Marano, J., Chen, C., & Boyadjieva, N. (2007). Alcohol exposure during the developmental period induces beta-endorphin neuronal death and causes alteration in the opioid control of stress axis function. *Endocrinology, 148*, 2828–2834.

Schulz, C., Gomez Perdiguero, E., Chorro, L., Szabo-Rogers, H., Cagnard, N., Kierdorf, K., et al. (2012). A lineage of myeloid cells independent of Myb and hematopoietic stem cells. *Science, 336*, 86–90.

Sierra, A., Encinas, J. M., Deudero, J. J., Chancey, J. H., Enikolopov, G., Overstreet-Wadiche, L. S., et al. (2010). Microglia shape adult hippocampal neurogenesis through apoptosis-coupled phagocytosis. *Cell Stem Cell, 7*, 483–495.

Sullivan, E. V., & Pfefferbaum, A. (2005). Neurocircuitry in alcoholism: A substrate of disruption and repair. *Psychopharmacology, 180*, 583–594.

Takeuchi, H. (2010). Neurotoxicity by microglia: Mechanisms and potential therapeutic strategy. *Clinical and Experimental Neuroimmunology, 1*, 12–21.

Tobar, N., Villar, V., & Santibanez, J. F. (2010). ROS-NFkappaB mediates TGF-beta1-induced expression of urokinase-type plasminogen activator, matrix metalloproteinase-9 and cell invasion. *Molecular and Cellular Biochemistry, 340*, 195–202.

Town, T., Nikolic, V., & Tan, J. (2005). The microglial "activation" continuum: From innate to adaptive responses. *Journal of Neuroinflammation, 2*, 24.

Tremblay, M. E., Lowery, R. L., & Majewska, A. K. (2010). Microglial interactions with synapses are modulated by visual experience. *PLoS Biology, 8*, e1000527.

Tremblay, M. E., Stevens, B., Sierra, A., Wake, H., Bessis, A., & Nimmerjahn, A. (2011). The role of microglia in the healthy brain. *The Journal of Neuroscience, 31*, 16064–16069.

Ward, R. J., Colivicchi, M. A., Allen, R., Schol, F., Lallemand, F., de Witte, P., et al. (2009). Neuro-inflammation induced in the hippocampus of 'binge drinking' rats may be mediated by elevated extracellular glutamate content. *Journal of Neurochemistry, 111*, 1119–1128.

Watari, H., Born, D. E., & Gleason, C. A. (2006). Effects of first trimester binge alcohol exposure on developing white matter in fetal sheep. *Pediatric Research, 59*, 560–564.

WHO (Producer). (2011). *Global status report on alcohol and health.* 3/25/2014, http://www.who.int/substance_abuse/publications/global_alcohol_report/msbgsruprofiles.pdf?ua=1.

Zhan, Y., Paolicelli, R. C., Sforazzini, F., Weinhard, L., Bolasco, G., Pagani, F., et al. (2014). Deficient neuron-microglia signaling results in impaired functional brain connectivity and social behavior. *Nature Neuroscience, 17*, 400–406.

Zhang, X., Lan, N., Bach, P., Nordstokke, D., Yu, W., Ellis, L., et al. (2012). Prenatal alcohol exposure alters the course and severity of adjuvant-induced arthritis in female rats. *Brain, Behavior, and Immunity, 26*, 439–450.

Zhao, Y. N., Wang, F., Fan, Y. X., Ping, G. F., Yang, J. Y., & Wu, C. F. (2013). Activated microglia are implicated in cognitive deficits, neuronal death, and successful recovery following intermittent ethanol exposure. *Behavioural Brain Research, 236*, 270–282.

CHAPTER FIVE

Functions of the Chemokine Receptor CXCR4 in the Central Nervous System and Its Regulation by μ-Opioid Receptors

Bradley Nash*, Olimpia Meucci*,†,1
*Department of Pharmacology and Physiology, Drexel University College of Medicine, Philadelphia, Pennsylvania, USA
†Department of Microbiology and Immunology, Drexel University College of Medicine, Philadelphia, Pennsylvania, USA
[1]Corresponding author: e-mail address: omeucci@drexelmed.edu

Contents

Abstract

Activation of the G protein-coupled receptor CXCR4 by its chemokine ligand CXCL12 regulates a number of physiopathological functions in the central nervous system, during development as well as later in life. In addition to the more classical roles of the CXCL12/CXCR4 axis in the recruitment of immune cells or migration and proliferation of neural precursor cells, recent studies suggest that CXCR4 signaling also modulates synaptic function and neuronal survival in the mature brain, through direct and indirect effects on neurons and glia. These effects, which include regulation of glutamate receptors and uptake, and of dendritic spine density, can significantly alter the ability of neurons to face excitotoxic insults. Therefore, they are particularly relevant to

ISSN 0074-7742
http://dx.doi.org/10.1016/B978-0-12-801284-0.00005-1

neurodegenerative diseases featuring alterations of glutamate neurotransmission, such as HIV-associated neurocognitive disorders. Importantly, CXCR4 signaling can be dysregulated by HIV viral proteins, host HIV-induced factors, and opioids. Potential mechanisms of opioid regulation of CXCR4 include heterologous desensitization, transcriptional regulation and changes in receptor expression levels, opioid–chemokine receptor dimer or heteromer formation, and the newly described modulation by the protein ferritin heavy chain—all leading to inhibition of CXCR4 signaling. After reviewing major effects of chemokines and opioids in the CNS, this chapter discusses chemokine–opioid interactions in neuronal and immune cells, focusing on their potential contribution to HIV-associated neurocognitive disorders.

1. CHEMOKINE SYSTEM OVERVIEW

In order for cells to communicate they must employ a language of sorts that allows them to respond to threats and to routine duties. Chemokines act as a part of this natural language and their physiological effects are myriad. Chemokine ligands are mostly secreted small proteins, although two chemokines, CX3CL1 and CXCL16, also exist in a membrane-bound form that allows their signaling events to happen specifically in nearby cells (Clark, Staniland, & Malcangio, 2011; La Porta, 2012). The chemokine superfamily is divided into different classes based on the order of four conserved cysteine residues. In alpha chemokines, the first two conserved cysteines are separated by any amino acid. Therefore, this class is denoted as CXC. The receptor or ligand designation (L/R) follows, and then a numerical identifier (Zlotnik & Yoshie, 2000). Other chemokine classes include CC, which has adjacent conserved cysteines, (X)C, which has only two conserved cysteines, and CX3C, which has three amino acids separating the first two conserved cysteines. Typically, chemokines in a particular class may only stimulate receptors of the same class, but this does not eliminate natural redundancy from the system as many chemokine ligands display promiscuous binding to receptors within their family (Zlotnik & Yoshie, 2012). Chemokine receptors are seven-transmembrane G protein-coupled receptors (GPCRs) that mostly signal through Gαi proteins (Réaux-Le Goazigo, Van Steenwinckel, Rostène, & Mélik Parsadaniantz, 2013) and thus are subject to GPCR–GPCR interactions that can modulate intracellular signals after ligand binding. In some cases, chemokine receptors can regulate the strength of an external signal by forming dimeric complexes (Mellado et al., 2001). Both homo- and heterodimers seem to occur within the chemokine receptor family, and heterodimers composed of chemokine/opioid receptors are thought to play an important role in signaling modulation of immune and neural cells (Mellado et al., 2001). Chemokine and opioid interactions at the receptor level will be covered in an upcoming section.

Chemokines have also been characterized on the basis of their function as inflammatory or homeostatic (Moser, Wolf, Walz, & Loetscher, 2004). Inflammatory chemokines are upregulated in damaged tissues and activated immune cells and have the ability to recruit immune effector cells to an area of infection or inflammation. Although there are a large number of potentially inflammatory chemokines, these proteins are typically more promiscuous in their binding, and many of them are located on the same areas of chromosomes 4 and 17 (Nomiyama, Osada, & Yoshie, 2011). This redundancy ensures that a proper immune response can be mounted in tissues that may possess different chemokine secretion profiles. Inflammatory chemokines also promote angiogenesis and help to activate the blood vessel endothelium to become leakier and express anchor proteins that allow circulating immune cells to more easily enter an inflamed area (Strieter, Burdick, Gomperts, Belperio, & Keane, 2005). The second functional classification comes from the discovery that chemokines are necessary for normal homeostatic processes to occur. These chemokine receptor pairs are usually located on distinct chromosomal sites and have much less redundancy compared with the inflammatory variety of chemokines (Zlotnik & Yoshie, 2012). CXCR4, which is the focus of this review, exerts several homeostatic effects throughout the body, including the CNS (Lazarini, Tham, Casanova, Arenzana-Seisdedos, & Dubois-Dalcq, 2003; Réaux-Le Goazigo et al., 2013). CXCL12 is the only chemokine ligand known to bind CXCR4; downstream effects of the CXCL12/CXCR4 axis in the CNS include regulation of the retinoblastoma cell-cycle protein, ensuring cell survival for postmitotic cells (Khan et al., 2008), migration of neuronal precursor cells (Stumm et al., 2003), neurogenesis (Réaux-Le Goazigo et al., 2013), protection against neurotoxic insults (Khan et al., 2004; Meucci et al., 1998; Shepherd, Loo, & Mohapatra, 2013), and regulation of dendritic spine density (Pitcher et al., 2014). These effects depend on the receptor and ligand being constitutively expressed in these tissues, which is in stark contrast to the highly inducible expression of inflammatory chemokines.

The homeostatic and inflammatory characterization is not a firm boundary, as many chemokines have activity in both facets, and still other chemokine receptors serve more diverse functions. CXCR7 was originally defined as an orphan receptor until CXCL12 was reported to bind it (Balabanian et al., 2005). Upon further investigation, CXCR7 was shown to lack a

DRY amino acid motif, implying that it was not capable of coupling to G proteins and activating the classical pathways associated with GPCRs (Graham, 2009). This receptor is important in development, as it acts as a chemokine sink for CXCL12, but it may have functionality that is cell-type specific and has yet to be uncovered through β-arrestin-mediated signals or other G protein-independent signals. Recent studies show that CXCR7's ability to act as a sink for CXCL12 prevents CXCR4 internalization in the presence of high-chemokine levels, thus preserving CXCR4 function and ensuring proper migration of interneurons in the developing CNS (Sanchez-Alcaniz et al., 2011). Involvement of CXCR7 in other physiological and developmental processes has yet to be fully characterized, although studies have highlighted its involvement in various cancers (Hattermann & Mentlein, 2013).

2. OPIOID SYSTEM OVERVIEW

The opioid system has been covered extensively in several past reviews (Corbett, Henderson, McKnight, & Paterson, 2006; Hutchinson et al., 2011; Waldhoer, Bartlett, & Whistler, 2004; Williams et al., 2013) and is discussed here only briefly. Opioids have a long history of usage as analgesics and pain relievers, and research today continues to reveal novel actions of these compounds. Opioid receptor ligands are typically referred to as *opiates* if they are obtained from natural substances such as the opium poppy and as *opioids* if they are either natural or made in a synthetic process. Opioid receptors (also GPCRs) exist in three major isoforms (μ, κ, and δ), which are expressed throughout the body (Wittert, Hope, & Pyle, 1996), and are sensitive to the general opioid receptor antagonist naloxone. Similarly to chemokine receptors, formation of receptor complexes with other opioid receptor subtypes, or other GPCRs can effect signal transduction to varying degrees. Opioid receptors have been reported to interact with chemokine receptors (described later) and cannabinoid receptors (Rios, Gomes, & Devi, 2006) resulting in decreased signal transduction for both opioid and cannabinoid ligands (Rios et al., 2006). Although opioid drugs are tightly regulated and are common drugs of abuse, natural opioid ligands produced by the body are important physiological regulators of many different processes, such as cell membrane homeostasis, cell proliferation, immune function, gastrointestinal function, and neuromodulation, among others (Feng et al., 2012). Importantly, endogenous opioids are peptide based, and are much larger than exogenous small-molecule opioids, although they are able to bind and activate the same receptors. Currently, 10 different endogenous opioid ligands have been discovered, which are all slightly different in their binding affinities and specificity for opioid receptor subtypes. Met and Leu enkephalin, the first endogenous opioid ligands discovered from purified brain extracts (Hughes et al., 1975), both derived from the precursor proenkephalin and have the greatest affinity for δ-opioid receptors. Dynorphins, the second group of endogenous opioids, are derived from the protein precursor prodynorphin and preferentially bind κ-opioid receptors (Goldstein, Tachibana, Lowney, Hunkapiller, & Hood, 1979). Endorphins are the third group, derived from pro-opiomelanocortin, and possess the highest affinity for the μ subtype (Li, Chung, & Doneen, 1976). Endogenous opioids have very diverse roles in many different tissues. For example, opioids typically have an inhibitory effect on neuronal activity, but the μ-opioid receptor agonist DAMGO (Tyr-D-Ala-Gly-*N*-methyl-Phe-Gly-ol) is able to regulate resting membrane potential in Purkinje neurons by its ability to increase intracellular calcium levels via a G protein-independent pathway, which results in increased neurotransmitter release (Iegorova, Fisyunov, & Krishtal, 2010). Painful stimuli including injuries and lipopolysaccharide administration are known to cause an upregulation of the μ-opioid receptor agonist β endorphin, which serves as a natural form of analgesia (Molina, 2002), and opioids are able to control feeding depending on which receptor subtype is bound (Gosnell & Levine, 2009).

Exogenous opioids may be both beneficial and problematic depending on the compound and dose taken. Common opioid compounds such as morphine and heroin are highly addictive and likely to be abused to achieve a sensation of euphoria. Many cellular processes that depend on tight regulation of the opioid system can be affected under conditions of abuse (covered later in this chapter). As opioid drugs, in particular morphine, are still the standard of care for chronic pain and other disorders involving pain, and their potential for abuse makes their use more widespread, understanding their total physiological influence will be important for determining potential effects of opioids in other pathophysiological conditions, such as neurocognitive disorders.

3. CXCR4 AND OPIOIDS ACTIONS IN THE CENTRAL NERVOUS SYSTEM

Although both the chemokine and opioid families are expressed in many different tissues, these proteins have been intensively investigated in the nervous and immune system where they exert important physiological and pathological effects. With regard to the CNS, both chemokines and opioids have been shown to act as neuromodulators (Rostene, Kitabgi, & Parsadaniantz, 2007) and as mediators of cell-to-cell communication (Sheridan & Murphy, 2013). The following sections cover the CXCL12/CXCR4 signaling axis and opioid family members in different CNS cell types, and their normal as well as pathological roles with a focus on HIV-associated neurocognitive disorders (HAND).

3.1. Physiological and pathological roles of CXCR4

CXCR4 is expressed in the brain and the spinal cord *in vitro* and *in vivo* in a vast variety of species (Meucci et al., 1998; Ohtani et al., 1998; Pitcher et al., 2014) and in all major CNS cell types, including neurons (Meucci et al., 1998; van der Meer, Ulrich, Gonzalez-Scarano, & Lavi, 2000), astroglia (Bajetto et al., 1999), microglia (Lipfert, Odemis, Wagner, Boltze, & Engele, 2013), and oligodendrocytes (Maysami et al., 2006). The receptor has the potential to activate several distinct signaling pathways and elicit various biological responses. The natural ligand CXCL12 binding results in inhibition of adenylate cyclase, via activation of Gαi proteins, and decreased protein kinase A activity (Zheng et al., 1999)—while also increasing intracellular Ca^{2+} levels and protein kinase C, via the phospholipase C pathway (Cali & Bezzi, 2010; Khan et al., 2004; Meucci et al., 1998). Additional downstream signals with direct effect on gene transcription include activation of the ERK and Akt cascade (Khan et al., 2004), the JAK/STAT, and the nuclear factor-κB pathways (Ganju et al., 1998). As a homeostatic chemokine/receptor pair, these proteins have much more varied roles than contributing to immune responses compared to their inflammatory counterparts. The signaling outcomes of the receptor are indeed similar to the classical immune chemotactic response, but the same signals can also occur on nonimmune cells that express CXCR4. CXCR4 and CXCL12 are expressed in both the developing and mature CNS, where they serve multiple vital functions. For instance, during development, CXCL12 guides developing interneurons to their proper cortical layer via a CXCL12 gradient that is produced by resident cells (Sanchez-Alcaniz et al., 2011; Stumm et al., 2003). As explained earlier, CXCR7 was shown to act as a chemokine sink, removing excess CXCL12 from the extracellular space and, in the process, preserving the developing interneurons' responsiveness to CXCL12-mediated chemotaxis (Sanchez-Alcaniz et al., 2011). Although this function does not involve the immune system, the signaling outcome of chemotaxis is similar to outcomes observed with inflammatory chemokines, suggesting that this chemokine/receptor pair is more ancient and more important because of its presence in organisms that do not have a functional immune system (Huising, Stet, Kruiswijk, Savelkoul, & Lidy Verburg-van Kemenade, 2003). Additionally, CXCR4 knockout animals do not survive past birth, further indicating the importance of this receptor during development (Ma et al., 1998). In the mature CNS, CXCL12 may also have other functions that are more diverse than chemotaxis. In periods of neuronal stress or excitotoxicity, CXCL12 can protect neurons by several different mechanisms, as discussed later.

The importance of the CXCL12/CXCR4 signaling axis becomes much more apparent in pathological states where the signaling is dysregulated, including HIV infection. HIV can infect peripheral immune cells and use them as a liaison to enter the CNS (Gonzalez-Scarano & Martin-Garcia, 2005). Upon entering, the virus can establish a CNS reservoir and by extension cause activation of CNS immune cells and an inflammatory response (Kraft-Terry, Buch, Fox, & Gendelman, 2009). Both cells that are infected and uninfected can contribute to the inflammatory/excitotoxic state, and the HIV proteins themselves can have detrimental effects on many different CNS cell types (recently reviewed by Gonzalez-Scarano & Martin-Garcia, 2005; Lindl, Marks, Kolson, & Jordan-Sciutto, 2010). The HIV-envelope protein from X4 viruses uses CXCR4 for entry into cells, but it can also cause cell damage via CXCR4-dependent signaling events (Pandey & Bolsover, 2000). Depending on the cell type expressing CXCR4, this binding event can result in different outcomes. For example, gp120-induced activation of CXCR4 in glia can cause secretion of several inflammatory mediators, including tumor necrosis factor-α and interleukin-1β (Bezzi et al., 2001). These mediators can activate nearby uninfected immune cells and precipitate inflammatory responses in the CNS (Kraft-Terry et al., 2009). Astrocytes are the main support cells in the CNS, and their typical functions can be altered in HIV infection, even though these cells are not the primary target of HIV in the CNS. Inflammatory mediators can alter their ability to prevent excessive stimulation of glutamate receptors in the tripartite synapse, resulting in the formation of an excitotoxic environment (Okamoto, Wang, & Baba, 2005). Additionally, gp120 can directly bind CXCR4 on neurons, which contributes to neuronal demise and simplification (Bardi, Sengupta, Khan, Patel, & Meucci, 2006; Ellis, Langford, & Masliah, 2007; Hesselgesser et al., 1998; Meucci et al., 1998). In summary, while activation of CXCR4 by its natural chemokine ligand, CXCL12, is generally neuroprotective, abnormal stimulation of this receptor during HIV infection can have opposite effects in the CNS.

Patients infected by HIV who also abuse opioid drugs often show enhanced disease progression at the periphery and in the CNS, so an examination of the interaction between the opioid and chemokine system in the brain can provide valuable insights for understanding this comorbid pathology. Interestingly, DAMGO, a potent and selective μ-opioid receptor agonist, was shown to inhibit neuroprotection afforded by CXCL12 treatment in NMDA-treated neuronal cultures (Patel et al., 2006). This effect was associated with DAMGO's ability to prevent ERK and Akt phosphorylation via CXCL12 signaling. CXCL12 is also able to directly modulate NMDA receptor subunit composition on cortical neurons, by preventing the production (and likely insertion) of the NR2B receptor subunit at extrasynaptic sites (Khan et al., 2008; Nicolai, Burbassi, Rubin, & Meucci, 2010). This subunit is associated with enhanced Ca^{2+} currents and overstimulation of the neuron, resulting in a disruption of ionic homeostasis, and potential activation of caspases (Leveille et al., 2008).

3.2. Effects of opioids on neuronal and non-neuronal cells

The opioid system has varied roles in the CNS, but this section emphasizes how opioids can affect cellular processes that are dysregulated in neurocognitive disorders. Opioids can change CNS physiology by two main mechanisms, first by their neuromodulatory actions and second through their effects on immune cells. Because endogenous opioids can be released in response to a painful stimulus or an infection, these molecules are thought to be helpful in restoring homeostasis during stressful episodes. For instance, in normoxic conditions, neuronal Na^+ and K^+ homeostasis is not affected by μ or δ agonists in mouse cortical slices (Chao, Bazzy-Asaad, Balboni, & Xia, 2007). However, under hypoxic conditions, δ-opioid receptor stimulation restored normal Na^+ and K^+ ion composition in the same cortical brain slices, suggesting the receptor signal has potential neuroprotective effects (Chao et al., 2007). Exogenous opioids are typically prescribed for pain and the endogenous system can be activated under similar circumstances. Ligand-bound opioid receptors function via Gαi proteins to inhibit adenylyl cyclase and Ca^{2+} channels, which may contribute to decreased neuronal excitability. Inhibiting neurons involved in ascending pain pathways, and activating inhibitory descending pathways in specific CNS areas including the periaqueductal gray, amygdala, insula, and spinal cord are important for achieving opioid analgesia (Mansour, Fox, Akil, & Watson, 1995). However, expression of opioid receptors in other tissues, such as the gastrointestinal tract, can contribute to unwanted side effects (Ketwaroo, Cheng, & Lembo, 2013). One very interesting set of side effects of opioid treatment concerns the regulation of the immune system. In addition to neuronal opioid receptor expression, several CNS cells that exert immune-like functions also express opioid receptors and can be modulated by both endogenous and exogenous opioids. Astrocytes express all three classical isoforms of opioid receptors (Hutchinson et al., 2011), but their individual expression can change depending on developmental status and their CNS localization (Ruzicka et al., 1995). The μ-opioid receptor is most abundantly expressed in the cortex, whereas κ- and δ-opioid receptors are abundantly expressed in the cortex and hypothalamus, among other areas (Ruzicka et al., 1995). Immature astrocytes express more κ-opioid receptors, and the expression of the other opioid receptor subtypes changes as these cells enter the cell cycle (Persson, Thorlin, Ronnback, Hansson, & Eriksson, 2000). Rat cortical microglia express all three classical opioid receptor isoforms (Turchan-Cholewo et al., 2008), but information about how the distribution of these receptors changes over cellular development is lacking. Oligodendrocytes express μ and κ isoforms, and the overall opioid receptor expression in these cells decreases as they mature (Hauser, Fitting, Dever, Podhaizer, & Knapp, 2012; Knapp, Maderspach, & Hauser, 1998).

Opioids can have effects on CNS cells via classical opioid receptors or via nonclassical receptors, such as toll-like receptor 4 (TLR4; Watkins, Hutchinson, Rice, & Maier, 2009). This pattern recognition receptor recognizes foreign bodies such as bacterial lipopolysaccharide and initiates an inflammatory response by activating its respective immune cells and facilitating the release of inflammatory cytokines and chemokines. Triple opioid receptor knockout mice are not protected from opioid-induced hyperalgesia, suggesting that another system is responsive to these compounds and may work to thwart normal opioid signaling (van Dorp et al., 2009). TLR4 may be a part of this system, but other receptors likely play a role as well. However, the effects of classical opioid receptor activation on central immune cells have been more thoroughly classified. Morphine and endogenous opioids are known to reduce proliferation of astrocytes in many different brain regions, even in the presence of epidermal growth factor (Belcheva, Tan, Heaton, Clark, & Coscia, 2003). This may contribute to a general suppression of immune activity in individuals who abuse opioids. Some reports suggest that opioids have proliferative effects as well, depending on what receptor subtype they bind and on the region in which the target cells are located. For instance, young astrocytes that highly express κ-opioid receptors are sensitive to reduced proliferation after treatment with a κ agonist, while mature astrocytes located in other CNS areas have been reported to proliferate after κ or δ stimulation (Bunn, Hanley, & Wilkin, 1985; Xu, Bruchas, Ippolito, Gendron, & Chavkin, 2007). In line with the general theme of immune suppression via opioids, morphine is able to induce apoptosis in microglial cells by activating caspase 3 and the p38 pathway (Hu, Sheng, Lokensgard, & Peterson, 2002). However, the same group demonstrated that astrocytes do not undergo apoptosis after morphine administration, suggesting integral differences in μ-opioid receptor signaling between these two cell types (Hu et al., 2002). Many interesting findings have been reported regarding opioids' ability to modulate immune cells, but further characterization is necessary to understand how specific opioids can modulate CNS cells in different CNS areas.

Generally, neurocognitive disorders are accompanied by a low-level chronic inflammation in the CNS, which can be detrimental to cellular homeostasis and result in faster disease progression. As HIV-positive patients who have abused opioids are important members of this group, understanding how opioids can enhance progression of disease is paramount in the development of future adjunctive therapies. The following section outlines what is currently known about how opioids can interact with chemokine receptors, mostly CXCR4, and the potential clinical outcomes that arise as a consequence of this interaction.

4. CXCR4 INTERACTIONS WITH OPIOIDS

The scientific literature describes many ways that the chemokine and opioid systems can differentially regulate each other in immune and neuronal cells, especially with regard to the chemokine receptor CXCR4 (please see Table 5.1 for recent examples). Classical modes of regulation, such as heterologous desensitization and transcriptional changes, have been described in addition to novel and unexpected regulatory pathways.

4.1. Heterologous desensitization

Heterologous desensitization is described as a broad desensitization of similar receptors after repeated exposure to a specific ligand (i.e., activation of receptor A by its ligand desensitizes receptor B) and it is typically a rapid event. As opioid and chemokine receptors are both G protein-coupled seven-transmembrane receptors that use the same cellular machinery to

Table 5.1 Examples of chemokine–opioid receptor interactions

Receptor	Regulation	Involved receptors	Cell type	References
CXCR4	Heterologous desensitization	μOR and δOR	Rat PAG	Chen, Geller, Rogers, and Adler (2007)
	Heterodimer formation	δOR	Murine PBMCs	Pello et al. (2008)
CCR5	Heterologous desensitization	μOR and δOR	Rat PAG	Chen et al. (2007)
CX_3CR1	Heterologous desensitization	μOR and δOR	Rat PAG	Heinisch, Palma, and Kirby (2011)
κOR	Heterologous desensitization	CXCR4	Jurkat cells	Finley et al. (2008)
	mRNA expression inhibition	CXCR4	Human PBMCs, CHME-3	Finley, Steele, Cornwell, and Rogers (2011)
μOR	Heterologous desensitization	CCR1, 2, and 5	Human PBMCs	Szabo et al. (2003)
	Cell surface downregulation	CXCR4	Rat DRG	Wilson, Jung, Ripsch, Miller, and White (2011)
	Cell surface upregulation	Chronic μOR > CXCR4	Rat DRG	Wilson et al. (2011)
	FHC upregulation	μOR > CXCR4	Human/monkey/rat cortical neurons	Sengupta et al. (2009), Pitcher et al. (2014)
δOR	Heterologous desensitization	CCR1, 2, and 5	Human PBMCs	Szabo et al. (2003)

broadcast their signals, stimulation of either has been shown to reduce the subsequent signaling of the other in some cases. Moreover, these interactions have been shown to be cell- and receptor subtype- specific. For instance, in immune cells μ- and δ-opioid receptor stimulation is able to desensitize CCR1, 2, and 5 but has no effect on CXCR4 signaling (Finley et al., 2008; Szabo et al., 2003). In contrast, CXCR4 activation by CXCL12 desensitizes signals from both μ and δ receptors (Chen et al., 2007;

Heinisch et al., 2011). The κ-opioid receptor is the only classical opioid receptor subtype that has been shown to desensitize subsequent CXCR4 signaling via heterologous desensitization, and Ca^{2+} signaling experiments suggest that bidirectional desensitization occurs within seconds of κ-opioid receptor or CXCR4 activation in a dose-dependent manner (Finley et al., 2008). Although total CXCR4 surface expression does not change acutely after κ stimulation by U50488H, CXCL12-mediated chemotaxis is disrupted in Jurkat T cells after pretreatment with the κ agonist. Surface expression of CXCR4 can be reduced over a longer period, but the initial desensitization of the receptor does not result from internalization. *In vivo*, dynorphin-induced analgesia is blocked by CXCL12 administration in rat periaqueductal gray, providing further evidence of these receptors' abilities to cross-desensitize each other (Finley et al., 2008). Desensitization is not limited to the receptors listed here or to the CNS, as CCL2, CCL3, CCL5, and CXCL8 can desensitize μ-opioid receptor signals in peripheral sensory neurons (Zhang, Rogers, Caterina, & Oppenheim, 2004).

4.2. Transcriptional regulation/changes in expression

Peripheral blood mononuclear cells have been shown to reduce total CXCR4 protein and transcript levels after a 30-min treatment with the κ agonist U50488H, and this effect was associated with reduced X4 HIV infection in these cells (Finley et al., 2011). This process was shown to be dependent on the JAK/STAT pathway and on interferon regulatory factor 2 (IRF2), as blocking the phosphorylation of JAK2 or STAT3 prevented the activation of IRF1 and 2 and their subsequent binding to the CXCR4 gene promoter. Although both isoforms of IRF are activated by κ stimulation, and both bind to the CXCR4 promoter, only IRF2 is necessary for CXCR4 downregulation, an interaction confirmed *in vivo* by chromatin immunoprecipitation analysis (Finley et al., 2011). Other opioids also have effects on the translation and expression of both chemokine receptors and ligands in various CNS cell types. In rat dorsal root ganglion (DRG), morphine administration for 5 days at 10 mg/kg resulted in marked CXCR4 downregulation in neurons and satellite glia cells, an effect lasting for at least 21 days after the last treatment (Wilson et al., 2011). However, the expression of CXCL12 actually increases after chronic morphine administration in the DRG, although it is unclear whether this is a direct effect of morphine treatment or an indirect feedback mechanism (Wilson et al., 2011). Additionally, this morphine dosing regimen led to opioid-induced hyperalgesia in some of the animals.

Treatment with the CXCR4 antagonist AMD3100 caused an upregulation of total CXCR4 expression in these animals that was accompanied by reduced hyperalgesia as measured by von Frey filaments (Wilson et al., 2011). Again, this interaction is mechanistically unclear, as it could be influenced by other receptors that bind morphine at high doses, such as TLR4.

4.3. Receptor dimerization

As both CXCR4 and all three classical opioid receptors are seven-transmembrane GPCRs, they have the ability to interact with each other on the cell surface. These interactions can dramatically change the way that both receptors normally bind their ligands and therefore regulate intracellular signaling. One example involves heterodimer formation between CXCR4 and δ-opioid receptors, which is dynamically regulated by the ligands of both receptors (Pello et al., 2008). Cell lines and primary monocytes that express both receptors display heterodimerization when treated with CXCL12 and D-penicillamine(2,5)-enkephalin (DPDPE) simultaneously, and this effect is associated with negative regulation of both receptors but not with heterologous desensitization (Pello et al., 2008). Fluorescence resonance energy transfer studies suggest that heterodimers can form in cell lines and primary monocytes in the absence of ligands as well. Formation of these heterodimers is disruptive to CXCR4 homodimer formation, but upon treatment with DPDPE alone, CXCR4 homodimer formation is restored (Pello et al., 2008). As homodimer formation is thought to occur during CXCR4 signaling (Toth, Ren, & Miller, 2004), this interaction demonstrated an additional avenue by which the opioid system can regulate chemokine receptor signaling. The μ-opioid receptor may also form heterodimers with CXCR4, but this has not been investigated as extensively as the δ-opioid receptor-CXCR4 heterodimer formation. In 3-week-old rats, guanosine 5′-O-[gamma-thio]triphosphate (GTPγS) studies show that μ-opioid receptors and CXCR4 are co-expressed in several different brain areas, including the medial and lateral cortex and the hippocampus (Burbassi, Aloyo, Simansky, & Meucci, 2008). Pretreatment with morphine or the μ-opioid receptor agonist DAMGO resulted in reduced GTPγS coupling after CXCL12 treatment (Burbassi et al., 2008). Interestingly, in μ-opioid receptor knockout glia, CXCL12-induced CXCR4 G protein coupling and other downstream signals are reduced, suggesting that regulatory mechanisms between these systems may differ depending on the cell type and the opioid receptor subtype (Burbassi, Sengupta, & Meucci, 2010).

4.4. Novel regulatory mechanisms: Ferritin heavy chain

Additional exploration of opioid–chemokine interactions has revealed that the heavy chain subunit of the ubiquitously expressed iron-binding protein ferritin (FHC) can act as a negative regulator of CXCR4 signaling (Festa & Meucci, 2012; Pitcher et al., 2014; Sengupta et al., 2009). This interaction is somewhat unusual, as ferritin has classically been described for its ability to sequester iron and for its role in iron homeostasis (Wang, Knovich, Coffman, Torti, & Torti, 2010). Unlike heterologous desensitization and dimer formation, FHC-mediated inhibition occurs over a long-time period, i.e., hours, from receptor stimulation rather than within minutes or fractions of seconds and can persist for days. The first evidence of the FHC–CXCR4 interaction was provided by Li and colleagues, who showed that this interaction was dependent on ligand binding to CXCR4 in transfected HEK293 and HeLa cells (Li, Luo, Mines, Zhang, & Fan, 2006). They also showed that upregulation of FHC in these cells prior to treatment with the natural ligand CXCL12 was able to inhibit CXCR4 downstream signals, including ERK1/2 activation, and chemotaxis (Li et al., 2006). FHC was also shown to negatively regulate CXCR4 in Jurkat T cells, which natively express the receptor, and was able to affect other chemokine receptor signals, such as ERK1/2 activation via CXCR2 (Li et al., 2006).

In primary rat cortical neurons, and *ex vivo* rat cortical slices, morphine, and DAMGO treatment increased protein levels of FHC. This is associated with decreased activation of CXCR4 by CXCL12 and decreased activation of downstream signals, including ERK1/2 and Akt (Sengupta et al., 2009). Interestingly, FHC co-immunoprecipitates with CXCR4 after morphine treatment (Sengupta et al., 2009), which corresponds to reduced coupling of CXCR4 to G proteins (Burbassi et al., 2008). These findings suggest that FHC may interfere with G protein-mediated signals from CXCR4. The specific μ-opioid receptor antagonist CTAP (D-Phe-Cys-Tyr-D-Trp-Arg-Thr-Pen-Thr-NH2) is able to block upregulation of FHC in morphine-treated neurons and restore ERK1/2 and Akt activity, suggesting that other receptors that can potentially bind opioids are not involved in this mechanism. This is also supported by studies in μ-opioid receptor-deficient animals (Burbassi et al., 2010). Interestingly Rhesus macaques chronically treated with morphine and opioid-abusing human patients show increased FHC protein levels in cortical neurons (Pitcher et al., 2014), which is in agreement with current *in vitro* data. Of note, increased expression of FHC in both human and macaque neurons positively correlates with reduced CXCR4 activation and with neurocognitive impairment in human patients, suggesting that CXCR4 activation is crucial for maintaining cognitive function (Pitcher et al., 2014). Additionally, morphine-treated rats show decreased dendritic spine density in cortical neurons compared with controls, whereas RNAi-mediated knockdown of FHC was able to maintain dendritic spines at control levels in the presence of morphine (Pitcher et al., 2014). Cognitive decline is often associated with reduced dendritic spine density (Ellis et al., 2007; Morrison & Baxter, 2012), and ferritin may be an important player in this process through its regulation of CXCR4.

5. CHEMOKINE AND OPIOID INTERACTIONS IN HAND

HAND affect the prognosis and quality of life of HIV-positive patients to varying degrees (McArthur, Steiner, Sacktor, & Nath, 2010). With the advent of combination antiretroviral therapy, the incidence of the more severe forms of the disease has declined, but the overall prevalence of HAND has increased (Heaton et al., 2010). HIV-positive patients who use drugs, and opioids in particular, seem to show faster disease progression, as demonstrated by higher viral loads, decreased immune function, and potential noncompliance with standard therapies (Nath, 2010). Several studies support the capability of opioids to enhance HIV progression in general and HAND in particular, even in patients who are already on combined antiretroviral therapy (Fitting et al., 2010; Hauser et al., 2012; Hu et al., 2012; Malik, Khalique, Buch, & Seth, 2011; Pitcher et al., 2014). As opioid drug abuse is a major mechanism of transmission of HIV (10% of all diagnoses in 2011), a large population of HIV-infected individuals will be faced with increasing cognitive impairment, resulting in increased healthcare costs, and poorer overall prognosis of the disease (Centers for Disease Control, 2013).

The chemokine receptors CCR5 and CXCR4 (also known as HIV co-receptors) play a major role in HIV infection of target cells. CCR5-using viruses are usually predominant in the brain and responsible for initial CNS invasion, which is due to the ability of specific subsets of peripheral monocytes to normally penetrate the brain parenchyma (Gonzalez-Scarano & Martin-Garcia, 2005). These cells can then spread infection to resident CNS cells (mainly microglia and perivascular macrophages). CXCR4-using viruses can also enter the brain, particularly at more advanced stages of disease as other immune cells can easily cross the damaged BBB. While both host and viral factors are implicated in HIV neuropathology, different mechanisms may prevail in different groups of patients (Gelman et al., 2013) thus explaining, at least in part, the limited success of antiretrovirals in reverting neurocognitive impairment. Although neurons are not infected by HIV, the neurological deficits are ultimately consequence of neuronal alterations (mainly synaptodendritic damage; Ellis et al., 2007) or loss. Direct and indirect actions of viral proteins (mostly gp120 and tat) as well as immune/inflammatory mediators that trigger oxidative stress and excitotoxicity lead to this outcome. In line with this, many studies have shown that gp120s from either X4 or R5 viruses are highly neurotoxic. This effect is likely mediated by abnormal engaging of the respective chemokine receptor and competition with endogenous chemokine ligands. However, as discussed above, chemokine receptors exert a number of physiological functions in the CNS (Li & Ransohoff, 2008). CXCR4, in particular, is essential to CNS homeostasis and activation of neuroprotective signals. Through its ability to regulate cell-cycle proteins (Khan et al., 2003, 2008), dendritic spines (Pitcher et al., 2014), excitatory and inhibitory neurotransmission (Miller et al., 2008; Nicolai et al., 2010), and neuronal glial communication (Réaux-Le Goazigo et al., 2013), among other pathways, CXCR4 may significantly contribute to natural repair mechanisms and the ability of neurons to overcome toxic insults. Therefore, alteration of the neuronal CXCL12/CXCR4 axis is thought to be involved in neurodegenerative conditions. For instance, excessive cleavage of the endogenous CXCR4 ligand CXCL12 (which leads to inability of the cleaved product to properly engage CXCR4) has been implicated in HIV-induced neuronal damage (Vergote et al., 2006; Zhang et al., 2003), while deficits in CXCR4 signaling caused by opioids are linked to HAND (Pitcher et al., 2014; Sengupta et al., 2009). On the other hand, opioids usually enhance HIV infection of target cells, which support a neurotoxic environment. Examples of CXCL12/CXCR4 alterations in the context of other CNS disorders also exist (Parachikova & Cotman, 2007; Patel, McCandless, Dorsey, & Klein, 2010).

6. GAPS AND FUTURE CHALLENGES

Many important questions remain in regards to elucidating opioid–chemokine system interactions and their potential clinical relevance. Recently developed biased agonist opioid ligands can be used to further describe mechanistic details for opioid–chemokine regulation by their ability to only activate G protein-mediated signals from the μ-opioid receptor. Because some of the clinically relevant side effects of opioid treatments, such as constipation, are mediated through β-arrestin signaling pathways, perhaps the negative regulation of CXCR4 and subsequent pathology can be precisely localized to one arm of the μ-opioid receptor signaling pathway (DeWire et al., 2013). Many gaps remain regarding the regulation of CXCR4 by FHC as well as the mechanism of μ-opioid receptor-induced FHC upregulation. Unpublished data from our group (Ponnuru P. et al., unpublished data) suggest that the latter process may be driven by morphine-induced changes in cellular iron levels and posttranscriptional regulation of FHC; preliminary findings also suggest that morphine may interfere with iron homeostasis in select neuronal populations. Elucidation of the molecular events implicated in these steps is required to better understand morphine control of FHC. Furthermore, identification of the neuronal subpopulations and other populations of CNS cells that are susceptible to increased FHC after morphine remains an important avenue of research that has yet to be fully characterized. Rat cortical astroglia do not appear to upregulate FHC after morphine treatment, while some but not all of the rat cortical neurons are susceptible (Nash B. et al., unpublished data). This suggests different mechanisms of opioid-induced chemokine regulation in the distinct CNS cell types, which is in agreement with data reported in other non-neuronal cells, and brain regions. Examination of morphine-induced changes in iron-related proteins in different CNS cells may also provide important clues regarding changes in iron metabolism in these cells. These studies may be instrumental in finding new drug targets for HAND and other neurocognitive disorders and help elucidate the pathophysiological regulatory mechanisms between these systems.

In closure, understanding how the chemokine and opioid systems interact with each other in neuronal and immune cells will provide insight into potential dysregulated signaling pathways in HAND patients that abuse opioids. Chemokine and opioid interactions seem to be complex and receptor/cells specific, but the general trend that emerges from the work summarized above suggests that these interactions generally are inhibitory towards one another. Therefore, opioids may disrupt homeostatic chemokine signals in the CNS that are crucial for neuronal protection and repair mechanisms (including neurogenesis), which are likely implicated in cognitive decline.

ACKNOWLEDGMENTS

The authors thank present and past members of the Meucci Lab for their contribution to the studies discussed here and NIH/NIDA for continuous and generous support (DA15014 and DA32444 to O. M.)

REFERENCES

Bajetto, A., Bonavia, R., Barbero, S., Piccioli, P., Costa, A., Florio, T., et al. (1999). Glial and neuronal cells express functional chemokine receptor CXCR4 and its natural ligand stromal cell-derived factor 1. *Journal of Neurochemistry*, *73*(6), 2348–2357.

Balabanian, K., Lagane, B., Infantino, S., Chow, K. Y., Harriague, J., Moepps, B., et al. (2005). The chemokine SDF-1/CXCL12 binds to and signals through the orphan receptor RDC1 in T lymphocytes. *Journal of Biological Chemistry*, *280*(42), 35760–35766. http://dx.doi.org/10.1074/jbc.M508234200.

Bardi, G., Sengupta, R., Khan, M. Z., Patel, J. P., & Meucci, O. (2006). Human immunodeficiency virus gp120-induced apoptosis of human neuroblastoma cells in the absence of CXCR4 internalization. *Journal of Neurovirology*, *12*(3), 211–218. http://dx.doi.org/10.1080/13550280600848373.

Belcheva, M. M., Tan, Y., Heaton, V. M., Clark, A. L., & Coscia, C. J. (2003). Mu opioid transactivation and down-regulation of the epidermal growth factor receptor in astrocytes: Implications for mitogen-activated protein kinase signaling. *Molecular Pharmacology*, *64*(6), 1391–1401. http://dx.doi.org/10.1124/mol.64.6.1391.

Bezzi, P., Domercq, M., Brambilla, L., Galli, R., Schols, D., De Clercq, E., et al. (2001). CXCR4-activated astrocyte glutamate release via TNFalpha: Amplification by microglia triggers neurotoxicity. *Nature Neuroscience*, *4*(7), 702–710. http://dx.doi.org/10.1038/89490.

Bunn, S. J., Hanley, M. R., & Wilkin, G. P. (1985). Evidence for a kappa-opioid receptor on pituitary astrocytes: An autoradiographic study. *Neuroscience Letters*, *55*(3), 317–323.

Burbassi, S., Aloyo, V. J., Simansky, K. J., & Meucci, O. (2008). GTPgammaS incorporation in the rat brain: A study on mu-opioid receptors and CXCR4. *Journal of Neuroimmune Pharmacology*, *3*(1), 26–34. http://dx.doi.org/10.1007/s11481-007-9083-1.

Burbassi, S., Sengupta, R., & Meucci, O. (2010). Alterations of CXCR4 function in mu-opioid receptor-deficient glia. *European Journal of Neuroscience*, *32*(8), 1278–1288. http://dx.doi.org/10.1111/j.1460-9568.2010.07402.x.

Cali, C., & Bezzi, P. (2010). CXCR4-mediated glutamate exocytosis from astrocytes. *Journal of Neuroimmunology*, *224*(1–2), 13–21. http://dx.doi.org/10.1016/j.jneuroim.2010.05.004.

Centers for Disease Control. (2013). *Rates of diagnoses of HIV infection among adults and adolescents, by area of residence, 2011—United States and 6 dependent areas,* HIV Surveillance Report, 23.

Chao, D., Bazzy-Asaad, A., Balboni, G., & Xia, Y. (2007). delta-, but not mu-, opioid receptor stabilizes K(+) homeostasis by reducing Ca(2+) influx in the cortex during acute hypoxia. *Journal of Cellular Physiology*, *212*(1), 60–67. http://dx.doi.org/10.1002/jcp.21000.

Chen, X., Geller, E. B., Rogers, T. J., & Adler, M. W. (2007). Rapid heterologous desensitization of antinociceptive activity between mu or delta opioid receptors and chemokine receptors in rats. *Drug and Alcohol Dependence*, *88*(1), 36–41. http://dx.doi.org/10.1016/j.drugalcdep.2006.09.010.

Clark, A. K., Staniland, A. A., & Malcangio, M. (2011). Fractalkine/CX3CR1 signalling in chronic pain and inflammation. *Current Pharmaceutical Biotechnology*, *12*(10), 1707–1714.

Corbett, A. D., Henderson, G., McKnight, A. T., & Paterson, S. J. (2006). 75 Years of opioid research: The exciting but vain quest for the Holy Grail. *British Journal of Pharmacology*, *147*(Suppl. 1), S153–S162. http://dx.doi.org/10.1038/sj.bjp.0706435.

DeWire, S. M., Yamashita, D. S., Rominger, D. H., Liu, G., Cowan, C. L., Graczyk, T. M., et al. (2013). A G protein-biased ligand at the mu-opioid receptor is potently analgesic with reduced gastrointestinal and respiratory dysfunction compared with morphine. *The Journal of Pharmacology and Experimental Therapeutics*, *344*(3), 708–717. http://dx.doi.org/10.1124/jpet.112.201616.

Ellis, R., Langford, D., & Masliah, E. (2007). HIV and antiretroviral therapy in the brain: Neuronal injury and repair. *Nature Reviews. Neuroscience*, *8*(1), 33–44. http://dx.doi.org/10.1038/nrn2040.

Feng, Y., He, X., Yang, Y., Chao, D., Lazarus, L. H., & Xia, Y. (2012). Current research on opioid receptor function. *Current Drug Targets*, *13*(2), 230–246.

Festa, L., & Meucci, O. (2012). Effects of opiates and HIV proteins on neurons: The role of ferritin heavy chain and a potential for synergism. *Current HIV Research*, *10*(5), 453–462.

Finley, M. J., Chen, X., Bardi, G., Davey, P., Geller, E. B., Zhang, L., et al. (2008). Bi-directional heterologous desensitization between the major HIV-1 co-receptor CXCR4 and the kappa-opioid receptor. *Journal of Neuroimmunology*, *197*(2), 114–123. http://dx.doi.org/10.1016/j.jneuroim.2008.04.021.

Finley, M. J., Steele, A., Cornwell, W. D., & Rogers, T. J. (2011). Transcriptional regulation of the major HIV-1 coreceptor, CXCR4, by the kappa opioid receptor. *Journal of Leukocyte Biology*, *90*(1), 111–121. http://dx.doi.org/10.1189/jlb.1010546.

Fitting, S., Xu, R., Bull, C., Buch, S. K., El-Hage, N., Nath, A., et al. (2010). Interactive comorbidity between opioid drug abuse and HIV-1 Tat: Chronic exposure augments spine loss and sublethal dendritic pathology in striatal neurons. *American Journal of Pathology*, *177*(3), 1397–1410. http://dx.doi.org/10.2353/ajpath.2010.090945.

Ganju, R. K., Brubaker, S. A., Meyer, J., Dutt, P., Yang, Y., Qin, S., et al. (1998). The alpha-chemokine, stromal cell-derived factor-1alpha, binds to the transmembrane G-protein-coupled CXCR-4 receptor and activates multiple signal transduction pathways. *Journal of Biological Chemistry*, *273*(36), 23169–23175.

Gelman, J. S., Sironi, J., Berezniuk, I., Dasgupta, S., Castro, L. M., Gozzo, F. C., et al. (2013). Alterations of the intracellular peptidome in response to the proteasome inhibitor bortezomib. *PLoS One*, *8*(1), e53263. http://dx.doi.org/10.1371/journal.pone.0053263.

Goldstein, A., Tachibana, S., Lowney, L. I., Hunkapiller, M., & Hood, L. (1979). Dynorphin-(1–13), an extraordinarily potent opioid peptide. *Proceedings of the National Academy of Sciences of the United States of America*, *76*(12), 6666–6670.

Gonzalez-Scarano, F., & Martin-Garcia, J. (2005). The neuropathogenesis of AIDS. *Nature Reviews. Immunology*, *5*(1), 69–81. http://dx.doi.org/10.1038/nri1527.

Gosnell, B. A., & Levine, A. S. (2009). Reward systems and food intake: Role of opioids. *International Journal of Obesity (London)*, *33*(Suppl. 2), S54–S58. http://dx.doi.org/10.1038/ijo.2009.73.

Graham, G. J. (2009). D6 and the atypical chemokine receptor family: Novel regulators of immune and inflammatory processes. *European Journal of Immunology*, *39*(2), 342–351. http://dx.doi.org/10.1002/eji.200838858.

Hattermann, K., & Mentlein, R. (2013). An infernal trio: The chemokine CXCL12 and its receptors CXCR4 and CXCR7 in tumor biology. *Annals of Anatomy*, *195*(2), 103–110. http://dx.doi.org/10.1016/j.aanat.2012.10.013.

Hauser, K. F., Fitting, S., Dever, S. M., Podhaizer, E. M., & Knapp, P. E. (2012). Opiate drug use and the pathophysiology of neuroAIDS. *Current HIV Research*, *10*(5), 435–452.

Heaton, R. K., Clifford, D. B., Franklin, D. R., Jr., Woods, S. P., Ake, C., Vaida, F., et al. (2010). HIV-associated neurocognitive disorders persist in the era of potent antiretroviral therapy: CHARTER Study. *Neurology*, *75*(23), 2087–2096. http://dx.doi.org/10.1212/WNL.0b013e318200d727.

Heinisch, S., Palma, J., & Kirby, L. G. (2011). Interactions between chemokine and mu-opioid receptors: Anatomical findings and electrophysiological studies in the rat

periaqueductal grey. *Brain, Behavior, and Immunity*, *25*(2), 360–372. http://dx.doi.org/10.1016/j.bbi.2010.10.020.
Hesselgesser, J., Taub, D., Baskar, P., Greenberg, M., Hoxie, J., Kolson, D. L., et al. (1998). Neuronal apoptosis induced by HIV-1 gp120 and the chemokine SDF-1 alpha is mediated by the chemokine receptor CXCR4. *Current Biology*, *8*(10), 595–598.
Hu, S., Sheng, W. S., Lokensgard, J. R., & Peterson, P. K. (2002). Morphine induces apoptosis of human microglia and neurons. *Neuropharmacology*, *42*(6), 829–836.
Hu, G., Yao, H., Chaudhuri, A. D., Duan, M., Yelamanchili, S. V., Wen, H., et al. (2012). Exosome-mediated shuttling of microRNA-29 regulates HIV Tat and morphine-mediated neuronal dysfunction. *Cell Death & Disease*, *3*, e381. http://dx.doi.org/10.1038/cddis.2012.114.
Hughes, J., Smith, T. W., Kosterlitz, H. W., Fothergill, L. A., Morgan, B. A., & Morris, H. R. (1975). Identification of two related pentapeptides from the brain with potent opiate agonist activity. *Nature*, *258*(5536), 577–580.
Huising, M. O., Stet, R. J., Kruiswijk, C. P., Savelkoul, H. F., & Lidy Verburg-van Kemenade, B. M. (2003). Molecular evolution of CXC chemokines: Extant CXC chemokines originate from the CNS. *Trends in Immunology*, *24*(6), 307–313.
Hutchinson, M. R., Shavit, Y., Grace, P. M., Rice, K. C., Maier, S. F., & Watkins, L. R. (2011). Exploring the neuroimmunopharmacology of opioids: An integrative review of mechanisms of central immune signaling and their implications for opioid analgesia. *Pharmacological Reviews*, *63*(3), 772–810. http://dx.doi.org/10.1124/pr.110.004135.
Iegorova, O., Fisyunov, A., & Krishtal, O. (2010). G-protein-independent modulation of P-type calcium channels by mu-opioids in Purkinje neurons of rat. *Neuroscience Letters*, *480*(2), 106–111. http://dx.doi.org/10.1016/j.neulet.2010.06.015.
Ketwaroo, G. A., Cheng, V., & Lembo, A. (2013). Opioid-induced bowel dysfunction. *Current Gastroenterology Reports*, *15*(9), 344. http://dx.doi.org/10.1007/s11894-013-0344-2.
Khan, M. Z., Brandimarti, R., Musser, B. J., Resue, D. M., Fatatis, A., & Meucci, O. (2003). The chemokine receptor CXCR4 regulates cell-cycle proteins in neurons. *Journal of Neurovirology*, *9*(3), 300–314. http://dx.doi.org/10.1080/13550280390201010.
Khan, M. Z., Brandimarti, R., Patel, J. P., Huynh, N., Wang, J., Huang, Z., et al. (2004). Apoptotic and antiapoptotic effects of CXCR4: Is it a matter of intrinsic efficacy? Implications for HIV neuropathogenesis. *AIDS Research and Human Retroviruses*, *20*(10), 1063–1071. http://dx.doi.org/10.1089/aid.2004.20.1063.
Khan, M. Z., Brandimarti, R., Shimizu, S., Nicolai, J., Crowe, E., & Meucci, O. (2008). The chemokine CXCL12 promotes survival of postmitotic neurons by regulating Rb protein. *Cell Death and Differentiation*, *15*(10), 1663–1672. http://dx.doi.org/10.1038/cdd.2008.95.
Knapp, P. E., Maderspach, K., & Hauser, K. F. (1998). Endogenous opioid system in developing normal and jimpy oligodendrocytes: Mu and kappa opioid receptors mediate differential mitogenic and growth responses. *Glia*, *22*(2), 189–201.
Kraft-Terry, S. D., Buch, S. J., Fox, H. S., & Gendelman, H. E. (2009). A coat of many colors: Neuroimmune crosstalk in human immunodeficiency virus infection. *Neuron*, *64*(1), 133–145. http://dx.doi.org/10.1016/j.neuron.2009.09.042.
La Porta, C. A. (2012). CXCR6: The role of environment in tumor progression. Challenges for therapy. *Stem Cell Reviews*, *8*(4), 1282–1285. http://dx.doi.org/10.1007/s12015-012-9383-6.
Lazarini, F., Tham, T. N., Casanova, P., Arenzana-Seisdedos, F., & Dubois-Dalcq, M. (2003). Role of the alpha-chemokine stromal cell-derived factor (SDF-1) in the developing and mature central nervous system. *Glia*, *42*(2), 139–148. http://dx.doi.org/10.1002/glia.10139.
Leveille, F., El Gaamouch, F., Gouix, E., Lecocq, M., Lobner, D., Nicole, O., et al. (2008). Neuronal viability is controlled by a functional relation between synaptic and

extrasynaptic NMDA receptors. *FASEB Journal*, *22*(12), 4258–4271. http://dx.doi.org/10.1096/fj.08-107268.

Li, C. H., Chung, D., & Doneen, B. A. (1976). Isolation, characterization and opiate activity of beta-endorphin from human pituitary glands. *Biochemical and Biophysical Research Communications*, *72*(4), 1542–1547.

Li, R., Luo, C., Mines, M., Zhang, J., & Fan, G. H. (2006). Chemokine CXCL12 induces binding of ferritin heavy chain to the chemokine receptor CXCR4, alters CXCR4 signaling, and induces phosphorylation and nuclear translocation of ferritin heavy chain. *Journal of Biological Chemistry*, *281*(49), 37616–37627. http://dx.doi.org/10.1074/jbc.M607266200.

Li, M., & Ransohoff, R. M. (2008). Multiple roles of chemokine CXCL12 in the central nervous system: A migration from immunology to neurobiology. *Progress in Neurobiology*, *84*(2), 116–131. http://dx.doi.org/10.1016/j.pneurobio.2007.11.003.

Lindl, K. A., Marks, D. R., Kolson, D. L., & Jordan-Sciutto, K. L. (2010). HIV-associated neurocognitive disorder: Pathogenesis and therapeutic opportunities. *Journal of Neuroimmune Pharmacology*, *5*(3), 294–309. http://dx.doi.org/10.1007/s11481-010-9205-z.

Lipfert, J., Odemis, V., Wagner, D. C., Boltze, J., & Engele, J. (2013). CXCR4 and CXCR7 form a functional receptor unit for SDF-1/CXCL12 in primary rodent microglia. *Neuropathology and Applied Neurobiology*, *39*(6), 667–680. http://dx.doi.org/10.1111/nan.12015.

Ma, Q., Jones, D., Borghesani, P. R., Segal, R. A., Nagasawa, T., Kishimoto, T., et al. (1998). Impaired B-lymphopoiesis, myelopoiesis, and derailed cerebellar neuron migration in CXCR4- and SDF-1-deficient mice. *Proceedings of the National Academy of Sciences of the United States of America*, *95*(16), 9448–9453.

Malik, S., Khalique, H., Buch, S., & Seth, P. (2011). A growth factor attenuates HIV-1 Tat and morphine induced damage to human neurons: Implication in HIV/AIDS-drug abuse cases. *PLoS One*, *6*(3), e18116. http://dx.doi.org/10.1371/journal.pone.0018116.

Mansour, A., Fox, C. A., Akil, H., & Watson, S. J. (1995). Opioid-receptor mRNA expression in the rat CNS: Anatomical and functional implications. *Trends in Neurosciences*, *18*(1), 22–29.

Maysami, S., Nguyen, D., Zobel, F., Pitz, C., Heine, S., Hopfner, M., et al. (2006). Modulation of rat oligodendrocyte precursor cells by the chemokine CXCL12. *Neuroreport*, *17*(11), 1187–1190. http://dx.doi.org/10.1097/01.wnr.0000227985.92551.9a.

McArthur, J. C., Steiner, J., Sacktor, N., & Nath, A. (2010). Human immunodeficiency virus-associated neurocognitive disorders: Mind the gap. *Annals of Neurology*, *67*(6), 699–714. http://dx.doi.org/10.1002/ana.22053.

Mellado, M., Rodriguez-Frade, J. M., Vila-Coro, A. J., Fernandez, S., Martin de Ana, A., Jones, D. R., et al. (2001). Chemokine receptor homo- or heterodimerization activates distinct signaling pathways. *The EMBO Journal*, *20*(10), 2497–2507. http://dx.doi.org/10.1093/emboj/20.10.2497.

Meucci, O., Fatatis, A., Simen, A. A., Bushell, T. J., Gray, P. W., & Miller, R. J. (1998). Chemokines regulate hippocampal neuronal signaling and gp120 neurotoxicity. *Proceedings of the National Academy of Sciences of the United States of America*, *95*(24), 14500–14505.

Miller, R. J., Rostene, W., Apartis, E., Banisadr, G., Biber, K., Milligan, E. D., et al. (2008). Chemokine action in the nervous system. *The Journal of Neuroscience*, *28*(46), 11792–11795. http://dx.doi.org/10.1523/JNEUROSCI.3588-08.2008.

Molina, P. E. (2002). Stress-specific opioid modulation of haemodynamic counter-regulation. *Clinical and Experimental Pharmacology and Physiology*, *29*(3), 248–253.

Morrison, J. H., & Baxter, M. G. (2012). The ageing cortical synapse: Hallmarks and implications for cognitive decline. *Nature Reviews. Neuroscience*, *13*(4), 240–250. http://dx.doi.org/10.1038/nrn3200.

Moser, B., Wolf, M., Walz, A., & Loetscher, P. (2004). Chemokines: Multiple levels of leukocyte migration control. *Trends in Immunology*, *25*(2), 75–84. http://dx.doi.org/10.1016/j.it.2003.12.005.
Nath, A. (2010). Human immunodeficiency virus-associated neurocognitive disorder: Pathophysiology in relation to drug addiction. *Annals of the New York Academy of Sciences*, *1187*, 122–128. http://dx.doi.org/10.1111/j.1749-6632.2009.05277.x.
Nicolai, J., Burbassi, S., Rubin, J., & Meucci, O. (2010). CXCL12 inhibits expression of the NMDA receptor's NR2B subunit through a histone deacetylase-dependent pathway contributing to neuronal survival. *Cell Death & Disease*, *1*, e33. http://dx.doi.org/10.1038/cddis.2010.10.
Nomiyama, H., Osada, N., & Yoshie, O. (2011). A family tree of vertebrate chemokine receptors for a unified nomenclature. *Developmental and Comparative Immunology*, *35*(7), 705–715. http://dx.doi.org/10.1016/j.dci.2011.01.019.
Ohtani, Y., Minami, M., Kawaguchi, N., Nishiyori, A., Yamamoto, J., Takami, S., et al. (1998). Expression of stromal cell-derived factor-1 and CXCR4 chemokine receptor mRNAs in cultured rat glial and neuronal cells. *Neuroscience Letters*, *249*(2–3), 163–166.
Okamoto, M., Wang, X., & Baba, M. (2005). HIV-1-infected macrophages induce astrogliosis by SDF-1alpha and matrix metalloproteinases. *Biochemical and Biophysical Research Communications*, *336*(4), 1214–1220. http://dx.doi.org/10.1016/j.bbrc.2005.08.251.
Pandey, V., & Bolsover, S. R. (2000). Immediate and neurotoxic effects of HIV protein gp120 act through CXCR4 receptor. *Biochemical and Biophysical Research Communications*, *274*(1), 212–215. http://dx.doi.org/10.1006/bbrc.2000.3113.
Parachikova, A., & Cotman, C. W. (2007). Reduced CXCL12/CXCR4 results in impaired learning and is downregulated in a mouse model of Alzheimer disease. *Neurobiology of Disease*, *28*(2), 143–153. http://dx.doi.org/10.1016/j.nbd.2007.07.001.
Patel, J. R., McCandless, E. E., Dorsey, D., & Klein, R. S. (2010). CXCR4 promotes differentiation of oligodendrocyte progenitors and remyelination. *Proceedings of the National Academy of Sciences of the United States of America*, *107*(24), 11062–11067. http://dx.doi.org/10.1073/pnas.1006301107.
Patel, J. P., Sengupta, R., Bardi, G., Khan, M. Z., Mullen-Przeworski, A., & Meucci, O. (2006). Modulation of neuronal CXCR4 by the micro-opioid agonist DAMGO. *Journal of Neurovirology*, *12*(6), 492–500. http://dx.doi.org/10.1080/13550280601064798.
Pello, O. M., Martinez-Munoz, L., Parrillas, V., Serrano, A., Rodriguez-Frade, J. M., Toro, M. J., et al. (2008). Ligand stabilization of CXCR4/delta-opioid receptor heterodimers reveals a mechanism for immune response regulation. *European Journal of Immunology*, *38*(2), 537–549. http://dx.doi.org/10.1002/eji.200737630.
Persson, P. A., Thorlin, T., Ronnback, L., Hansson, E., & Eriksson, P. S. (2000). Differential expression of delta opioid receptors and mRNA in proliferating astrocytes during the cell cycle. *Journal of Neuroscience Research*, *61*(4), 371–375.
Pitcher, J., Abt, A., Myers, J., Han, R., Snyder, M., Graziano, A., et al. (2014). Neuronal ferritin heavy chain and drug abuse affect HIV-associated cognitive dysfunction. *Journal of Clinical Investigation*, *124*(2), 656–669. http://dx.doi.org/10.1172/JCI70090, Epub 2014 Jan 9.
Réaux-Le Goazigo, A., Van Steenwinckel, J., Rostène, W., & Mélik Parsadaniantz, S. (2013). Current status of chemokines in the adult CNS. *Progress in Neurobiology*, *104*, 67–92. http://dx.doi.org/10.1016/j.pneurobio.2013.02.001.
Rios, C., Gomes, I., & Devi, L. A. (2006). mu opioid and CB1 cannabinoid receptor interactions: Reciprocal inhibition of receptor signaling and neuritogenesis. *British Journal of Pharmacology*, *148*(4), 387–395. http://dx.doi.org/10.1038/sj.bjp.0706757.
Rostene, W., Kitabgi, P., & Parsadaniantz, S. M. (2007). Chemokines: A new class of neuromodulator? *Nature Reviews. Neuroscience*, *8*(11), 895–903. http://dx.doi.org/10.1038/nrn2255.

Ruzicka, B. B., Fox, C. A., Thompson, R. C., Meng, F., Watson, S. J., & Akil, H. (1995). Primary astroglial cultures derived from several rat brain regions differentially express mu, delta and kappa opioid receptor mRNA. *Brain Research. Molecular Brain Research*, *34*(2), 209–220.

Sanchez-Alcaniz, J. A., Haege, S., Mueller, W., Pla, R., Mackay, F., Schulz, S., et al. (2011). Cxcr7 controls neuronal migration by regulating chemokine responsiveness. *Neuron*, *69*(1), 77–90. http://dx.doi.org/10.1016/j.neuron.2010.12.006.

Sengupta, R., Burbassi, S., Shimizu, S., Cappello, S., Vallee, R. B., Rubin, J. B., et al. (2009). Morphine increases brain levels of ferritin heavy chain leading to inhibition of CXCR4-mediated survival signaling in neurons. *Journal of Neuroscience*, *29*(8), 2534–2544. http://dx.doi.org/10.1523/JNEUROSCI.5865-08.2009.

Shepherd, A. J., Loo, L., & Mohapatra, D. P. (2013). Chemokine co-receptor CCR5/CXCR4-dependent modulation of Kv2 1 channel confers acute neuroprotection to HIV-1 glycoprotein gp120 exposure. *PLoS One*, *8*(9), e76698. http://dx.doi.org/10.1371/journal.pone.0076698.

Sheridan, G. K., & Murphy, K. J. (2013). Neuron-glia crosstalk in health and disease: Fractalkine and CX3CR1 take centre stage. *Open Biology*, *3*(12), 130181. http://dx.doi.org/10.1098/rsob.130181.

Strieter, R. M., Burdick, M. D., Gomperts, B. N., Belperio, J. A., & Keane, M. P. (2005). CXC chemokines in angiogenesis. *Cytokine & Growth Factor Reviews*, *16*(6), 593–609. http://dx.doi.org/10.1016/j.cytogfr.2005.04.007.

Stumm, R. K., Zhou, C., Ara, T., Lazarini, F., Dubois-Dalcq, M., Nagasawa, T., et al. (2003). CXCR4 regulates interneuron migration in the developing neocortex. *Journal of Neuroscience*, *23*(12), 5123–5130.

Szabo, I., Wetzel, M. A., Zhang, N., Steele, A. D., Kaminsky, D. E., Chen, C., et al. (2003). Selective inactivation of CCR5 and decreased infectivity of R5 HIV-1 strains mediated by opioid-induced heterologous desensitization. *Journal of Leukocyte Biology*, *74*(6), 1074–1082. http://dx.doi.org/10.1189/jlb.0203067.

Toth, P. T., Ren, D., & Miller, R. J. (2004). Regulation of CXCR4 receptor dimerization by the chemokine SDF-1alpha and the HIV-1 coat protein gp120: A fluorescence resonance energy transfer (FRET) study. *Journal of Pharmacology and Experimental Therapeutics*, *310*(1), 8–17. http://dx.doi.org/10.1124/jpet.103.064956.

Turchan-Cholewo, J., Dimayuga, F. O., Ding, Q., Keller, J. N., Hauser, K. F., Knapp, P. E., et al. (2008). Cell-specific actions of HIV-Tat and morphine on opioid receptor expression in glia. *Journal of Neuroscience Research*, *86*(9), 2100–2110. http://dx.doi.org/10.1002/jnr.21653.

van der Meer, P., Ulrich, A. M., Gonzalez-Scarano, F., & Lavi, E. (2000). Immunohistochemical analysis of CCR2, CCR3, CCR5, and CXCR4 in the human brain: Potential mechanisms for HIV dementia. *Experimental and Molecular Pathology*, *69*(3), 192–201. http://dx.doi.org/10.1006/exmp.2000.2336.

van Dorp, E. L., Kest, B., Kowalczyk, W. J., Morariu, A. M., Waxman, A. R., Arout, C. A., et al. (2009). Morphine-6beta-glucuronide rapidly increases pain sensitivity independently of opioid receptor activity in mice and humans. *Anesthesiology*, *110*(6), 1356–1363. http://dx.doi.org/10.1097/ALN.0b013e3181a105de.

Vergote, D., Butler, G. S., Ooms, M., Cox, J. H., Silva, C., Hollenberg, M. D., et al. (2006). Proteolytic processing of SDF-1alpha reveals a change in receptor specificity mediating HIV-associated neurodegeneration. *Proceedings of the National Academy of Sciences of the United States of America*, *103*(50), 19182–19187. http://dx.doi.org/10.1073/pnas.0604678103.

Waldhoer, M., Bartlett, S. E., & Whistler, J. L. (2004). Opioid receptors. *Annual Review of Biochemistry*, *73*, 953–990. http://dx.doi.org/10.1146/annurev.biochem.73.011303.073940.

Wang, W., Knovich, M. A., Coffman, L. G., Torti, F. M., & Torti, S. V. (2010). Serum ferritin: Past, present and future. *Biochimica et Biophysica Acta, 1800*(8), 760–769. http://dx.doi.org/10.1016/j.bbagen.2010.03.011.

Watkins, L. R., Hutchinson, M. R., Rice, K. C., & Maier, S. F. (2009). The "toll" of opioid-induced glial activation: Improving the clinical efficacy of opioids by targeting glia. *Trends in Pharmacological Sciences, 30*(11), 581–591. http://dx.doi.org/10.1016/j.tips.2009.08.002.

Williams, J. T., Ingram, S. L., Henderson, G., Chavkin, C., von Zastrow, M., Schulz, S., et al. (2013). Regulation of mu-opioid receptors: Desensitization, phosphorylation, internalization, and tolerance. *Pharmacological Reviews, 65*(1), 223–254. http://dx.doi.org/10.1124/pr.112.005942.

Wilson, N. M., Jung, H., Ripsch, M. S., Miller, R. J., & White, F. A. (2011). CXCR4 signaling mediates morphine-induced tactile hyperalgesia. *Brain, Behavior, and Immunity, 25*(3), 565–573. http://dx.doi.org/10.1016/j.bbi.2010.12.014.

Wittert, G., Hope, P., & Pyle, D. (1996). Tissue distribution of opioid receptor gene expression in the rat. *Biochemical and Biophysical Research Communications, 218*(3), 877–881. http://dx.doi.org/10.1006/bbrc.1996.0156.

Xu, M., Bruchas, M. R., Ippolito, D. L., Gendron, L., & Chavkin, C. (2007). Sciatic nerve ligation-induced proliferation of spinal cord astrocytes is mediated by kappa opioid activation of p38 mitogen-activated protein kinase. *Journal of Neuroscience, 27*(10), 2570–2581. http://dx.doi.org/10.1523/JNEUROSCI.3728-06.2007.

Zhang, K., McQuibban, G. A., Silva, C., Butler, G. S., Johnston, J. B., Holden, J., et al. (2003). HIV-induced metalloproteinase processing of the chemokine stromal cell derived factor-1 causes neurodegeneration. *Nature Neuroscience, 6*(10), 1064–1071. http://dx.doi.org/10.1038/nn1127.

Zhang, N., Rogers, T. J., Caterina, M., & Oppenheim, J. J. (2004). Proinflammatory chemokines, such as C–C chemokine ligand 3, desensitize mu-opioid receptors on dorsal root ganglia neurons. *Journal of Immunology, 173*(1), 594–599.

Zheng, J., Thylin, M. R., Ghorpade, A., Xiong, H., Persidsky, Y., Cotter, R., et al. (1999). Intracellular CXCR4 signaling, neuronal apoptosis and neuropathogenic mechanisms of HIV-1-associated dementia. *Journal of Neuroimmunology, 98*(2), 185–200.

Zlotnik, A., & Yoshie, O. (2000). Chemokines: A new classification system and their role in immunity. *Immunity, 12*(2), 121–127.

Zlotnik, A., & Yoshie, O. (2012). The chemokine superfamily revisited. *Immunity, 36*(5), 705–716. http://dx.doi.org/10.1016/j.immuni.2012.05.008.

CHAPTER SIX

Discovery of a Novel Site of Opioid Action at the Innate Immune Pattern-Recognition Receptor TLR4 and its Role in Addiction

Jonathan Henry W. Jacobsen*, Linda R. Watkins†, Mark R. Hutchinson‡,1

*Discipline of Pharmacology, School of Medical Sciences, University of Adelaide, Adelaide, South Australia, Australia
†Department of Psychology and Centre for Neuroscience, University of Colorado at Boulder, Boulder, Colorado, USA
‡Discipline of Physiology, School of Medical Sciences, University of Adelaide, Adelaide, South Australia, Australia
[1]Corresponding author: e-mail address: mark.hutchinson@adelaide.edu.au

Contents

Abstract

Opioids have historically, and continue to be, an integral component of pain management. However, despite pharmacokinetic and dynamic optimization over the past 100 years, opioids continue to produce many undesirable side effects such as tolerance, reward, and dependence. As such, opioids are liable for addiction. Traditionally, opioid addiction was viewed as a solely neuronal process, and while substantial headway has been made into understanding the molecular and cellular mechanisms mediating this process, research has however, been relatively ambivalent to how the rest of the central

International Review of Neurobiology, Volume 118
ISSN 0074-7742
http://dx.doi.org/10.1016/B978-0-12-801284-0.00006-3

nervous system (CNS) responds to opioids. Evidence over the past 20 years has clearly demonstrated the importance of the immunocompetent cells of the CNS (glia) in many aspects of opioid pharmacology. Particular focus has been placed on microglia and astrocytes, who in response to opioids, become activated and release inflammatory mediators. Importantly, the mechanism underlying immune activation is beginning to be elucidated. Evidence suggests an innate immune pattern-recognition receptor (toll-like receptor 4) as an integral component underlying opioid-induced glial activation. The subsequent proinflammatory response may be viewed akin to neurotransmission creating a process termed central immune signaling. Translationally, we are beginning to appreciate the importance of central immune signaling as it contributes to many behavioral actions of addiction including reward, withdrawal, and craving. As such, the aim of this chapter is to review and integrate the neuronal and central immune signaling perspective of addiction.

1. INTRODUCTION

Opioid addiction is second only to amphetamines in terms of illicit drug dependence worldwide and represents a major medical, social, and economic burden (Degenhardt et al., 2013). Interestingly, however, there has been a shift in the type of opioids associated with addiction. Abuse of illicit opioids such as heroin now represents a small proportion of the opioid abusers with prescription pain medications, such as OxyContin® (oxycodone) and Vicodin® (hydrocodone) now among the most-abused opioid-based drug (National Institute of Drug Abuse, 2005, 2011). Classically, addiction research has focused on cellular and molecular alterations within neurons and their circuits. As such, an overwhelming proportion of pharmacotherapies for opioid addiction are designed to target these neuronally based processes known to be affected by drug intake. Despite the pivotal role of neurons in the initiation, transition, and maintenance of opioid addiction (Koob & Volkow, 2009), these are not the only cells within the brain affected by, and reacting to, opioids (Hutchinson et al., 2007). Of particular importance are glial cells (microglia and astrocytes). Traditionally, these cells were considered passive accessories to neurons. However, recent research has demonstrated that these cells actively participate in synaptogenesis, neuronal excitability, and neurotransmission (Ben Achour & Pascual, 2010; Watkins, Hutchinson, Johnston, & Maier, 2005). Additionally, glia demonstrate robust changes in their physiology and morphology in response to opioid exposure within key brain regions identified as contributing to addiction (Miguel-Hidalgo, 2009). Many studies have identified potential mechanisms by which these cells contribute to the state of addiction (see Coller & Hutchinson, 2012, for review). However, evidence implicates glial-derived inflammatory signaling (central immune signaling) as a pivotal component to opioid addiction-like behaviors (Hutchinson & Watkins, 2014). This chapter aims to integrate current neurobiological theories of addiction in-line with the emerging importance of central immune signaling.

2. INNATE IMMUNITY OF THE CENTRAL NERVOUS SYSTEM

Glia are by far the most numerous cell type within the CNS accounting for approximately 70–90% of all cell populations (Banati, 2003; Temburni & Jacob, 2001). However, addiction research, and neuroscience in general has often underappreciated the importance of these cells viewing them as passive elements. Furthermore, the notion that these cells were capable of contributing to the intricate dynamics of neurons and their circuits was not considered until recently.

Glial cells were traditionally viewed as the primary immune effector cells within the CNS (Jensen, Massie, & Keyser, 2013; Ransohoff & Perry, 2009). Both microglia and astrocytes, two subclasses of glia, are capable of reacting to perturbations in CNS homeostasis such as injury or infection. Under these conditions, glia become activated exhibiting alterations to their morphology, gene expression, and proliferation (Sofroniew & Vinters, 2009). From a morphological standpoint, immune activation changes glia from a ramified (resting) to an amoeboid (activated) state with increased expression of cell surface markers—cluster of differentiation 11b (CD11b) and ionized calcium-binding adapter 1 for microglia and glial fibrillary acid protein (GFAP) for astrocytes (Graeber, 2010; Sofroniew & Vinters, 2009). Additionally, immune activation increases the expression of inflammatory mediators such as interleukin-1β (IL-1β), IL-6, tumor necrosis factor α (TNF-α), monocyte chemoattractant protein-1 (MCP-1 or CCL2), and inducible nitric oxide synthase (iNOS) (Jensen et al., 2013). These mediators serve two main functions: to eliminate the pathology and to enhance repair (Ransohoff & Brown, 2012). It is now recognized, however, that the immune response is context specific, varies according to insult, occurs along a graded continuum, and is tightly regulated (Graeber, 2010). Such a gain in immune activity can result in a loss of housekeeping functions which can be both beneficial and detrimental to neuronal function (Sofroniew, 2009). Notably, important immunological differences exist between glia. For example, microglia are extremely sensitive and react quickly to alterations in the CNS. By contrast, astrocytes are slower and exhibit more control over their immune activation (Ransohoff & Brown, 2012). Interestingly, after the initial stimuli has resolved, rather than returning to a basal state, microglia and astrocytes can remain in a sensitized state termed "primed." In the absence of an activation stimulus, sensitized glia appear morphologically active. However, they do not over produce inflammatory mediators under basal conditions. Upon subsequent challenges, primed cells react quicker and exhibit an exacerbated immune response. The factors leading to glial cells remaining in a primed state remain to be fully elucidated (Bilbo & Schwarz, 2012).

In addition to their immunological role, it is becoming increasingly apparent that glia modulate neuronal physiology and function. Both anatomical and molecular evidence support this concept. For example, a single astrocyte can potentially envelope up to 100,000 neuronal synapses (Bushong, Martone, Jones, & Ellisman, 2002; Halassa, Fellin, Takano, Dong, & Haydon, 2007). Both astrocytes and microglia are extremely dynamic under basal settings, constantly surveying synapses for any alterations in homeostasis (Nimmerjahn, 2005). Additionally, both these cells express a breadth of neurotransmitters and their corresponding receptors allowing bidirectional communication between neurons and glia (Fellin, 2009). This dynamic interaction between astrocytes and microglia with pre- and postsynaptic neurons has been termed the tetrapartite synapse (De Leo, Tawfik, & LaCroix-Fralish, 2006).

3. PARALLELS BETWEEN AN IMMUNE RESPONSE TO LIPOPOLYSACCHARIDE AND OPIOIDS

Parallels can be drawn between glial responses to typical immunogen, for example, lipopolysaccharide (LPS), an endotoxin expressed by Gram negative bacteria, and opioids (Wang et al., 2012). It has been clearly demonstrated that opioids activate brain and spinal glia (reviewed by Hutchinson et al., 2007; Hutchinson et al., 2011; Watkins, Hutchinson, Rice, & Maier, 2009). Opioids increase the expression of GFAP and CD11b and alter cellular morphology from a ramified to an amoeboid state—mimicking a response to LPS (Hutchinson et al., 2008; Wang et al., 2012; Yoon, Patel, & Dougherty, 2012). Importantly, just as neurons exhibit areas of increased and decreased activity in response to opioids, glial cells also display regional heterogeneity, indicating specificity in their response (Hutchinson et al., 2009). For example, chronic morphine increases the expression of both CD11b and GFAP in the ventral tegmental area, dentate gyrus, and dorsal and ventral periaqueductal gray. However, GFAP expression was additionally increased in the nucleus accumbens and *cornu ammonis* of the hippocampus, while CD11b was not (Hutchinson et al., 2009). It is worth highlighting all the areas that are implicated in the pathogenesis of addiction (Koob & Le Moal, 2008).

Consistent with the concept of opioids inducing a central immune response paralleling LPS, inflammatory cytokines, and chemokines such as IL-1β, TNF-α, and MCP-1 have been observed to be upregulated in the brain following acute and chronic opioid exposure (see Coller & Hutchinson, 2012, for review). Similarly, signaling pathways and transcription factors, for example, extracellular signal-regulated kinase (ERK)/mitogen-activated protein kinase, c-Jun N-terminal kinase/signal transducer and activator of transcription, and nuclear factor kappa-light-chain-enhancer of activated B cells (NF-κB) and activator protein-1 (AP-1), have been observed in response to acute and chronic opioids as well (see Coller & Hutchinson, 2012, for review). Importantly, administration of glial attenuators such as minocycline, fluorocitrate, or ibudilast (was AV411, now MN166) reverse the LPS- and opioid-induced inflammatory response of glia (Hutchinson et al., 2009; Song & Zhoa, 2001; Zhang et al., 2012).

However, a key point of contrast between LPS and opioids is the nature of the immune response (Stevens, Aravind, Das, & Davis, 2013). An immune response to LPS is extremely proinflammatory resulting in high circulating levels of inflammatory products and relies upon peripheral immune signals to activate glia due to its poor brain penetrance at subseptic doses (Quan & Banks, 2007). By contrast, opioids readily permeate the brain-activating glia directly. However, the degree to which the immune system is engaged is far less than LPS (Stevens et al., 2013). As a consequence, the inflammatory mediators, may in fact be acting at subinflammatory levels in a manner more akin to neurotransmission than neuroinflammation (Hutchinson & Watkins, 2014) (Fig. 6.1).

The mechanism underlying opioid-induced glial activation is beginning to be elucidated. Increasing evidence suggests that toll-like receptors (TLRs), a class of an innate immune receptor, are key initiators to opioid-induced activation (Hutchinson et al., 2007). Each TLR is a single transmembrane-spanning protein containing a toll/interleukin receptor domain and leucine-rich repeats. As such TLRs belong to the interleukin-1/TLR superfamily. TLRs are pattern-recognition receptors

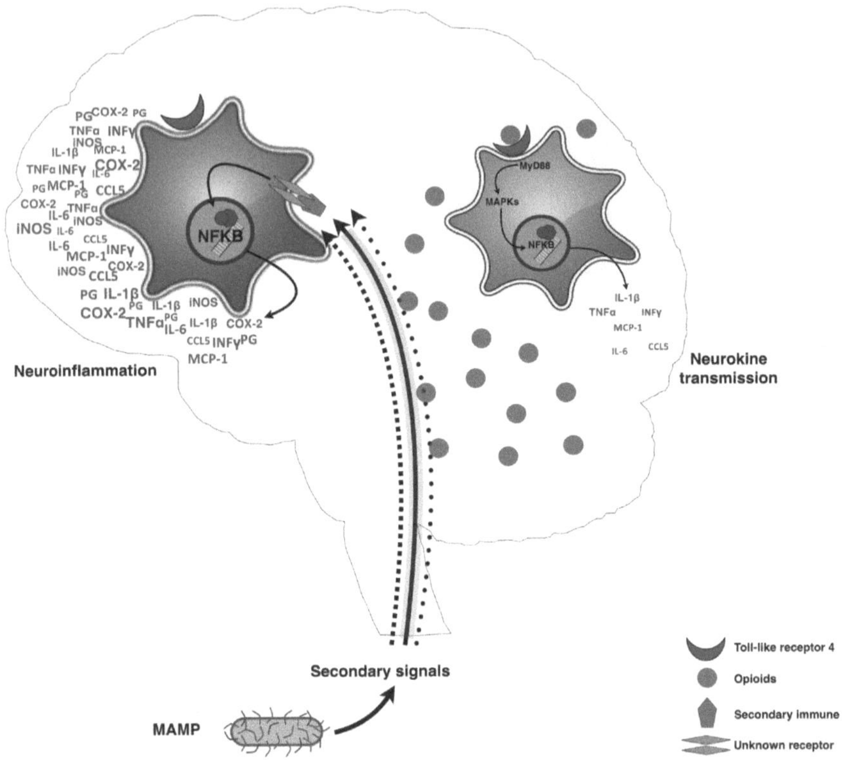

Figure 6.1 Comparisons between an opioid- and microbe-associated molecular pattern (MAMP) (LPS)-induced glial response. In order for a peripheral MAMP (LPS) challenge to trigger a response by glial cells within the CNS, it must be converted to a secondary immune signal due to the poor brain penetrance of MAMPs. Importantly, multiple parallel pathways are required to elicit this response (as outlined below). The secondary signal activates glia within the brain eliciting a proinflammatory immune response. The response is characterized by higher concentrations of inflammatory mediators such as cytokines, chemokines, and inflammatory proteins: creating a neuroinflammatory environment. By contrast, opioids trigger glial responses directly, in neuroanatomical areas associated with addiction. The immune response is relatively smaller and has a different neuroanatomical distribution when compared to that of peripherally induced innate immune response. As such, it has been hypothesized that the immune response is acting in a manner similar to neurotransmission than neuroinflammation (central immune signaling or neurokine transmission). CCL5, chemokine ligand 5; COX-2, cyclooxygenase 2; iNOS, inducible nitric oxide synthase; IL-1β, interleukin-1β; IL-6, interleukin-6; MCP-1, monocyte chemoattractant protein-1; PG, prostaglandin; TNF-α, tumor necrosis factor α.

capable of responding to broad range of molecular motifs expressed on exogenous (microbe-associated molecular patterns, MAMPs) and endogenous molecules (danger-associated molecular patterns, DAMPs) (Akira & Takeda, 2004). Recently, xenobiotics (xenobiotic molecular patterns, XAMPs), which include foreign chemicals, such as alcohol, methamphetamine, and cocaine, have also found to signal through this receptor class (Hutchinson & Watkins, 2014). Within the CNS, the expression of TLRs is limited mainly to glial cells, with limited neuronal expression under basal conditions. However, following an inflammatory challenge, the expressions of TLRs are significantly increased on all cells of the CNS (Bsibsi et al., 2002). Immune activation of TLRs results in the downstream activation of inflammatory transcription factors such as NF-κB and AP-1 leading to the subsequent transcription and translation of inflammatory products, such as cytokines and chemokines, for example, TNF-α, IL-1β, IL-6, IFNγ, and MCP-1 (Akira & Takeda, 2004).

In the context of opioids, however, particular emphasis has been placed upon a specific TLR: Toll-like receptor 4 (TLR4). TLR4, along with its coreceptors MD-2 and CD14, was originally identified as the receptor-recognizing LPS (Poltorak, 1998). However, *in vitro*, *in vivo*, and *in silico* experiments have demonstrated the pivotal role of TLR4 in opioid-induced activation of glia leading to a neuroinflammatory response (Hutchinson et al., 2012, 2010). Importantly, Hutchinson et al. (2010) uncovered key pharmacological characteristics of opioid-induced TLR4 signaling:

- TLR4 signaling occurs in response to natural, semisynthetic, and fully synthetic opioids
- It is nonstereoselective for opioids
- TLR4 exhibits some selectivity among opioids, as morphine-3-glucuronide, an opioid receptor inactive metabolite increases TLR4 signaling whereas morphine-6-glucuronide, an opioid receptor active metabolite, does not
- Natural and unnatural isomers of some μ-opioid receptor antagonists, including naloxone and naltrexone, nonstereoselectively inhibit TLR4 signaling
- *In silico* docking predicts that opioids preferentially bind to the LPS-binding pocket in MD-2 rather than TLR4 itself
- Opioid-induced TLR4 signaling increases the expression of inflammatory cytokines and proteins and increases the expression of glial activation markers (Wang et al., 2012).

Preclinical research has implicated TLR4-dependent signaling in a number of behavioral actions of opioids such as hyperalgesia, tolerance, dependence, and addiction (Watkins et al., 2009). However, as will be discussed below, in most instances how an initially inflammatory TLR4 signal is subsequently converted to a behavioral response is presently unclear.

It is worth highlighting, however, TLR2, 8, and 9 and other receptors such as the purinergic P2X4 and α2-adrenocreptor have also been implicated in the response to opioids (Coller & Hutchinson, 2012).

4. HOW IMMUNE MEDIATORS CAN ALTER BEHAVIOR

It is now established that inflammatory mediators (e.g., cytokines and chemokines) contribute to brain function and behavior (Besedovsky & Rey, 2007; Capuron & Miller, 2011; Wrona, 2006). These mediators contribute to learning, memory, neuroplasticity, neurogenesis, and many other homeostatic brain functions (Yirmiya & Goshen, 2011). Additionally, they have the capability to influence neurocircuitries and neurotransmitter systems in turn altering behavior (Adler & Rogers, 2005). The classic example highlighting the importance of inflammatory mediators on producing behavior is the sickness response. Systemic or central administration of cytokines, or innate immune activation via LPS, for example, induces a behavioral response known as sickness behavior (Dantzer, 2001). Sickness behavior is characterized by anhedonia, malaise, anxiety, social withdrawal, fever, pain, and depression (Hart, 1988). Interestingly, similar symptoms are seen during periods of opioid withdrawal in addicts (Shurman, Koob, & Gutstein, 2010). Sickness behavior is hypothesized to benefit an organism by promoting conservation of energy, thereby reallocating energetic resources to recover from illness (Kelley et al., 2003). However, under chronic conditions of immune stimulation, constant elevation of inflammatory cytokines produce profound alterations in neuronal function and behavior and is hypothesized to contribute to the development of psychiatric conditions (Felger & Lotrich, 2013) and possibly contributing to addiction-related neuroadaptations.

In the context of the sickness response, peripherally released inflammatory mediators mediate the translation from an immune response to altered behavior. However, the large size of inflammatory mediators such as cytokines and proteases results in an inability to readily permeate the blood brain barrier. However, peripheral immune mediators can access the CNS by several mechanisms such as entry through the leaky regions in the blood brain barrier (circumventricular organs); active transport across the brain endothelium by specific cytokine transporters; activation of peripheral nerve fibers such as the vagal nerve, which transmit the cytokine signal to specific brain regions; activation of endothelial cells and peripheral macrophages within the cerebral vasculature to produce inflammatory mediators; and recruitment of peripheral immune cells into the brain parenchyma (Quan & Banks, 2007). Within the brain, these inflammatory signals are amplified by microglia and astrocytes. The subsequent release of inflammatory mediators within the CNS can have profound effects on neuronal function. For example, cytokines alter the metabolism of serotonin, noradrenalin, dopamine, and many other neurotransmitters in brain regions associated with addiction (Song, Merali, & Anisman, 1999), such as the amygdala, hippocampus, and nucleus accumbens. In turn, these areas regulate emotion, motivation, and reward (Dantzer, 2001).

It is worth highlighting that the release of inflammatory products within the CNS during acute immune challenges is beneficial. However, under periods of chronic immune activation, these inflammatory mediators alter neuronal plasticity, transmission, and integrity. For example, inflammatory mediators decrease brain-derived neurotrophic factor, an important neurotrophin for regulating neuroplasticity and neurogenesis (Felger & Lotrich, 2013). As previously mentioned, inflammatory mediators regulate glutamate, dopamine, and serotonin neurotransmission. Cytokines, among other inflammatory products activate indoleamine-2,3-dioxygenase, an enzyme present in astrocytes, which catabolizes tryptophan, a precursor for serotonin and dopamine, into kynurenine. Kynurenine is subsequently metabolized into kynurenic acid and quinolinic acid. Both metabolites influence glutamate neurotransmission. For example, quinolinic acid is a *N*-methyl-D-aspartate receptor (NMDA) agonist and inhibits glutamate reuptake by astrocyte excitatory amino acid transporter (EAAT). The subsequent increase in synaptic glutamate contributes to excitotoxicity eventuating in neuronal apoptosis (Felger & Lotrich, 2013). Furthermore, the shift away from serotonin and dopamine to kynurenic acid may contribute to many behavioral characteristics of the sickness response such as anxiety, depression, and anhedonia.

There is evidence to suggest the opioid-induced inflammatory response is responsible for many of the undesirable behavioral effects of opioids such as tolerance, dependence, reduced analgesia, and addiction. This has been inferred from research demonstrating that inhibition of glia or their proinflammatory signaling pathways, transcription factors or cytokines results in enhanced morphine analgesia, attenuated tolerance and dependence, and, importantly, reduced addiction (Hutchinson et al., 2007; Watkins et al., 2005).

5. STRESS AND THE CENTRAL NERVOUS SYSTEM'S INNATE IMMUNE SYSTEM

Stress is an integral component of opioid addiction contributing to the withdrawal and craving stages (Koob, 2013). Similar to immune challenges, stress can activate central immune cells (Frank, Watkins, & Maier, 2011). In brief, a stress response is mediated by the release of corticotrophin-releasing hormone (CRH), a primary regulator of the hypothalamic-pituitary-adrenal axis (HPA axis). In response to stress, CRH activates the HPA axis, releasing glucocorticoids. Glucocorticoids negatively regulate the expression of CRH, and are anti-inflammatory when exposure follows an immune challenge. However, glucocorticoids have recently been recognized as exerting delayed proinflammatory effects via glial priming (Frank et al., 2011). The system's complexity is increased as the HPA axis receives a number of inputs from brain regions associated with addiction, such as the amygdala, locus coeruleus, and the medial prefrontal cortex (Tsigos & Chrousos, 2002). These brain regions are crucially involved in the processing of fear, pain, and anxiety and contribute to the development of withdrawal and craving for opioids (Koob & Le Moal, 2008).

The CNS immune response to stress is hypothesized to consist of two phases. Importantly, the response parallels allostasis and opponent theory, as will be discussed in detail later. The type of immune response is dependent upon the nature of the stress and the duration (acute vs. chronic). In general, the first response to stress is inflammatory, rapid, and transient. For example, inflammatory mediators are elevated after 2 h of stress and persist for several hours after the stressor has concluded (O'Connor et al., 2003). The acute inflammatory response is mediated primarily by glia with several mechanisms proposed to underlie stress-induced activation. For example, an increase in both endogenous danger signals (DAMPS) released by neurons, and neuronal noradrenergic processes from the locus coeruleus are hypothesized to be immunostimulatory (Iwata, Ota, & Duman, 2013; Johnson et al., 2005). The second immune response to stress is sluggish in onset, slow to build up, and decay. The second immune response is suggested to be a sensitization event, in which glia are primed to subsequent stimuli (similar to a MAMP-induced sensitization event) (Frank et al., 2011). For example, systemic LPS administered 24 h following a chronic stressor resulted in a potentiated inflammatory response within the brain. Inhibiting TLR2 and

TLR4 activation prior to the stressor inhibited immune potentiation (Weber, Frank, Sobesky, Watkins, & Maier, 2013). The sensitization event appears to be in part mediated by glucocorticoids. However, the precise mechanism is unknown. Importantly, TLR activation is necessary for both the acute response to and subsequent sensitization response to stress. Behaviorally, stress-induced immune activation has been implicated in the development of psychiatric conditions such as anxiety and depression (Najjar, Pearlman, Alper, Najjar, & Devinsky, 2013) and as will be outlined in future sections (Section 6).

However, communication between stress and neuroimmune cells is not unidirectional as evidence suggests that immune mediators can influence the stress response (Raison, Capuron, & Miller, 2006). For example, heat-shock proteins (some of which are TLR4 agonists) are released during periods of stress and initiate an immune response culminating in the induction of the NLRP3 inflammasome—a multimeric protein complex responsible for the maturation and release of IL-1β (Latz, Xiao, & Stutz, 2013). Induction of IL-1β expression and release can subsequently induce the expression of additional inflammatory mediators. Under acute stress conditions, the glucocorticoid response can inhibit ongoing inflammatory signaling. However, under prolonged immune activation or chronic stress, the glucocorticoid response can no longer attenuate this inflammatory response, resulting in an exacerbated stress response (Raison et al., 2006). Importantly, these cytokines can further stimulate the production of CRH, thereby worsening the stress response. Elevated CRH levels are strongly implicated in many behavioral responses of addiction, such as dysphoria, fear, and anxiety (Koob, 2013).

As such, it could be hypothesized that the chronic stress of drug addiction, which typically occurs during periods of abstinence, could be creating a sensitized neuroimmune environment. Subsequent exposure to opioids could then produce an exacerbated immune response within the brain, potentially driving addiction to a greater extent.

6. ADDICTION

It is important to highlight that drug addiction is clinically and neurobiologically distinct from an initial drug-taking experience. While these two events share similar neurobiological substrates, addiction is a complex, relapsing disorder resulting from neuroadaptations over prolonged drug exposure. These adaptations are complex and persist beyond acute withdrawal and are thought to involve environmental (stress, learning, and conditioning) and genetic elements (predisposition to drug taking) (Koob & Volkow, 2009). Furthermore, in humans, most drug users do not become addicts with a similar scenario observed in animals. Stable drug intake can be observed within animals without indications of addiction even in paradigms designed to recapitulate addictive scenarios (limited access, intravenous administration). This reflects the complex etiology of drug addiction, which is influenced by genetics, drug-use history, and the environment (Koob, 1997). Importantly, many of the mechanistic- and system-based schemas for addiction were developed in a model devoid of glia. Nevertheless, it is apparent that these theories have served, and continue to serve, as foundational intellectual stepping-stones upon which modern addiction research has been based. Therefore, to progress in our understanding of addiction, we need to integrate the emerging importance of glia with the current established models of addiction (as outlined in Section 6.1).

Both the International Classification of Diseases 10th Revision (ICD-10) and the Diagnostic and Statistical Manual of Mental Disorders V (DSM-V) recognize drug addiction as a chronic disorder which is characterized by an impaired control over, and an increased craving for, taking the drug of abuse, a withdrawal syndrome upon cessation, tolerance—with a larger dose required to achieve the desired psychological effect, a disproportionate time seeking and consuming the drug and persistence of drug-use despite negative consequences (The Diagnostic and Statistical Manual of Mental Disorders V., 2013; World Health Organization, 1993). Drug addiction is thought to reflect a shift from impulsive to compulsive behavior (Koob & Le Moal, 2001). Impulsivity reflects an increased sense of tension or arousal before committing the act, with intense pleasure and gratification at the time of committing the act followed by guilt and self-regret. Impulsivity is associated with positive reinforcement (the stimulus increases the probability of a response). By contrast, compulsivity requires anxiety and stress before committing the compulsive behavior, with relief from the stress after the event. Compulsivity reflects negative reinforcement (removal of the stimulus will increase the probability of a response) (American Psychiatric Association, 1994, 2013). The shift from impulsivity to compulsivity reflects a change in motivation state from positive to negative reinforcement and is comprised of three interrelated stages of binge/intoxication, withdrawal/negative affect, and preoccupation/craving. These three stages interact with one another, become more intense, and ultimately lead to addiction (Koob & Le Moal, 2001).

6.1. Opponent theory and allostasis

Several theories have been applied to drug addiction, often reflecting different aspects of the disorder. And, although many theories account for some aspects of drug addiction, they cannot in their entirety account for the entire disorder. Therefore, while we acknowledge the presence and importance of other theories, for example, incentive sensitization (Robinson & Berridge, 1993) and associative-learning disorder (Di Chiara, 1999), we have chosen to examine and apply one leading theory of addiction.

The opponent-process theory (Solomon & Corbit, 1974) is particularly relevant to drug addiction. Opponent-process theory states, hedonic, affective, or emotional states are automatically opposed by CNS mechanisms, which reduce the intensity of these states, both pleasant and aversive in order to maintain homeostasis. In brief, the theory consists of two processes termed A and B. A-process consists of the initial positive or negative hedonic response to a stimuli. The degree of response correlates with the quality and duration of the stimuli and exhibits tolerance. B-process occurs after the A-process has finished, is sluggish in onset, is slow to build up and decay, and is opposite in response to A. However, if the stimuli is consistently repeated, A-process becomes weaker, and the B-process becomes stronger and long lasting (allostasis) (Koob, 1997) (Fig. 6.2).

In the context of addiction, an opioid produces an initial hedonic experience (euphoria), which serves as a positive reinforcer (A-process). However, the user subsequently experiences negative hedonic affect (anhedonia and dysphoria), which acts a negative reinforcer (B-process) (Solomon & Corbit, 1974). Koob furthered this concept by theorizing as occasional drug-use transitions into a compulsive habit; this homeostatic process becomes dysregulated and breaks. The initial euphoric sensation begins to weaken (tolerance, desensitization), and the negative hedonic experiences occur more rapidly, is enduring, more intense, and fails to return to base line, creating a new allostatic state. Consequently, the body must actively compensate for the dysregulation of the reward pathway by creating a new-set point that is now below the original reward value (allostatic point). Collectively, these processes alter the overall drug-taking experience. What was once considered rewarding to take is now taken in an attempt to return to a baseline state. Continual resetting contributes to the pathology of addiction (Koob, 1997; Koob & Le Moal, 2001).

Two key cellular processes are hypothesized to underlie allostasis in opioid addiction—within and between-system adaptations (Koob, 1997). In the presence of constant drug exposure, a within-system adaptation

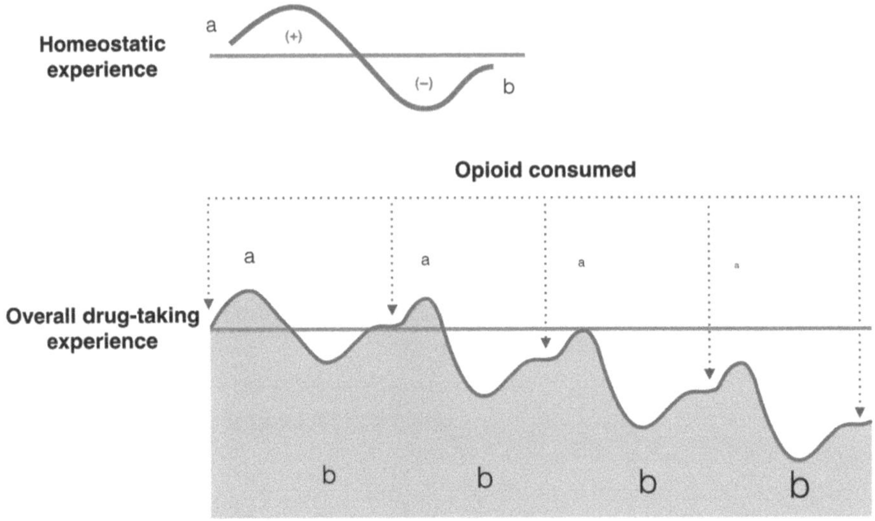

Figure 6.2 The transition of opponent-process theory to allostasis. The opponent-process theory (top) states that an affect response consists of two opposing process termed A and B. In the context of opioid addiction, A-process represents the initial positive hedonic (euphoric) sensation upon drug intake. Process-B, by contrast, represents the negative hedonic experience (dysphoria), which occurs after A; it is sluggish on onset and is enduring. The affect response (mood) is the sum of these two processes and defines the overall drug-taking experience. However, as drug addiction develops (bottom) and there becomes less time between opioid uses; the CNS cannot maintain this homeostatic state. A-process becomes weaker and B-process becomes more intense and long lasting (feed-forward mechanism). B-process does not return to the initial homeostatic level as such the body recreates a new-set point around which to operate (allostatic state). As addiction progresses, the body continually attempts to achieve stability through change (allostasis), ultimately, however, this process becomes more damaging (Koob & Le Moal, 2001).

refers to the brain's attempt to neutralize the drug's effect by altering the primary responding element which is producing the desired effect such as reward. In the context of opioid addiction, a within-system adaptation is hypothesized to include cellular changes within the mesolimbic dopamine pathway (primary responding element). A within-system adaptation contributes to drug tolerance—a higher dose and drug-taking frequency is required to achieve the same effect. However, the persistence of the within-system adaptations in the absence of a drug may contribute to withdrawal and craving. By contrast, a between-system adaptation is a different cellular system, which has opposing effects to the primary response element (antireward). A between-system adaptation is aimed to limit the reward by induce a stressful or dysphoric response. However, in the absence of opioids, the system still persists, exacerbating withdrawal, and craving.

7. INTEGRATING ADDICTION NEUROSCIENCE WITH IMMUNOLOGY

Koob's theory of addiction provides an excellent understanding of the neuronal alterations underlying addiction. However, in its current state it cannot accurately portray the emerging importance of the neuroimmune system and central immune signaling in drug addiction. Furthermore, it does not account for the way pathways interact with one another on a molecular level. Therefore, in order to create a neuroimmune perspective of addiction, we must borrow and merge tenants from Koob's neuroscience perspective and an important immunology tenant proposed by Nish and Medzhitov (2011), as described below.

Similar to how different brain pathways work together to form complex behavior (e.g., the interaction between CRF and noradrenalin to create withdrawal behavior), innate host defense pathways also work together to produce controlled and regulated immunological signaling. In brief, innate host defense pathways consist of microbial sensors, their signaling pathways, and the effector mechanisms they induce. TLRs are one of the best-characterized microbial sensors, which as previously mentioned recognize PAMPs, DAMPs, and XAMPs. TLRs demonstrate functional redundancy such that, pathogens can be detected by more than one receptor. For example, *Helicobacter pylori* (*H. pylori*), a Gram negative bacterium, is detected by both TLR2 and TLR5 (Smith et al., 2003). Activation of these receptors results in signal transduction via intracellular messengers, which culminate in the activation of several transcription factors such as NF-κB, AP-1, and interferon regulatory factors. These transcription factors result in the induction of inflammatory and innate immune responses (effector mechanisms) such as cytokines, chemokines, and inflammatory peptides (Kawai & Akira, 2007).

Host defense pathways do not act independently from one another and can engage in three types of interactions: cooperation, complementation, and compensation (Fig. 6.3) (Nish & Medzhitov, 2011). These types of molecular interactions are not limited to immunological signaling and can be applied to interactions occurring within the neurocircuitries governing addiction. For example:

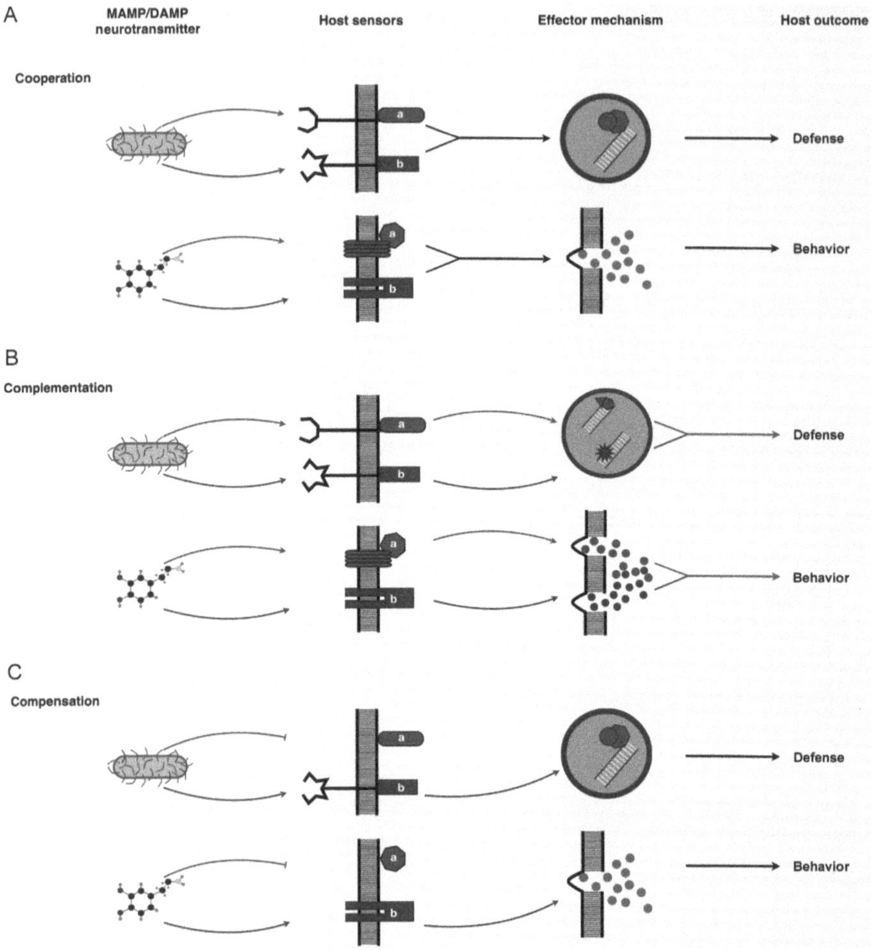

Figure 6.3 Interactions between host defense pathways. Three forms of interactions exist between host defense pathways: cooperation, complementation, and compensation. (A) Cooperation: an effector mechanism is optimally induced when two pathways are activated simultaneously. (B) Complementation: two host pathways engage individual effector mechanisms that synergise with each other to form one functional unit of host defense. (C) Compensation: only exists when one host defense pathway is deficient. The remaining host defense pathway can compensate to produce a functional effector mechanism. Compensation can occur at any stage of the host defense pathway. Importantly, this tenant governing host defense pathways can be applied to the neurotransmission systems during addiction. MAMP, microbe-associated molecular pattern; DAMP, danger-associated molecular pattern; A and B, host sensors.

- Cooperation: a molecular pattern activates two- or more receptors and signaling pathways that induce the same effect mechanism. For example, TNF-α and IL-1β cooperate to optimally induce IL-6 secretion (McGee, Bamberg, Vitkus, & McGhee, 1995). Similarly, CRF acts via $CRFR_1$ and $CRFR_2$ to induce anxiety behavior (Risbrough, 2004).
- Complementation: a molecular pattern activates two distinct pathways which complement each other to form a functional unit (synergy). For example, an infection can induce both an innate and adaptive immune response, collectively generating host defense. Within the limbic system, cooperation exists between both dynorphin and GABA which optimally inhibit dopamine release when acting together compared to either neurotransmitter alone (Krebs, Gauchy, Desban, Glowinski, & Kemel, 1994).
- Compensation: is a form of functional redundancy. If one of two pathways becomes inactivated (either the receptor or signaling pathway), the remaining pathway can compensate for the missing signal. For example, viral RNA or DNA can be recognized by either TLRs or Rig-1-like receptors (Szabo & Rajnavolgyi, 2013). If either sensor is inactivated, an antiviral response can still be mounted. Within the limbic system compensation occurs in dopamine deficient mice, which are still capable of producing morphine-conditioned place preference (Hnasko, Sotak, & Palmiter, 2005). This indicates another pathway, such as endogenous opioids, is compensating for dopamine.

The interactions exhibited by host defense pathways increase the heterogeneity of an immune response resulting in extremely diverse and individually tailored responses. Similarly, the existence of these molecular interactions within the neurocircuitries of addiction increases the drug-induced behavior allowing for individual drugs to cause unique behavioral responses.

Koob's (Koob, 1997) theory of drug addiction is composed of three interrelated phases: binge and intoxication, withdrawal and negative affect, and preoccupation and craving. The aim of this section is to integrate current neuroimmune interactions at both a global (cellular) and local (molecular) level with current neuronal aspects of addiction. We have selected to integrate both glia and TLR4 into Koob's theory of drug addiction as it provides a clear scaffold for incorporating central immune signaling involvement in drug addiction.

7.1. Binge and intoxication

The binge and intoxication stage of addiction is governed primarily by the rewarding (euphoric) properties of opioids (positive reinforcement, A-process). This phase is characterized by the impulsive desire to consume opioids (Koob & Le Moal, 2001). Research over the past 25 years has demonstrated a pivotal role of the midbrain's mesolimbic dopamine pathway in the rewarding effects of opioids (Di Chiara & Imperato, 1988). The mesolimbic pathway consists of the ventral tegmental area, nucleus accumbens, amygdala, hippocampus, and the bed nucleus of the stria terminalis (BNST). Of particular importance to reward are the ventral tegmental area and the nucleus accumbens. Electrophysiological studies have consistently reported that intravenous, systemic, or local administration of morphine increases the firing rate of dopaminergic neurons originating in the ventral tegmental area (Gysling & Wang, 1983; Matthews & German, 1984). The increase in firing rate is associated with an increase in the release of dopamine within the nucleus accumbens that is thought to contribute to the rewarding effects of opioids. The molecular mechanisms underlying the increased firing of dopaminergic neurons relates to a disinhibitory effect mediated by GABAergic interneurons (Johnson & North, 1992). Within the ventral tegmental area, GABAergic interneurons express high levels of μ-opioid receptors. Direct binding of exogenous opioids to the μ-opioid receptors on the interneurons results in membrane hyperpolarization, decreasing their activity, and the subsequent release of GABA (Matsui & Williams, 2011). The reduction of inhibitory tone onto the dopaminergic neurons increases their activity thereby increasing the release of dopamine within the nucleus accumbens. However, opioids are unique in terms of their reward pathway as evidence implicates dopamine-independent pathways in reward-like behaviors (Bozarth & Wise, 1981; Phillips & LePiane, 1980). For example, mice deficient in dopamine (Th−/−; DbhTh/+) develop conditioned place preference to morphine when caffeine or L-DOPA are provided shortly after morphine administration (Hnasko et al., 2005), an effect potentially mediated by the brain's endogenous opioid peptides (Koob, 1992).

Interestingly, immune involvement in the binge and intoxication stages of addiction can be seen at both a global (glia) and local (TLR4) level. For example, administration of glial attenuators propentofylline or minocycline, significantly reduced morphine-induced conditioned place preference, with control experiments illustrating glial attenuators are not inherently aversive (Hutchinson et al., 2008; Narita et al., 2006) Importantly, microdialysis revealed that ibudilast significantly reduces the concentration of extracellular dopamine within the shell of the nucleus accumbens, thereby providing a direct link between glia and reward (Bland, Hutchinson, Maier, Watkins, & Johnson, 2009). Interestingly, a similar effect can be seen at a more local level. Hutchinson et al. (2012) have clearly demonstrated the importance of TLR4 in generating opioid-induced reward. In a series of experiments they showed that mice deficient in TLR4 or MyD88 (a TLR4 adaptor molecule) did not develop conditioned place preference to oxycodone; (+)-naloxone, a nonstereoselective pharmacological TLR4 antagonist (a) inhibited morphine-induced conditioned place preference, (b) inhibited remifentanil self-administration, and (c) suppressed morphine-induced dopamine release in the shell of the nucleus accumbens. Importantly, the experiments used natural, semisynthetic, and fully synthetic opioids, clearly illustrating that opioid-induced TLR4 signaling is not limited to morphine. The results suggest that TLR4 is cooperating with traditional neuronal process of reward (thereby contributing to A-process of opponent theory). Opioids are preferentially activating different receptors on glia and neurons, which in turn act via different signaling pathways. However, the mechanism underlying this phenomenon remains to be determined. The end products of glial and neuronal effects all converge to increase the release of dopamine with the shell of the nucleus accumbens (Fig. 6.4). The neuroanatomical area in which these immune mediators are released and act remain to be fully elucidated.

Several mechanisms underlying TLR4-induced opioid reward have been proposed. For example, Narita et al. (2006) have demonstrated that administration of astrocyte-conditioned media (ACM) directly into the nucleus accumbens or intracingulate cortex, key areas relating to the development of opioid addiction, potentiates morphine-induced conditioned place preference. Furthermore, this ACM response was area specific, as intracorpus striatum administration of ACM did not influence reward behavior. The ACM induced several inflammatory mediators, for example, MCP-5. It was hypothesized that these inflammatory mediators would function akin to neurotransmitters and act upon neurons to potentiate reward. Interesting nucleus accumbens administration of microglial conditioned media failed to potentiate reward behavior. This supports observations by Hutchinson et al. (2009) who demonstrated that microglia are not activated by morphine within the nucleus accumbens. However, these data contrast those of Schwarz and colleagues (Schwarz & Bilbo, 2013; Schwarz, Hutchinson, & Bilbo, 2011) who found increased activation under similar

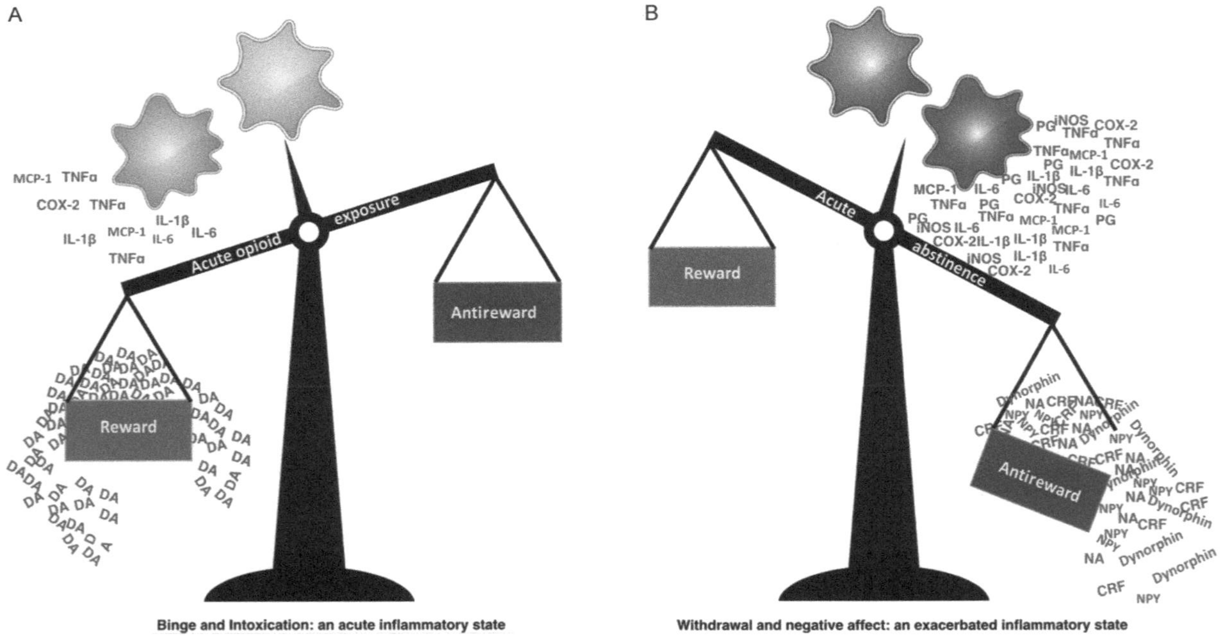

Figure 6.4 Neuroimmune perspective of binge and intoxication and withdrawal and negative affect. (A) Compulsive administration of opioids during the binge and intoxication stage of addiction results in the activation of glial responses within ventral tegmental area and the nucleus accumbens. Opioid-induced glial responses increase the expression and release of inflammatory mediators such as IL-1β and TNF-α. It is hypothesized that these inflammatory mediators act in a similar manner to traditional neurotransmitters (neurokines) which enhance the activity of dopaminergic neurons, thereby increasing the release of dopamine within the nucleus accumbens. The cooperation between these inflammatory mediators and dopamine is thought to drive opioid-induced reward to a greater extent. (B) Repeated cycles of binge and intoxication of opioids sensitize glia. In the absence of opioids, these cells remained in a primed state capable of responding to minor perturbations in homeostasis. During withdrawal, there are high levels of stress hormones, which are hypothesized to interact with glia, generating an inflammatory

circumstances. Inflammatory mediators such as those found in the media act via several mechanisms to modulate neuronal function. For example, TNF-α alters the composition of AMPA receptors expressed on postsynaptic neurons (Stellwagen, 2005) and produces a sustained downregulation of the surface expression of $GABA_A$ receptors collectively reducing inhibitory tone (Pribiag & Stellwagen, 2013). In addition, TNF-α is a key factor regulating synaptic scaling and neurogenesis (Savin, Triesch, & Meyer-Hermann, 2009). Furthermore, prolonged administration of chemokine, CCL2 (MCP-1), increases burst and pacemaker firing in dopaminergic neurons via closure of selective potassium channels. This in turn increases the extracellular release of dopamine in the substantia nigra resulting in an increase in locomotor activity (circling behavior) in rats (Guyon et al., 2009). Alternatively, Frank et al. (2011) suggest that opioid-induced central immune signaling molecules, such as IL-1β, downregulate the expression of astrocytic glutamate-reuptake transporter, EAAT. Consequently, the extracellular concentration of glutamate would increase enhancing glutamatergic neurotransmission. If this build up occurred in the ventral tegmental area or nucleus accumbens, dopaminergic neurotransmission would be augmented.

7.2. Withdrawal and negative affect

From a clinical perspective, opioid withdrawal is one of the most powerful factors driving addiction (Kosten & George, 2002). Withdrawal contributes to opioid addiction, as repeated dosing is required to avoid and alleviate the withdrawal symptoms (B-process of opponent theory). Consequently, tolerance and addiction develop (Koob, 1997). Withdrawal and negative affect manifests as chronic irritability, pain, anhedonia, depression, anxiety, dysphoria, and hyperalgesia (Cushman & Dole, 1973). Importantly, many of

response. This inflammatory response is greater and more prolonged than the response that occurred during the initial reward phase. As such, this inflammatory response may be viewed more akin to a peripheral immune challenge capable of generating high levels of inflammatory mediators within the CNS. The outcomes of this are twofold. First, the high levels of inflammatory mediators act in a feed-forward mechanism to induce the expression of more stress hormones. This may contribute to the high levels of stress and anxiety observed during withdrawal. Second, central immune signaling may be capable of creating a form of opioid-induced sickness response that contributes to some of the symptoms observed during withdrawal such as depression, anhedonia, and dysphoria. Furthermore these inflammatory mediators may contribute to the development of tolerance within the reward pathway. Collectively, central immune signaling is cooperating with current antireward neurons to exacerbate withdrawal and negative affect.

these symptoms are similar to those experienced during cytokine- or innate immune-induced sickness behavior. Similarly, animals exhibit an increase in anxiety- and dysphoric-like behaviors and an increase in reward thresholds (which correlates with an escalation in drug intake). As such there is strong predicative validity between the two species (Morgan & Christie, 2012).

Opioid withdrawal typically accompanies prolonged drug use and it results from repeated cycles of neuroadaptations within the mesolimbic dopamine pathway resulting in tolerance as well as recruitment of the "antireward" pathways. Antireward is based on a concept that brain systems are in place to limit reward triggered by excessive activity within the reward system (B-process) (Koob & Le Moal, 2008) and manifests as the withdrawal and negative affect experienced by opioid addicts. Several anatomical sites have been proposed which mediate the antireward properties of opioids such as the locus coeruleus, the extended amygdala (BNST, central nucleus of the amygdala (CeA), and the shell of the nucleus accumbens; Koob & Le Moal, 2008) and the periaqueductal gray (Ouyang et al., 2012). It is important to highlight that these brain regions are involved in arousal, stress, fear, anxiety, and the emotional component of pain.

The association between glia, TLR4, and withdrawal has predominately been studied using spontaneous or precipitated naloxone-induced withdrawal. Similar to reward, the glial attenuators minocycline and ibudilast decreased the symptoms of naloxone-precipitated withdrawal in rats, which was accompanied by an immediate decrease in accumbal dopamine (Bland et al., 2009; Hutchinson et al., 2009). Furthermore, ibudilast significantly attenuated morphine withdrawal-induced increases in GFAP and CD11b and the induction of proinflammatory cytokines IL-1β, MCP-1, MIP-2 in the periaqueductal gray and amygdala, two key brain regions involved in opioid withdrawal (Hutchinson et al., 2009). Similarly, the genetic (TLR4 K.O.) or pharmacological ((+)-naloxone) attenuation of TLR4 significantly reduces the somatic symptoms of withdrawal in morphine-dependent rodents following (−)-naloxone-precipitated withdrawal (Liu, Coller, Watkins, Somogyi, & Hutchinson, 2011). Therefore, glia and TLR4 are actively contributing to the development of negative affect of opioid addiction (B-process).

The finding that glia and TLR4 contribute to both hedonic (A-process) and dysphoric (B-process) sensations associated with opioid addiction seems paradoxical. How can an initially rewarding process subsequently become altered so to assist in developing negative affect? Parallels can be drawn between this behavior and a stress response. For example, an initial stress response may prove beneficial to an organism. However, chronic stress often proves detrimental (hormesis). Similarly, an acute inflammatory response to opioids may prove beneficial. However, a prolonged and chronic response may be detrimental, as will be outlined in the next section.

7.2.1 Within system adaptations—A desensitized reward pathway

As drug use becomes more frequent, the brain attempts to counteract these effects. As such, the same brain regions and molecular process mediating positive reinforcement may also be contributing to withdrawal and the transition to addiction (within-system adaptations). During withdrawal and subsequently craving, there is an increase in brain reward threshold and, as a consequence, larger and more frequent doses are required to achieve the same desired response. This is attributable to receptor desensitization and tolerance at both a molecular and cell level. In the case of opioids, long-term exposure decreases the cell surface expression of the μ-opioid receptor, reduced receptor coupling to g-proteins, and dysregulation of intracellular signaling pathways (see Dang & Christie, 2012; Williams, Ingram, Henderson, & Chavkin, 2012, for review). Interestingly, these same actions are thought to underlie analgesic tolerance. Collectively, these actions reduce the ability of opioids to inhibit the GABAergic tone. Consequently, the same dose of an opioid can no longer induce VTA firing and dopamine release. However, in the absence of the opioids, this inhibitory tone remains, reducing basal dopamine levels. This is hypothesized to contribute to the negative affect and withdrawal behaviors (Koob, 2013).

Interestingly, glia have been implicated in the development of analgesic tolerance (Eidson & Murphy, 2013). Both analgesic and reward-induced tolerance are hypothesized to occur via a similar molecular mechanism as outlined above. Therefore, as glia and their inflammatory mediators are known to contribute to the development of analgesic tolerance (Watkins et al., 2009), it is possible that the same glial-derived tolerance mechanisms are also occurring under the context of addiction (as reviewed by Hutchinson et al., 2011).

7.2.2 Between system adaptations—Recruitment of antireward pathway

In addition to the decreased activity within the mesolimbic dopamine pathway observed during withdrawal, the brain's antireward systems (CeA, locus coeruleus, and periaqueductal gray) are activated. These systems have been hypothesized to overcome the chronic presence of the drug and restore homeostasis. However, in the absence of the drug, these systems remain active thereby producing the features of withdrawal (aversive, anhedonic, dysphoric, or stress-like behavior). For example, CRF, a key mediator of both the brain stress system and the HPA axis, mediates the manifestation of anxiety and aversive-like behavior in rodents (Shekhar, Truitt, Rainnie, & Sajdyk, 2005). During opioid withdrawal, CRF is elevated in the amygdala and is hypothesized to contribute to the withdrawal symptoms experienced by drug users (Koob, 2013). Mice deficient in CRF-1 receptor exhibit decreased signs of morphine withdrawal. Additionally, CRF-1 or -2 receptor antagonism decreases morphine withdrawal behavior (Navarro-Zaragoza, Nunez, Laorden, & Milanes, 2010; Navarro-Zaragoza et al., 2011). These findings are furthered as suppression of amygdala CRF reduces conditioned place aversion to morphine withdrawal (Heinrichs, Menzaghi, Schulteis, Koob, & Stinus, 1995).

As previously mentioned, bidirectional communication exists between stress and glia. For example, chronic stress upregulates TLR4, its coreceptor MD-2 and downstream inflammatory mediators including NF-κB, IL-1β, COX-2, and PGE2 within the prefrontal cortex of mice coinciding with elevated levels of corticosterone in rats (Gárate et al., 2011, 2013). Continual exposure to opioids during the binge and intoxication stage of addiction has primed or sensitized glial cells (Schwarz et al., 2011). These cells will readily react to minor perturbations in their microenvironment. During periods of withdrawal, levels of CRF are high. It is therefore hypothesized that CRF reactivates glia to an inflammatory state. The subsequent release of inflammatory mediators then feedback stimulating the production of more CRF (Kageyama et al., 2010). This feed-forward mechanism between inflammatory mediators and CRF may contribute to some of the symptoms observed during withdrawal. As such, TLR4 and glia are likely cooperating with CRF to generate a stress response (B-process) (Fig. 6.4).

As previously mentioned, withdrawal and negative affect share many similar behavioral symptoms to the cytokine-induced sickness response. This raises the possibility that glially derived cytokines may be contributing to the symptoms of withdrawal (Hutchinson et al., 2009). Similar to a sickness response, inflammatory cytokines in key brain regions mediating stress, pain, and depression are elevated following morphine withdrawal (Hutchinson et al., 2009). Furthermore, both the sickness response and withdrawal symptoms are attenuated *in vivo* when TLR4 antagonists or glial attenuators are administered prior to the event (Henry et al., 2008; Hutchinson, Zhang, et al., 2008). Recent research highlights the importance of TNF-α, an inflammatory mediator whose expression is regulated by TLR4 activation. Previous studies have illustrated opioid-induced TLR4 activation increases TNF-α within the brain which has the potential to influence synaptic plasticity and transmission (Hutchinson et al., 2009). Hao et al. (2010) demonstrated morphine withdrawal-induced astrocyte activation which causes the subsequent release of TNF-α into the periaqueductal gray. The rise in TNF-α exacerbated withdrawal-like behaviors. Importantly, these results were verified as a similar behavioral outcome was achieved by applying exogenous TNF-α to naïve mice. Similar behaviors are observed with TNF-α and LPS-induced sickness behavior (Dantzer, O'Connor, Freund, Johnson, & Kelley, 2008; Palin et al., 2007). However, the precise mechanism underlying the TNF-α effect remains to be determined.

Furthermore, the withdrawal and negative affect stage is characterized by alterations in brain plasticity. Glially derived opioid-induced TNF-α is hypothesized to contribute to alterations in brain plasticity. For example, TNF-α binds to TNFR1 receptor expressed by neurons. This results in the subsequent activation of NMDA receptor and ERK. ERK activation results in the phosphorylation of CREB, a transcription factor associated with brain plasticity and is activated during periods of drug addiction. Similarly, activation of NMDA receptors results in calcium influx, activating calmodulin and adenylyl cyclase, leading to activation of protein kinase A, a regulator of both CREB and FOS (Ouyang et al., 2012). These findings suggest that glia are pivotally involved in opioid-induced neuroplasticity, further contributing to the pathogenesis of addiction.

The association between glia and TLR4 in withdrawal is interesting, given that it is also implicated in the rewarding effects of opioids. The different effects can be explained, however, by drawing parallels to the opponent theory and allostasis. During the early phases of addiction, opioid-induced glial activation can be viewed as cooperating with the reward pathway (A-process). However, after repeated rounds of opioid use, glia are primed to over produce inflammatory mediators in response to opioids, which may contribute to tolerance observed in the reward pathway. Additionally, these inflammatory mediators are now interacting with the newly activated stress pathway (B-process). As such, the subsequent inflammatory response will be complementing the antireward pathway, while continue to desensitize the reward pathway. Glia and TLRs can be viewed as a within-system which cooperates with neuronal mechanisms to result in the desensitization of the primary responding element and complements the brain's antireward system.

7.3. Preoccupation and craving

A distinguishing feature of opioid addiction is the high rates of relapse following periods of self-imposed or forced abstinence which is commonly observed among opioid addicts (Bossert, Ghitza, Lu, Epstein, & Shaham, 2005). Craving typically presents before the symptoms of withdrawal occur. However, craving also exists after long periods of abstinence indicating long-term plastic changes in brain regions governing motivation. Craving and relapse to opioids is associated with the anticipation of their rewarding effects and the removal of the aversive effects of withdrawal (Shurman et al., 2010). Preclinical and clinical studies have clearly illustrated that drug craving or relapse to drug taking can occur as a result of exposure to the self-administered drug, drug-associated cues, or context and stress.

It is important to highlight that during this stage of addiction, there has been integration (neuroplasticity) in brain regions mediating reward, stress, and memory and a loss from drive to the prefrontal cortex. Collectively, these may be a trigger for impulsivity and contribute to the loss of control associated with drug use. Due to the within- and between-system adaptations (sensitization of the reward pathway and recruitment of the stress pathway), individuals may become particularly sensitive to drug priming, drug cues, and stressors which may create drug-craving behavior observed during the preoccupation and anticipation of the addiction cycle (Koob, 2013).

Research over the past 20 years has highlighted key brain regions involved in the reinstatement of drug craving. Evidence suggests drug-induced reinstatement is dependent on both dopaminergic neurons from the VTA to the nucleus accumbens, and glutamtergic neurons from the medial prefrontal cortex to the nucleus accumbens and ventral pallidum (Bossert, Marchant, Calu, & Shaham, 2013). Cue (discreet, contextual, or discriminative)-induced reinstatement involves the basolateral amygdala projections to the medial prefrontal cortex (Feltenstein & See, 2009; Rogers, Ghee, & See, 2008). Stress-induced reinstatement is thought to occur via the BNST and the CeA with CRF and noradrenalin playing a pivotal role (Koob & Volkow, 2009). In addition, CRF-sensitive serotonergic projections from the dorsal raphe nuclei also play a role in stress-induced reinstatement (Lunden & Kirby, 2013).

Few studies have examined the influence of central immune signaling on the preoccupation and craving stage of opioid addiction. Schwarz et al.

(2011) illustrated morphine-induced glial activation within the nucleus accumbens, as determined by increased expression of cell surface markers, cytokines, and chemokines, predicts future reinstatement to morphine-conditioned place preference. Importantly, this study illustrated reduced early life stress (increase in the quantity and quality of maternal care), increases IL-10, attenuates morphine-induced glial activation, and the subsequent reinstatement to morphine-conditioned place preference in adulthood—an effect mimicked by the glial activation attenuator, ibudilast. Furthermore, using a rat model of heroin-induced relapse, Theberge et al. (2013) demonstrated that chronic administration of (+)-naltrexone, a TLR4 antagonist, for 13 days during a craving period significantly attenuated relapse to heroin. However, chronic delivery of (+)-naltrexone or (+)-naloxone had no effect on the acquisition or maintenance of heroin self-administration—contrasting Hutchinson et al. (2012) findings. Collectively, these results suggest that glia and TLR4 are required for the craving, preoccupation, and relapse to opioids.

Craving is driven by the memory of opioid-induced reward and negative emotional states during periods of protracted abstinence (Koob & Le Moal, 2008). As previously discussed, both glia and TLR4 are required for the generation of the negative emotional state and the rewarding effects of opioids. Furthermore, the nature of Theberge et al. (2013) and Schwarz et al. (2011) experiments (reinstatement models) clearly demonstrated that TLR4 and glia are involved in generating opioid reward-based contextual memories. Interestingly, TLR4's downstream immune molecules are intricately involved in the memory facilitation and consolidation process (Yirmiya & Goshen, 2011). For example, IL-1β is involved conditioned-fear induced, spatial memory, and context-specific memory facilitation. Additionally, NF-κB is required for reconsolidation of morphine reward memory and the reconsolidation of fear memories (Si et al., 2012; Yang et al., 2011). Importantly, NF-κB's reconsolidation of fear memory was constrained to the amygdala. Therefore, during periods of protracted abstinence, it is hypothesized that glial-derived central immune signaling may be consolidating and reactivating opioid-reward memories and creating negative emotional states. This in turn, drives opioid-craving to a greater extent. If addicts were overcome by craving and relapsed, the primed glia will rapidly reactive in brain regions associated with reward and negative affect. Thereby, restarting the cycle of drug addiction (Fig. 6.5).

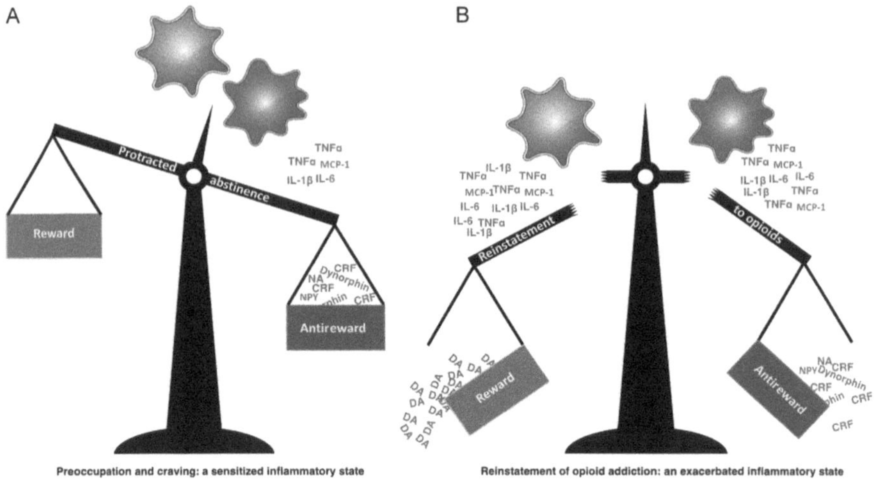

Figure 6.5 Neuroimmune perspective of craving, preoccupation, and reinstatement. (A) During periods of protracted abstinence, glial cells remained in a primed state. Similar to withdrawal, stress hormones reactivate glia in brain regions governing motivation and seeking behavior, for example, the nucleus accumbens. Reactivated glia may release inflammatory mediators consolidating or reactivating memories of opioids—increasing the desire to consume the drug. The desire to consume the drug is associated with high levels of stress and activation of the brain's antireward system (negative affect) that further increases the craving behavior. (B) Reinstatement to opioids breaks the system. Opioid-induced glial activation is concurrently driving reward while increasing the activity of the brain's antireward pathway creating negative affect. Collectively, central immune signaling is cooperating with all circuits underlying addiction: perpetuating the cycle.

8. CONCLUSION

This chapter has clearly demonstrated that glial-derived central immune signaling contributes to the binge and intoxication, withdrawal and negative affect, and preoccupation and craving stages of opioid addiction. However, it must be stressed that glia do not function alone in creating this complex behavior and either complement or cooperate with existing neuronal circuits. However, glia's importance cannot be underemphasized. The above sections clearly illustrated the importance of glia and TLR4 in A- and B-processes of opponent theory and allostasis. This finding is of significance as it demonstrates glial- or TLR4-based interventions are disease modifying. It has the potential to alleviate the symptoms (withdrawal and craving behavior) and the underlying pathology (tolerance,

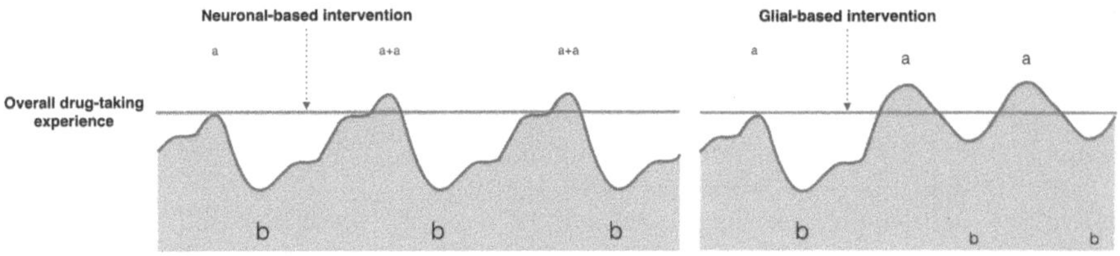

Figure 6.6 Managing parts or treating the whole: a central immune signaling perspective of allostasis. It is now apparent that both glia and TLR4 contribute to opioid addiction. The hedonic (A-process) and dysphoric (B-process) experiences of opioid addiction involve a complex interaction between TLR4, glia, and neurons. Traditional neuronally based interventions for treating opioid addiction inhibit many key events relating to hedonic and dysphoric experiences. However, they fail to treat the underlying glial-mediated events, which may contribute to the reinstatement of addiction. A glial-based intervention targets both A- and B-processes of opponent theory and in turn drug addiction.

reward-deficiencies, and reward-memories) of opioid addiction. By contrast, current neuronally based therapies are mostly symptom modifying and do not address the underlying processes of addiction and allostasis (Fig. 6.6). As such, the high rates of relapse to opioids using neuronally targeted therapies are not surprising. This research is still in its infancy and many questions still remain. Research must clarify how central immune signaling contributes to addiction, and whether it is stage and drug specific. Furthermore, biomarkers indicating which patient may benefit from immune-based therapies needs to be determined. However, the future is bright for glial-based therapies as clinical trials are currently underway evaluating the effects of ibudilast on oxycodone self-administration and opioid withdrawal (trial identifier: NCT01740414 and NCT00723177, respectively).

REFERENCES

Adler, M. W., & Rogers, T. J. (2005). Are chemokines the third major system in the brain? *Journal of Leukocyte Biology*, *78*(6), 1204–1209.

Akira, S., & Takeda, K. (2004). Toll-like receptor signalling. *Nature Reviews Immunology*, *4*(7), 499–511.

American Psychiatric Association. (1994). *Diagnostic and Statistical Manual of Mental Disorders* (4th ed.). Washington, D.C.: American Psychiatry Press.

American Psychiatric Association. (2013). *Diagnostic and Statistical Manual of Mental Disorders* (5th ed.). Washington, D.C.: American Psychiatric Association.

Banati, R. B. (2003). Neuropathological imaging: In vivo detection of glial activation as a measure of disease and adaptive change in the brain. *British Medical Bulletin*, *65*, 121–131.

Ben Achour, S., & Pascual, O. (2010). Glia: The many ways to modulate synaptic plasticity. *Neurochemistry International*, *57*(4), 440–445.
Besedovsky, H. O., & Rey, A. D. (2007). Physiology of psychoneuroimmunology: A personal view. *Brain, Behavior, and Immunity*, *21*(1), 34–44.
Bilbo, S. D., & Schwarz, J. M. (2012). The immune system and developmental programming of brain and behavior. *Frontiers in Neuroendocrinology*, *33*(3), 267–286.
Bland, S. T., Hutchinson, M. R., Maier, S. F., Watkins, L. R., & Johnson, K. W. (2009). The glial activation inhibitor AV411 reduces morphine-induced nucleus accumbens dopamine release. *Brain, Behavior, and Immunity*, *23*(4), 492–497.
Bossert, J. M., Ghitza, U. E., Lu, L., Epstein, D. H., & Shaham, Y. (2005). Neurobiology of relapse to heroin and cocaine seeking: An update and clinical implications. *European Journal of Pharmacology*, *526*(1–3), 36–50.
Bossert, J. M., Marchant, N. J., Calu, D. J., & Shaham, Y. (2013). The reinstatement model of drug relapse: Recent neurobiological findings, emerging research topics, and translational research. *Psychopharmacology*, *229*(3), 453–476.
Bozarth, M. A., & Wise, R. A. (1981). Intracranial self-administration of morphine into the ventral tegmental area in rats. *Life Sciences*, *28*(5), 551–555.
Bsibsi, M., Ravid, R., Gveric, D., & van Noort, J. M. (2002). Broad expression of Toll-like receptors in the human central nervous system. *Journal of Neuropathology and Experimental Neurology*, *61*(11), 1013–1021.
Bushong, E. A., Martone, M. E., Jones, Y. Z., & Ellisman, M. H. (2002). Protoplasmic astrocytes in CA1 stratum radiatum occupy separate anatomical domains. *Journal of Neuroscience*, *22*(1), 183–192.
Capuron, L., & Miller, A. H. (2011). Immune system to brain signaling: Neuropsychopharmacological implications. *Pharmacology and Therapeutics*, *130*(2), 226–238.
Coller, J. K., & Hutchinson, M. R. (2012). Implications of central immune signaling caused by drugs of abuse: Mechanisms, mediators and new therapeutic approaches for prediction and treatment of drug dependence. *Pharmacology and Therapeutics*, *134*(2), 219–245.
Cushman, P., & Dole, V. P. (1973). Detoxification of rehabilitated methadone-maintained patients. *JAMA*, *226*(7), 747–752.
Dang, V. C., & Christie, M. J. (2012). Mechanisms of rapid opioid receptor desensitization, resensitization and tolerance in brain neurons. *British Journal of Pharmacology*, *165*(6), 1704–1716.
Dantzer, R. (2001). Cytokine-induced sickness behavior: Mechanisms and implications. *Annals of the New York Academy of Sciences*, *933*, 222–234.
Dantzer, R., O'Connor, J. C., Freund, G. G., Johnson, R. W., & Kelley, K. W. (2008). From inflammation to sickness and depression: When the immune system subjugates the brain. *Nature Reviews Neuroscience*, *9*(1), 46–56.
De Leo, J. A., Tawfik, V. L., & LaCroix-Fralish, M. L. (2006). The tetrapartite synapse: Path to CNS sensitization and chronic pain. *Pain*, *122*(1–2), 17–21.
Degenhardt, L., Whiteford, H. A., Ferrari, A. J., Baxter, A. J., Charlson, F. J., Hall, W. D., et al. (2013). Global burden of disease attributable to illicit drug use and dependence: Findings from the Global Burden of Disease Study 2010. *The Lancet*, *382*(9904), 1564–1574.
Di Chiara, G. (1999). Drug addiction as dopamine-dependent associative learning disorder. *European Journal of Pharmacology*, *375*(1–3), 13–30.
Di Chiara, G., & Imperato, A. (1988). Drugs abused by humans preferentially increase synaptic dopamine concentrations in the mesolimbic system of freely moving rats. *Proceedings of the National Academy of Sciences*, *85*(14), 5274–5278.
Eidson, L. N., & Murphy, A. Z. (2013). Blockade of toll-like receptor 4 attenuates morphine tolerance and facilitates the pain relieving properties of morphine. *Journal of Neuroscience*, *33*(40), 15952–15963.

Felger, J. C., & Lotrich, F. E. (2013). Inflammatory cytokines in depression: Neurobiological mechanisms and therapeutic implications. *Neuroscience*, *246*, 199–229.
Fellin, T. (2009). Communication between neurons and astrocytes: Relevance to the modulation of synaptic and network activity. *Journal of Neurochemistry*, *108*(3), 533–544.
Feltenstein, M. W., & See, R. E. (2009). The neurocircuitry of addiction: An overview. *British Journal of Pharmacology*, *154*(2), 261–274.
Frank, M. G., Watkins, L. R., & Maier, S. F. (2011). Stress- and glucocorticoid-induced priming of neuroinflammatory responses: Potential mechanisms of stress-induced vulnerability to drugs of abuse. *Brain, Behavior, and Immunity*, *25*(S1), S21–S28.
Gárate, I., García-Bueno, B., Madrigal, J. L., Bravo, L., Berrocoso, E., Caso, J. R., et al. (2011). Origin and consequences of brain toll-like receptor 4 pathway stimulation in an experimental model of depression. *Journal of Neuroinflammation*, *8*(1), 151.
Gárate, I., Garcia-Bueno, B., Madrigal, J. L. M., Caso, J. R., Alou, L., Gomez-Lus, M. L., et al. (2013). Stress-induced neuroinflammation: Role of the toll-like receptor-4 pathway. *Biological Psychiatry*, *73*(1), 32–43.
Graeber, M. B. (2010). Changing face of microglia. *Science*, *330*(6005), 783–788.
Guyon, A., Skrzydelski, D., De Giry, I., Rovère, C., Conductier, G., Trocello, J. M., et al. (2009). Long term exposure to the chemokine Ccl2 activates the nigrostriatal dopamine system: A novel mechanism for the control of dopamine release. *Neuroscience*, *162*(4), 1072–1080.
Gysling, K., & Wang, R. Y. (1983). Morphine-induced activation of A10 dopamine neurons in the rat. *Brain Research*, *277*(1), 119–127.
Halassa, M. M., Fellin, T., Takano, H., Dong, J. H., & Haydon, P. G. (2007). Synaptic islands defined by the territory of a single astrocyte. *Journal of Neuroscience*, *27*(24), 6473–6477.
Hao, S., Liu, S., Zheng, X., Zheng, W., Ouyang, H., Mata, M., et al. (2010). The role of TNF. *Neuropsychopharmacology*, *36*(3), 664–676.
Hart, B. L. (1988). Biological basis of the behavior of sick animals. *Neuroscience and Biobehavioral Reviews*, *12*(2), 123–137.
Heinrichs, S. C., Menzaghi, F., Schulteis, G., Koob, G. F., & Stinus, L. (1995). Suppression of corticotropin-releasing factor in the amygdala attenuates aversive consequences of morphine withdrawal. *Behavioural Pharmacology*, *6*(1), 74–80.
Henry, C. J., Huang, Y., Wynne, A., Hanke, M., Himler, J., Bailey, M. T., et al. (2008). Minocycline attenuates lipopolysaccharide (LPS)-induced neuroinflammation, sickness behavior, and anhedonia. *Journal of Neuroinflammation*, *5*(1), 1–15.
Hnasko, T. S., Sotak, B. N., & Palmiter, R. D. (2005). Morphine reward in dopamine-deficient mice. *Nature*, *438*(7069), 854–857.
Hutchinson, M. R., Bland, S. T., Johnson, K. W., Rice, K. C., Maier, S. F., & Watkins, L. R. (2007). Opioid-induced glial activation: Mechanisms of activation and implications for opioid analgesia, dependence, and reward. *The Scientific World Journal*, 7, 98–111.
Hutchinson, M. R., Lewis, S. S., Coats, B. D., Skyba, D. A., Crysdale, N. Y., Berkelhammer, D. L., et al. (2009). Reduction of opioid withdrawal and potentiation of acute opioid analgesia by systemic AV411 (Ibudilast). *Brain, Behavior, and Immunity*, *23*(2), 240–250.
Hutchinson, M. R., Northcutt, A. L., Chao, L. W., Kearney, J. J., Zhang, Y., Berkelhammer, D. L., et al. (2008). Minocycline suppresses morphine-induced respiratory depression, suppresses morphine-induced reward, and enhances systemic morphine-induced analgesia. *Brain, Behavior, and Immunity*, *22*(8), 1248–1256.
Hutchinson, M. R., Northcutt, A. L., Hiranita, T., Wang, X., Lewis, S. S., Thomas, J., et al. (2012). Opioid activation of toll-like receptor 4 contributes to drug reinforcement. *Journal of Neuroscience*, *32*(33), 11187–11200.

Hutchinson, M. R., Shavit, Y., Grace, P. M., Rice, K. C., Maier, S. F., & Watkins, L. R. (2011). Exploring the neuroimmunopharmacology of opioids: An integrative review of mechanisms of central immune signaling and their implications for opioid analgesia. *Pharmacological Reviews*, *63*(3), 772–810.

Hutchinson, M. R., & Watkins, L. R. (2014). Why is neuroimmunopharmacology crucial for the future of addiction research? *Neuropharmacology*, *76*, 218–227.

Hutchinson, M. R., Zhang, Y., Brown, K., Coats, B. D., Shridhar, M., Sholar, P. W., et al. (2008). Non-stereoselective reversal of neuropathic pain by naloxone and naltrexone: Involvement of toll-like receptor 4 (TLR4). *European Journal of Neuroscience*, *28*(1), 20–29.

Hutchinson, M. R., Zhang, Y., Shridhar, M., Evans, J. H., Buchanan, M. M., Zhao, T. X., et al. (2010). Evidence that opioids may have toll-like receptor 4 and MD-2 effects. *Brain, Behavior, and Immunity*, *24*(1), 83–95.

Iwata, M., Ota, K. T., & Duman, R. S. (2013). The inflammasome: Pathways linking psychological stress, depression and systemic illness. *Brain, Behavior, and Immunity*, *31*(C), 105–114.

Jensen, C. J., Massie, A., & Keyser, J. (2013). Immune players in the CNS: The astrocyte. *Journal of Neuroimmune Pharmacology*, *8*(4), 824–839.

Johnson, J. D., Campisi, J., Sharkey, C. M., Kennedy, S. L., Nickerson, M., Greenwood, B. N., et al. (2005). Catecholamines mediate stress-induced increases in peripheral and central inflammatory cytokines. *Neuroscience*, *135*(4), 1295–1307.

Johnson, S. W., & North, R. A. (1992). Opioids excite dopamine neurons by hyperpolarization of local interneurons. *The Journal of Neuroscience*, *12*(2), 483–488.

Kageyama, K., Kagaya, S., Takayasu, S., Hanada, K., Iwasaki, Y., & Suda, T. (2010). Cytokines induce NF-κB, Nurr1 and corticotropin-releasing factor gene transcription in hypothalamic 4B cells. *Neuroimmunomodulation*, *17*(5), 305–313.

Kawai, T., & Akira, S. (2007). Signaling to NF-κB by toll-like receptors. *Trends in Molecular Medicine*, *13*(11), 460–469.

Kelley, K. W., Bluthé, R.-M., Dantzer, R., Zhou, J.-H., Shen, W.-H., Johnson, R. W., et al. (2003). Cytokine-induced sickness behavior. *Brain, Behavior, and Immunity*, *17*(Suppl. 1), S112–S118.

Koob, G. F. (1992). Neural mechanisms of drug reinforcement. *Annals of the New York Academy of Sciences*, *654*, 171–191.

Koob, G. F. (1997). Drug abuse: Hedonic homeostatic dysregulation. *Science*, *278*(5335), 52–58.

Koob, G. F. (2013). Addiction is a reward deficit and stress surfeit disorder. *Frontiers in Psychiatry*, *4*, 1–18.

Koob, G. F., & Le Moal, M. (2001). Drug addiction, dysregulation of reward, and allostasis. *Neuropsychopharmacology*, *24*(2), 97–129.

Koob, G. F., & Le Moal, M. (2008). Addiction and the brain antireward system. *Annual Review of Psychology*, *59*(1), 29–53.

Koob, G. F., & Volkow, N. D. (2009). Neurocircuitry of addiction. *Neuropsychopharmacology*, *35*(1), 217–238.

Kosten, T. R., & George, T. P. (2002). The neurobiology of opioid dependence: Implications for treatment. *Science & Practice Perspectives*, *1*(1), 13–20.

Krebs, M. O., Gauchy, C., Desban, M., Glowinski, J., & Kemel, M. L. (1994). Role of dynorphin and GABA in the inhibitory regulation of NMDA-induced dopamine release in striosome- and matrix-enriched areas of the rat striatum. *Journal of Neuroscience*, *14*(4), 2435–2443.

Latz, E., Xiao, T. S., & Stutz, A. (2013). Activation and regulation of the inflammasomes. *Nature Review Immunology*, *13*(6), 397–411.

Liu, L., Coller, J. K., Watkins, L. R., Somogyi, A. A., & Hutchinson, M. R. (2011). Naloxone-precipitated morphine withdrawal behavior and brain IL-1β expression: Comparison of different mouse strains. *Brain, Behavior, and Immunity*, *25*(6), 1223–1232.

Lunden, J. W., & Kirby, L. G. (2013). Opiate exposure and withdrawal dynamically regulate mRNA expression in the serotonergic dorsal raphe nucleus. *Neuroscience*, *254*(C), 160–172.

Matsui, A., & Williams, J. T. (2011). Opioid-sensitive GABA inputs from rostromedial tegmental nucleus synapse onto midbrain dopamine neurons. *Journal of Neuroscience*, *31*(48), 17729–17735.

Matthews, R. T., & German, D. C. (1984). Electrophysiological evidence for excitation of rat ventral tegmental area dopamine neurons by morphine. *Neuroscience*, *11*(3), 617–625.

McGee, D. W., Bamberg, T., Vitkus, S. J., & McGhee, J. R. (1995). A synergistic relationship between TNF-alpha, IL-1 beta, and TGF-beta 1 on IL-6 secretion by the IEC-6 intestinal epithelial cell line. *Immunology*, *86*(1), 6–11.

Miguel-Hidalgo, J. J. (2009). The role of glial cells in drug abuse. *Current Drug Abuse Reviews*, *2*(1), 76–82.

Morgan, M. M., & Christie, M. J. (2012). Analysis of opioid efficacy, tolerance, addiction and dependence from cell culture to human. *British Journal of Pharmacology*, *164*(4), 1322–1334.

Najjar, S., Pearlman, D. M., Alper, K., Najjar, A., & Devinsky, O. (2013). Neuroinflammation and psychiatric illness. *Journal of Neuroinflammation*, *10*(43), 1–24.

Narita, M., Miyatake, M., Narita, M., Shibasaki, M., Shindo, K., Nakamura, A., et al. (2006). Direct evidence of astrocytic modulation in the development of rewarding effects induced by drugs of abuse. *Neuropsychopharmacology*, *31*(11), 2476–2488.

National Institute of Drug Abuse. (2005). *Heroin: Abuse and addiction (National Institute of Drug Abuse Research Report Series)*. Bethesda: National Institute of Health.

National Institute of Drug Abuse. (2011). *Commonly abused prescription drugs (National Institute of Drug Abuse)*. Bethesda: National Institute of Drug Abuse.

Navarro-Zaragoza, J., Nunez, C., Laorden, M. L., & Milanes, M. V. (2010). Effects of corticotropin-releasing factor receptor-1 antagonists on the brain stress system responses to morphine withdrawal. *Molecular Pharmacology*, 77(5), 864–873.

Navarro-Zaragoza, J., Núñez, C., Ruiz-Medina, J., Laorden, M. L., Valverde, O., & Milanés, M. V. (2011). CRF2 mediates the increased noradrenergic activity in the hypothalamic paraventricular nucleus and the negative state of morphine withdrawal in rats. *British Journal of Pharmacology*, *162*(4), 851–862.

Nimmerjahn, A. (2005). Resting microglial cells are highly dynamic surveillants of brain parenchyma in vivo. *Science*, *308*(5726), 1314–1318.

Nish, S., & Medzhitov, R. (2011). Host defense pathways: Role of redundancy and compensation in infectious disease phenotypes. *Immunity*, *34*(5), 629–636.

O'Connor, K. A., Johnson, J. D., Hansen, M. K., Wieseler Frank, J. L., Maksimova, E., Watkins, L. R., et al. (2003). Peripheral and central proinflammatory cytokine response to a severe acute stressor. *Brain Research*, *991*(1–2), 123–132.

Ouyang, H., Liu, S., Zeng, W., Levitt, R. C., Candiotti, K. A., & Hao, S. (2012). An emerging new paradigm in opioid withdrawal: A critical role for glia-neuron signaling in the periaqueductal gray. *The Scientific World Journal*, *2012*(24), 1–9.

Palin, K., Bluthé, R.-M., McCusker, R. H., Moos, F., Dantzer, R., & Kelley, K. W. (2007). TNFα-induced sickness behavior in mice with functional 55 kD TNF receptors is blocked by central IGF-I. *Journal of Neuroimmunology*, *187*(1–2), 55–60.

Phillips, A. G., & LePiane, F. G. (1980). Reinforcing effects of morphine microinjection into the ventral tegmental area. *Pharmacology, Biochemistry and Behavior*, *12*(6), 965–968.

Poltorak, A. (1998). Defective LPS signaling in C3H/HeJ and C57BL/10ScCr mice: Mutations in Tlr4 gene. *Science*, *282*(5396), 2085–2088.

Pribiag, H., & Stellwagen, D. (2013). TNF-downregulates inhibitory neurotransmission through protein phosphatase 1-dependent trafficking of GABAA receptors. *Journal of Neuroscience*, *33*(40), 15879–15893.

Quan, N., & Banks, W. A. (2007). Brain-immune communication pathways. *Brain, Behavior, and Immunity*, *21*(6), 727–735.

Raison, C. L., Capuron, L., & Miller, A. H. (2006). Cytokines sing the blues: Inflammation and the pathogenesis of depression. *Trends in Immunology*, *27*(1), 24–31.

Ransohoff, R. M., & Brown, M. A. (2012). Innate immunity in the central nervous system. *Journal of Clinical Investigation*, *122*(4), 1164–1171.

Ransohoff, R. M., & Perry, V. H. (2009). Microglial physiology: Unique stimuli, specialized responses. *Annual Review of Immunology*, *27*(1), 119–145.

Risbrough, V. B. (2004). Corticotropin-releasing factor receptors CRF1 and CRF2 exert both additive and opposing influences on defensive startle behavior. *Journal of Neuroscience*, *24*(29), 6545–6552.

Robinson, T. E., & Berridge, K. C. (1993). The neural basis of drug craving: An incentive-sensitization theory of addiction. *Brain Research Reviews*, *18*(3), 247–291.

Rogers, J. L., Ghee, S., & See, R. E. (2008). The neural circuitry underlying reinstatement of heroin-seeking behavior in an animal model of relapse. *Neuroscience*, *151*(2), 579–588.

Savin, C., Triesch, J., & Meyer-Hermann, M. (2009). Epileptogenesis due to glia-mediated synaptic scaling. *Journal of the Royal Society Interface*, *6*(37), 655–668.

Schwarz, J. M., & Bilbo, S. D. (2013). Adolescent morphine exposure affects long-term microglial function and later-life relapse liability in a model of addiction. *Journal of Neuroscience*, *33*(3), 961–971.

Schwarz, J. M., Hutchinson, M. R., & Bilbo, S. D. (2011). Early-life experience decreases drug-induced reinstatement of morphine CPP in adulthood via microglial-specific epigenetic programming of anti-inflammatory IL-10 expression. *Journal of Neuroscience*, *31*(49), 17835–17847.

Shekhar, A., Truitt, W., Rainnie, D., & Sajdyk, T. (2005). Role of stress, corticotrophin releasing factor (CRF) and amygdala plasticity in chronic anxiety. *Stress*, *8*(4), 209–219.

Shurman, J., Koob, G. F., & Gutstein, H. B. (2010). Opioids, pain, the brain, and hyperkatifeia: A framework for the rational use of opioids for pain. *Pain Medicine*, *11*(7), 1092–1098.

Si, J., Yang, J., Xue, L., Yang, C., Luo, Y., Shi, H., et al. (2012). Activation of NF-κB in basolateral amygdala is required for memory reconsolidation in auditory fear conditioning. *PloS One*, 7(9), e43973.

Smith, M. F., Mitchell, A., Li, G., Ding, S., Fitzmaurice, A.-M., Ryan, K., et al. (2003). Toll-like receptor (TLR) 2 and TLR5, but not TLR4, are required for Helicobacter pylori-induced NF-κB activation and chemokine expression by epithelial cells. *Journal of Biological Chemistry*, *278*(35), 32552–32560.

Sofroniew, M. V. (2009). Molecular dissection of reactive astrogliosis and glial scar formation. *Trends in Neurosciences*, *32*(12), 638–647.

Sofroniew, M. V., & Vinters, H. V. (2009). Astrocytes: Biology and pathology. *Acta Neuropathologica*, *119*(1), 7–35.

Solomon, R. L., & Corbit, J. D. (1974). An opponent-process theory of motivation. I. Temporal dynamics of affect. *Psychological Review*, *81*(2), 119–145.

Song, C., Merali, Z., & Anisman, H. (1999). Variations of nucleus accumbens dopamine and serotonin following systemic Interleukin-1, Interleukin-2 or Interleukin-6 treatment. *Neuroscience*, *88*(3), 823–836.

Song, P., & Zhoa, Z.-Q. (2001). The involvement of glial cells in the development of morphine tolerance. *Neuroscience Research*, *39*, 281–286.

Stellwagen, D. (2005). Differential regulation of AMPA receptor and GABA receptor trafficking by tumor necrosis factor. *Journal of Neuroscience*, *25*(12), 3219–3228.

Stevens, C. W., Aravind, S., Das, S., & Davis, R. L. (2013). Pharmacological characterization of LPS and opioid interactions at the toll-like receptor 4. *British Journal of Pharmacology*, *168*(6), 1421–1429.

Szabo, A., & Rajnavolgyi, E. (2013). Collaboration of toll-like and RIG-1-like receptors in human dendritic cells: Triggering antiviral innate immune responses. *American Journal of Clinical and Experimental Immunology*, *2*(3), 195–207.

Temburni, M. K., & Jacob, M. H. (2001). New Functions for Glia in the Brain. *Proceedings of the National Academy of Sciences*, *98*(7), 3631–3632.

Theberge, F. R., Li, X., Kambhampati, S., Pickens, C. L., Laurent, R. S., & Bossert, J. M. (2013). Effect of chronic delivery of the toll-like receptor 4 antagonist ()-Naltrexone on incubation of Heroin craving. *Biological Psychiatry*, *73*(8), 729–737.

Tsigos, C., & Chrousos, G. P. (2002). Hypothalamic-pituitary-adrenal axis, neuroendocrine factors and stress. *Journal of Psychosomatic Research*, *53*(4), 865–871.

Wang, X., Loram, L. C., Ramos, K., de Jesus, A. J., Thomas, J., Cheng, K., et al. (2012). Morphine activates neuroinflammation in a manner parallel to endotoxin. *Proceedings of the National Academy of Sciences*, *109*(16), 6325–6330.

Watkins, L. R., Hutchinson, M. R., Johnston, I. N., & Maier, S. F. (2005). Glia: Novel counter-regulators of opioid analgesia. *Trends in Neurosciences*, *28*(12), 661–669.

Watkins, L. R., Hutchinson, M. R., Rice, K. C., & Maier, S. F. (2009). The 'Toll' of opioid-induced glial activation: Improving the clinical efficacy of opioids by targeting glia. *Trends in Pharmacological Sciences*, *30*(11), 581–591.

Weber, M. D., Frank, M. G., Sobesky, J. L., Watkins, L. R., & Maier, S. F. (2013). Blocking toll-like receptor 2 and 4 signaling during a stressor prevents stress-induced priming of neuroinflammatory responses to a subsequent immune challenge. *Brain, Behavior, and Immunity*, *32*(C), 112–121.

Williams, J. T., Ingram, S. L., Henderson, G., Chavkin, C., von, M, Zastrow, Schulz, S., et al. (2012). Regulation of -opioid receptors: Desensitization, phosphorylation, internalization, and tolerance. *Pharmacological Reviews*, *65*(1), 223–254.

World Health Organization (1993). *The ICD-10 classification of mental and behavioural disorders: Diagnostic criteria for research.* Geneva, Switzerland: World Health Organization.

Wrona, D. (2006). Neural-immune interactions: An integrative view of the bidirectional relationship between the brain and immune systems. *Journal of Neuroimmunology*, *172*(1–2), 38–58.

Yang, J., Yu, J., Jia, X., Zhu, W., Zhao, L., Li, S., et al. (2011). Inhibition of nuclear factor-kB impairs reconsolidation of morphine reward memory in rats. *Behavioural Brain Research*, *216*(2), 592–596.

Yirmiya, R., & Goshen, I. (2011). Immune modulation of learning, memory, neural plasticity and neurogenesis. *Brain, Behavior, and Immunity*, *25*(2), 181–213.

Yoon, S. Y., Patel, D., & Dougherty, P. M. (2012). Minocycline blocks lipopolysaccharide induced hyperalgesia by suppression of microglia but not astrocytes. *Neuroscience*, *221*(C), 214–224.

Zhang, X.-Q., Cui, Y., Cui, Y., Chen, Y., Na, X.-D., Chen, F.-Y., et al. (2012). Activation of P38 signaling in the microglia in the nucleus accumbens contributes to the acquisition and maintenance of morphine-induced conditioned place preference. *Brain, Behavior, and Immunity*, *26*(2), 318–325.

CHAPTER SEVEN

Neuroimmune Basis of Methamphetamine Toxicity

Jennifer M. Loftis[*,†,‡,1], **Aaron Janowsky**[*,†,‡,§]
[*]Research & Development Service, Portland VA Medical Center, Portland, Oregon, USA
[†]Department of Psychiatry, Oregon Health & Science University, School of Medicine, Portland, Oregon, USA
[‡]Methamphetamine Abuse Research Center, Oregon Health & Science University, Portland, Oregon, USA
[§]Department of Behavioral Neuroscience, Oregon Health & Science University, Portland, Oregon, USA
[1]Corresponding author: e-mail address: loftisj@ohsu.edu

Contents

Abstract

Although it is not known which antigen-specific immune responses (or if antigen-specific immune responses) are relevant or required for methamphetamine's neurotoxic effects, it is apparent that methamphetamine exposure is associated with significant effects on adaptive and innate immunity. Alterations in lymphocyte activity and number, changes in cytokine signaling, impairments in phagocytic functions, and glial activation and gliosis have all been reported. These drug-induced changes in immune response, particularly within the CNS, are now thought to play a critical role in the addiction process for methamphetamine dependence as well as for other substance use disorders. In Section 2, methamphetamine's effects on glial cell (e.g., microglia and astrocytes) activity and inflammatory signaling cascades are summarized, including how alterations in immune cell function can induce the neurotoxic and addictive effects of methamphetamine. Section 2 also describes neurotransmitter involvement in the

International Review of Neurobiology, Volume 118
ISSN 0074-7742
http://dx.doi.org/10.1016/B978-0-12-801284-0.00007-5

modulation of methamphetamine's inflammatory effects. Section 3 discusses the very recent use of pharmacological and genetic animal models which have helped elucidate the behavioral effects of methamphetamine's neurotoxic effects and the role of the immune system. Section 4 is focused on the effects of methamphetamine on blood–brain barrier integrity and associated immune consequences. Clinical considerations such as the combined effects of methamphetamine and HIV and/or HCV on brain structure and function are included in Section 4. Finally, in Section 5, immune-based treatment strategies are reviewed, with a focus on vaccine development, neuroimmune therapies, and other anti-inflammatory approaches.

1. INTRODUCTION

The toxic effects of methamphetamine have been recognized for decades. Only recently, however, the role of the immune system in methamphetamine's neurotoxic effects has been examined in detail. A number of molecular and cellular mechanisms are triggered following exposure of cells or animals to methamphetamine, and the cascade of events from exposure to neurotoxicity involves cellular components from receptors to immune system activation and inflammation, to energy metabolism. The term "neurotoxicity" can be ambiguous due to the array of methods and perspectives that are used to address methamphetamine's effects. Here, the term is used to describe a condition that follows exposure to methamphetamine, which initiates a cascade of events resulting in altered behavior or cellular function *in the absence of drug*. This distinguishes neurotoxicity from intoxication and effects that occur simultaneously with drug exposure. This chapter reviews some of the more recent findings related to the mechanisms involved in the neuroimmune basis for methamphetamine toxicity.

2. PART 1

2.1. Microglia and astrocytes

Methamphetamine affects glial cell (e.g., microglia and astrocytes) activity, and glial cell activity can modulate the neurotoxic and addictive effects of methamphetamine. Microglia are the major antigen-presenting cells in brain and when activated, they secrete an array of signaling molecules (e.g., proinflammatory cytokines and chemokines) that can cause neuronal damage. Astrocytes are the most numerous and diverse glial cells in the CNS, with a variety of functions including, but not limited to, maintenance of brain homeostasis, storage and distribution of energy substrates, synaptogenesis, and brain defense. Like microglia, astrocytes can also secret a number of signaling molecules that play a putative role in methamphetamine-induced neurotoxicity, such as pro- and anti-inflammatory cytokines including interleukins (ILs), interferons (IFNs), and tumor necrosis factors (TNFs), as well as chemokines (e.g., macrophage inflammatory proteins and monocyte chemoattractant proteins) (Whitney, Eidem, Peng, Huang, & Zheng, 2009). Activation of microglia and astrocytes are normal compensatory reactions to brain injury, but excess neuroinflammation can lead to further brain damage. Indeed, repeated or neurotoxic (i.e., high dose) methamphetamine exposure induces alterations in glial cell functions that contribute to a complex cascade of events, leading to neuroinflammation, neuronal damage, and behavioral impairments (Fig. 7.1). Table 7.1 provides a review of studies that describe acute and chronic effects of methamphetamine exposure on peripheral and central immune function.

2.1.1 Microglia

The role of microglia in mediating methamphetamine's effects has been appreciated since the late 1990s when it was reported that following methamphetamine exposure microglia participate more in necrosis–phagocytosis of neuronal debris than in restoration of neuronal processes (Escubedo et al., 1998). Subsequently, glial cell activation associated with methamphetamine exposure has been well documented with significant contributions made from a number of laboratories (e.g., Asanuma, Miyazaki, Higashi, Tsuji, & Ogawa, 2004; Fantegrossi et al., 2008; Guilarte, Nihei, McGlothan, & Howard, 2003; Hebert & O'Callaghan, 2000; Thomas, Francescutti-Verbeem, Liu, & Kuhn, 2004; Thomas, Walker, Benjamins, Geddes, & Kuhn, 2004). Microglial activation appears to be an early step in the process of methamphetamine-induced neurotoxicity, especially following exposure to high doses of methamphetamine (Thomas, Francescutti-Verbeem, & Kuhn, 2008b; Thomas, Francescutti-Verbeem, et al., 2004; Thomas, Walker, et al., 2004). Although, in some brain regions (e.g., caudate–putamen) microglial activation may not occur until 1 day or more after exposure to high-dose methamphetamine (Bowyer, Robinson, Ali, & Schmued, 2008). Neurotoxic regimens of methamphetamine exposure result in reactive microglia and increased inflammatory gene expression, particularly in striatum—but evidence of methamphetamine-induced microgliosis has also been observed in other brain regions [e.g., cerebellum and hippocampus (Escubedo et al., 1998); somatosensory and piriform cortices and periaqueductal gray (LaVoie, Card, & Hastings, 2004)].

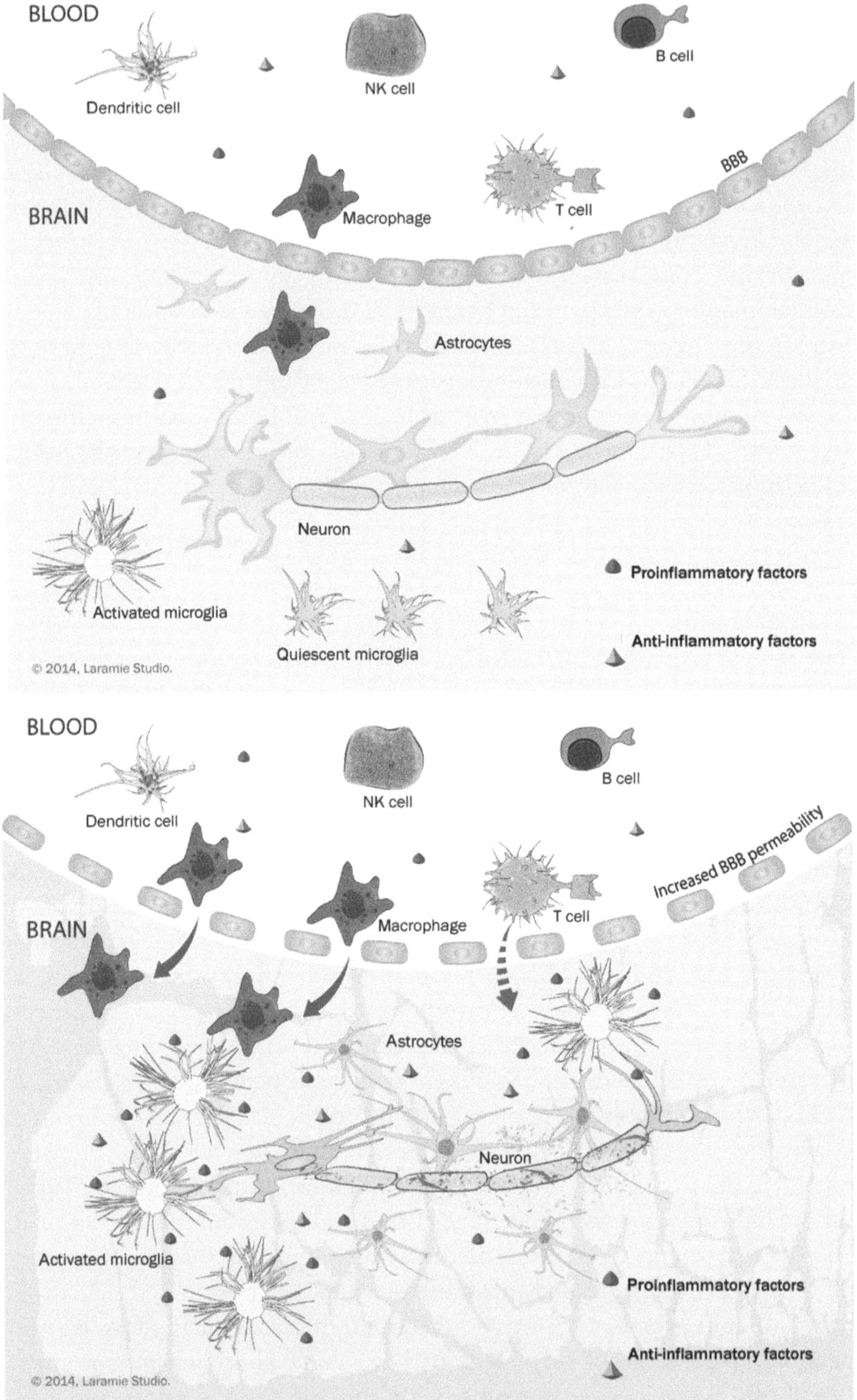

Figure 7.1 Neuroimmune mechanisms of methamphetamine-induced CNS toxicity. The simplified schematic of brain and periphery illustrates the blood–brain barrier

Methamphetamine activates microglia in a dose-dependent manner and via a time course that is generally concurrent with damage to the dopaminergic system (Thomas, Walker, et al., 2004). As a possible target for methamphetamine's effects on microglia, the microglial-specific fractalkine receptor (CX3CR1) [a mediator of 1-methyl-4-phenyl-1,2,3,4-tetrahydropyridine (MPTP)-induced neurodegeneration of dopamine neurons (Cardona et al., 2006)] was evaluated for a role in methamphetamine-induced microglial activation and neurotoxicity. Using mice in which the *CX3CR1* gene was deleted (CX3CR1 knock-out mice), Thomas, Francescutti-Verbeem, and Kuhn (2008a) determined that CX3CR1 signaling does not modulate methamphetamine-induced neurotoxicity or microglial activation. Specifically, methamphetamine exposure had similar effects in both the CX3CR1 knock-out mice and in the wild-type control mice (e.g., microglial activation, increases in body temperature, and reductions in dopamine) (Thomas et al., 2008a).

Once activated, microglia contribute to and potentially perpetuate methamphetamine-induced neuroinflammation and neurodegeneration through inflammatory processes, including the production of proinflammatory cytokines (e.g., TNF-α, IL-1β, and IL-6), or through oxidative mechanisms (Clark, Wiley, & Bradberry, 2013; Yamamoto & Raudensky, 2008) (Fig. 7.1). For example, the excess dopamine resulting from methamphetamine exposure produces dopamine quinones (DAQs) which can activate microglia. Kuhn, Francescutti-Verbeem, and Thomas (2006) demonstrated that DAQs cause time-dependent activation of cultured microglial cells. Importantly, microarray analysis of the effects of DAQs on microglial gene

(BBB), immune and neural cells, and pro- and anti-inflammatory factors before (panel on left) and after (panel on right) methamphetamine. Following exposure to methamphetamine disruption of the BBB is observed (reviewed in Kousik, Napier, & Carvey, 2012) as well as activation of brain microglia and astrocytes, infiltration of macrophages/monocytes, alterations in the expression and balance of proinflammatory factors (e.g., IL-1β), myelin damage, and other indicators of neurodegeneration. Although not depicted in this figure, methamphetamine-induced neurotoxicity is accompanied not only by changes in immunity but also by alterations in neurotransmitter expression and function (such as excitotoxicity from glutamatergic overactivation) and changes in cellular bioenergetics (including oxidative and nitrosative stress). Collectively, these pathological processes contribute to neurotoxicity (e.g., increased BBB permeability, inflammation, neuronal degeneration, cell death) and neuropsychiatric impairments (e.g., cognitive deficits, mood disorders)—processes that can perpetuate addictive behaviors and hinder recovery efforts.

Table 7.1 Effects of methamphetamine exposure on immunologic outcomes in human and animal models

Immune effects measured	Results	References
Splenocyte immune factor production, following *in vivo* methamphetamine exposure and *in vitro* antigen stimulation	Decreased ConA and LPS-induced production of IL-2 and IFN-γ; normal production of IL-4 and IL-6; increased production of TNF-α	Yu et al. (2002)
Peritoneal macrophage function, following *in vivo* methamphetamine exposure and *in vitro* antigen exposure	Reduced macrophage counts and reduced phagocytosis. Following exposure to LPS, reduced cytotoxic activity against melanoma cells; reduced antiviral activity against poly I:C; reduced NO_2^- production; reduced TNF-α, IL-1, and IL-6 production; reduced expression of CD14 receptors	In, Son, Rhee, and Pyo (2004)
Lymphocyte proliferation and NK cell activity, following *in vivo* methamphetamine exposure and *in vitro* antigen exposure	In monkeys, following ConA and PHA exposure, increased lymphocyte proliferation and NK cell activity	Saito et al. (2006)
Dendritic cell immune factor production, following *in vitro* methamphetamine exposure and *in vitro* stimulation	Within a large gene microarray, altered IL-4/GMCSF induced production of functional classes of genes involved in chemokine and cytokine regulation, signal transduction mechanisms, apoptosis, and cell cycle regulation	Mahajan et al. (2006)
Splenocyte T-cell proliferation and macrophage and dendritic cell function, following *in vitro* methamphetamine exposure and *in vitro* antigen exposure	Following OVA exposure, reduced spleen T-cell proliferation; following exposure to *Candida albicans* and *Cryptococcus neoformans*, reduced macrophage phagocytosis, and increased fungal replication; within macrophages, collapsed pH gradients and increased accumulation of autophagosomes	Talloczy et al. (2008)

Splenocyte antibody and immune factor production, following *in vivo* methamphetamine exposure and *in vivo* antigen sensitization	Following sensitzation to OVA, reduced splenocyte production of OVA-specific IgM, IgG1, and IgG2a; reduced IL-4 and IFN-γ production upon *ex vivo* restimulation with OVA	Wey, Wu, Chang, and Jan (2008)
Dendritic cell immune factor production, following *in vitro* methamphetamine exposure and *in vitro* stimulation	Within a large protein array, altered IL-4/GMCSF induced production of a number of functional classes of proteins that modulate apoptosis, protein folding, protein kinase activity, metabolism, and intracellular signal transduction	Reynolds et al. (2009)
Splenocyte T-cell proliferation, and peritoneal macrophage function, following *in vivo* methamphetamine exposure and *in vitro* antigen exposure/Disease progression, lung immune factor expression, and antibody production, following *in vivo* methamphetamine exposure and *in vivo* antigen exposure	Following exposure to histoplasmosis, reduced splenocyte T-cell proliferation and reduced macrophage function (e.g., reduced phagocytosis), resulting in increased fungal load and infection/ Following exposure to histoplasmosis, increased disease progression (e.g., to death); increased lung expression of TNF-α, IFN-γ, IL-4, IL-10, and TGF-β; increased IgG2b levels	Martinez, Mihu, Gacser, Santambrogio, and Nosanchuk (2009)
Neuroinflammatory responses to a subsequent peripheral immune stimulus in mice administered a neurotoxic methamphetamine treatment regimen	Methamphetamine exacerbated the LPS-induced increase in central cytokine mRNA. Methamphetamine alone increased microglial Iba1 expression (a marker for microglial activation) and expression was further increased when mice were exposed to both methamphetamine and LPS	Buchanan, Sparkman, and Johnson (2010a, 2010b)
Peripheral blood T-cell proliferation, following *in vitro* methamphetamine exposure and *in vitro* T-cell stimulation	Reduced anti-CD3/CD28 induced T-cell proliferation and IL-2 production	Potula et al., (2010)

Continued

Table 7.1 Effects of methamphetamine exposure on immunologic outcomes in human and animal models—cont'd

Immune effects measured	Results	References
Peripheral and central immune factor levels, following *in vivo* methamphetamine exposure in humans and mice during different periods of withdrawal/remission	A number of significant methamphetamine-induced changes in cytokines, chemokines, and adhesion factors were observed. Of particular interest were monocyte chemoattractant protein 1 (MCP-1; a.k.a., CCL2) and intercellular adhesion molecule (ICAM-1; a.k.a. CD54), which were similarly increased in the plasma of methamphetamine exposed mice as well as humans. In human participants, methamphetamine-induced changes in the cytokine and chemokine milieu were accompanied by increased cognitive impairments	Loftis, Choi, Hoffman, and Huckans (2011)
Phenotypic changes in leukocyte subsets (e.g., NK cells, T cells, macrophages) from spleen and lymph node of mice treated with an escalating methamphetamine dose schedule	Methamphetamine exposure was associated with an overall decrease in both proportion and number of splenic NK cells, dendritic cells, and a subset of monocytes/macrophages [i.e., monocytes expressing high levels of granulocyte antigen 1 ($Gr\text{-}1^{high}$)]	Harms, Morsey, Boyer, Fox, and Sarvetnick (2012)
Astrocyte immune factor production, following *in vitro* methamphetamine exposure	In human fetal astrocytes, methamphetamine increased IL-6 and IL-8 production (blocked by a metabotropic glutamate receptor-5 inhibitor)	Shah, Silverstein, Singh, and Kumar (2012)

CCR5, C-C chemokine receptor type 5; ConA, concanavalin A; CXCR4, C-X-C chemokine receptor 4; DA, dopamine; GMCSF, granulocyte–macrophage colonly stimulating factorl; HCV, hepatitis C virus; HIV, human immunodeficiency virus; ICAM, intracellular adhesion molecule; IFN, interferon; Ig, immunoglobulin; IL, interleukin; LPS, lipopolysaccharide; MAPK, mitogen-activated protein kinase; MCP, monocyte chemotactic protein; MIP, macrophage inflammatory protein; mRNA, messenger ribonucleic acid; NK, natural killer; OVA, ovalbumin; PHA, phytochemagglutinin-P; STAT1, signal transducer and activator of transcription1; TGF, transforming growth factor; TNF, tumor necrosis factor.

expression indicated that many of the genes differentially regulated by DAQs were those associated with inflammation and neurotoxicity, including cytokines, chemokines, and prostaglandins. Thus, following methamphetamine exposure, the generation of DAQs may induce early activation of microglial cells and increased expression of inflammatory signaling cascades. Of note, one study reported a global pattern of microglial activation and microgliosis in individuals with a history of methamphetamine addiction, which appeared to persist for at least 2 years into abstinence (Sekine et al., 2008).

2.1.2 Astrocytes

For astrocytes, methamphetamine's effects are mediated, in part, by changes in: (1) transcription factor pathways, (2) astrocytic cytokine receptors, (3) excitatory amino acid transporters (EAATs), and (4) glucose uptake mechanisms (Abdul Muneer, Alikunju, Szlachetka, & Haorah, 2011). Methamphetamine can activate astrocytes and induce astrogliosis (e.g., in striatum) via activation of the Janus kinase 2 (JAK2)/signal transducer and activator of transcription 3 (STAT3) signaling cascade (Hebert & O'Callaghan, 2000; Robson et al., 2014)—a pathway that is similarly thought to contribute to astrogliosis following exposure to other neurotoxic substances (MPTP) (e.g., Sriram, Benkovic, Hebert, Miller, & O'Callaghan, 2004) and one that may promote the persistence of reactive gliosis following toxicant exposure (Hebert & O'Callaghan, 2000). For example, Friend and Keefe (2013) reported that astrocytes (but not microglia) remain reactive for at least 30 days following methamphetamine exposure. Consistent with a role for inflammatory signaling in maintaining methamphetamine's activation of astrocytes, mice treated with a neurotoxic regimen of methamphetamine (i.e., four doses of 5 mg/kg administered every 2 h) show significant increases in oncostatin M receptor expression (an astrocytic receptor, Tamura, Morikawa, & Senba, 2003, for the proinflammatory IL-6-type cytokine oncostatin M) (Robson et al., 2014). Methamphetamine also alters EAATs on astrocytes [particularly type 2, EAAT-2 (reviewed in Cisneros & Ghorpade, 2012)] putatively via the production of reactive oxygen species (ROS) (Vaarmann, Gandhi, & Abramov, 2010). These changes in transporter expression could compromise the ability of EAATs to remove extracellular glutamate from the synapse (Cisneros & Ghorpade, 2012; Lau & Tymianski, 2010), thus contributing to methamphetamine-induced excitotoxicity and neurotoxic immune activation [a.k.a. inflammotoxicity (Wilhelm, Hashimoto, Roberts, & Loftis, 2014)].

2.2. Inflammatory signaling

A growing literature demonstrates that methamphetamine exposure and activation of microglia and astrocytes (as well as other cell types) alters peripheral and central immune functions (In, Son, Rhee, & Pyo, 2005; In et al., 2004; Liang et al., 2008; Martinez et al., 2009; Ye et al., 2008) and that immune factors such as cytokines (e.g., IL-1β, IL-6, and TNF-α), chemokines [e.g., monocyte chemotactic protein-1 (MCP-1)], and adhesion molecules play a critical role in the development and persistence of methamphetamine-induced neuronal injury and neuropsychiatric impairments (Clark et al., 2013; Loftis et al., 2011; Yamamoto, Moszczynska, & Gudelsky, 2010; Yamamoto & Raudensky, 2008). In an astrocytic cell line, methamphetamine exposure for 3 days increases IL-6 and IL-8 RNA levels by 4.6 ± 0.2 fold and 3.5 ± 0.2 fold, respectively (Shah et al., 2012). Consistent with these findings, mice treated with methamphetamine show brain region-specific increases in the expression of proinflammatory cytokines (e.g., IL-1β) up to 3 weeks after methamphetamine exposure (Loftis et al., 2011). In a rigorous study designed to evaluate whether or not deletion of the TNF-α gene in mice affects addictive-like behavior, including methamphetamine self-administration, motivation to self-administer methamphetamine, and cue-induced reinstatement of methamphetamine-seeking behavior, Yan, Nitta, Koseki, Yamada, and Nabeshima (2012) measured methamphetamine self-administration and reinstatement of drug-seeking behavior in TNF-α knock-out and wild-type mice. The authors observed an upward shift of dose responses to methamphetamine self-administration under a fixed ratio schedule of reinforcement in the TNF-α knockout as compared with wild-type mice, indicating that mice lacking TNF-α administered more methamphetamine than controls. Similarly, TNF-α knock-out mice also had a higher breaking point under a progressive ratio schedule of reinforcement, suggesting that TNF-α can also influence motivation to self-administer drug. Taken together, these findings demonstrate that TNF-α signaling affects methamphetamine self-administration and motivation to obtain the drug. However, TNF-α did not appear to contribute to methamphetamine-associated cue-induced relapsing behavior (Yan et al., 2012).

Interestingly, IFN-γ (another proinflammatory cytokine) injected systemically prior to repeated methamphetamine exposure protects against methamphetamine-induced neurotoxicity (as measured by a reduction in striatal dopamine transporters) and hyperthermia, putatively through intracerebral molecular pathways (Hozumi et al., 2008). Thus, of the altered immune factors, some cytokines appear to facilitate or promote methamphetamine-induced toxicities but others may slow or prevent the development of adverse drug effects. Taken together, these observations describe a critical role for CNS immune signaling in methamphetamine neurotoxicity and dependence. More research is needed to elucidate the specific signaling pathways [e.g., STAT3, nuclear factor kappa-light-chain-enhancer of activated B cells (NF-κB) (Wires et al., 2012)] and cell types (e.g., microglia, astrocytes, and vascular endothelial cells) that induce and perpetuate these neuroinflammatory processes—particularly because these immune effects can impact addictive and related behavioral responses.

2.3. Molecular/cellular: receptors and transporters

2.3.1 Neurotransmitter modulation of methamphetamine's inflammatory effects

Methamphetamine alters the disposition of a number of neurotransmitters and high doses cause reductions in a number of cell types, especially in the striatum. Loss of dopaminergic, GABAergic, and cholinergic cells has been described (Zhu, Xu, & Angulo, 2006). However, glial cells are also involved in neurotoxic reactions to methamphetamine, and receptors for neurotransmitters have been found on various cell types that are involved in immune response. Methamphetamine-induced alterations in neurotransmitter availability can affect receptor-mediated second messenger systems and downstream events inside the cell. Recent evidence for possible roles of methamphetamine in altering neurotransmitter-linked changes in intermediates of the immune response is described here.

As opposed to its indirect effects on neurotransmitter receptors via its actions as a serotonin, norepinephrine, dopamine, and vesicular monoamine (2) transporter substrate (Eshleman, Calligaro, & Eldefrawi, 1993; Sulzer et al., 1995; Wilhelm, Johnson, Lysko, Eshleman, & Janowsky, 2004), methamphetamine and its analogs, including methylenedioxymethamphetamine (MDMA), are direct agonists at trace amine associated receptor 1 (TAAR1). Endogenous agonists for the TAAR1 include tyramine and octopamine, as well as other neurotransmitter precursors and metabolites (Borowsky et al., 2001; Bunzow et al., 2001). TAAR1 mRNA and mRNA for other TAARs have been identified on mouse B and NK cells, and in stimulated human peripheral blood lymphocytes (Nelson, Tolbert, Singh, & Bost, 2007). In addition, TAAR1 protein has been characterized in rhesus monkey peripheral blood mononuclear cells (Panas et al., 2012). This latter report also characterized the effects of methamphetamine on phosphorylation of protein kinase A (PKA) and protein kinase C (PKC) activity in peripheral blood mononuclear cells as well as in rhesus monkey immortalized B lymphocytes. Generally, methamphetamine increased expression of the phosphorylated enzymes in PHA-treated cells. TAAR1-mediated stimulation of PKA and PKC phosphorylation appears to be necessary but not sufficient, i.e., PHA is also required for phosphorylation. In fact, it appears that PHA is required to increase the expression of TAAR1, which precedes methamphetamine stimulation of TAAR1 and subsequent phosphorylation of PKA and PKC. The authors also reported methamphetamine-induced increases in nuclear factor of activated T cells, part of the calcium response pathway, and an increase in cyclic AMP-response element binding protein (CREB) pathway in rhesus monkey TAAR1-transfected cells. Thus, methamphetamine stimulation of TAAR1 in lymphocytes appears to play an important role in lymphocyte function preceding immune response. Babusyte, Kotthoff, Fiedler, and Krautwurst (2013) subsequently demonstrated that activation of T-cell and B-cell TAAR1 (and 2) by endogenous ligands such as phenethylamine results in altered cytokine expression. Although the effects of methamphetamine were not examined, similar results could be expected. Reports also indicate that malignant B cells, including a Burkitt's lymphoma cell line are sensitive to TAAR1 agonists, which reduce their viability (Wasik, Millan, Scanlan, & Barnes, 2012). Clearly, TAAR1 receptors on lymphocytes are involved in cell function and cellular response. However, it is not clear that TAAR1-mediated cellular mechanisms in neurons and lymphocytes are shared.

Glutamate receptors have also been implicated in the behavioral effects of methamphetamine (Achat-Mendes, Platt, & Spealman, 2012; Crawford, Roberts, & Beveridge, 2013; Herrold, Voigt, & Napier, 2013), and there are some data to indicate the role of specific glutamate receptors in neurotoxicity. Metabotropic glutamate receptor 5 (mGluR5) appears to play a role in the methamphetamine-induced change in immune response signaling. Shah et al. (2012) reported that exposure of the SVGA astrocyte cell line to methamphetamine caused a dose-dependent increase in mRNA and protein for the proinflammatory cytokines IL-6 and IL-8, as well as translocation of the nuclear subunit of NFκB, and the effects are most likely due to activation of Akt/PI3K pathways. Importantly, the effects of methamphetamine on IL-6 and IL-8 expression were reduced in the presence of 2-methyl-6-(phenylethynyl)pyridine (MPEP), an mGluR5 antagonist. The authors proposed a model consistent with direct stimulation of mGluR5 by methamphetamine. Additional studies suggest that mGluR5 regulates methamphetamine-induced dopamine release and the neurotoxic effects of methamphetamine on dopaminergic neurons, as MPEP also blocks the reduction in dopamine that is caused by high-dose methamphetamine, as well as methamphetamine-induced dopamine release (Gołembiowska, Konieczny, Wolfarth, & Ossowska, 2003).

The involvement of ionotropic *N*-methyl-D-aspartate (NMDA) and alpha-amino-3-hydroxy-5-methyl-4-isoxazolepropionic acid (AMPA) receptor subtypes of glutamate receptors have also been implicated in methamphetamine-induced neurotoxicity by experimental results indicating that the NMDA receptor antagonist (dl)-amino-5-phosphonovaleric acid (AP5) or the AMPA receptor antagonist dinitroquinoxaline-2,3-dione (DNQX) prevent methamphetamine-induced c-fos expression and dopamine depletion, possibly as a result of reducing corticostriatal input (Gross, Duncker, & Marshall, 2011). Calcium release has been implicated in the mechanism of action of glutamate and methamphetamine neurotoxicity: NMDA + methamphetamine-induced neurotoxicity could be reduced by exposure of hippocampal slices in culture to the endoplasmic reticulum-bound sigma-1 receptor antagonist BD1047 which is coupled to calcium-mediated signal cascades (Smith, Butler, & Prendergast, 2010). Thus, evidence suggests both direct and indirect roles for glutamate receptors in methamphetamine-induced neurotoxic events.

A number of reports implicate the α7 subtype of nicotinic cholinergic receptor in the neurotoxic effects of methamphetamine. In the rat, 6-hydroxydopamine model of Parkinson's disease, the α7 receptor agonist, 3-[(2,4-dimethoxy)benzylidene]-anabaseine dihydrochloride, reduces 6-hydroxydopamine-induced cell loss and the elevation of markers of neurotoxicity (Suzuki et al., 2013). The possibility that α7 receptors play a similar role in methamphetamine-induced neurotoxicity is supported by evidence that the α7 antagonist methyllycaconitine (MLA) prevents methamphetamine-induced neurotoxic events including decreases in cellular dopamine and in dopamine transporter expression, and methamphetamine-induced glutamate release (Northrop, Smith, Yamamoto, & Eyerman, 2011). The effect may be mediated by ROS, since MLA blocks methamphetamine-induced ROS production (Escubedo, Camarasa, Chipana, García-Ratés, & Pubill, 2009; Escubedo, Chipana, Pérez-Sánchez, Camarasa, & Pubill, 2005; Pubill et al., 2005). Therefore, the data support the hypothesis that methamphetamine indirectly causes release of the glutamate that is involved in methamphetamine-induced neurotoxicity. Although methamphetamine affects a number of receptors, including specific nicotinic cholinergic receptor subtypes, in its toxic effects, the role of muscarinic cholinergic receptors in methamphetamine's cellular effects has not been described in detail.

Methamphetamine is a substrate for the dopamine transporter and causes neurotransmitter release (Eshleman et al., 2013), which may trigger drug-induced changes in behavior, but may not exert its neurotoxic effects via direct interaction with dopamine receptors. However, chronic changes in neurotransmitter interactions with receptors that accompany drug abuse can alter cellular function, and lymphocytes express a number of receptors that might be affected by methamphetamine (see also Table 7.1). In Section 3.1, effects of methamphetamine on signaling systems in brain are discussed.

3. PART 2

3.1. Animal models

Behavioral effects of methamphetamine's neurotoxic effects and the role of the immune system have been characterized using pharmacological and genetic models. Many of the genetic models involve knock-out animals. For instance, dopamine D1 knock-out mice have less hyperthermia, gliosis, and inducible nitric oxide synthase in the striatum and less dopaminergic cell loss in the substantia nigra in response methamphetamine, as compared to wild-type control mice (Ares-Santos et al., 2012), and dopamine D2 receptor knock-out mice have reduced hyperthermia, loss of striatal dopaminergic markers, as well as a reduction in other markers of neurotoxicity (Granado, Ares-Santos, et al., 2011). Nrf2 is a neuroprotective gene that attenuates the effects of oxidative stress and is involved in innate immune response in the CNS. Nrf2 knock-out mice have an exaggerated response to methamphetamine, including increased glial activation, and expression of TNF-α and IL-15 (Granado, Lastres-Becker, et al., 2011). Prodynorphin-deficient mice were less sensitive to the dopamine depleting effects of methamphetamine as compared to wild-type mice. The mechanism for this effect may be via anti-inflammatory effects of dynorphin, which is found in very high concentrations in the nigrostriatal pathway (Wang et al., 2012). Likewise, protein kinase Cδ knock-out animals have reduced lipid peroxidation, protein oxidation, and behavioral deficits in response to methamphetamine. These studies were corroborated using pharmacological interventions, which also indicate increased tyrosine hydroxylase phosphorylation in response to methamphetamine (Shin et al., 2011, 2012).

Using pharmacological manipulations to reduce serotonin, Thomas, Angoa Pérez, Francescutti-Verbeem, Shah, and Kuhn (2010) demonstrated that serotonin may not play an essential role in methamphetamine neurotoxicity (reductions in dopaminergic cell markers such as dopamine, tyrosine hydroxylase, and the dopamine transporter). Importantly, this study also demonstrated that hyperthermia is not a prerequisite for toxicity, since depletion of serotonin is associated with increased hyperthermia. In experiments connecting the pharmacological effects of methamphetamine to its immunological effects, Buchanan et al. (2010b) demonstrated that a high-dose methamphetamine regimen increased the CNS cytokine response to subsequent i.p. administration of LPS, and also activated microglia.

4. PART 3

4.1. Blood–brain barrier and clinical considerations

4.1.1 Blood–brain barrier

The blood–brain barrier (BBB) is comprised of endothelial cells that line brain capillaries and contribute to supplying the brain with nutrients, regulating the transport of essential molecules (e.g., glucose and amino acids) to the brain, and protecting neural tissue from toxic substances—functions that collectively serve to maintain a stable environment. The BBB endothelial cells are surrounded by astrocytic perivascular endfeet which release specific factors, including transforming growth factor-β (TGF-β) and glial-derived neurotrophic factor (GDNF), important for the development and function of the BBB. Continuous tight junctions between the BBB endothelial cells as well as metabolic barriers (e.g., enzymes, transport systems) regulate transport across the BBB (e.g., limit the diffusion of molecules across the BBB and control the entry of neurotransmitters into the brain) (Abbott, Rönnbäck, & Hansson, 2006). Disturbances of brain homeostasis as a result of BBB damage are evident in a number of pathological conditions such as stroke, multiple sclerosis, Alzheimer's disease, HIV, and substance use disorders (including methamphetamine dependence). Studies show that although the initial etiologies for BBB disruption are different among these pathologies, characteristic alterations and common intracellular pathways are involved. NFκB activation and proinflammatory signaling [e.g., via IL-6 and matrix metalloproteinases] often mediate dysfunction of the BBB's structural and functional integrity (Abbott et al., 2006; Persidsky, Ramirez, Haorah, & Kanmogne, 2006). For example, proinflammatory cytokine-mediated release of matrix metalloproteinases (e.g., MMP-9) can increase BBB permeability by activating NF-κB, regulating microfilament reorganization and reducing the expression of tight junction proteins (or redistributing tight junction proteins from the membrane to the cytosol) (Aghajanian, Wittchen, Allingham, Garrett, & Burridge, 2008; Bogatcheva & Verin, 2008; Brown et al., 2003; Lu et al., 2009).

4.1.2 Methamphetamine exposure and BBB integrity

Psychostimulant exposure, and in particular methamphetamine abuse, is associated with disruption of the BBB (reviewed in Kousik et al., 2012) as well as with increased expression of proinflammatory factors. Pre-clinical *in vivo* studies and cultured human brain microvascular endothelial cell *in vitro* experiments show that methamphetamine contributes to inflammation-induced BBB dysfunction in a dose- and time-dependent manner by activating NF-κB, upregulating inflammatory gene expression (TNF-α), increasing the expression and activity of MMP-9, reducing the expression of tight junction proteins [e.g., zonula occludens (ZO)-1, claudin-5, and occludin], increasing the production of ROS (Lee, Hennig, Yao, & Toborek, 2001; Martins et al., 2011; Ramirez et al., 2009). Collectively, these adverse effects on BBB integrity contribute to neurodegeneration and may perpetuate toxic neural-immune interactions. For example, following exposure to high-dose methamphetamine, disruption of the BBB [as measured using immunoglobulin G (IgG) immunoreactivity to identify extravasated IgG in brain] is associated with neurodegeneration and activation of brain microglia and/or infiltration of macrophages. Methamphetamine-induced neurodegeneration is evident in brain regions with notable IgG immunoreactivity (i.e., hippocampus and amygdala) and during hyperthermic conditions (mouse body temperatures >40.5 °C). It was additionally reported that the hippocampal damage induced by methamphetamine and BBB disruption is likely sufficient to compromise cognitive function (Bowyer & Ali, 2006).

In a follow-up study, Bowyer et al. (2008) investigated BBB integrity to determine whether barrier disruption also plays a role in methamphetamine-induced neurotoxicity in the caudate–putamen—as it did in other brain regions (i.e., amygdala and hippocampus). Mice were administered high-dose methamphetamine and evaluated at 90 min, 4 h, 12 h, 1 and 3 days following drug exposure. The results showed that the BBB was only modestly altered (as indicated by IgG immunoreactivity) in the caudate–putamen at time points and subsequent to observation of neurodegeneration, and similar to previous work, these changes were dependent on hyperthermia (Bowyer et al., 2008; but see discussion of hyperthermia and serotonin involvement; Thomas et al., 2010).

Methamphetamine exposure, in combination with stress, can synergistically exacerbate BBB damage (i.e., reductions in occludin and claudin-5 immunoreactivity) via inflammatory mechanisms that persist for at least 7 days following methamphetamine (Northrop & Yamamoto, 2012). Collectively, these methamphetamine-induced effects on BBB contribute to immune dysfunction, such as increased leukocyte/monocyte transmigration across the endothelium and into the CNS (Buch et al., 2012; Ramirez et al., 2009, Fig. 7.1) as well as increased invasion of peripheral bacteria and viruses into the brain.

4.1.3 Clinical picture: immune signaling in addictive and related behaviors

In a classic positron emission tomography imaging study, Volkow et al. (2001) demonstrate that protracted abstinence can reverse some methamphetamine-induced brain insults (e.g., reductions in brain dopamine terminals) but suggest that other neurotoxic processes may persist and contribute to neuropsychiatric impairments during and following drug exposure. Research is beginning to link immune factor signaling with neural and behavioral aspects of addiction, such as impaired cognitive function (Loftis et al., 2011), drug seeking behaviors, and resilience to relapse (Blednov, Mayfield, Belknap, & Harris, 2012; Blednov, Ponomarev, et al., 2012; Schwarz, Hutchinson, & Bilbo, 2011; Zhang et al., 2012). For example, in a pre-clinical study propentofylline (a phosphodiesterase inhibitor and glial cell modulator) significantly reduced the activation of astrocytes induced by methamphetamine, and treatment with propentofylline suppressed the rewarding effects of methamphetamine, as measured using conditioned place preference testing (Narita et al., 2006). These findings suggest that methamphetamine-induced astrocyte activation and the subsequent release of proinflammatory factors can play a role in the development of the rewarding effects of methamphetamine.

In addition to contributing to the addictive properties of methamphetamine, cytokines and other inflammatory factors can also be potent modulators of mood and cognitive function. This is important because neuropsychiatric impairments can persist following abstinence and are associated with poorer treatment outcomes, such as increased relapse rates, lower treatment retention rates, and reduced daily functioning (e.g., Zorick et al., 2010). Repeated methamphetamine exposure induces alterations in peripheral and central immune factor expression, and in clinical samples of methamphetamine-dependent humans, these peripheral alterations in immune factor expression are associated with cognitive impairments and mood disturbances (Letendre et al., 2005; Loftis et al., 2011). The neuroinflammatory effects of methamphetamine appear to be brain region specific and can lead to differential effects on cognitive function (Chang, Cloak, et al., 2005; Jernigan et al., 2005). Thus, interventions that target immune mechanisms for repairing persistent methamphetamine-induced CNS injury and neuropsychiatric impairments, both vital for successful recovery from methamphetamine dependence, may help reduce relapse rates and improve treatment outcomes in adults recovering from methamphetamine dependence and other substance use disorders (see Section 5.2).

4.1.4 Chronic infection with human immunodeficiency virus and/or hepatitis C virus

Exposure to methamphetamine and other substances of abuse can increase the invasion of peripheral bacteria and viruses, such as human immunodeficiency virus (HIV) and hepatitis C virus (HCV), into the brain (e.g., Gavrilin, Mathes, & Podell, 2002). Further, individuals who are infected with one virus have an increased risk of coinfection with the other. Given the high rates of comorbid viral infection with substance use disorders (Fuller, Loftis, Rodriguez, McQuesten, & Hauser, 2009; Loftis, Matthews, & Hauser, 2006), particularly psychostimulant dependence (e.g., Buchacz et al., 2005), the combined effects of substance use disorders and HIV and/or HCV on neural-immune interactions are of scientific and clinical interest (Cadet & Krasnova, 2007). In addition, the effects of chronic viral infection on brain are likely exacerbated by the adverse effects of comorbid methamphetamine dependence on BBB integrity and function (Liang et al., 2008) (Fig. 7.1).

HIV infection of glial cells results in their activation and release of cytokines, ROS, HIV-related proteins, and other neurotoxic mediators (Nath & Geiger, 1998). Reports indicate that HIV-induced neurotoxicity may preferentially affect brain regions rich with dopaminergic transporters (e.g., basal ganglia)—brain areas that also show damage following methamphetamine abuse (Nath, Maragos, Avison, Schmitt, & Berger, 2001). In fact, methamphetamine use has been linked to worse neuropsychiatric functioning in acute and early HIV infection, a critical period of immune activation (Weber et al., 2013).

One prominent neurotoxin associated with HIV is Tat (released from HIV-infected glia). Tat has been suggested to be pathogenically relevant in HIV-1-induced neuronal injury via diverse mechanisms. For example, coexposure to Tat and methamphetamine causes a synergistic loss of striatal dopamine and binding to the dopamine transporter, suggesting a loss of dopamine terminals (Theodore, Stolberg, Cass, & Maragos, 2006). In addition, methamphetamine and Tat can activate microglia and induce cytokine production. Interestingly, the interaction of Tat and methamphetamine is prevented in MCP-1-deficient mice and this also attenuates methamphetamine + Tat neurotoxicity (Theodore, Cass, & Maragos, 2006). Follow-up studies demonstrate that Tat contributes to increased TNF-α levels that exacerbates dopamine transporter damage from methamphetamine (Theodore, Cass, Nath, et al., 2006). Similarly, methamphetamine use in the presence of HIV produces additive interneuron loss (Chana et al., 2006) and abnormal brain metabolism in frontostriatal pathways (Chang, Ernst, Speck, & Grob, 2005).

The combination of HIV and methamphetamine is also associated with higher rates of neuropsychiatric impairment than either of the independent conditions (Rippeth et al., 2004). However, not all reports have been consistent regarding additive or synergistic effects of HIV and methamphetamine on brain structure and function (e.g., Archibald et al., 2012). The mechanisms of such injury are currently under investigation and inflammation, oxidative damage, and glutamate imbalance appear to contribute to neurotoxicity associated with methamphetamine dependence and comorbid HIV.

A few research groups are investigating CNS mechanisms associated with the neurotoxic effects of HCV and methamphetamine dependence. For example, using neuroimaging techniques, Taylor et al. (2004) found that *N*-acetylaspartate (NAA), a marker of neuronal integrity, was lower in the white matter of patients with HCV and a history of methamphetamine abuse, as compared to participants with HCV and no history of methamphetamine abuse. Importantly, this reduction in NAA was correlated with worse global neuropsychological deficit scores in the individuals with a history of methamphetamine addiction. However, at least one study of methamphetamine users (without HCV or HIV) found that impairments in cognitive function were unrelated to prior drug use histories (including self-reported age at first use, total years of use, route of consumption, or length of abstinence) (Cherner et al., 2010). Thus, more research is needed, as the additive or synergic effects of methamphetamine and viral infection on specific neurotoxicities and neuropsychiatric impairments are yet to be defined.

5. PART 4

5.1. Vaccine development

Anti-addiction or anti-drug vaccines are currently among the most developed immunotherapeutic approaches for substance use disorders [see National Institute on Drug Abuse (NIDA)-sponsored anti-drug vaccine animation at http://www.drugabuse.gov/news-events/nida-notes/anti-drug-vaccine-animation]. These vaccines are designed to generate and attract antibodies to a substance so that it is too large to pass through the BBB (Cerny & Cerny, 2009; Gentry, Rüedi-Bettschen, & Owens, 2009; Kinsey, Jackson, & Orson, 2009; Li et al., 2011; Orson, Kinsey, Singh, Wu, & Kosten, 2009), thereby blocking its CNS actions and rewarding effects. To date, vaccines have been created against nicotine, morphine/heroin, cocaine, and methamphetamine, and a number of compounds are undergoing clinical trials or are in preclinical development (Goniewicz & Delijewski, 2013; Kosten, Domingo, Orson, & Kinsey, 2014; Shen et al., 2013). Researchers recently completed the first clinical study to establish the safety of a human-mouse chimeric monoclonal antibody (mAb) (i.e., ch-mAb7F9) designed to bind methamphetamine with high affinity and specificity in humans (Stevens et al., 2014). Results from the Phase 1a, double-blind, randomized, placebo-controlled clinical trial are not yet published (https://ClinicalTrials.gov Identifier: NCT01603147).

5.2. Neuroimmune-based and other anti-inflammatory treatment strategies

Ibudilast (AV-411 or MN-166) is a nonspecific phosphodiesterase inhibitor which suppresses glial cell activation and causes other anti-inflammatory effects. Prior to its development as a potential substance abuse treatment (e.g., for methamphetamine, alcohol, and opioid dependence), ibudilast had been used clinically for asthma, pulmonary, and cardiovascular diseases. Preclinically, administration of ibudilast following exposure to methamphetamine reduces: (1) the acute, chronic, and sensitization effects of the drug's locomotor activity, (2) stress-induced methamphetamine relapse, and (3) methamphetamine self-administration—suggesting that glial cell activity can modulate methamphetamine's behavioral effects (Beardsley, Shelton, Hendrick, & Johnson, 2010; Snider, Hendrick, & Beardsley, 2013; Snider et al., 2012). To date, a Phase I clinical trial of ibudilast with methamphetamine dependence has been completed. This initial human study was done in order to test the safety of the drug taken in combination with methamphetamine. Results from the Phase I safety interaction trial of ibudilast with methamphetamine are not yet publicly available (https://ClinicalTrials.gov Identifier: NCT01217970). Currently, the research is into its second phase, and results from the 12-week Phase II clinical trial are expected to be released in early 2015 (https://ClinicalTrials.gov Identifier: NCT01860807).

In addition to the development of therapeutic strategies for preventing and reducing methamphetamine use, interventions that target immune mechanisms for repairing persistent methamphetamine-induced CNS injury and neuropsychiatric impairments may also help reduce relapse rates and improve treatment outcomes in adults recovering from methamphetamine dependence and other substance use disorders. Given that a partial major histocompatibility complex (MHC)/neuroantigen peptide construct [pI-A^b/mMOG-35-55; a.k.a. recombinant T-cell receptor ligand (RTL)] effectively reduces the inflammatory and behavioral effects of experimental models of multiple sclerosis and stroke (Sinha et al., 2007; Subramanian et al., 2009; Vandenbark et al., 2003; Wang et al., 2006), it is possible that partial MHC/neuroantigen constructs could also effectively address the neuropsychiatric effects of chronic methamphetamine addiction. In a preclinical study, RTL-containing MHC coupled to myelin peptide [mouse myelin oligodendrocyte glycoprotein (mMOG)] improves the learning and memory impairments and CNS inflammation induced by repeated methamphetamine exposure in mouse models of chronic methamphetamine addiction (Loftis, Wilhelm, Vandenbark, & Huckans, 2013).

In addition to vaccine and neuroimmune-based strategies (e.g., ibudilast and partial MHC/neuroantigen peptide constructs), pharmacotherapies that regulate and reduce inflammation and oxidative stress are also under investigation. For example, minocycline (a broad-spectrum tetracycline antibiotic) has anti-inflammatory and antioxidant properties and has demonstrated some efficacy in the treatment of psychiatric disorders and neuropsychiatric symptoms. Preclinically, minocycline improves deficits in novel object recognition (a measure of cognitive function) induced by phencyclidine and methamphetamine treatment in mice (Fujita et al., 2008; Mizoguchi et al., 2008) and reduces the behavioral sensitization induced by methamphetamine and cocaine (Chen, Uz, & Manev, 2009; Zhang et al., 2006). In clinical studies, the results vary, with findings supporting the use of minocycline in schizophrenia, but showing less benefit for nicotine dependence (reviewed in Dean, Data-Franco, Giorlando, & Berk, 2012). Nonsteroidal anti-inflammatory drugs [(e.g., ketoprofen), traditional and selective inhibitors of cyclooxygenase (COX)-2] have also been evaluated for the treatment of psychiatric illness (Berk et al., 2013). For example, in one preclinical study, ketoprofen pretreatment (but not aspirin) dose-dependently attenuated methamphetamine-induced neuroxicity, as measured by reduction of dopamine transporters and accumulation of microglial cells in the striatum (Asanuma, Tsuji, Miyazaki, Miyoshi, & Ogawa, 2003).

Taken together, these preclinical and initial clinical findings indicate that immunotherapeutic strategies which target specific inflammatory pathways, reduce neurotoxicity, promote neuronal repair, and improve neuropsychiatric function may have potential as treatments for methamphetamine dependence and other substances of abuse.

5.3. Future directions

Genetic variation commonly influences psychiatric responses to drugs and medications, and this variation has implications for substance abuse treatment and the use of personalized therapeutic strategies. Thus, future research on methamphetamine dependence could investigate how drug use patterns as well as different treatment approaches may be influenced by genetic polymorphisms, as in alcohol dependence for example [naltrexone treatment seems more effective in carriers of a specific variant of the μ-opioid receptor gene (i.e., G polymorphism of SNP Rs1799971 in the gene OPRM1)] (Anton et al., 2008). Immunologic genes may be good candidates for affecting vulnerability to substance abuse treatment outcomes (Table 7.2). To date, the single nucleotide polymorphisms (SNPs) shown in Table 7.2 have not been directly studied in the context of methamphetamine addiction, but there is preliminary evidence implicating their possible involvement in neuropsychiatric

Table 7.2 Candidate polymorphisms with preliminary evidence for possible involvement in adverse neuropsychiatric symptoms relevant to addiction

Gene symbol	Gene name	ID number	Role in neuropsychiatric symptoms	Select references
IDO1	Indoleamine 2,3-dioxygenase 1	rs9657182	Homozygosity for the risk allele (CC) is associated with moderate–severe interferon (IFN)-α-induced depressive symptoms[a]	Smith et al. (2012)
IL28B	Interleukin 28B (interferon, lambda 3)	rs8099917, rs1297860	In adults with HCV, the allele (C) is associated with both better viral clearance and more subjective appetite, energy, and sleep complaints	Ge et al. (2009); Lotrich, Loftis, Ferrell, Rabinovitz, and Hauser (2010)
IL6	Interleukin 6	rs1800795	A functional G > C SNP in the promoter region is associated with differential IL-6 expression and IL-6 plasma concentrations. The C allele is associated with lower plasma IL-6 during immune activation than the G allele. The "low IL-6 synthesizing" genotype (CC) is associated with significantly fewer symptoms of IFN-α therapy induced depression	Bull et al. (2009); Fishman et al. (1998)
IFNG	Interferon-gamma	rs2430561	IFNG (+874)T/A polymorphism. The presence of T alleles represents a genetic risk factor for the development of IFN-α therapy induced depression	Oxenkrug et al. (2011)
TNF	Tumor necrosis factor-alpha	rs1800630, rs1800629	A decreased frequency of the −863CC TNF-α promoter genotype (involved in high production of this proinflammatory cytokine) is present in patients recovered from HCV infection; the G-308A polymorphism of the TNF gene may be independently associated with hypertension, leptin levels, and hypercholesterolemia, leading to metabolic syndrome	Gupta et al. (2012); Lio et al. (2003)

[a]IFN-γ and IFN-α transcriptionally induce IDO, the rate-limiting enzyme of the kynurenine pathway of tryptophan metabolism.
It is hypothesized that IFN-induced upregulation of IDO contributes to depression, in part, by shifting tryptophan metabolism from the formation of serotonin to the production of neuroactive/toxic kynurenines (Loftis & Turner, 2010).

impairments, particularly depression. A polymorphism near the *IL28B* gene encoding IFN lambda 3 is associated with depressive symptoms, including subjective appetite, energy, and sleep complaints (Lotrich et al., 2010). In addition to *IL28B*, the risk for developing adverse neuropsychiatric symptoms during or following methamphetamine abuse may be influenced by genetic polymorphisms in other cytokine genes, including IL-6 (*IL6*) (Bull et al., 2009), tumor necrosis factor-alpha (*TNF*) (Dogra, Chakravarti, Kar, & Chawla, 2011; Thio et al., 2004), and interferon-γ (*IFNG*) (Oxenkrug et al., 2011)—genetic variations which may have implications for the management and treatment of methamphetamine dependence and other substance use disorders (with or without comorbid viral infection).

ACKNOWLEDGMENTS

This work was in part supported by National Institutes of Health Grant DA018165 to the Methamphetamine Abuse Research Center (MARC) in Portland, Oregon. This material is the result of work supported with resources and the use of facilities at the Portland Veterans Affairs Medical Center and Oregon Health & Science University. Jennifer M. Loftis, Ph.D., and Aaron Janowsky, Ph.D., are Research Scientists at the Portland Veterans Affairs Medical Center. The authors would like to thank Jason A. Laramie, Certified Medical Illustrator, for help in preparing Fig. 7.1. The authors acknowledge Elsevier for the use of Table 7.1, a version of which was previously published in *Pharmacology & Therapeutics* (i.e., Loftis and Huckans, 2013).

Conflict of interest statement: The Department of Veterans Affairs and Oregon Health & Science University own a technology referenced in this review (a partial MHC/neuroantigen peptide construct). The Department of Veterans Affairs, OHSU, and Dr. Loftis have rights to the royalties from the licensing agreement with Artielle (the company that has licensed the technology). These potential conflicts of interest have been reviewed and managed by the Conflict of Interest Committees at the Portland Veterans Affairs Medical Center and Oregon Health & Science University.

REFERENCES

Abbott, N. J., Rönnbäck, L., & Hansson, E. (2006). Astrocyte-endothelial interactions at the blood-brain barrier. *Nature Reviews Neuroscience*, 7, 41–53.

Abdul Muneer, P. M., Alikunju, S., Szlachetka, A. M., & Haorah, J. (2011). Methamphetamine inhibits the glucose uptake by human neurons and astrocytes: Stabilization by acetyl-L-carnitine. *PLoS One*, *6*, e19258.

Achat-Mendes, C., Platt, D. M., & Spealman, R. D. (2012). Antagonism of metabotropic glutamate 1 receptors attenuates behavioral effects of cocaine and methamphetamine in squirrel monkeys. *The Journal of Pharmacology and Experimental Therapeutics*, *343*, 214–224.

Aghajanian, A., Wittchen, E. S., Allingham, M. J., Garrett, T. A., & Burridge, K. (2008). Endothelial cell junctions and the regulation of vascular permeability and leukocyte transmigration. *Journal of Thrombosis and Haemostasis*, *6*, 1453–1460.

Anton, R., Oroszi, G., O'Malley, S., Couper, D., Swift, R., Pettinati, H., et al. (2008). An evaluation of opioid receptor (OPRM1) as a predictor of naltrexone response in the treatment of alcohol dependence. *Archives of General Psychiatry, 65*, 135–144.

Archibald, S. L., Jacobson, M. W., Fennema-Notestine, C., Ogasawara, M., Woods, S. P., Letendre, S., et al. (2012). Functional interactions of HIV-infection and methamphetamine dependence during motor programming. *Psychiatry Research, 202*, 46–52.

Ares-Santos, S., Granado, N., Oliva, I., O'Shea, E., Martin, E. D., Colado, M. I., et al. (2012). Dopamine D(1) receptor deletion strongly reduces neurotoxic effects of methamphetamine. *Neurobiology of Disease, 45*, 810–820.

Asanuma, M., Miyazaki, I., Higashi, Y., Tsuji, T., & Ogawa, N. (2004). Specific gene expression and possible involvement of inflammation in methamphetamine-induced neurotoxicity. *Annals of the New York Academy of Sciences, 1025*, 69–75.

Asanuma, M., Tsuji, T., Miyazaki, I., Miyoshi, K., & Ogawa, N. (2003). Methamphetamine-induced neurotoxicity in mouse brain is attenuated by ketoprofen, a non-steroidal anti-inflammatory drug. *Neuroscience Letters, 352*, 13–16.

Babusyte, A., Kotthoff, M., Fiedler, J., & Krautwurst, D. (2013). Biogenic amines activate blood leukocytes via trace amine-associated receptors TAAR1 and TAAR2. *Journal of Leukocyte Biology, 93*, 387–394.

Beardsley, P. M., Shelton, K. L., Hendrick, E., & Johnson, K. W. (2010). The glial cell modulator and phosphodiesterase inhibitor, AV411 (ibudilast), attenuates prime- and stress-induced methamphetamine relapse. *European Journal of Pharmacology, 637*, 102–108.

Berk, M., Dean, O., Drexhage, H., McNeil, J. J., Moylan, S., O'Neil, A., et al. (2013). Aspirin: A review of its neurobiological properties and therapeutic potential for mental illness. *BMC Medicine, 11*, 74.

Blednov, Y. A., Mayfield, R. D., Belknap, J., & Harris, R. A. (2012). Behavioral actions of alcohol: Phenotypic relations from multivariate analysis of mutant mouse data. *Genes, Brain, and Behavior, 11*, 424–435.

Blednov, Y. A., Ponomarev, I., Geil, C., Bergeson, S., Koob, G. F., & Harris, R. A. (2012). Neuroimmune regulation of alcohol consumption: Behavioral validation of genes obtained from genomic studies. *Addiction Biology, 17*, 108–120.

Bogatcheva, N. V., & Verin, A. D. (2008). The role of cytoskeleton in the regulation of vascular endothelial barrier function. *Microvascular Research, 76*, 202–207.

Borowsky, B., Adham, N., Jones, K. A., Raddatz, R., Artymyshyn, R., Ogozalek, K. L., et al. (2001). Trace amines: Identification of a family of mammalian G protein-coupled receptors. *Proceedings of the National Academy of Sciences of the United States of America, 98*, 8966–8971.

Bowyer, J. F., & Ali, S. (2006). High doses of methamphetamine that cause disruption of the blood-brain barrier in limbic regions produce extensive neuronal degeneration in mouse hippocampus. *Synapse, 60*, 521–532.

Bowyer, J. F., Robinson, B., Ali, S., & Schmued, L. C. (2008). Neurotoxic-related changes in tyrosine hydroxylase, microglia, myelin, and the blood-brain barrier in the caudate-putamen from acute methamphetamine exposure. *Synapse, 62*, 193–204.

Brown, R. C., Mark, K. S., Egleton, R. D., Huber, J. D., Burroughs, A. R., & Davis, T. P. (2003). Protection against hypoxia-induced increase in blood-brain barrier permeability: Role of tight junction proteins and NFkappaB. *Journal of Cell Science, 116*, 693–700.

Buch, S., Yao, H., Guo, M., Mori, T., Mathias-Costa, B., Singh, V., et al. (2012). Cocaine and HIV-1 interplay in CNS: Cellular and molecular mechanisms. *Current HIV Research, 10*, 425–428.

Buchacz, K., McFarland, W., Kellogg, T. A., Loeb, L., Holmberg, S. D., Dilley, J., et al. (2005). Amphetamine use is associated with increased HIV incidence among men who have sex with men in San Francisco. *AIDS, 19*, 1423–1424.

Buchanan, J. B., Sparkman, N. L., & Johnson, R. W. (2010a). A neurotoxic regimen of methamphetamine exacerbates the febrile and neuroinflammatory response to a subsequent peripheral immune stimulus. *Journal of Neuroinflammation*, *7*, 82.

Buchanan, J. B., Sparkman, N. L., & Johnson, R. W. (2010b). Methamphetamine sensitization attenuates the febrile and neuroinflammatory response to a subsequent peripheral immune stimulus. *Brain, Behavior, and Immunity*, *24*, 502–511.

Bull, S. J., Huezo-Diaz, P., Binder, E. B., Cubells, J. F., Ranjith, G., Maddock, C., et al. (2009). Functional polymorphisms in the interleukin-6 and serotonin transporter genes, and depression and fatigue induced by interferon-alpha and ribavirin treatment. *Molecular Psychiatry*, *14*, 1095–1104.

Bunzow, J. R., Sonders, M. S., Arttamangkul, S., Harrison, L. M., Zhang, G., Quigley, D. I., et al. (2001). Amphetamine, 3,4-methylenedioxymethamphetamine, lysergic acid diethylamide, and metabolites of the catecholamine neurotransmitters are agonists of a rat trace amine receptor. *Molecular Pharmacology*, *60*, 1181–1188.

Cadet, J. L., & Krasnova, I. N. (2007). Interactions of methamphetamine and HIV: Cellular and molecular mechanisms of toxicity potentiation. *Neurotoxicity Research*, *112*, 181–204.

Cardona, A. E., Pioro, E. P., Sasse, M. E., Kostenko, V., Cardona, S. M., Dijkstra, I. M., et al. (2006). Control of microglial neurotoxicity by the fractalkine receptor. *Nature Neuroscience*, *9*, 917–924.

Cerny, E. H., & Cerny, T. (2009). Vaccines against nicotine. *Human Vaccines*, *5*, 200–205.

Chana, G., Everall, I. P., Crews, L., Langford, D., Adame, A., Grant, I., et al. (2006). Cognitive deficits and degeneration of interneurons in HIV+methamphetamine users. *Neurology*, *67*, 1486–1489.

Chang, L., Cloak, C., Patterson, K., Grob, C., Miller, E. N., & Ernst, T. (2005). Enlarged striatum in abstinent methamphetamine abusers: A possible compensatory response. *Biological Psychiatry*, *57*(9), 967–974.

Chang, L., Ernst, T., Speck, O., & Grob, C. S. (2005). Additive effects of HIV and chronic methamphetamine use on brain metabolite abnormalities. *The American Journal of Psychiatry*, *162*, 361–369.

Chen, H., Uz, T., & Manev, H. (2009). Minocycline affects cocaine sensitization in mice. *Neuroscience Letters*, *452*, 258–261.

Cherner, M., Suarez, P., Casey, C., Deiss, R., Letendre, S., Marcotte, T., et al. (2010). Methamphetamine use parameters do not predict neuropsychological impairment in currently abstinent dependent adults. *Drug and Alcohol Dependence*, *106*, 154–163.

Cisneros, I. E., & Ghorpade, A. (2012). HIV-1, methamphetamine and astrocyte glutamate regulation: Combined excitotoxic implications for neuro-AIDS. *Current HIV Research*, *10*, 392–406.

Clark, K. H., Wiley, C. A., & Bradberry, C. W. (2013). Psychostimulant abuse and neuroinflammation: Emerging evidence of their interconnection. *Neurotoxicity Research*, *23*, 174–188.

Crawford, J. T., Roberts, D. C., & Beveridge, T. J. (2013). The group II metabotropic glutamate receptor agonist, LY379268, decreases methamphetamine self-administration in rats. *Drug and Alcohol Dependence*, *132*, 414–419.

Dean, O. M., Data-Franco, J., Giorlando, F., & Berk, M. (2012). Minocycline: Therapeutic potential in psychiatry. *CNS Drugs*, *26*, 391–401.

Dogra, G., Chakravarti, A., Kar, P., & Chawla, Y. K. (2011). Polymorphism of tumor necrosis factor-α and interleukin-10 gene promoter region in chronic hepatitis C virus patients and their effect on pegylated interferon-α therapy response. *Human Immunology*, *72*, 935–939.

Escubedo, E., Camarasa, J., Chipana, C., García-Ratés, S., & Pubill, D. (2009). Involvement of nicotinic receptors in methamphetamine- and MDMA-induced neurotoxicity: Pharmacological implications. *International Review of Neurobiology*, *88*, 121–166.

Escubedo, E., Chipana, C., Pérez-Sánchez, M., Camarasa, J., & Pubill, D. (2005). Methyllycaconitine prevents methamphetamine-induced effects in mouse striatum: Involvement of alpha7 nicotinic receptors. *The Journal of Pharmacology and Experimental Therapeutics*, *315*, 658–667.

Escubedo, E., Guitart, L., Sureda, F. X., Jiménez, A., Pubill, D., Pallàs, M., et al. (1998). Microgliosis and down-regulation of adenosine transporter induced by methamphetamine in rats. *Brain Research*, *814*, 120–126.

Eshleman, A. J., Calligaro, D. O., & Eldefrawi, M. E. (1993). Allosteric regulation by sodium of the binding of [3H]cocaine and [3H]GBR 12935 to rat and bovine striata. *Membrane Biochemistry*, *10*, 129–144.

Eshleman, A. J., Wolfrum, K. M., Hatfield, M. G., Johnson, R. A., Murphy, K. V., & Janowsky, A. (2013). Substituted methcathinones differ in transporter and receptor interactions. *Biochemical Pharmacology*, *85*, 1803–1815.

Fantegrossi, W. E., Ciullo, J. R., Wakabayashi, K. T., De La Garza, R., 2nd., Traynor, J. R., & Woods, J. H. (2008). A comparison of the physiological, behavioral, neurochemical and microglial effects of methamphetamine and 3,4-methylenedioxymethamphetamine in the mouse. *Neuroscience*, *151*, 533–543.

Fishman, D., Faulds, G., Jeffery, R., Mohamed-Ali, V., Yudkin, J. S., Humphries, S., et al. (1998). The effect of novel polymorphisms in the interleukin-6 (IL-6) gene on IL-6 transcription and plasma IL-6 levels, and an association with systemic-onset juvenile chronic arthritis. *The Journal of Clinical Investigation*, *102*, 1369–1376.

Friend, D. M., & Keefe, K. A. (2013). Glial reactivity in resistance to methamphetamine-induced neurotoxicity. *Journal of Neurochemistry*, *125*, 566–574.

Fujita, Y., Ishima, T., Kunitachi, S., Hagiwara, H., Zhang, L., Iyo, M., et al. (2008). Phencyclidine-induced cognitive deficits in mice are improved by subsequent subchronic administration of the antibiotic drug minocycline. *Progress in Neuro-Psychopharmacology & Biological Psychiatry*, *32*, 336–339.

Fuller, B. E., Loftis, J. M., Rodriguez, V. L., McQuesten, M. J., & Hauser, P. (2009). Psychiatric and substance use disorders comorbidities in veterans with hepatitis C virus and HIV coinfection. *Current Opinion in Psychiatry*, *22*, 401–408.

Gavrilin, M. A., Mathes, L. E., & Podell, M. (2002). Methamphetamine enhances cell-associated feline immunodeficiency virus replication in astrocytes. *Journal of Neurovirology*, *8*, 240–249.

Ge, D., Fellay, J., Thompson, A. J., Simon, J. S., Shianna, K. V., Urban, T. J., et al. (2009). Genetic variation in IL28B predicts hepatitis C treatment-induced viral clearance. *Nature*, *461*, 399–401.

Gentry, W. B., Rüedi-Bettschen, D., & Owens, S. M. (2009). Development of active and passive human vaccines to treat methamphetamine addiction. *Human Vaccines*, *5*, 206–213.

Gołembiowska, K., Konieczny, J., Wolfarth, S., & Ossowska, K. (2003). Neuroprotective action of MPEP, a selective mGluR5 antagonist, in methamphetamine-induced dopaminergic neurotoxicity is associated with a decrease in dopamine outflow and inhibition of hyperthermia in rats. *Neuropharmacology*, *45*, 484–492.

Goniewicz, M. L., & Delijewski, M. (2013). Nicotine vaccines to treat tobacco dependence. *Human Vaccines & Immunotherapeutics*, *9*, 13–25.

Granado, N., Ares-Santos, S., Oliva, I., O'Shea, E., Martin, E. D., Colado, M. I., et al. (2011). Dopamine D2-receptor knockout mice are protected against dopaminergic neurotoxicity induced by methamphetamine or MDMA. *Neurobiology of Disease*, *42*, 391–403.

Granado, N., Lastres-Becker, I., Ares-Santos, S., Oliva, I., Martin, E., Cuadrado, A., et al. (2011). Nrf2 deficiency potentiates methamphetamine-induced dopaminergic axonal damage and gliosis in the striatum. *Glia*, *59*, 1850–1863.

Gross, N. B., Duncker, P. C., & Marshall, J. F. (2011). Cortical ionotropic glutamate receptor antagonism protects against methamphetamine-induced striatal neurotoxicity. *Neuroscience*, *199*, 272–283.

Guilarte, T. R., Nihei, M. K., McGlothan, J. L., & Howard, A. S. (2003). Methamphetamine-induced deficits of brain monoaminergic neuronal markers: Distal axotomy or neuronal plasticity. *Neuroscience*, *122*, 499–513.

Gupta, V., Gupta, A., Jafar, T., Gupta, V., Agrawal, S., Srivastava, N., et al. (2012). Association of TNF-α promoter gene G-308A polymorphism with metabolic syndrome, insulin resistance, serum TNF-α and leptin levels in Indian adult women. *Cytokine*, *57*, 32–36.

Harms, R., Morsey, B., Boyer, C. W., Fox, H. S., & Sarvetnick, N. (2012). Methamphetamine administration targets multiple immune subsets and induces phenotypic alterations suggestive of immunosuppression. *PLoS One*, 7, e49897.

Hebert, M. A., & O'Callaghan, J. P. (2000). Protein phosphorylation cascades associated with methamphetamine-induced glial activation. *Annals of the New York Academy of Sciences*, *914*, 238–262.

Herrold, A. A., Voigt, R. M., & Napier, T. C. (2013). mGluR5 is necessary for maintenance of methamphetamine-induced associative learning. *European Neuropsychopharmacology*, *23*, 691–696.

Hozumi, H., Asanuma, M., Miyazaki, I., Fukuoka, S., Kikkawa, Y., Kimoto, N., et al. (2008). Protective effects of interferon-gamma against methamphetamine-induced neurotoxicity. *Toxicology Letters*, *177*, 123–129.

In, S. W., Son, E. W., Rhee, D. K., & Pyo, S. (2004). Modulation of murine macrophage function by methamphetamine. *Journal of Toxicology and Environmental Health. Part A*, *67*, 1923–1937.

In, S. W., Son, E. W., Rhee, D. K., & Pyo, S. (2005). Methamphetamine administration produces immunomodulation in mice. *Journal of Toxicology and Environmental Health*, *68*, 2133–2145.

Jernigan, T. L., Gamst, A. C., Archibald, S. L., Fennema-Notestine, C., Mindt, M. R., Marcotte, T. D., et al. (2005). Effects of methamphetamine dependence and HIV infection on cerebral morphology. *The American Journal of Psychiatry*, *162*, 1461–1472.

Kinsey, B. M., Jackson, D. C., & Orson, F. M. (2009). Anti-drug vaccines to treat substance abuse. *Immunology and Cell Biology*, *87*, 309–314.

Kosten, T., Domingo, C., Orson, F., & Kinsey, B. (2014). Vaccines against stimulants: Cocaine and MA. *British Journal of Clinical Pharmacology*, *77*, 368–374.

Kousik, S. M., Napier, T. C., & Carvey, P. M. (2012). The effects of psychostimulant drugs on blood brain barrier function and neuroinflammation. *Frontiers in Pharmacology*, *3*, 121.

Kuhn, D. M., Francescutti-Verbeem, D. M., & Thomas, D. M. (2006). Dopamine quinones activate microglia and induce a neurotoxic gene expression profile: Relationship to methamphetamine-induced nerve ending damage. *Annals of the New York Academy of Sciences*, *1074*, 31–41.

Lau, A., & Tymianski, M. (2010). Glutamate receptors, neurotoxicity and neurodegeneration. *Pflügers Archiv*, *460*, 525–542.

LaVoie, M. J., Card, J. P., & Hastings, T. G. (2004). Microglial activation precedes dopamine terminal pathology in methamphetamine-induced neurotoxicity. *Experimental Neurology*, *187*, 47–57.

Lee, Y. W., Hennig, B., Yao, J., & Toborek, M. (2001). Methamphetamine induces AP-1 and NF-kappaB binding and transactivation in human brain endothelial cells. *Journal of Neuroscience Research*, *66*, 583–591.

Letendre, S. L., Cherner, M., Ellis, R. J., Marquie-Beck, J., Gragg, B., Marcotte, T., et al. (2005). The effects of hepatitis C, HIV, and methamphetamine dependence on neuropsychological performance: Biological correlates of disease. *AIDS*, *19*(Suppl. 3), S72–S78.
Li, Q. Q., Luo, Y. X., Sun, C. Y., Xue, Y. X., Zhu, W. L., Shi, H. S., et al. (2011). A morphine/heroin vaccine with new hapten design attenuates behavioral effects in rats. *Journal of Neurochemistry*, *119*, 1271–1281.
Liang, H., Wang, X., Chen, H., Song, L., Ye, L., Wang, S. H., et al. (2008). Methamphetamine enhances HIV infection of macrophages. *The American Journal of Pathology*, *172*, 1617–1624.
Lio, D., Caruso, C., Di Stefano, R., Colonna Romano, G., Ferraro, D., Scola, L., et al. (2003). IL-10 and TNF-alpha polymorphisms and the recovery from HCV infection. *Human Immunology*, *64*, 674–680.
Loftis, J. M., Choi, D., Hoffman, W., & Huckans, M. S. (2011). Methamphetamine causes persistent immune dysregulation: A cross-species, translational report. *Neurotoxicity Research*, *20*, 59–68.
Loftis, J. M., & Huckans, M. (2013). Substance use disorders: Psychoneuroimmunological mechanisms and new targets for therapy. *Pharmacology & Therapeutics*, *139*, 289–300.
Loftis, J. M., Matthews, A. M., & Hauser, P. (2006). Psychiatric and substance use disorders in individuals with hepatitis C: Epidemiology and management. *Drugs*, *66*, 155–174.
Loftis, J. M., & Turner, E. H. (2010). Novel treatment strategies for depression in patients with hepatitis C. *Psychosomatics*, *51*, 357–358.
Loftis, J. M., Wilhelm, C. J., Vandenbark, A. A., & Huckans, M. (2013). Partial MHC/neuroantigen peptide constructs: A potential neuroimmune-based treatment for methamphetamine addiction. *PLoS One*, *8*, e56306.
Lotrich, F. E., Loftis, J. M., Ferrell, R. E., Rabinovitz, M., & Hauser, P. (2010). IL28B polymorphism is associated with both side effects and clearance of hepatitis c during interferon-alpha therapy. *Journal of Interferon & Cytokine Research*, *31*(3), 331–336.
Lu, D. Y., Yu, W. H., Yeh, W. L., Tang, C. H., Leung, Y. M., Wong, K. L., et al. (2009). Hypoxia-induced matrix metalloproteinase-13 expression in astrocytes enhances permeability of brain endothelial cells. *Journal of Cellular Physiology*, *220*, 163–173.
Mahajan, S. D., Hu, Z., Reynolds, J. L., Aalinkeel, R., Schwartz, S. A., & Nair, M. P. (2006). Methamphetamine modulates gene expression patterns in monocyte derived mature dendritic cells: Implications for HIV-1 pathogenesis. *Molecular Diagnosis & Therapy*, *10*, 257–269.
Martinez, L. R., Mihu, M. R., Gacser, A., Santambrogio, L., & Nosanchuk, D. (2009). Methamphetamine enhances histoplasmosis by immunosuppression of the host. *The Journal of Infectious Diseases*, *200*, 131–141.
Martins, T., Baptista, S., Goncalves, J., Leal, E., Milhazes, N., Borges, F., et al. (2011). Methamphetamine transiently increases the blood-brain barrier permeability in the hippocampus: Role of tight junction proteins and matrix metalloproteinase-9. *Brain Research*, *1411*, 28–40.
Mizoguchi, H., Takuma, K., Fukakusa, A., Ito, Y., Nakatani, A., Ibi, D., et al. (2008). Improvement by minocycline of methamphetamine-induced impairment of recognition memory in mice. *Psychopharmacology*, *196*, 233–241.
Narita, M., Miyatake, M., Narita, M., Shibasaki, M., Shindo, K., Nakamura, A., et al. (2006). Direct evidence of astrocytic modulation in the development of rewarding effects induced by drugs of abuse. *Neuropsychopharmacology*, *31*, 2476–2488.
Nath, A., & Geiger, J. (1998). Neurobiological aspects of human immunodeficiency virus infection: Neurotoxic mechanisms. *Progress in Neurobiology*, *54*, 19–33.
Nath, A., Maragos, W. F., Avison, M. J., Schmitt, F. A., & Berger, J. R. (2001). Acceleration of HIV dementia with methamphetamine and cocaine. *Journal of Neurovirology*, *7*, 66–71.
Nelson, D. A., Tolbert, M. D., Singh, S. J., & Bost, K. L. (2007). Expression of neuronal trace amine-associated receptor (Taar) mRNAs in leukocytes. *Journal of Neuroimmunology*, *192*, 21–30.

Northrop, N. A., Smith, L. P., Yamamoto, B. K., & Eyerman, D. J. (2011). Regulation of glutamate release by α7 nicotinic receptors: Differential role in methamphetamine-induced damage to dopaminergic and serotonergic terminals. *The Journal of Pharmacology and Experimental Therapeutics, 336*, 900–907.

Northrop, N. A., & Yamamoto, B. K. (2012). Persistent neuroinflammatory effects of serial exposure to stress and methamphetamine on the blood-brain barrier. *Journal of Neuroimmune Pharmacology, 7*, 951–968.

Orson, F. M., Kinsey, B. M., Singh, R. A., Wu, Y., & Kosten, T. R. (2009). Vaccines for cocaine abuse. *Human Vaccines, 5*, 194–199.

Oxenkrug, G., Perianayagam, M., Mikolich, D., Requintina, P., Shick, L., Ruthazer, R., et al. (2011). Interferon-gamma (+874) T/A genotypes and risk of IFN-alpha-induced depression. *Journal of Neural Transmission, 118*, 271–274.

Panas, M. W., Xie, Z., Panas, H. N., Hoener, M. C., Vallender, E. J., & Miller, G. M. (2012). Trace amine associated receptor 1 signaling in activated lymphocytes. *Journal of Neuroimmune Pharmacology, 7*, 866–876.

Persidsky, Y., Ramirez, S. H., Haorah, J., & Kanmogne, G. D. (2006). Blood-brain barrier: Structural components and function under physiologic and pathologic conditions. *Journal of Neuroimmune Pharmacology, 1*, 223–236.

Potula, R., Hawkins, B. J., Cenna, J. M., Fan, S., Dykstra, H., Ramierez, S. H., et al. (2010). Methamphetamine causes mitochondrial oxidative damage in human T lymphocytes leading to functional impairment. *Journal of Immunology, 185*, 2867–2876.

Pubill, D., Chipana, C., Camins, A., Pallàs, M., Camarasa, J., & Escubedo, E. (2005). Free radical production induced by methamphetamine in rat striatal synaptosomes. *Toxicology and Applied Pharmacology, 204*, 57–68.

Ramirez, S. H., Potula, R., Fan, S., Eidem, T., Papugani, A., Reichenbach, N., et al. (2009). Methamphetamine disrupts blood-brain barrier function by induction of oxidative stress in brain endothelial cells. *Journal of Cerebral Blood Flow and Metabolism, 29*, 1933–1945.

Reynolds, J. L., Mahajan, S. D., Aalinkeel, R., Nair, B., Sykes, D. E., & Schwartz, S. A. (2009). Proteomic analyses of the effects of drugs of abuse on monocyte-derived mature dendritic cells. *Immunological Investigations, 38*, 526–550.

Rippeth, J. D., Heaton, R. K., Carey, C. L., Marcotte, T. D., Moore, D. J., Gonzalez, R., et al. (2004). Methamphetamine dependence increases risk of neuropsychological impairment in HIV infected persons. *Journal of the International Neuropsychological Society, 10*, 1–14.

Robson, M. J., Turner, R. C., Naser, Z. J., McCurdy, C. R., O'Callaghan, J. P., Huber, J. D., et al. (2014). SN79, a sigma receptor antagonist, attenuates methamphetamine-induced astrogliosis through a blockade of OSMR/gp130 signaling and STAT3 phosphorylation. *Experimental Neurology, 254*, 180–189.

Saito, M., Yamaguchi, T., Kawata, T., Ito, H., Kanai, T., Terada, M., et al. (2006). Effects of methamphetamine on cortisone concentration, NK cell activity and mitogen response of T-lymphocytes in female cynomolgus monkeys. *Experimental Animals, 55*, 477–481.

Schwarz, J. M., Hutchinson, M. R., & Bilbo, S. D. (2011). Early-life experience decreases drug-induced reinstatement of morphine CPP in adulthood via microglial-specific epigenetic programming of anti-inflammatory IL-10 expression. *The Journal of Neuroscience, 31*, 17835–17847.

Sekine, Y., Ouchi, Y., Sugihara, G., Takei, N., Yoshikawa, E., Nakamura, K., et al. (2008). Methamphetamine causes microglial activation in the brains of human abusers. *The Journal of Neuroscience, 28*, 5756–5761.

Shah, A., Silverstein, P. S., Singh, D. P., & Kumar, A. (2012). Involvement of metabotropic glutamate receptor 5, AKT/PI3K signaling and NF-κB pathway in methamphetamine-mediated increase in IL-6 and IL-8 expression in astrocytes. *Journal of Neuroinflammation, 9*, 52.

Shen, X. Y., Kosten, T. A., Lopez, A. Y., Kinsey, B. M., Kosten, T. R., & Orson, F. M. (2013). A vaccine against methamphetamine attenuates its behavioral effects in mice. *Drug and Alcohol Dependence, 129*, 41–48.

Shin, E. J., Duong, C. X., Nguyen, T. X., Bing, G., Bach, J. H., Park, D. H., et al. (2011). PKCδ inhibition enhances tyrosine hydroxylase phosphorylation in mice after methamphetamine treatment. *Neurochemistry International, 59*, 39–50.

Shin, E. J., Duong, C. X., Nguyen, X. K., Li, Z., Bing, G., Bach, J. H., et al. (2012). Role of oxidative stress in methamphetamine-induced dopaminergic toxicity mediated by protein kinase Cδ. *Behavioural Brain Research, 232*, 98–113.

Sinha, S., Subramanian, S., Proctor, T. M., Kaler, L. J., Grafe, M., Dahan, R., et al. (2007). A promising therapeutic approach for multiple sclerosis: Recombinant T-cell receptor ligands modulate experimental autoimmune encephalomyelitis by reducing interleukin-17 production and inhibiting migration of encephalitogenic cells into the CNS. *The Journal of Neuroscience, 27*, 12531–12539.

Smith, K. J., Butler, T. R., & Prendergast, M. A. (2010). Inhibition of sigma-1 receptor reduces N-methyl-D-aspartate induced neuronal injury in methamphetamine-exposed and-naive hippocampi. *Neuroscience Letters, 13*(481), 144–148.

Smith, A. K., Simon, J. S., Gustafson, E. L., Noviello, S., Cubells, J. F., Epstein, M. P., et al. (2012). Association of a polymorphism in the indoleamine- 2,3-dioxygenase gene and interferon-α-induced depression in patients with chronic hepatitis C. *Molecular Psychiatry, 17*, 781–789.

Snider, S. E., Hendrick, E. S., & Beardsley, P. M. (2013). Glial cell modulators attenuate methamphetamine self-administration in the rat. *European Journal of Pharmacology, 701*, 124–130.

Snider, S. E., Vunck, S. A., van den Oord, E. J., Adkins, D. E., McClay, J. L., & Beardsley, P. M. (2012). The glial cell modulators, ibudilast and its amino analog, AV1013, attenuate methamphetamine locomotor activity and its sensitization in mice. *European Journal of Pharmacology, 679*, 75–80.

Sriram, K., Benkovic, S. A., Hebert, M. A., Miller, D. B., & O'Callaghan, J. P. (2004). Induction of gp130-related cytokines and activation of JAK2/STAT3 pathway in astrocytes precedes up-regulation of glial fibrillary acidic protein in the 1-methyl-4-phenyl-1,2,3,6-tetrahydropyridine model of neurodegeneration: Key signaling pathway for astrogliosis in vivo? *The Journal of Biological Chemistry, 279*, 19936–19947.

Stevens, M. W., Tawney, R. L., West, C. M., Kight, A. D., Henry, R. L., Owens, S. M., et al. (2014). Preclinical characterization of an anti-methamphetamine monoclonal antibody for human use. *MAbs, 6*, 547–555.

Subramanian, S., Zhang, B., Kosaka, Y., Burrows, G. G., Grafe, M. R., Vandenbark, A. A., et al. (2009). Recombinant T cell receptor ligand treats experimental stroke. *Stroke, 40*, 2539–2545.

Sulzer, D., Chen, T. K., Lau, Y. Y., Kristensen, H., Rayport, S., & Ewing, A. (1995). Amphetamine redistributes dopamine from synaptic vesicles to the cytosol and promotes reverse transport. *The Journal of Neuroscience, 15*, 4102–4108.

Suzuki, S., Kawamata, J., Matsushita, T., Matsumura, A., Hisahara, S., Takata, K., et al. (2013). 3-[(2,4-Dimethoxy)benzylidene]-anabaseine dihydrochloride protects against 6-hydroxydopamine-induced parkinsonian neurodegeneration through α7 nicotinic acetylcholine receptor stimulation in rats. *Journal of Neuroscience Research, 91*, 462–471.

Talloczy, Z., Martinez, J., Joset, D., Ray, Y., Gacser, A., Toussi, S., et al. (2008). Methamphetamine inhibits antigen processing, presentation, and phagocytosis. *PLoS Pathogens, 4*, e28.

Tamura, S., Morikawa, Y., & Senba, E. (2003). Localization of oncostatin M receptor beta in adult and developing CNS. *Neuroscience, 119*, 991–997.

Taylor, M. J., Letendre, S. L., Schweinsburg, B. C., Alhassoon, O. M., Brown, G. G., Gongvatana, A., et al. (2004). Hepatitis C virus infection is associated with reduced white matter N-acetylaspartate in abstinent methamphetamine users. *Journal of the International Neuropsychological Society, 10*, 110–113.

Theodore, S., Cass, W. A., & Maragos, W. F. (2006). Involvement of cytokines in human immunodeficiency virus-1 protein Tat and methamphetamine interactions in the striatum. *Experimental Neurology, 199*, 490–498.

Theodore, S., Cass, W. A., Nath, A., Steiner, J., Young, K., & Maragos, W. F. (2006). Inhibition of tumor necrosis factor-alpha signaling prevents human immunodeficiency virus-1 protein Tat and methamphetamine interaction. *Neurobiology of Disease, 23*, 663–668.

Theodore, S., Stolberg, S., Cass, W. A., & Maragos, W. F. (2006). Human immunodeficiency virus-1 protein tat and methamphetamine interactions. *Annals of the New York Academy of Sciences, 1074*, 178–190.

Thio, C. L., Goedert, J. J., Mosbruger, T., Vlahov, D., Strathdee, S. A., O'Brien, S. J., et al. (2004). An analysis of tumor necrosis factor alpha gene polymorphisms and haplotypes with natural clearance of hepatitis C virus infection. *Genes and Immunity, 5*, 294–300.

Thomas, D. M., Angoa Pérez, M., Francescutti-Verbeem, D. M., Shah, M. M., & Kuhn, D. M. (2010). The role of endogenous serotonin in methamphetamine-induced neurotoxicity to dopamine nerve endings of the striatum. *Journal of Neurochemistry, 115*, 595–605.

Thomas, D. M., Francescutti-Verbeem, D. M., & Kuhn, D. M. (2008a). Methamphetamine-induced neurotoxicity and microglial activation are not mediated by fractalkine receptor signaling. *Journal of Neurochemistry, 106*, 696–705.

Thomas, D. M., Francescutti-Verbeem, D. M., & Kuhn, D. M. (2008b). The newly synthesized pool of dopamine determines the severity of methamphetamine-induced neurotoxicity. *Journal of Neurochemistry, 105*, 605–616.

Thomas, D. M., Francescutti-Verbeem, D. M., Liu, X., & Kuhn, D. M. (2004). Identification of differentially regulated transcripts in mouse striatum following methamphetamine treatment–an oligonucleotide microarray approach. *Journal of Neurochemistry, 88*, 380–393.

Thomas, D. M., Walker, P. D., Benjamins, J. A., Geddes, T. J., & Kuhn, D. M. (2004). Methamphetamine neurotoxicity in dopamine nerve endings of the striatum is associated with microglial activation. *The Journal of Pharmacology and Experimental Therapeutics, 311*, 1–7.

Vaarmann, A., Gandhi, S., & Abramov, A. Y. (2010). Dopamine induces Ca2+ signaling in astrocytes through reactive oxygen species generated by monoamine oxidase. *The Journal of Biological Chemistry, 285*, 25018–25023.

Vandenbark, A. A., Rich, C., Mooney, J., Zamora, A., Wang, C., Huan, J., et al. (2003). Recombinant TCR ligand induces tolerance to myelin oligodendrocyte glycoprotein 35–55 peptide and reverses clinical and histological signs of chronic experimental autoimmune encephalomyelitis in HLA-DR2 transgenic mice. *Journal of Immunology, 171*, 127–133.

Volkow, N. D., Chang, L., Wang, G. J., Fowler, J. S., Franceschi, D., Sedler, M., et al. (2001). Loss of dopamine transporters in methamphetamine abusers recovers with protracted abstinence. *The Journal of Neuroscience, 21*, 9414–9418.

Wang, C., Gold, B. G., Kaler, L. J., Yu, X., Afentoulis, M. E., Burrows, G. G., et al. (2006). Antigen-specific therapy promotes repair of myelin and axonal damage in established EAE. *Journal of Neurochemistry, 98*, 1817–1827.

Wang, Q., Shin, E. J., Nguyen, X. K., Li, Q., Bach, J. H., Bing, G., et al. (2012). Endogenous dynorphin protects against neurotoxin-elicited nigrostriatal dopaminergic neuron damage and motor deficits in mice. *Journal of Neuroinflammation, 9*, 124.

Wasik, A. M., Millan, M. J., Scanlan, T., & Barnes, N. M. (2012). Gordon evidence for functional trace amine associated receptor-1 in normal and malignant B cells. *Leukemia Research, 36*, 245–249.

Weber, E., Morgan, E. E., Iudicello, J. E., Blackstone, K., Grant, I., Ellis, R. J., et al. (2013). Substance use is a risk factor for neurocognitive deficits and neuropsychiatric distress in acute and early HIV infection. *Journal of Neurovirology*, *19*, 65–74.

Wey, S. P., Wu, H. Y., Chang, F. C., & Jan, T. R. (2008). Methamphetamine and diazepam suppress antigen-specific cytokine expression and antibody production in ovalbumin-sensitized BALB/c mice. *Toxicology Letters*, *181*, 157–162.

Whitney, N. P., Eidem, T. M., Peng, H., Huang, Y., & Zheng, J. C. (2009). Inflammation mediates varying effects in neurogenesis: Relevance to the pathogenesis of brain injury and neurodegenerative disorders. *Journal of Neurochemistry*, *108*, 1343–1359.

Wilhelm C. J., Hashimoto J. G., Roberts M. L., Loftis J. M., Wiren K. M. (2014) Distinct glucocorticoid signaling associated with ethanol-induced inflammotoxicity. 37th annual research society on alcoholism (RSA) scientific meeting, June 2014 (abstract accepted).

Wilhelm, C. J., Johnson, R. A., Lysko, P. G., Eshleman, A. J., & Janowsky, A. (2004). Effects of methamphetamine and lobeline on vesicular monoamine and dopamine transporter-mediated dopamine release in a cotransfected model system. *The Journal of Pharmacology and Experimental Therapeutics*, *310*, 1142–1151.

Wires, E. S., Alvarez, D., Dobrowolski, C., Wang, Y., Morales, M., Karn, J., et al. (2012). Methamphetamine activates nuclear factor kappa-light-chain-enhancer of activated B cells (NF-κB) and induces human immunodeficiency virus (HIV) transcription in human microglial cells. *Journal of Neurovirology*, *18*, 400–410.

Yamamoto, B. K., Moszczynska, A., & Gudelsky, G. A. (2010). Amphetamine toxicities: Classical and emerging mechanisms. *Annals of the New York Academy of Sciences*, *1187*, 101–121.

Yamamoto, B. K., & Raudensky, J. (2008). The role of oxidative stress, metabolic compromise, and inflammation in neuronal injury produced by amphetamine-related drugs of abuse. *Journal of Neuroimmune Pharmacology*, *3*, 203–217.

Yan, Y., Nitta, A., Koseki, T., Yamada, K., & Nabeshima, T. (2012). Dissociable role of tumor necrosis factor alpha gene deletion in methamphetamine self-administration and cue-induced relapsing behavior in mice. *Psychopharmacology*, *221*, 427–436.

Ye, L., Peng, J. S., Wang, X., Wang, Y. J., Luo, G. X., & Ho, W. Z. (2008). Methamphetamine enhances Hepatitis C virus replication in human hepatocytes. *Journal of Viral Hepatitis*, *15*, 261–270.

Yu, Q., Zhang, D., Walston, M., Zhang, J., Liu, Y., & Watson, R. R. (2002). Chronic methamphetamine exposure alters immune function in normal and retrovirus-infected mice. *International Immunopharmacology*, *2*, 951–962.

Zhang, X. Q., Cui, Y., Cui, Y., Chen, Y., Na, X. D., Chen, F. Y., et al. (2012). Activation of p38 signaling in the microglia in the nucleus accumbens contributes to the acquisition and maintenance of morphine-induced conditioned place preference. *Brain, Behavior, and Immunity*, *26*, 318–325.

Zhang, L., Kitaichi, K., Fujimoto, Y., Nakayama, H., Shimizu, E., Iyo, M., et al. (2006). Protective effects of minocycline on behavioral changes and neurotoxicity in mice after administration of methamphetamine. *Progress in Neuro-Psychopharmacology & Biological Psychiatry*, *30*, 1381–1393.

Zhu, J. P., Xu, W., & Angulo, J. A. (2006). Methamphetamine-induced cell death: Selective vulnerability in neuronal subpopulations of the striatum in mice. *Neuroscience*, *140*, 607–622.

Zorick, T., Nestor, L., Miotto, K., Sugar, C., Hellemann, G., Scanlon, G., et al. (2010). Withdrawal symptoms in abstinent methamphetamine-dependent subjects. *Addiction*, *105*, 1809–1818.

CHAPTER EIGHT

Marijuana Use and Brain Immune Mechanisms

Guy A. Cabral[1], Melissa Jamerson
Department of Microbiology and Immunology, School of Medicine, Virginia Commonwealth University, Richmond, Virginia, USA
[1]Corresponding author: e-mail address: gacabral@vcu.edu

Contents

Abstract

The recreational smoking of marijuana, or *Cannabis sativa*, has become widespread, including among adolescents. Marijuana contains a class of compounds known as phytocannabinoids that include cannabidiol (CBD) and Δ^9-tetrahydrocannabinol (THC). THC is the major psychoactive component in marijuana, but also exhibits immunosuppressive activity. CBD, while not psychotropic, also modulates immune function, but its mechanism of action appears to differ from that of THC. Since both compounds are highly lipophilic, they readily passage the blood–brain barrier and access the central nervous system. Since CBD is not psychotropic, it has been considered as a candidate therapeutic compound for ablating neuropathological processes characterized by hyperinflammation. However, an unresolved question centers around the impact of these compounds on immune-competent cells within the CNS in relation to susceptibility to infection. There are accumulating data indicating that THC inhibits the migratory capability of macrophage-like cells resident in the CNS, such as microglia, toward nodes of microbial invasion. Furthermore, phytocannabinoids have been reported to exert developmental and long-term effects on the immune system suggesting that exposure to these substances during an early stage in life has the potential to alter the fundamental neuroimmune response to select microbial agents in the adult.

International Review of Neurobiology, Volume 118
ISSN 0074-7742
http://dx.doi.org/10.1016/B978-0-12-801284-0.00008-7

1. INTRODUCTION

The recreational smoking of *Cannabis sativa*, mainly its resin (hashish) or the chopped flowering tops of the plant (marijuana), has become widespread. Because of its lack of acute life-threatening effects, marijuana has been often thought of as a "soft drug." However, this perception may have to be revised in face of mounting scientific data that it impairs memory and learning and has residual effects on lung, immune, and reproductive function. Products derived from smoking marijuana vary in the content of Δ^9-tetrahydrocannabinol (THC), the major psychoactive component (reviewed in Nahas and Latour, 1993; Fig. 8.1). The flowering tops of the plant may contain 1–6% THC total weight, although the percentage by weight of this component may reach 8% in hashish and exceed 50% in hash oil. However, the fiber type of marijuana that is used for the manufacture of rope and twine contains little THC. In addition to THC, over 100 other cannabinoids have been identified in the particulate phase of the marijuana plant, mainly cannabidiol (CBD) and cannabinol (CBN) that are not psychoactive but are biologically active (Fig. 8.1). The term phytocannabinoid has been used to distinguish these plant-derived compounds from cannabinoids that have been synthesized in the laboratory. Upon heating, phytocannabinoids rapidly decarboxylate and at the temperature of pyrolysis (200–400 °C) undergo aromatization. Polycydic aromatic hydrocarbons have been identified in marijuana smoke for which the proportions of the higher molecular weight compounds, particularly the carcinogens benzo(α)pyrene and benz(α)anthracene, are greater than in tobacco smoke. The gas phase of marijuana smoke includes toxic substances such as carbon monoxide, hydrogen cyanide, and nitrosamines, although these are present in equivalent concentrations in tobacco smoke. THC and other cannabinoids are lipophilic and are stored in liver, lung, spleen, and neutral fat. Because THC has a half-life of 8 days in fat, it may take up to 1 month for complete elimination of a single dose. Furthermore, since THC is a polar compound it is slowly metabolized into more water-soluble, nonpsychoactive metabolites. The bioavailability of inhaled and ingested THC is 20% and 6%, respectively. Less than 1% of the bioavailable THC reaches the brain, a fact that illustrates the psychoactive potency of this compound (Chesher & Jackson, 1985).

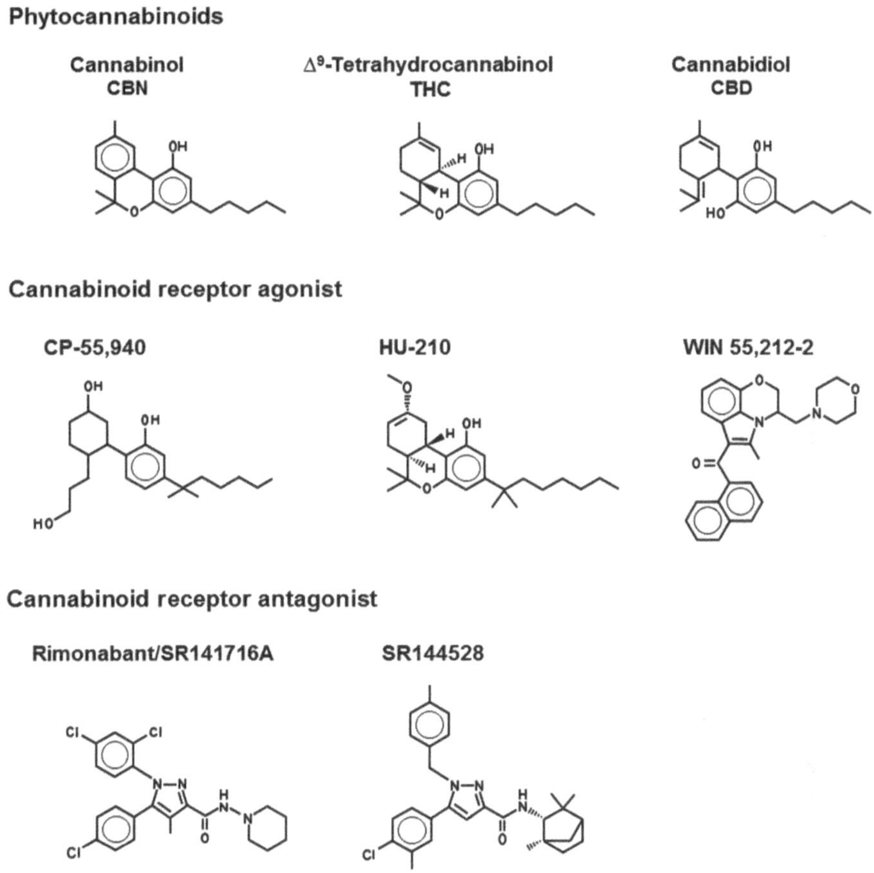

Figure 8.1 Structures of representative cannabinoids (phytocannabinoids, cannabinoid receptor agonists, and cannabinoid receptor antagonists). Cannabinol/CBN: 6,6,9-trimethyl-3-pentylbenzo[c]chromen-1-ol, Δ^9 tetrahydrocannabinol/THC; (6aR,10aR)-6,6,9-trimethyl-3-pentyl-6a,7,8,10a-tetrahydro-6*H*-benzo[c]chromen-1-ol, Cannabidiol/CBD; 2-[(1R,6R)-6-Isopropenyl-3-methyl-2-cyclohexen-1-yl]-5-pentyl-1,3-benzenediol, CP-55,940; 2-[(1R,2R,5R)-5-hydroxy-2-(3-hydroxypropyl) cyclohexyl]-5-(2-methyloctan-2-yl)phenol, HU-210; (6aS,10aR)-9-(hydroxymethyl)-6,6-dimethyl-3-(2-methyloctan-2-yl)-6a,7,10,10a-tetrahydrobenzo[c]chromen-1-ol, WIN 55,212-2; [(3R)-5-methyl-3-(morpholin-4-ylmethyl)-2,3-dihydro[1,4]oxazino[2,3,4-hi]indol-6-yl](naphthalen-1-yl) methanone, Rimonabant/SR141716A; 5-(4-chlorophenyl)-1-(2,4-dichlorophenyl)-4-methyl-*N*-(piperidin-1-yl)-1*H*-pyrazole-3-carboxamide and SR144528; 5-(4-chloro-3-methylphenyl)-1-[(4-methylphenyl)methyl]-*N*-[(1R,3S,4S)-2,2,4-trimethyl-3-biccyclo[2.2.1] heptanyl]pyrazole-3-carboxamide. CBD, CBN, and THC are found natively in *Cannabis sativa*.

2. PHYTOCANNABINOIDS AND IMMUNE FUNCTION

THC also has been shown to alter immune function, most of its effects being immune suppressive (reviewed in Cabral & Staab, 2005). The purification and structural characterization of THC (Gaoni & Mechoulam, 1971) led to the chemical synthesis of various cannabinoid analogs that have been used extensively in structure–activity studies to characterize cannabinoid-mediated effects *in vitro* and *in vivo* related to the central nervous and immune systems and contributed to the definition of the mechanisms by which cannabinoids exert their effects. THC has been reported to suppress the antibody response of humans and animals (Klein, Friedman, & Specter, 1998) and to suppress a variety of activities of T lymphocytes (Kaminski, 1998; Klein et al., 2004). Administration of THC to mice also resulted in inhibition of natural killer (NK) cytolytic activity and in reduction of interferon-gamma (IFNγ) levels (Massi, Fuzio, Vigano, Sacerdote, & Parolaro, 2000). In addition, THC has been reported to abolish the functional activities of macrophages and macrophage-like cells, including macrophage-like cell contact-dependent cytolysis of tumor cells and the processing of antigens (Burnette-Curley, Marciano-Cabral, Fischer-Stenger, & Cabral, 1993; Klein, Kawakami, Newton, & Friedman, 1991; McCoy, Matveyeva, Carlisle, & Cabral, 1999). It has been reported also that THC alters the production of chemokines and cytokines, leading to a perturbation in the homeostatic balance between proinflammatory (Th1) and antiinflammatory (Th2) activities. Proinflammatory cytokines such as interleukin-1 (IL-1) and tumor necrosis factor-α (TNF-α) promote systemic inflammation. Antiinflammatory cytokines such as IL-4 and IL-10, on the other hand, play an immunoregulatory role in controlling the proinflammatory response. Klein, Newton, Nakachi, and Friedman (2000) showed that treatment of BALB/c mice with THC resulted in a decrease in levels of the proinflammatory (Th1) cytokines IFNγ and IL-12 and an increase in the level of the antiinflammatory cytokine IL-4 in response to infection with *Legionella pneumophila*. Conversely, through the use of a mouse lung tumor model in which THC mediated a decrease in tumor immunogenicity, Zhu et al. (2000) demonstrated that levels of the immune inhibitory Th2 cytokines IL-10 and transforming growth factor were augmented, while the level of the immune stimulatory Th1 cytokine IFNγ was downregulated. These collective results suggested that THC exposure alters the Th1/Th2 cytokine profile in experimental animals such that the balance of chemokine and cytokine functional activity shifts from that of a Th1-type proinflammatory cytokine profile to that of a Th2-type antiinflammatory cytokine profile. Such a shift in the Th1/Th2 cytokine profile could contribute to altered proinflammatory responsiveness to infection with select biological agents such as bacteria and viruses. Although the preponderance of scientific reports indicates that THC inhibits immune functional activities *in vitro* and *in vivo*, it is now recognized that other phytocannabinoids such as CBD and CBN may alter the functional activities of the immune system. However, whether the mechanism by which these phytocannabinoids alter immune function is comparable to that exerted by THC remains to be defined.

3. IMMUNE MODULATION AND CANNABINOID RECEPTORS

Studies in which synthetic cannabinoid compounds such as CP-55,940, HU-210, WIN 55,212, JWH-015, ACEA, SR141716A, and SR144528 (Fig. 8.1) have been used have provided important insights into the functional relevance and mechanism of action by which THC exerts its effects on the immune system. Furthermore, these collective studies have served as a basis for the identification of specific binding sites in mammalian brain and peripheral nonneuronal tissues which are now recognized as representing cannabinoid receptors (Matsuda, Lolait, Brownstein, Young, & Bonner, 1990; Munro, Thomas, & Abu-Shaar, 1993). To date, two cannabinoid receptors that meet stringent pharmacological and molecular criteria have been identified (Fig. 8.2). The first of these, the cannabinoid receptor type 1 (CBR1), is a serpentine, seven-transmembranal, G protein-coupled receptor that has been found by radioligand binding and *in situ* mRNA hybridization to be distributed throughout the brain and localized predominantly in the cerebellum, cerebral cortex, hippocampus, basal ganglia, and spinal cord (Herkenham et al., 1990; Matsuda et al., 1990; Westlake, Howlett, Bonner, Matsuda, & Herkenham, 1994). The CB1R also has been found in testis (Gerard, Mollereau, Vassart, & Parmentier, 1991) and, to a lesser extent, in other tissues and cells. The psychotropic effects attributed to THC are due to activation of this receptor (Agarwal et al., 2007; Kunos, Osei-Hyiaman, Batkai, Sharkey, & Makriyannis, 2009).

A second serpentine, seven-transmembranal, G protein-coupled cannabinoid receptor, the cannabinoid receptor type 2 (CB2R), has been identified in cells and tissues of the immune system (Munro et al., 1993). This receptor possesses approximately 44% amino acid similarity to the CB1R.

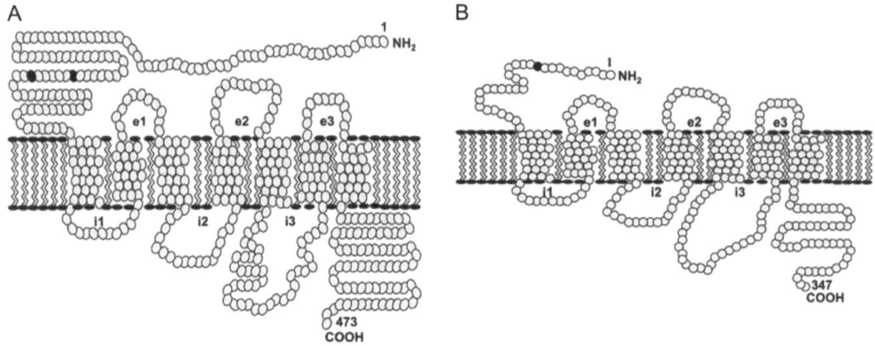

Figure 8.2 Diagrammatic representation of the CB1R and CB2R. (A) Mouse CB1R: The solid circles denote putative glycosylation sites at asparagine residues 78 and 84. The extracellular and intracellular loops are denoted as e1, e2, and e3, and i1, i2, and i3, respectively. (B) Mouse CB2R: The solid circle denotes a putative glycosylation site at asparagine residue 11 of CB2. The extracellular and intracellular loops are denoted as e1, e2, and e3, and i1, i2, and i3, respectively. The CB1Rs of different mammalian species exhibit a high level of similarity. The CB2R exhibits greater interspecies differences with the deduced amino acid sequence of the mouse CB2R differing from that of human in 60 residues (82% similarity) (Shire et al., 1996). *Panel (B) From Cabral and Griffin-Thomas (2008).*

The level of the CB2R varies among different immune cell populations, with B lymphocytes expressing the highest levels followed by macrophages, monocytes, NK cells, and polymorphonuclear cells, in that order (Galiegue et al., 1995; Schatz, Lee, Condie, Pulaski, & Kaminski, 1997). Early studies indicated that the distribution of the CB2R was confined to peripheral non-neuronal sites. However, it is now recognized that this receptor is expressed by different subsets of immune-competent cells found in the CNS (Cabral & Marciano-Cabral, 2005; Carlisle, Marciano-Cabral, Staab, Ludwick, & Cabral, 2002; Carrier et al., 2004; Fernandez-Ruiz et al., 2007; Nunez et al., 2004; Ramirez, Blazquez, Gomez del Pulgar, Guzman, & de Ceballos, 2005). Most of the immunomodulatory effects attributed to THC have been linked to activation of the CB2R.

There is evidence for the existence of additional cannabinoid receptors, based primarily on *in vivo* studies in which cannabinoid receptor knock-out mice have been used to investigate the pharmacology and pharmacokinetics of cannabinoids (Breivogel, Griffin, Di Marzo, & Martin, 2001; Di Marzo et al., 2000; Jarai et al., 1999). Included among these candidate cannabinoid receptors is GPR55, a seven-transmembranal G protein-coupled receptor first cloned and identified *in silico* from an expressed sequence tags database

(Baker, Pryce, Davies, & Hiley, 2006; Pertwee, 2007; Sawzdargo et al., 1999). GPR55 is activated by the phytocannabinoids THC and CBD, the synthetic cannabinoids abnormal CBD, HU-210, and CP-55,940, and the endogenous cannabinoids anandamide, 2-arachidonoylglycerol, and noladin ether (Ryberg et al., 2007). However, unlike the CB1R and the CB2R, GPR55 is coupled to a G-alpha (Gα) protein instead of a Gi/o protein (Ryberg et al., 2007), is not activated by the synthetic agonist WIN 55212-2, and increases intracellular calcium levels upon activation (Lauchner et al., 2008). Nevertheless, to date, a novel non-CB1, non-CB2 cannabinoid receptor (non-CB1R, non-CB2R) that meets rigid pharmacological and functional criteria as selective for cannabinoid ligands has yet to be cloned and characterized at the molecular level (Breivogel et al., 2001; Di Marzo et al., 2000; Jarai et al., 1999; Wiley & Martin, 2002).

4. MARIJUANA AND NEUROIMMUNITY

The blood–brain barrier (BBB) is a selectively permeable continuous layer of microvascular endothelial cells that is interconnected by tight junctions. It serves to regulate the traffic into the CNS of substances and cells that are present in the circulatory system. Perivascular macrophages are found in the region surrounding the endothelium of brain capillaries and are replenished through the migration of circulating monocytes. In addition, astrocytes are found in this region and play a critical homeostatic role by maintaining contact with endothelial cells and regulating permeability of the endothelial layer through the release of soluble factors. Within the CNS are found immune-competent cells such as microglia, perivascular macrophages, and pericytes. These cells, once activated, release a plethora of inflammatory factors that secondarily activate astrocytes, resulting in a cascade of chemokines and cytokines that promote expansion of the neuroinflammatory response. The recognition that these resident immune cells within the CNS also express the CB2R, and that they were responsive to activation by cannabinoids, suggested that exogenously introduced cannabinoids had the capacity to modulate their functional activities. Furthermore, they could do so in a fashion selectively differential from that resulting from activation of the CB1R that was linked to elicitation of psychotropic effects. Indeed, cannabinoids that target the CB2R have been considered as therapeutic agents for management of multiple sclerosis (Maresz et al., 2007; Zhang, Adler, et al., 2009), ischemic/perfusion injury following an induced stroke (Ni et al., 2004; Zhang et al., 2007; Zhang, Martin et al., 2009), rheumatoid arthritis (Sumariwalla et al., 2004), inflammatory bowel disease (Storr et al., 2008, 2009), inflammatory autoimmune diabetes (Li, Kaminski, & Fischer, 2001), spinal cord injury (Adhikary et al., 2011; Baty et al., 2008), sepsis (Tschöp et al., 2009), autoimmune uveoretinitis (Xu et al., 2007), osteoporosis (Ofek et al., 2006), and systemic sclerosis (Servettaz et al., 2010). This potential for selectively targeting the CB2R is especially relevant since cannabinoids are highly lipophilic molecules, readily cross the BBB, and have the capacity to interact selectively with target cells at concentrations that do not engender overt cytotoxic effects.

However, while it has been proposed that the CB2R may have therapeutic potential, especially when considered for dampening of untoward immune responses associated with a variety of neuropathological conditions, exogenously introduced cannabinoids have the potential to place a host at greater susceptibility to infection. This potential could be realized under conditions in which THC or other exogenous cannabinoid is introduced into the CNS through the medium of marijuana smoke. However, the acquisition of direct data that unequivocally links marijuana use in humans to immune dysfunction within the CNS has been difficult to obtain. Marijuana users may partake of other drugs that affect immune function and dissecting the relative contribution of distinct phytocannabinoids has been difficult. Furthermore, because marijuana contains a plethora of phytocannabinoids, attributing a specified action as linked functionally to THC, CBD, or CBN divorced from their potential interactive action is difficult. The complexity is further amplified by the emerging scientific data that distinct phytocannabinoids may activate immune cells by receptor-mediated as well as by nonreceptor-mediated modes. Thus, in order to garner insight as to the potential linkage between marijuana use in humans and compromised neuroimmune function, investigators have resorted to the use of purified synthetic phytocannabinoid preparations in cell culture models and in experimental animals. Monnet-Tschudi et al. (2008) assessed THC accumulation, metabolism, and cell-type-specific adverse effects in aggregating brain cell cultures. Mixed-cell aggregate cultures of fetal rat telencephalon were used as an *in vitro* model, as well as aggregates enriched either in neurons or in glial cells. It was found that THC accumulated preferentially in neurons, and that glia–neuron interactions decreased THC accumulation. The quantification of 11-OH–THC and of THC–COOH showed that brain aggregates were capable of THC metabolism. No cell-type difference was found for the metabolite 11-OH–THC, whereas the THC–COOH content was higher in mixed-cell cultures. Neurons, and particularly GABAergic neurons, were most sensitive to THC. JWH-015, a CB2R agonist, showed effects similar to THC, whereas ACEA, a CB1R agonist, had no effect. The expression of IL-6 was upregulated after treatment with THC or JWH-015, whereas the expression of TNF-α remained unchanged. These results suggested that the adverse effects of THC were related either to THC accumulation or to CBR activation and associated with IL-6 upregulation. Nevertheless, while data derivative from the use of experimental models must be interpreted with appropriate caveats, they converge on the outcome that phytocannabinoids such as THC and CBD when used at biologically relevant concentrations alter immune function.

5. EFFECT OF PHYTOCANNABINOIDS ON MICROGLIA

The preponderance of the data that are available concerning the effects of phytocannabinoids on CNS immune functional activity involves their action on microglia. Microglia are a population of glial cells that act as resident macrophages of the CNS and represent 10–15% of the total glial population. Once considered to be scavenger cells in this compartment, it is now recognized that they exert a major role in CNS remodeling and regeneration (reviewed in Moore & Thanos, 1996). Chronic activation of microglia, however, can exacerbate pathologies of the CNS including AIDS dementia and Alzheimer's disease (Dickson, Lee, Mattiace, Yen, & Brosnan, 1993). Microglia may contribute to the neurodegeneration by secreting nitric oxide (NO), IL-1β, IL-6, and TNF-α (Lee, Liu, Dickson, Brosnan, & Berman, 1993; Merrill, Ignarro, Sherman, Melinek, & Lane, 1993). In this context, select cannabinoids have been proposed as having therapeutic potential in that they may downmodulate the excessive inflammatory response. However, there is evidence that phytocannabinoids may decrease inflammatory responses that are consequent of bacterial infection with a CNS dimension. Puffenbarger, Boothe, and Cabral (2000) examined the effect of cannabinoids on the induction of cytokine messenger RNAs (mRNAs) by rat microglia in response to bacterial lipopolysaccharide (LPS), a component that is found in the outer membrane of Gram-positive bacteria. Exposure of neonatal rat cortical microglial cells to THC resulted in reduced amounts of LPS-induced mRNAs for IL-1α, IL-1β, IL-6, and TNF-α. Of these proinflammatory cytokine mRNAs, the response of that for IL-6 was exquisitely sensitive to THC. When the paired enantiomers CP-55,940 or CP-56,667 were used as a measure of enantiomeric selectivity as a hallmark of a receptor-mediated action, a similar inhibition of LPS-induced cytokine mRNA expression was obtained. A comparable inhibitory outcome was obtained when the paired enantiomers levonantradol and dextronantradol were employed. Neither the CB1R-selective antagonist SR141716A nor the CB2R-selective antagonist SR144528 was able to reverse the inhibition of cytokine mRNA. The collective results indicated that THC down-modulated levels of proinflammatory cytokine mRNAs in rat microglia, but did so by a mode that did not involve a known cannabinoid receptor. Cutando et al. (2013) reported that microglial activation served as an underlying basis for cerebellar deficits produced by repeated use of marijuana. It was found that subchronic administration of THC to mice activated cerebellar microglia and increased the expression of neuroinflammatory markers, including that of IL-1β. The neuroinflammatory phenotype correlated with deficits in cerebellar conditioned learning and fine motor coordination. The neuroinflammatory phenotype was readily detectable in the cerebellum of mice with global loss of the CB1R (i.e., CB1R, Cb1 (−/−) mice) and in mice lacking CB1R in the cerebellar parallel fibers, suggesting that CB1R downregulation in the cerebellar molecular layer played a key role in THC-induced cerebellar deficits. Expression of CB2R and IL-1β mRNA was increased under neuroinflammatory conditions in activated CD11b-positive microglia. Furthermore, administration of the immunosuppressant minocycline or an inhibitor of IL-1β receptor signaling prevented the deficits in cerebellar function in Cb1(−/−) and THC-withdrawn mice. The results suggested that cerebellar microglial activation played a crucial role in the cerebellar deficits induced by repeated cannabis exposure. Thus, while THC affects microglial responsiveness, the mechanism through which it alters this responsiveness, and whether the altered activity is attributed to activation of a specified receptor, remains to be resolved.

Studies conducted on the effects of THC on microglia have been extended to include those involving CBD. The rationale for examining the effects of this phytocannabinoid on microglial function is based on recognition that it has immunomodulatory activity divorced from elicitation of psychotropic effects. Carrier, Auchampach, and Hillard (2006) examined the effects of both THC and CBD on microglial proliferation and found that these compounds potently inhibited [^{3}H]thymidine incorporation into a mouse microglial cell line while having no effect on cell cycle. Treatment with THC and CBD decreased [^{3}H]thymidine uptake into microglia. CBD and, less potently, THC decreased uptake of [^{3}H]adenosine to a similar extent as [^{3}H]thymidine in both mouse microglia and mouse RAW264.7

macrophage-like cells. Binding studies confirmed that CBD bound to the equilibrative nucleoside transporter 1. Because adenosine agonists have antiinflammatory effects, and because uptake of adenosine is a primary process that is involved in terminating adenosine signaling, these investigators tested the hypothesis that CBD is immunosuppressive since it enhanced endogenous adenosine signaling. *In vivo* treatment with CBD decreased TNF-α production in LPS-treated mice, an effect that was reversed with an A2A adenosine receptor antagonist and abolished in A2A receptor knock-out mice. This G protein-coupled receptor has been reported to negatively regulate overreactive immune cells so as to protect tissues from collateral inflammatory damage (Ohta & Sitkovsky, 2001). It was suggested that CBD had the ability to enhance adenosine signaling through inhibition of its uptake and that a non-cannabinoid receptor-mediated mechanism accounted for the CBD-induced decrease in inflammation. Kozela et al. (2010) reported that THC and CBD differentially inhibited the LPS-activated NF-κB and interferon-β/STAT proinflammatory pathways in BV-2 microglial-like cells. It was found that THC and CBD, the two major cannabinoids present in marijuana, decreased the production and release of proinflammatory cytokines, including IL-1β, IL-6, and IFN-β from LPS-activated microglial cells. The cannabinoid antiinflammatory action did not appear to involve the CB1R and the CB2R or the abn-CBD-sensitive receptor, a putative receptor that is sensitive to a synthetic regioisomer of CBD (abn-CBD). In addition, it was found that THC and CBD acted through different, although partially overlapping, mechanisms. CBD, but not THC, reduced the activity of the NF-κB pathway, a primary pathway regulating the expression of proinflammatory genes. Moreover, CBD, but not THC, upregulated the activation of the STAT3 transcription factor, an element of homeostatic mechanism(s) inducing antiinflammatory events. Following CBD treatment, a decreased level of mRNA for the suppressor of cytokine signaling (SOC3) gene, a main negative regulator of STATs and particularly of STAT3, was observed. However, both CBD and THC decreased the activation of the LPS-induced STAT1 transcription factor, a key player in IFN-β-dependent proinflammatory processes. Thus, CBD and THC appeared to vary in their effects on antiinflammatory pathways, including the NF-κB and IFN-β-dependent pathways. Wu et al. (2012) investigated the proapoptotic effect of CBD on primary microglial cells. Treatment of mouse primary microglial cultures with CBD resulted in a time- and concentration-dependent induction of apoptosis, as shown by increase in hypodiploid cells and DNA strand breaks, and marked activation of both caspase-8 and -9. Mechanistic studies revealed that antioxidants such as *N*-acetyl-L-cysteine and glutathione, that the putative GPR55 agonist abn-CBD, and that specific antagonists for the vanilloid receptor, the CB1R, and the CB2R did not counteract the apoptosis induced by CBD. In contrast, methyl-β-cyclodextrin (MCD), a lipid raft disruptor, potently attenuated CBD-induced microglial apoptosis and caspase activation. Furthermore, CBD-induced lipid raft coalescence and augmented the expression of GM1 ganglioside and caveolin-1, all of which were attenuated by MCD. The results suggested that CBD induced a proapoptotic effect in primary microglia through lipid raft coalescence and the elevation of expression of GM1 ganglioside and caveolin-1.

The anti-neuroinflammatory effects of phytocannabinoids may be exerted differentially in male versus female animals and may be linked to age differences. For example, adolescent exposure to THC may exert sex-dependent long-term effects on neuroinflammation, serotoninergic, and cannabinoid systems. Lopez-Rodriguez, Llorente-Berzal, Garcia-Segura, and Viveros (2014) examined the long-term effects of THC and 3,4-methylenedioxymethamphetamine (MDMA) on diverse neuroinflammation and neurotoxic markers in male and female Wistar rats. Rats were chronically treated with increasing doses of THC and/or MDMA during adolescence. The effects of THC and/or MDMA on glial reactivity and on serotoninergic and cannabinoid systems were assessed by immunohistochemistry in the hippocampus and parietal cortex. THC was shown to significantly increase the glial fibrillary acid protein (GFAP), or astrocytic, stained area in both sexes. In males, both drugs either separately or in combination induced a significant increase of the percentage of reactive microglial cells based on staining for the microglial marker-ionized calcium adapter molecule 1(Iba-1, also referred to as allograft inflammatory factor 1). In contrast, in females, each drug, when administered alone produced a significant decrease of this percentage, whereas the combination of both drugs resulted in a "normalization" to control values. In males, MDMA reduced the number of serotonin transporter-positive (SERT+) fibers. THC induced an opposite effect and the group receiving both drugs did not differ significantly from the controls. In females, MDMA reduced the number of SERT+ fibers and the combination of both drugs counteracted this effect. THC induced a significant reduction of the CB1R in females, an effect that was aggravated by the combination with MDMA. The collective results indicated that adolescent exposure to THC and/or MDMA induced long-term, sex-dependent neurochemical and glial alterations. Furthermore, the results indicated that combination drug exposure may lead to additive or synergistic immunomodulatory effects in the affected host.

6. MARIJUANA AND ASTROCYTES

Astrocytes are the most abundant cells in the CNS and provide biochemical support of endothelial cells that form the BBB, nutrients to nervous tissue, maintenance of extracellular ion balance, and repair of neuronal tissue. Astrocytes are sensitive to factors released from microglia and macrophage-like cells and, in response, elicit inflammatory factors that have the potential to generate a "cytokine storm." Because astrocytes, coexpress CB1R and CB2R (Stella, 2010), they have been considered as candidate targets for therapeutic manipulation of untoward inflammatory responses. However, by virtue of the fact that they express the CB1R, they have limited therapeutic potential for immune modulation using ligands that target the CB1R, the activation of which would engender psychotropic effects. Nevertheless, various investigators have conducted experiments to determine the level to which this receptor plays a functionally relevant role in cannabinoid-mediated modulation of astrocyte activities. Sanchez, Galve-Roperh, Canova, Brachet, and Guzman (1998) reported that THC-induced apoptosis in rat C6.9 glioma cells, as determined by DNA fragmentation and loss of plasma membrane asymmetry. THC stimulated sphingomyelin hydrolysis in these cells. However, THC and *N*-acetylsphingosine, a cell-permeable ceramide analog, induced apoptosis in several transformed neural cells but not in primary astrocytes or neurons. Neither the THC-induced apoptosis nor the THC-induced sphingomyelin breakdown was prevented by the CB1R-selective antagonist SR141716A. The investigators suggested that THC-induced apoptosis in glioma C6.9 cells relied on a CB1R-independent stimulation of sphingomyelin breakdown. Sanchez, Galve-Roperh, Rueda, and Guzman (1998) extended these studies to show that sphingomyelin hydrolysis and the mitogen-activated protein kinase (MAPK) cascade were involved in the THC-induced stimulation of glucose metabolism in primary rat astrocytes. THC increased the rate of glucose oxidation to CO_2 as well as the rate of glucose incorporation into phospholipids and glycogen. These effects of THC were mimicked by HU-210, a synthetic THC analog, and prevented by forskolin, pertussis toxin, and the CB1R antagonist SR141716A. THC did not affect basal cAMP levels but partially antagonized the forskolin-induced elevation of intracellular cAMP concentration. THC stimulated p42/p44 MAPK activity, Raf-1 phosphorylation, and Raf-1 translocation to the particulate cell fraction. In addition, the MAPK inhibitor PD 098095 and the phosphoinositide 3-kinase inhibitors wortmannin and LY 294002 were able to antagonize the THC-induced stimulation of glucose oxidation to CO_2, phospholipid synthesis, and glycogen synthesis. In addition, the involvement of sphingomyelin breakdown in the metabolic effects of THC was studied. THC produced a rapid stimulation of sphingomyelin hydrolysis that occurred concomitant to an elevation of intracellular ceramide levels. This effect was prevented by the CB1R antagonist SR141716A. Moreover, the cell-permeable ceramide analog D-erythro-*N*-octanoylsphingosine as well as exogenous sphingomyelinase were able to stimulate MAPK activity, to increase the amount of Raf-1 bound to the particulate cell fraction, and to stimulate glucose metabolism. The latter effect was prevented by PD 098059, a selective mitogen-activated protein (MAP) kinase inhibitor, and was not additive to that exerted by THC. The collective results suggested that THC produced a cannabinoid receptor-mediated stimulation of astrocyte metabolism that appeared to rely on sphingomyelin hydrolysis and MAPK stimulation. There is accumulating evidence that THC may exert a wide range of effects on a diverse array of gene products. Bindukumar et al. (2008) undertook genomic and proteomic analysis of the effects of cannabinoids on normal human astrocytes (NHA) to determine the global molecular effects of cannabinoids on NHA using genomic and proteomic analyses. NHA were treated with THC and assayed using gene microarrays and two-dimensional (2D) difference gel electrophoresis coupled with mass spectrometry to elucidate their genomic and proteomic profiles, respectively. The results showed that the expression of more than 20 translated protein gene products from NHA was differentially dysregulated by treatment with THC.

Pre- and perinatal exposure to THC may lead to long-term effects in both neurons and glial cells. Suárez, Bodega, Ramos, Fernandez-Ruiz, and Fernandez (2000) evaluated the responses of neurons and astroglial cells to pre- and perinatal exposure to THC in the substantia nigra (SN) of male and female rats, at three postnatal ages (PD21, PD30, and PD70), by immunohistochemical detection of tyrosine hydroxylase (TH) in dopaminergic neurons and of GFAP in astrocytes. The results showed that the effects of pre- and perinatal exposure to THC on neuronal and astroglial immunoreactivities in the SN (compacta and reticulata) varied with sex, with male rats being more susceptible than females. Prenatal exposure to THC decreased

TH immunoreactivity in the SN of males on PD21 when compared to both their controls and THC-exposed females of the same age. Furthermore, the TH expression decreased with age in THC-exposed males in the SN pars compacta, whereas it increased in controls. In contrast, TH expression was maintained at a stable level in the SN pars compacta of THC-exposed females from PD21. These differences in neuronal development caused by prenatal THC exposure were associated with significant differences in GFAP expression by astroglial cells in both sexes. On PD21, GFAP immunoreactivity decreased in the SN in THC-exposed male rats. Although GFAP expression increased in THC-exposed males with age, it did not reach control levels by PD70. In contrast, significantly increased GFAP expression in THC-exposed females on PD21 was observed, compared to controls and also to THC-exposed male rats. It was suggested that these THC-induced changes in glial development indicated that this cannabinoid accelerated the maturation of astrocytes in female rats, whereas astrocytic maturation was delayed in THC-exposed males. Suárez et al. (2002) also analyzed the responses of cerebellar astroglial cells to pre- and perinatal THC exposure in three postnatal ages and both sexes. To determine whether THC directly modified astroglial growth during development, the effects of this cannabinoid on astroglial morphological changes and on the expression of specific astroglial markers (GFAP and glutamine synthetase: GS) were investigated. It was demonstrated that the administration of THC during development had deleterious effects on astroglial maturation in the cerebellum. These results also confirmed that THC could interfere with astroglial differentiation in a mode that was dependent on sex. The data suggested that pre- and perinatal THC exposure directly interfered with astroglial maturation by disrupting normal cytoskeletal formation, as indicated by the irregular disposition of GFAP and the lower GFAP expression observed at all the ages studied. It was indicated that THC exposure during development could also modulate glutamatergic nervous activity since GS expression was reduced in THC-exposed brains. GS expression increased progressively after THC withdrawal, but this expression failed to reach control values even after 2 months following THC withdrawal. These results suggested that glutamate uptake was lower in glial cells exposed to THC. Furthermore, it was suggested that glutamatergic neurotransmission was affected by THC exposure during gestation and that cannabinoids exerted developmental toxicity, at least on astroglial cells, which could contribute to fetal brain growth retardation. The collective data suggest that prenatal and/or postnatal exposure to THC has potential to hinder maturation of astrocytes, an outcome that may predispose these cells to compromised responsiveness to neuroinvasive agents.

The effect of CBD on astrocyte function also has been examined. The data that are available suggest that CBD has a general protective effect within the CNS. Thus, CBD may act to counteract the potential deleterious effects of THC on astrocytes. For example, Lafuente et al. (2011) reported that CBD reduced brain damage and improved functional recovery after acute hypoxia-ischemia (HI) in newborn pigs. In their studies, newborn piglets exposed to acute HI received intravenous CBD (HI + CBD) or vehicle (HI + VEH). In HI + VEH, 72 h post-HI brain activity as assessed by amplitude-integrated EEG (aEEG), near-infrared spectroscopy (NIRS) parameters remained lower than normal, and neurobehavioral performance was abnormal. In the brain, there were fewer normal and more pyknotic neurons, while astrocytes were less numerous and swollen. Cerebrospinal fluid concentration of neuronal-specific enolase (NSE) and S100β protein (a multifunctional protein found in large amounts in astrocytes) and brain tissue percentage of TNF-α-positive cells were higher. In contrast, in HI +CBD, aEEG had recovered to $86 \pm 5\%$, NIRS parameters increased, and the neurobehavioral score normalized. HI-induced histological changes, and NSE and S100β concentration and TNF-α-positive cell increases were suppressed by CBD. It was concluded that post-HI administration of CBD-protected neurons and astrocytes, leading to histological, functional, biochemical, and neurobehavioral improvements. Nabissi, Morelli, Santoni, and Santoni (2013) suggested that CBD could have a potential therapeutic application in that triggering of the transient receptor potential vanilloid type 2 (TRPV2) channel by CBD-sensitized glioblastoma multiforme (GBM) cells to cytotoxic chemotherapeutic agents. Several chemotherapeutic agents such as temozolomide (TMZ), carmustine (BCNU), or doxorubicin (DOXO) have been employed for treatment of GBM, but display limited efficacy. Activation of the TRPV2 has been found to inhibit human GBM cell proliferation and overcome BCNU resistance of GBM cells. Thus, the involvement of CBD-induced TRPV2 activation, in the modulation of glioma cell chemosensitivity to TMZ, BCNU, and DOXO was evaluated. It was found that CBD increased TRPV2 expression and activity. CBD, by triggering TRPV2-dependent Ca(2+) influx, increased drug uptake and synergized with cytotoxic agents to induce apoptosis of glioma cells. Moreover, as the pore region of transient receptor potential channels is critical for ion channel permeation, it was demonstrated that deletion of the TRPV2 pore domain inhibited CBD-induced Ca(2+)

influx, drug uptake, and cytotoxic effects. Thus, coadministration of cytotoxic agents together with CBD increased drug uptake and potentiated cytotoxic activity in human glioma cells. These latter observations highlight the fact that distinct chemotherapeutic agents, and by implication cannabinoids, may interact to modulate glial cell functional activities. They also emphasize the difficulty of assignation of a particular mechanistic effect to a specified phytocannabinoid divorced from the milieu presented by other components in marijuana.

7. MARIJUANA AND INFECTIOUS AGENTS THAT TARGET THE CNS

The observation that select phytocannabinoids such as THC and CBD can alter immune function, including that attributed to immune-competent cells resident in the CNS, suggested that these compounds also had the capability to alter immune responsiveness to infectious neurotropic agents. In this context, by ablating the inflammatory response to an invasive microbial agent THC and CBD could place a host at increased risk of infection. The extent to which a phytocannabinoid mediates immune dysfunction may be linked to the nature of the immune response elicited against the invasive agent. This area of phytocannabinoid research has been understudied, principally due to the fact that cannabinoids that have immune modulatory capacity and little psychotropic activity, such as CBD, or selective agonists for the CB2R, have come under scrutiny as potential agents for moderating neuropathological processes that are characterized by hyperinflammatory responses. The data that are available converge on the outcome that THC compromises host resistance to protozoan infections of the CNS and effects migratory responses of CNS-related macrophage-like cells to select components of the human immunodeficiency virus type-1 (HIV-1). Studies by Marciano-Cabral, Ferguson, Bradley, and Cabral (2001) and Cabral and Marciano-Cabral (2004) demonstrated, using a mouse model of Granulomatous Amebic Encephalitis (GAE), that THC exacerbated CNS infection by the opportunistic ameba *Acanthamoeba culbertsoni* (*A. culbertsoni*). Mice administered THC and infected with these amebae exhibited dose-related higher mortalities than infected vehicle controls. The greater severity of disease for THC-treated mice was accompanied by decreased accumulation of macrophage-like cells at focal sites of infection in the brain (Fig. 8.3). Analysis of the migratory capability of macrophage-like cells toward extracellular products secreted by these amebae was inhibited by

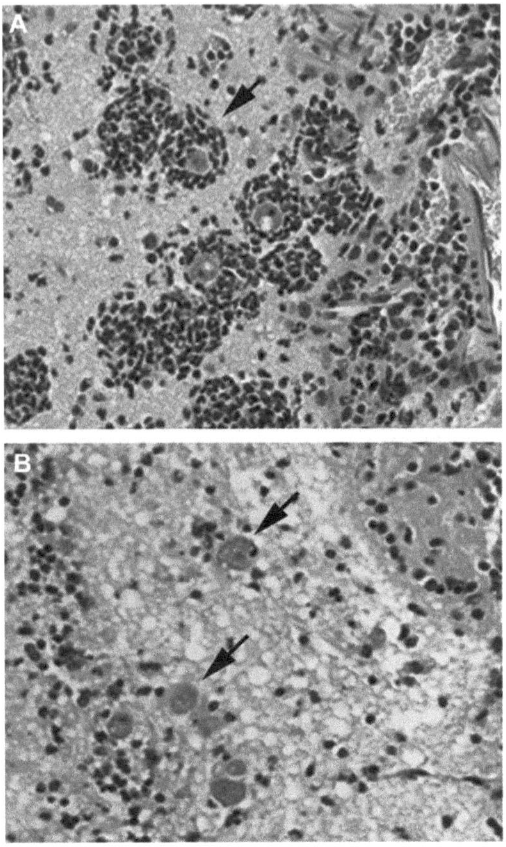

Figure 8.3 THC downregulates accumulation of macrophage-like cells at focal sites of *Acanthamoeba* in mouse brain. $(B_6C_3)F_1$ mice were treated once intraperitoneally with THC (25 mg kg^{-1}) or vehicle (ethanol:emulphor:saline, 1:1:18), inoculated intranasally with three 50% lethal doses (3 LD_{50}) of *A. culbertsoni*. Paraffin sections of brain were stained with hematoxylin and eosin. (A) Section from vehicle-treated mouse depicting accumulation of macrophage-like cells around *Acanthamoeba* (arrow). (B) Section from THC-treated mouse depicting *Acanthamoeba* in the brain in the absence of macrophage-like cell accumulation (arrows). THC, Δ^9-tetrahydrocannabinol. *From Cabral, Raborn, Griffin, Dennis, and Marciano-Cabral (2008).*

THC in a mode (Fig. 8.4) that was found to be linked pharmacologically to the CB2R (Fig. 8.5). In addition, THC administration resulted in decreased levels of mRNA for the proinflammatory cytokines IL-1α, IL-1β, and TNF-α for neonatal rat microglia co-cultured with the *Acanthamoeba*. However, whether these latter events are linked functionally to a cannabinoid receptor has not been resolved. These results indicated a potential for the

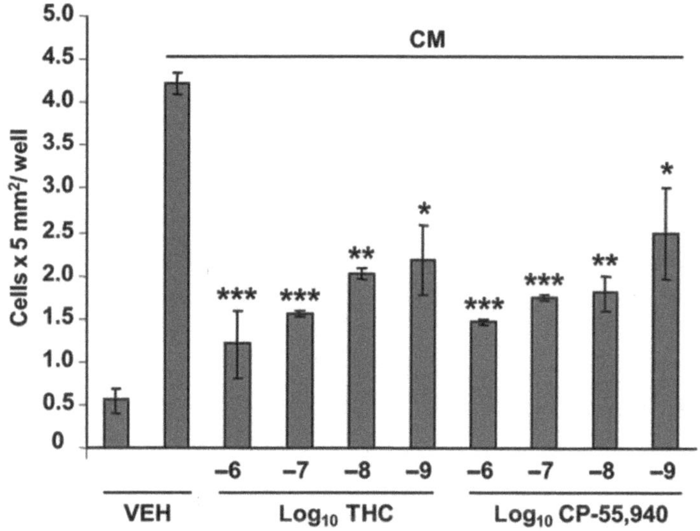

Figure 8.4 THC inhibits chemotaxis of microglia. Microglia were isolated from neonatal Sprague–Dawley rats and purified as described (Waksman, Olson, Carlisle, & Cabral, 1999), treated (3 h) with cannabinoid or vehicle (0.01% ethanol) and assessed (2 h) for migration against *Acanthamoeba*-conditioned medium (CM). The CB1R/CB2R partial agonist THC has a $K_i = 46$ nM at the CB2R, while the potent full agonist CP-55,940 has a $K_i = 0.9$ nM at the CB2R. $^*P<0.05$, $^{**}P<0.01$, $^{***}P<0.001$. $n=3$/group. CB1R, CB1 receptor; CB2R, CB2 receptor; CM, *Acanthamoeba*-conditioned medium; THC, Δ^9-tetrahydrocannabinol. *Acanthamoeba* were maintained in culture for 24 h to generate the *Acanthamoeba*-conditioned medium (CM) that harbors proteases and other factors released from amebae that serve as chemotactic stimuli for attracting microglia. The full agonist CP-55,940 elicited a concentration-related inhibition of chemotaxis comparable to that of the partial agonist THC. *From Cabral et al. (2008).*

phytocannabinoid THC to dampen the capacity of brain macrophage-like cells to mount a full complement of immune responsiveness to brain infection by these opportunistic amebae. Fraga, Raborn, Ferreira, and Cabral (2011) also reported that cannabinoids altered the responsiveness of microglial-like cells to the transactivating protein Tat of the HIV-1. This protein is an early RNA-binding protein that plays a crucial role in the replication of the HIV-1. At initiation of infection, large amounts of Tat, and other viral-regulatory proteins, are synthesized and released from infected cells (Ensoli et al., 1993). The Tat protein then can translocate across cell membranes and localize in the nucleus of cells where it drives virus replication (Watson & Edwards, 1999). In addition, Tat modulates expression of genes that regulate cellular activities related to survival and growth, inflammation, and angiogenesis (Chang, Gallo, & Ensoli, 1995); promotes chemotaxis and invasive behavior

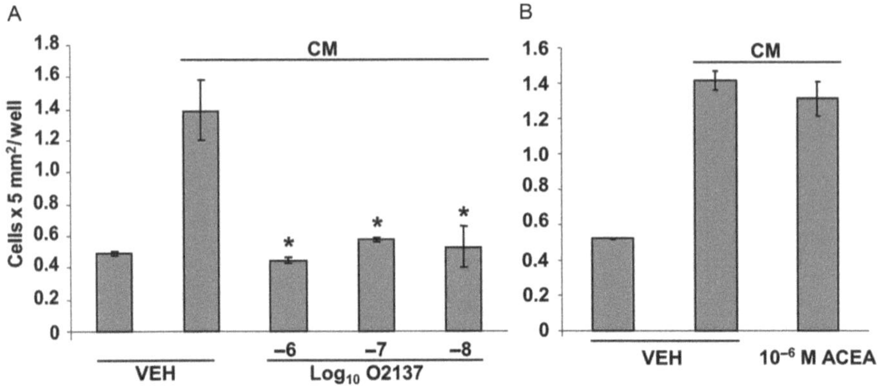

Figure 8.5 Inhibition of chemotaxis of microglia in response to *Acanthamoeba*-conditioned medium (CM) is linked pharmacologically to the CB2R. Microglia were treated (3 h) with cannabinoid or vehicle (0.01% ethanol) and assessed (2 h) for migration against CM. (A) O2137 has greater than 200-fold selectivity for the CB2R: CB1R $K_i = 2700$ nM, CB2R $K_i = 11$ nM; (B) ACEA CB1R $K_i = 1.4$ nM, >1400-fold selectivity over the CB2R. $^*P < 0.05$. $n = 3$/group. ACEA, *N*-(2-chloroethyl)-5Z,8Z,11Z,14Z-eicosatetraenamide; CB1R, CB1 receptor; CB2R, CB2 receptor; CM, ameba-conditioned medium. *From Cabral et al. (2008).*

by monocytes, T-helper lymphocytes, neutrophils, and microglia (Albini et al., 1998; Benelli et al., 1998; de Paulis et al., 2000; Pu et al., 2003); and induces production of chemokines (D'Aversa, Yu, & Berman, 2004) and cytokines (Nath, Conant, Chen, Scott, & Major, 1999; Pu et al., 2003).Tat also harbors a β-chemokine ligand motif articulating a mode by which it recruits uninfected immunocytes to focal areas of HIV infection (de Paulis et al., 2000). Thus, it is not surprising that HIV-1 infection is associated with brain damage (Gendelman et al., 1997) typified by cerebral atrophy, neuronal loss, gliosis, infiltration of inflammatory cells, and microglial activation (New, Maggirwar, Epstein, Dewhurst, & Gelbard, 1998; Orsini, Debouck, Webb, & Lysko, 1996). Fraga et al. (2011) used a mouse BV-2 microglial-like cell model to demonstrate that THC exerted a concentration-related reduction in the migration of BV-2 cells toward Tat. The CB2R antagonist SR144528, but not the CB1R antagonist SR141716A, blocked this inhibition of migration. Similarly, CB2R knockdown with small interfering RNA reversed the cannabinoid-mediated inhibition. In addition, the level of the β-chemokine receptor CCR-3 was reduced and its intracellular localization was altered. These collective results suggested that THC-mediated inhibition of BV-2 microglial-like cell migration to Tat was linked functionally to the CB2R and, in that fashion could ablate HIV-associated neuroinflammatory expansion.

While phytocannabinoids such as THC may have an effect on immune responsiveness *in vitro* to select gene products of the HIV-1, *in vivo* studies that have been conducted to date show no major effect on virus replication and/or associated disease progression. In fact, Molina et al. (2011) reported that THC administration attenuated the progression of simian immunodeficiency virus in the rhesus macaque that is used as a primate model of human HIV infection. Ramirez et al. (2013) implied that cannabinoids that activate the CB2R played a role in limiting HIV-1 infection in the CNS. It was speculated that the prevention of HIV-1 entry into target cells was not a central mechanism for CB2R-mediated suppression in viral replication but that activation of the CB2R was involved in HIV-1 replication. This speculation was supported by results from a single-round infection with a pseudotyped virus that revealed a marked decrease in HIV-1 LTR activation by the CB2R ligands. Williams et al. (2014) reported that THC treatment during human monocyte differentiation reduced macrophage susceptibility to HIV-1 Infection. THC treatment of primary human monocytes during differentiation reduced HIV-1 infection of subsequent macrophages. In contrast, treatment of macrophages with THC immediately prior to, or continuously following, HIV-1 exposure failed to alter infection. Through the use of selective receptor agonists, it was indicated that the THC effect during monocyte differentiation was mediated primarily through the CB2R. Based on these data, it was indicated that THC suppressed HIV-1 infection by a reduction in cell surface HIV receptor expression that diminished efficiency of entry.

Cells of monocyte/macrophage lineage are prime targets for infection with the HIV-1 and serve as replication-competent hosts for the HIV-1. In this capacity they act as "carriers" of virus to distal sites, and upon activation play a central role in dissolution of the BBB leading to expansion of disease into the CNS (Gendelman et al., 2012). HIV-infected monocytes/macrophages produce virus, release virus-specified gene products such as the transactivating protein Tat and the major envelope glycoprotein gp120, and emit a plethora of cell-specified inflammatory products such as NO, chemokines, and cytokines (Ensoli et al., 1993; Schneider, Kaaden, Copeland, Oroszlan, & Hunsmann, 1986; Zagury et al., 1998; reviewed in Haughey & Mattson, 2002; Li, Lee, Cheung, & Lau, 2005; Nath et al., 1999; Yeh et al., 2000; reviewed in Lee et al., 2003). These products also serve as chemotactic and/or inflammatory stimuli that engender inflammatory responses from bystander uninfected immunocytes, the concerted action of which results in breakdown of the BBB, enhancement of trafficking of immunocytes into the CNS, and expansion of the CNS inflammatory process (Mitola et al., 1997; Vené, Benelli, Noonan, & Albini, 2000; reviewed in Pugliese, Vidotto, Beltramo, & Torre, 2005). Indeed, it has been postulated that HIV-associated neuropathogenesis is due mainly to effects of cytokines and neurotoxins that are produced by activated monocytes, perivascular macrophages, and microglia in the CNS rather than to direct cytotoxic effects of the HIV (Genis et al., 1992; Nath et al., 2000; Wesselingh et al., 1993; Williams, Turchan, Lu, Nath, & Drachman, 2005; reviewed in Mattson, Haughey, & Nath, 2005; van de Bovenkamp, Nottet, & Pereira, 2002). These inflammatory events are salient in the post highly active antiretroviral therapy (HAART) era and have been attributed mostly to Tat, which is not targeted by protease inhibitor and deoxynucleoside analog composites of HAART, and/or to gp120 that is emitted from cells that serve as virus reservoirs (reviewed in Nath & Sacktor, 2006; Rumbaugh et al., 2006; reviewed in Navia & Rostasy, 2005; reviewed in Anthony & Bell, 2008; reviewed in Alexaki, Liu, & Wigdahl, 2008). Thus, phytocannabinoid dampening of hyperinflammatory responses that are associated with select neuropathological processes such as those engendered by HIV could prove beneficial to the human host. On the other hand, phytocannabinoid exposure could prove detrimental to the host by inhibiting macrophage-like cell migration to, and subsequent sequestration of, a pathogen such as *A. culbertsoni*. In this context, the nature of neuroimmune response to the invasive agent may predicate whether a phytocannabinoid plays a role in mediating disease progression. Nevertheless, the study of the effects of marijuana or its major phytocannabinoid components on the neuroimmune response to infectious agents is in its infancy.

8. SUMMARY AND FUTURE PROSPECTIVES

Marijuana is a complex substance that harbors a group of terpenoid-like compounds known as phytocannabinoids, of which the psychoactive compound THC and the nonpsychoactive compound CBD have been the most studied. These phytocannabinoids have been shown to alter the functional activities of a variety of immune cells at peripheral sites and in the CNS. While there is a large body of data from animal models that the effects of THC on immunity are linked to decreased resistance to infectious agents, a comparable linkage in humans has yet to be obtained, especially at least as relates to long-terms effects. A major confound in resolving this issue is that individuals who use marijuana in a recreational mode also partake of other substances that have immune-suppressing potential. On the other hand, individuals who use marijuana, or select cannabinoid formulations, for therapeutic purposes already have underlying health conditions that may render them immune compromised or susceptible to infection. Finally, the presence of CBD in marijuana may counteract the effects of THC and temper the overall immune dysfunctional outcome *in vitro* and in experimental animals. In this context, the effect of THC on a given immune functional system may depend on the marijuana formulation and its included concentration of THC. Many of the activities of THC are mediated through activation of cannabinoid receptors. The CB1R appears to be functionally relevant for the overall homeostatic balance and regulation of the CNS (Marsicano et al., 2003; reviewed in Cota, 2007). It has been suggested that the CB1R has potential as a molecular target for therapeutic attenuation of cognitive impairment and degeneration in select CNS disorders (Pryce et al., 2003; Pryce & Baker, 2007; Shen & Thayer, 1998). However, a drawback to this consideration is the recognition that activation of the CB1R also engenders psychotropic effects. On the other hand, while many neuropathogenic processes are characterized by progressive decline in cognitive functions, a major hallmark of CNS pathologies is inflammation. Since the CB2R has been linked to the modulation of immune responses, and its expression by macrophages and microglia appears to be upregulated in response to various stimuli (Carlisle et al., 2002), this receptor has potential to serve as a selective molecular target for ablating untoward inflammatory responses. Finally, the body of scientific data that is available suggests that THC has a major effect on macrophage-like cell migratory activity. This recognition may provide a rationale for distinctive THC-mediated effects on the responsiveness of macrophage-like cells to select pathogens that invade the CNS. For example, microglia play a critical role in sequestering invasive opportunistic amebae such as *A. culbertsoni* within the CNS. THC-mediated inhibition of the capacity of microglia to migrate to nodal sites of infection could compromise this process of sequestration and contribute to dissemination of amebae. On the other hand, microglia serve as replication-competent hosts for certain viruses such as the HIV-1. In this context, inhibition of microglial migration to sites of HIV infection within the CNS could hinder dissemination of the virus. Thus, the functional consequences of THC exposure on immune responsiveness within the CNS to a specified pathogen may be linked to the fundamental nature of the host–pathogen interaction. Resolution of these potential distinctive differences requires further investigation.

Finally, the question of whether select cannabinoids, including those that are native to marijuana, alters immune homeostatic balance within the CNS remains unresolved. This issue is particularly salient since use of marijuana may place a human at increased risk of infection with neurotropic infectious agents. The data that are available suggest that casual use of marijuana is not associated with overt perturbations of cellular and humoral immunity in humans. However, habitual marijuana smokers could experience a different outcome. Furthermore, comparatively little is known about the effects of marijuana or its included phytocannabinoids on immune responsiveness within the CNS. Infectious agents may engender disparate immune responses within different compartments that may be selectively responsive to the action of cannabinoids. Such a distinction may be manifest in terms of responsiveness to infection with an opportunistic ameba such as *A. culbertsoni* versus a viral agent such as the HIV-1. While it is becoming apparent that these infectious agents engender distinctive cellular and soluble factor responses within the CNS, a potential commonality of action on the part of cannabinoids is that they target cell migratory responses that may be linked to these agents rather than targeting replication of the agent. In addition, there is increasing recognition that cannabinoid exposure may exert long-term effects on immune responsiveness and that these may be linked to sex differences or developmental age. For example, Lombard, Hegde, Nagarkatti, and Nagarkatti (2011) reported that perinatal exposure to THC had a profound effect on the mouse fetus as evidenced by a decrease in thymic cellularity on gestational days 16, 17, and 18 and postgestational day 1 and marked alterations in T cell subpopulations. These outcomes were reversed by CB1R/CB2R antagonists, suggesting that the THC-mediated effects were due to activation of cannabinoid receptors. Thymic atrophy induced in the fetus correlated with caspase-dependent apoptosis in thymocytes. Thymic atrophy was the result of direct action of THC and not based on maternal factors since THC was able to induce T cell apoptosis *in vitro* in fetal thymic organ cultures. These results indicated that mouse perinatal exposure to THC-triggered T cell dysfunction, suggesting translationally that the offspring of marijuana abusers who have been exposed to THC *in utero* could be at a higher risk of exhibiting immune dysfunction and contracting infectious diseases. In addition, Moretti et al. (2014) reported recently that THC exposure has an impact on the immune system during adolescence. THC exposure during adolescence triggered immune dysfunction that lasted long after the end of exposure to this cannabinoid, a dysfunction that was characterized by induction of a proinflammatory macrophage phenotype. This dysfunction was characterized by a decrease in levels of both the Th1 and Th2 cytokines, indicating a general immune dysregulatory rather than one directed directly against Th1/Th2 bias. It was speculated that exposure to THC at a young age could place the affected individual at higher risk of developing autoimmune or chronic inflammatory disease. The recognition that phytocannabinoids can exert developmental and long-term effects on the immune system suggests that exposure to these substances, or marijuana, during an early stage in life has the capacity to alter host resistance to select microbial agents in the adult.

REFERENCES

Adhikary, S., Li, H., Heller, J., Skarica, M., Zhang, M., Ganea, D., et al. (2011). Modulation of inflammatory responses by a cannabinoid-2-selective agonist after spinal cord injury. *Journal of Neurotrauma, 28*, 2417–2427.

Agarwal, N., Pacher, P., Tegeder, I., Amaya, F., Constantin, C. E., Brenner, G. J., et al. (2007). Cannabinoids mediate analgesia largely via peripheral type 1 cannabinoid receptors in nociceptors. *Nature Neuroscience, 10*, 870–879.

Albini, A., Benelli, R., Giunciuglio, D., Cai, T., Mariani, G., Ferrini, S., et al. (1998). Identification of a novel domain of HIV tat involved in monocyte chemotaxis. *The Journal of Biological Chemistry, 273*, 15895–15900.

Alexaki, A., Liu, Y., & Wigdahl, B. (2008). Cellular reservoirs of HIV-1 and their role in viral persistence. *Current HIV Research, 6*, 388–400.

Anthony, I. C., & Bell, J. E. (2008). The neuropathology of HIV/AIDS. *International Review of Psychiatry, 20*, 15–24.

Baker, D., Pryce, G., Davies, W. L., & Hiley, C. R. (2006). In silico patent searching reveals a new cannabinoid receptor. *Trends in Pharmacological Sciences, 27*, 1–4.

Baty, D. E., Zhang, M., Li, H., Erb, C. J., Adler, M. W., Ganea, D., et al. (2008). Cannabinoid CB2 receptor activation attenuates motor and autonomic function deficits in a mouse model of spinal cord injury. *Clinical Neurosurgery, 55*, 172–177.

Benelli, R., Mortarini, R., Anichini, A., Giunciuglio, D., Noonan, D. M., Montalti, S., et al. (1998). Monocyte-derived dendritic cells and monocytes migrate to HIV-Tat RGD and basic peptides. *AIDS, 12*, 261–268.

Bindukumar, B., Mahajan, S. D., Reynolds, J. L., Hu, Z., Sykes, D. E., Aalinkeel, R., et al. (2008). Genomic and proteomic analysis of the effects of cannabinoids on normal human astrocytes. *Brain Research, 1191*, 1–11.

Breivogel, C. S., Griffin, G., Di Marzo, V., & Martin, B. R. (2001). Evidence for a new G protein-coupled cannabinoid receptor in mouse brain. *Molecular Pharmacology, 60*, 155–163.

Burnette-Curley, D., Marciano-Cabral, F., Fischer-Stenger, K., & Cabral, G. A. (1993). delta-9 Tetrahydrocannabinol inhibits cell contact-dependent cytotoxicity of Bacillus Calmette-Guerin-activated macrophages. *International Journal of Immunopharmacology, 15*, 371–382.

Cabral, G. A., & Griffin-Thomas, L. (2008). Cannabinoids as therapeutic agents for ablating neuroinflammatory disease. *Endocrine, Metabolic & Immune Disorders Drug Targets, 8*, 159–172.

Cabral, G. A., & Marciano-Cabral, F. (2004). Cannabinoid-mediated exacerbation of brain infection by opportunistic amebae. *Journal of Neuroimmunology, 147*, 127–130.

Cabral, G. A., & Marciano-Cabral, F. (2005). Cannabinoid receptors in microglia of the central nervous system: Immune functional relevance. *Journal of Leukocyte Biology*, *78*, 1192–1197.

Cabral, G. A., Raborn, E. S., Griffin, L., Dennis, J., & Marciano-Cabral, F. (2008). CB2 receptors in the brain: Role in central immune function. *British Journal of Pharmacology*, *153*, 240–251.

Cabral, G. A., & Staab, A. (2005). Effects on the immune system. *Handbook of Experimental Pharmacology*, *168*, 385–423.

Carlisle, S. J., Marciano-Cabral, F., Staab, A., Ludwick, C., & Cabral, G. A. (2002). Differential expression of the CB2 cannabinoid receptor by rodent macrophages and macrophage-like cells in relation to cell activation. *International Immunopharmacology*, *2*, 69–82.

Carrier, E. J., Auchampach, J. A., & Hillard, C. J. (2006). Inhibition of an equilibrative nucleoside transporter by cannabidiol: A mechanism of cannabinoid immunosuppression. *Proceedings of the National Academy of Sciences of the United States of America*, *103*, 7895–7900.

Carrier, E. J., Kearn, C. S., Barkmeier, A. J., Breese, N. M., Yang, W., Nithipatikom, K., et al. (2004). Cultured rat microglial cells synthesize the endocannabinoid 2-arachidonylglycerol, which increases proliferation via a CB2 receptor-dependent mechanism. *Molecular Pharmacology*, *65*, 999–1007.

Chang, H. K., Gallo, R. C., & Ensoli, B. (1995). Regulation of cellular gene expression and function by the human immunodeficiency virus type 1 Tat protein. *Journal of Biomedical Science*, *2*, 189–202.

Chesher, G. B., & Jackson, D. M. (1985). The quasi-morphine withdrawal syndrome: Effect of cannabinol, cannabidiol and tetrahydrocannabinol. *Pharmacology, Biochemistry, and Behavior*, *23*, 13–15.

Cota, D. (2007). CB1 receptors: Emerging evidence for central and peripheral mechanisms that regulate energy balance, metabolism, and cardiovascular health. *Diabetes/Metabolism Research and Reviews*, *23*, 507–517.

Cutando, L., Busquets-Garcia, A., Puighermanal, E., Gomis-Gonzalez, M., Delgado-Garcia, J. M., Gruart, A., et al. (2013). Microglial activation underlies cerebellar deficits produced by repeated cannabis exposure. *The Journal of Clinical Investigation*, *123*, 2816–2831.

D'Aversa, T. G., Yu, K. O., & Berman, J. W. (2004). Expression of chemokines by human fetal microglia after treatment with the human immunodeficiency virus type 1 protein Tat. *Journal of Neurovirology*, *10*, 86–97.

de Paulis, A., De Palma, R., Di Gioia, L., Carfora, M., Prevete, N., Tosi, G., et al. (2000). Tat protein is an HIV-1-encoded beta-chemokine homolog that promotes migration and up-regulates CCR3 expression on human Fc epsilon RI+ cells. *Journal of Immunology*, *165*, 7171–7179.

Di Marzo, V., Breivogel, C. S., Tao, Q., Bridgen, D. T., Razdan, R. K., Zimmer, A. M., et al. (2000). Levels, metabolism, and pharmacological activity of anandamide in CB(1) cannabinoid receptor knockout mice: Evidence for non-CB(1), non-CB(2) receptor-mediated actions of anandamide in mouse brain. *Journal of Neurochemistry*, *75*, 2434–2444.

Dickson, D. W., Lee, S. C., Mattiace, L. A., Yen, S. H., & Brosnan, C. (1993). Microglia and cytokines in neurological disease, with special reference to AIDS and Alzheimer's disease. *Glia*, *7*, 75–83.

Ensoli, B., Buonaguro, L., Barillari, G., Fiorelli, V., Gendelman, R., Morgan, R. A., et al. (1993). Release, uptake, and effects of extracellular human immunodeficiency virus type 1 Tat protein on cell growth and viral transactivation. *Journal of Virology*, *67*, 277–287.

Fernandez-Ruiz, J., Romero, J., Velasco, G., Tolon, R. M., Ramos, J. A., & Guzman, M. (2007). Cannabinoid CB2 receptor: A new target for controlling neural cell survival? *Trends in Pharmacological Sciences*, *28*, 39–45.

Fraga, D., Raborn, E. S., Ferreira, G. A., & Cabral, G. A. (2011). Cannabinoids inhibit migration of microglial-like cells to the HIV protein Tat. *Journal of Neuroimmune Pharmacology*, *6*, 566–577.

Galiegue, S., Mary, S., Marchand, J., Dussossoy, D., Carriere, D., Carayon, P., et al. (1995). Expression of central and peripheral cannabinoid receptors in human immune tissues and leukocyte subpopulations. *European Journal of Biochemistry*, *232*, 54–61.

Gaoni, Y., & Mechoulam, R. (1971). The isolation and structure of delta-1-tetrahydrocannabinol and other neutral cannabinoids from hashish. *Journal of the American Chemical Society*, *93*, 217–224.

Gendelman, H. E., Grant, I., Everall, I. A., Fox, H. A., Gelbard, H. A., Lipton, S. A., et al. (2012). *The neurobiology of AIDS* (3rd ed.). New York: Oxford Press.

Gendelman, H. E., Persidsky, Y., Ghorpade, A., Limoges, J., Stins, M., Fiala, M., et al. (1997). The neuropathogenesis of the AIDS dementia complex. *AIDS*, *11*(Suppl. A), S35–S45.

Genis, P., Jett, M., Bernton, E. W., Boyle, T., Gelbard, H. A., Dzenko, K., et al. (1992). Cytokines and arachidonic metabolites produced during human immunodeficiency virus (HIV)-infected macrophage-astroglia interactions: Implications for the neuropathogenesis of HIV disease. *The Journal of Experimental Medicine*, *176*, 1703–1718.

Gerard, C. M., Mollereau, C., Vassart, G., & Parmentier, M. (1991). Molecular cloning of a human cannabinoid receptor which is also expressed in testis. *Biochemical Journal*, *279*(Pt. 1), 129–134.

Haughey, N. J., & Mattson, M. P. (2002). Calcium dysregulation and neuronal apoptosis by the HIV-1 proteins Tat and gp120. *Journal of Acquired Immune Deficiency Syndromes*, *31*(Suppl. 2), S55–S61.

Herkenham, M., Lynn, A. B., Little, M. D., Johnson, M. R., Melvin, L. S., de Costa, B. R., et al. (1990). Cannabinoid receptor localization in brain. *Proceedings of the National Academy of Sciences of the United States of America*, *87*, 1932–1936.

Jarai, Z., Wagner, J. A., Varga, K., Lake, K. D., Compton, D. R., Martin, B. R., et al. (1999). Cannabinoid-induced mesenteric vasodilation through an endothelial site distinct from CB1 or CB2 receptors. *Proceedings of the National Academy of Sciences of the United States of America*, *96*, 14136–14141.

Kaminski, N. E. (1998). Regulation of the cAMP cascade, gene expression and immune function by cannabinoid receptors. *Journal of Neuroimmunology*, *83*, 124–132.

Klein, T. W., Friedman, H., & Specter, S. (1998). Marijuana, immunity and infection. *Journal of Neuroimmunology*, *83*, 102–115.

Klein, T. W., Kawakami, Y., Newton, C., & Friedman, H. (1991). Marijuana components suppress induction and cytolytic function of murine cytotoxic T cells in vitro and in vivo. *Journal of Toxicology and Environmental Health*, *32*, 465–477.

Klein, T. W., Newton, C., Larsen, K., Chou, J., Perkins, I., Lu, L., et al. (2004). Cannabinoid receptors and T helper cells. *Journal of Neuroimmunology*, *147*, 91–94.

Klein, T. W., Newton, C. A., Nakachi, N., & Friedman, H. (2000). Delta 9-tetrahydrocannabinol treatment suppresses immunity and early IFN-gamma, IL-12, and IL-12 receptor beta 2 responses to Legionella pneumophila infection. *The Journal of Immunology*, *164*, 6461–6466.

Kozela, E., Pietr, M., Juknat, A., Rimmerman, N., Levy, R., & Vogel, Z. (2010). Cannabinoids Delta(9)-tetrahydrocannabinol and cannabidiol differentially inhibit the lipopolysaccharide-activated NF-kappaB and interferon-beta/STAT proinflammatory pathways in BV-2 microglial cells. *The Journal of Biological Chemistry*, *285*, 1616–1626.

Kunos, G., Osei-Hyiaman, D., Batkai, S., Sharkey, K. A., & Makriyannis, A. (2009). Should peripheral CB(1) cannabinoid receptors be selectively targeted for therapeutic gain? *Trends in Pharmacological Sciences*, *30*, 1–7.

Lafuente, H., Alvarez, F. J., Pazos, M. R., Alvarez, A., Rey-Santano, M. C., Mielgo, V., et al. (2011). Cannabidiol reduces brain damage and improves functional recovery after acute hypoxia-ischemia in newborn pigs. *Pediatric Research*, *70*, 272–277.

Lauchner, J. E., Jensen, J. B., Chen, H. Y., Lu, H. C., Hille, B., & Mackie, K. (2008). GPR55 is a cannabinoid receptor that increases intracellular calcium and inhibits M current. *Proceedings of the National Academy of Sciences of the United States of America*, *105*, 2699–2704.

Lee, S. C., Liu, W., Dickson, D. W., Brosnan, C. F., & Berman, J. W. (1993). Cytokine production by human fetal microglia and astrocytes. Differential induction by lipopolysaccharide and IL-1 beta. *Journal of Immunology*, *150*, 2659–2667.

Lee, C., Liu, Q. H., Tomkowicz, B., Yi, Y., Freedman, B. D., & Collman, R. G. (2003). Macrophage activation through CCR5-CXCR4-mediated gp120-elicited signal pathways. *Journal of Leukocyte Biology*, *74*, 676–682.

Li, X., Kaminski, N. E., & Fischer, L. J. (2001). Examination of the immunosuppressive effect of delta9-tetrahydrocannabinol in streptozotocin-induced autoimmune diabetes. *International Immunopharmacology*, *1*, 699–712.

Li, J. C., Lee, D. C., Cheung, B. K., & Lau, A. S. (2005). Mechanisms for HIV Tat upregulation of IL-10 and other cytokine expression: Kinase signaling and PKR-mediated immune response. *FEBS Letters*, *579*, 3055–3062.

Lombard, C., Hegde, V. L., Nagarkatti, M., & Nagarkatti, P. S. (2011). Perinatal exposure to Delta9-tetrahydrocannabinol triggers profound defects in T cell differentiation and function in fetal and postnatal stages of life, including decreased responsiveness to HIV antigens. *The Journal of Pharmacology and Experimental Therapeutics*, *339*, 607–617.

Lopez-Rodriguez, A. B., Llorente-Berzal, A., Garcia-Segura, L. M., & Viveros, M. P. (2014). Sex-dependent long-term effects of adolescent exposure to THC and/or MDMA on neuroinflammation and serotoninergic and cannabinoid systems in rats. *British Journal of Pharmacology*, *171*, 1435–1447.

Marciano-Cabral, F., Ferguson, T., Bradley, S. G., & Cabral, G. (2001). Delta-9-tetrahydrocannabinol (THC), the major psychoactive component of marijuana, exacerbates brain infection by Acanthamoeba. *The Journal of Eukaryotic Microbiology*, *48*(Suppl.), 4S–5S.

Maresz, K., Pryce, G., Ponomarev, E. D., Marsicano, G., Croxford, J. L., Shriver, L. P., et al. (2007). Direct suppression of CNS autoimmune inflammation via the cannabinoid receptor CB1 on neurons and CB2 on autoreactive T cells. *Nature Medicine*, *13*, 492–497.

Marsicano, G., Goodenough, S., Monory, K., Hermann, H., Eder, M., Cannich, A., et al. (2003). CB1 cannabinoid receptors and on-demand defense against excitotoxicity. *Science*, *302*, 84–88.

Massi, P., Fuzio, D., Vigano, D., Sacerdote, P., & Parolaro, D. (2000). Relative involvement of cannabinoid CB(1) and CB(2) receptors in the Delta(9)-tetrahydrocannabinol-induced inhibition of natural killer activity. *European Journal of Pharmacology*, *387*, 343–347.

Matsuda, L. A., Lolait, S. J., Brownstein, M. J., Young, A. C., & Bonner, T. I. (1990). Structure of a cannabinoid receptor and functional expression of the cloned cDNA. *Nature*, *346*, 561–564.

Mattson, M. P., Haughey, N. J., & Nath, A. (2005). Cell death in HIV dementia. *Cell Death and Differentiation*, *12*(Suppl. 1), 893–904.

McCoy, K. L., Matveyeva, M., Carlisle, S. J., & Cabral, G. A. (1999). Cannabinoid inhibition of the processing of intact lysozyme by macrophages: Evidence for CB2 receptor participation. *The Journal of Pharmacology and Experimental Therapeutics*, *289*, 1620–1625.

Merrill, J. E., Ignarro, L. J., Sherman, M. P., Melinek, J., & Lane, T. E. (1993). Microglial cell cytotoxicity of oligodendrocytes is mediated through nitric oxide. *The Journal of Immunology, 151*, 2132–2141.
Mitola, S., Sozzani, S., Luini, W., Primo, L., Borsatti, A., Weich, H., et al. (1997). Tat-human immunodeficiency virus-1 induces human monocyte chemotaxis by activation of vascular endothelial growth factor receptor-1. *Blood, 90*, 1365–1372.
Molina, P. E., Winsauer, P., Zhang, P., Walker, E., Birke, L., Amedee, A., et al. (2011). Cannabinoid administration attenuates the progression of simian immunodeficiency virus. *AIDS Research and Human Retroviruses, 27*, 585–592.
Monnet-Tschudi, F., Hazekamp, A., Perret, N., Zurich, M. G., Mangin, P., Giroud, C., et al. (2008). Delta-9-tetrahydrocannabinol accumulation, metabolism and cell-type-specific adverse effects in aggregating brain cell cultures. *Toxicology and Applied Pharmacology, 228*, 8–16.
Moore, S., & Thanos, S. (1996). The concept of microglia in relation to central nervous system disease and regeneration. *Progress in Neurobiology, 48*, 441–460.
Moretti, S., Castelli, M., Franchi, S., Raggi, M. A., Mercolini, L., Protti, M., et al. (2014). Delta-tetrahydrocannabinol-induced anti-inflammatory responses in adolescent mice switch to pro-inflammatory in adulthood. *Journal of Leukocyte Biology, 96,* In Press.
Munro, S., Thomas, K. L., & Abu-Shaar, M. (1993). Molecular characterization of a peripheral receptor for cannabinoids. *Nature, 365*, 61–65.
Nabissi, M., Morelli, M. B., Santoni, M., & Santoni, G. (2013). Triggering of the TRPV2 channel by cannabidiol sensitizes glioblastoma cells to cytotoxic chemotherapeutic agents. *Carcinogenesis, 34*, 48–57.
Nahas, G. G., & Latour, C. (1993). *Cannabis physiopathology, epidemiology, and detection.* Boca Raton, Florida: CRC Press Inc.
Nath, A., Conant, K., Chen, P., Scott, C., & Major, E. O. (1999). Transient exposure to HIV-1 Tat protein results in cytokine production in macrophages and astrocytes. A hit and run phenomenon. *The Journal of Biological Chemistry, 274*, 17098–17102.
Nath, A., Haughey, N. J., Jones, M., Anderson, C., Bell, J. E., & Geiger, J. D. (2000). Synergistic neurotoxicity by human immunodeficiency virus proteins Tat and gp120: Protection by memantine. *Annals of Neurology, 47*, 186–194.
Nath, A., & Sacktor, N. (2006). Influence of highly active antiretroviral therapy on persistence of HIV in the central nervous system. *Current Opinion in Neurology, 19*, 358–361.
Navia, B. A., & Rostasy, K. (2005). The AIDS dementia complex: Clinical and basic neuroscience with implications for novel molecular therapies. *Neurotoxicity Research, 8*, 3–24.
New, D. R., Maggirwar, S. B., Epstein, L. G., Dewhurst, S., & Gelbard, H. A. (1998). HIV-1 Tat induces neuronal death via tumor necrosis factor-alpha and activation of non-N-methyl-D-aspartate receptors by a NFkappaB-independent mechanism. *The Journal of Biological Chemistry, 273*, 17852–17858.
Ni, X., Geller, E. B., Eppihimer, M. J., Eisenstein, T. K., Adler, M. W., & Tuma, R. F. (2004). Win 55212–2, a cannabinoid receptor agonist, attenuates leukocyte/endothelial interactions in an experimental autoimmune encephalomyelitis model. *Multiple Sclerosis, 10*, 158–164.
Nunez, E., Benito, C., Pazos, M. R., Barbachano, A., Fajardo, O., Gonzalez, S., et al. (2004). Cannabinoid CB2 receptors are expressed by perivascular microglial cells in the human brain: An immunohistochemical study. *Synapse, 53*, 208–213.
Ofek, O., Karsak, M., Leclerc, N., Fogel, M., Frenkel, B., Wright, K., et al. (2006). Peripheral cannabinoid receptor, CB2, regulates bone mass. *Proceedings of the National Academy of Sciences of the United States of America, 103*, 696–701.
Ohta, A., & Sitkovsky, M. (2001). Role of G-protein-coupled adenosine receptors in downregulation of inflammation and protection from tissue damage. *Nature, 414*, 916–920.

Orsini, M. J., Debouck, C. M., Webb, C. L., & Lysko, P. G. (1996). Extracellular human immunodeficiency virus type 1 Tat protein promotes aggregation and adhesion of cerebellar neurons. *The Journal of Neuroscience: The Official Journal of the Society for Neuroscience, 16*, 2546–2552.

Pertwee, R. G. (2007). GPR55: A new member of the cannabinoid clan? *British Journal of Pharmacology, 152*, 984–986.

Pryce, G., Ahmed, Z., Hankey, D. J., Jackson, S. J., Croxford, J. L., Pocock, J. M., et al. (2003). Cannabinoids inhibit neurodegeneration in models of multiple sclerosis. *Brain, 126*, 2191–2202.

Pryce, G., & Baker, D. (2007). Control of spasticity in a multiple sclerosis model is mediated by CB1, not CB2, cannabinoid receptors. *British Journal of Pharmacology, 150*, 519–525.

Pu, H., Tian, J., Flora, G., Lee, Y. W., Nath, A., Hennig, B., et al. (2003). HIV-1 Tat protein upregulates inflammatory mediators and induces monocyte invasion into the brain. *Molecular and Cellular Neurosciences, 24*, 224–237.

Puffenbarger, R. A., Boothe, A. C., & Cabral, G. A. (2000). Cannabinoids inhibit LPS-inducible cytokine mRNA expression in rat microglial cells. *Glia, 29*, 58–69.

Pugliese, A., Vidotto, V., Beltramo, T., & Torre, D. (2005). Phagocytic activity in human immunodeficiency virus type 1 infection. *Clinical and Diagnostic Laboratory Immunology, 12*, 889–895.

Ramirez, B. G., Blazquez, C., Gomez del Pulgar, T., Guzman, M., & de Ceballos, M. L. (2005). Prevention of Alzheimer's disease pathology by cannabinoids: Neuroprotection mediated by blockade of microglial activation. *The Journal of Neuroscience, 25*, 1904–1913.

Ramirez, S. H., Reichenbach, N. L., Fan, S., Rom, S., Merkel, S. F., Wang, X., et al. (2013). Attenuation of HIV-1 replication in macrophages by cannabinoid receptor 2 agonists. *Journal of Leukocyte Biology, 93*, 801–810.

Rumbaugh, J., Turchan-Cholewo, J., Galey, D., St. Hillaire, C., Anderson, C., Conant, K., et al. (2006). Interaction of HIV Tat and matrix metalloproteinase in HIV neuropathogenesis: A new host defense mechanism. *FASEB Journal, 20*, 1736–1738.

Ryberg, E., Larsson, N., Sjögren, S., Hjorth, S., Hermansson, N. O., Leonova, J., et al. (2007). The orphan receptor GPR55 is a novel cannabinoid receptor. *British Journal of Pharmacology, 152*, 1092–1101.

Sanchez, C., Galve-Roperh, I., Canova, C., Brachet, P., & Guzman, M. (1998). Delta9-tetrahydrocannabinol induces apoptosis in C6 glioma cells. *FEBS Letters, 436*, 6–10.

Sanchez, C., Galve-Roperh, I., Rueda, D., & Guzman, M. (1998). Involvement of sphingomyelin hydrolysis and the mitogen-activated protein kinase cascade in the Delta9-tetrahydrocannabinol-induced stimulation of glucose metabolism in primary astrocytes. *Molecular Pharmacology, 54*, 834–843.

Sawzdargo, M., Nguyen, T., Lee, D. K., Lynch, K. R., Cheng, R., Heng, H. H., et al. (1999). Identification and cloning of three novel human G protein-coupled receptor genes GPR52, GPR53, and GPR55: GPR55 is extensively expressed in human brain. *Brain Research. Molecular Brain Research, 5*, 193–198.

Schatz, A. R., Lee, M., Condie, R. B., Pulaski, J. T., & Kaminski, N. E. (1997). Cannabinoid receptors CB1 and CB2: A characterization of expression and adenylate cyclase modulation within the immune system. *Toxicology and Applied Pharmacology, 142*, 278–287.

Schneider, J., Kaaden, O., Copeland, T. D., Oroszlan, S., & Hunsmann, G. (1986). Shedding and interspecies type sero-reactivity of the envelope glycopolypeptide gp120 of the human immunodeficiency virus. *The Journal of General Virology, 67*(Pt. 11), 2533–2538.

Servettaz, A., Kavian, N., Nicco, C., Deveaux, V., Chereau, C., Wang, A., et al. (2010). Targeting the cannabinoid pathway limits the development of fibrosis and autoimmunity in a mouse model of systemic sclerosis. *The American Journal of Pathology, 177*, 187–196.

Shen, M., & Thayer, S. A. (1998). Cannabinoid receptor agonists protect cultured rat hippocampal neurons from excitotoxicity. *Molecular Pharmacology, 54*, 459–462.

Shire, D., Calandra, B., Rinaldi-Carmona, M., Oustric, D., Pessègue, B., Bonnin-Cabanne, O., et al. (1996). Molecular cloning, expression and function of the murine CB2 peripheral cannabinoid receptor. *Biochimica et Biophysica Acta, 1307*, 132–136.

Stella, N. (2010). Cannabinoid and cannabinoid-like receptors in microglia, astrocytes, and astrocytomas. *Glia, 58*, 1017–1030.

Storr, M. A., Keenan, C. M., Emmerdinger, D., Zhang, H., Yuce, B., Sibaev, A., et al. (2008). Targeting endocannabinoid degradation protects against experimental colitis in mice: Involvement of CB1 and CB2 receptors. *Journal of Molecular Medicine (Berlin, Germany), 86*, 925–936.

Storr, M. A., Keenan, C. M., Zhang, H., Patel, K. D., Makriyannis, A., & Sharkey, K. A. (2009). Activation of the cannabinoid 2 receptor (CB2) protects against experimental colitis. *Inflammatory Bowel Diseases, 15*, 1678–1685.

Suárez, I., Bodega, G., Fernandez-Ruiz, J. J., Ramos, J. A., Rubio, M., & Fernandez, B. (2002). Reduced glial fibrillary acidic protein and glutamine synthetase expression in astrocytes and Bergmann glial cells in the rat cerebellum caused by delta(9)-tetrahydrocannabinol administration during development. *Developmental Neuroscience, 24*, 300–312.

Suárez, I., Bodega, G., Ramos, J. A., Fernandez-Ruiz, J. J., & Fernandez, B. (2000). Neuronal and astroglial response to pre- and perinatal exposure to delta-9-tetra-hydrocannabinol in the rat substantia nigra. *Developmental Neuroscience, 22*, 253–263.

Sumariwalla, P. F., Gallily, R., Tchilibon, S., Fride, E., Mechoulam, R., & Feldmann, M. (2004). A novel synthetic, nonpsychoactive cannabinoid acid (HU-320) with antiinflammatory properties in murine collagen-induced arthritis. *Arthritis and Rheumatism, 50*, 985–998.

Tschöp, J., Kasten, K. R., Nogueiras, R., Goetzman, H. S., Cave, C. M., England, L. G., et al. (2009). The cannabinoid receptor 2 is critical for the host response to sepsis. *The Journal of Immunology, 183*, 499–505.

van de Bovenkamp, M., Nottet, H. S., & Pereira, C. F. (2002). Interactions of human immunodeficiency virus-1 proteins with neurons: Possible role in the development of human immunodeficiency virus-1-associated dementia. *European Journal of Clinical Investigation, 32*, 619–627.

Vené, R., Benelli, R., Noonan, D. M., & Albini, A. (2000). HIV-Tat dependent chemotaxis and invasion, key aspects of tat mediated pathogenesis. *Clinical & Experimental Metastasis, 18*, 533–538.

Waksman, Y., Olson, J. M., Carlisle, S. J., & Cabral, G. A. (1999). The central cannabinoid receptor (CB1) mediates inhibition of nitric oxide production by rat microglial cells. *The Journal of Pharmacology and Experimental Therapeutics, 288*, 1357–1366.

Watson, K., & Edwards, R. J. (1999). HIV-1-trans-activating (Tat) protein: Both a target and a tool in therapeutic approaches. *Biochemical Pharmacology, 58*, 1521–1528.

Wesselingh, S. L., Power, C., Glass, J. D., Tyor, W. R., McArthur, J. C., Farber, J. M., et al. (1993). Intracerebral cytokine messenger RNA expression in acquired immunodeficiency syndrome dementia. *Annals of Neurology, 33*, 576–582.

Westlake, T. M., Howlett, A. C., Bonner, T. I., Matsuda, L. A., & Herkenham, M. (1994). Cannabinoid receptor binding and messenger RNA expression in human brain: An in vitro receptor autoradiography and in situ hybridization histochemistry study of normal aged and Alzheimer's brains. *Neuroscience, 63*, 637–652.

Wiley, J. L., & Martin, B. R. (2002). Cannabinoid pharmacology: Implications for additional cannabinoid receptor subtypes. *Chemistry and Physics of Lipids, 121*, 57–63.

Williams, J. C., Appelberg, S., Goldberger, B. A., Klein, T. W., Sleasman, J. W., & Goodenow, M. M. (2014). Delta-tetrahydrocannabinol treatment during human monocyte differentiation reduces macrophage susceptibility to HIV-1 infection. *Journal of Neuroimmune Pharmacology*, In Press.

Williams, M. A., Turchan, J., Lu, Y., Nath, A., & Drachman, D. B. (2005). Protection of human cerebral neurons from neurodegenerative insults by gene delivery of soluble tumor necrosis factor p75 receptor. *Experimental Brain Research, 165*, 383–391.

Wu, H. Y., Goble, K., Mecha, M., Wang, C. C., Huang, C. H., Guaza, C., et al. (2012). Cannabidiol-induced apoptosis in murine microglial cells through lipid raft. *Glia, 60*, 1182–1190.

Xu, H., Cheng, C. L., Chen, M., Manivannan, A., Cabay, L., Pertwee, R. G., et al. (2007). Anti-inflammatory property of the cannabinoid receptor-2-selective agonist JWH-133 in a rodent model of autoimmune uveoretinitis. *Journal of Leukocyte Biology, 82*, 532–541.

Yeh, M. W., Kaul, M., Zheng, J., Nottet, H. S., Thylin, M., Gendelman, H. E., et al. (2000). Cytokine-stimulated, but not HIV-infected, human monocyte-derived macrophages produce neurotoxic levels of l-cysteine. *The Journal of Immunology, 164*, 4265–4270.

Zagury, D., Lachgar, A., Chams, V., Fall, L. S., Bernard, J., Zagury, J. F., et al. (1998). Interferon alpha and Tat involvement in the immunosuppression of uninfected T cells and C-C chemokine decline in AIDS. *Proceedings of the National Academy of Sciences of the United States of America, 95*, 3851–3856.

Zhang, M., Adler, M. W., Abood, M. E., Ganea, D., Jallo, J., & Tuma, R. F. (2009). CB2 receptor activation attenuates microcirculatory dysfunction during cerebral ischemic/reperfusion injury. *Microvascular Research, 78*, 86–94.

Zhang, M., Martin, B. R., Adler, M. W., Razdan, R. K., Jallo, J. I., & Tuma, R. F. (2007). Cannabinoid CB(2) receptor activation decreases cerebral infarction in a mouse focal ischemia/reperfusion model. *Journal of Cerebral Blood Flow and Metabolism, 27*, 1387–1396.

Zhang, M., Martin, B. R., Adler, M. W., Razdan, R. J., Kong, W., Ganea, D., et al. (2009). Modulation of cannabinoid receptor activation as a neuroprotective strategy for EAE and stroke. *Journal of Neuroimmune Pharmacology, 4*, 249–259.

Zhu, L. X., Sharma, S., Stolina, M., Gardner, B., Roth, M. D., Tashkin, D. P., et al. (2000). Delta-9-tetrahydrocannabinol inhibits antitumor immunity by a CB2 receptor-mediated, cytokine-dependent pathway. *The Journal of Immunology, 165*, 373–380.

CHAPTER NINE

Interactions of HIV and Drugs of Abuse: The Importance of Glia, Neural Progenitors, and Host Genetic Factors

Kurt F. Hauser[*,1], **Pamela E. Knapp**[*,†]
[*]Department of Pharmacology & Toxicology, Institute for Drug and Alcohol Studies, Virginia Commonwealth University, Richmond, Virginia, USA
[†]Department of Anatomy & Neurobiology, Institute for Drug and Alcohol Studies, Virginia Commonwealth University, Richmond, Virginia, USA
[1]Corresponding author: e-mail address: kfhauser@vcu.edu

Contents

International Review of Neurobiology, Volume 118
ISSN 0074-7742
http://dx.doi.org/10.1016/B978-0-12-801284-0.00009-9

Abstract

Considerable insight has been gained into the comorbid, interactive effects of HIV and drug abuse in the brain using experimental models. This review, which considers opiates, methamphetamine, and cocaine, emphasizes the importance of host genetics and glial plasticity in driving the pathogenic neuron remodeling underlying neuro-acquired immunodeficiency syndrome and drug abuse comorbidity. Clinical findings are less concordant than experimental work, and the response of individuals to HIV and to drug abuse can vary tremendously. Host-genetic variability is important in determining viral tropism, neuropathogenesis, drug responses, and addictive behavior. However, genetic differences alone cannot account for individual variability in the brain "connectome." Environment and experience are critical determinants in the evolution of synaptic circuitry throughout life. Neurons and glia both exercise control over determinants of synaptic plasticity that are disrupted by HIV and drug abuse. Perivascular macrophages, microglia, and to a lesser extent astroglia can harbor the infection. Uninfected bystanders, especially astroglia, propagate and amplify inflammatory signals. Drug abuse by itself derails neuronal and glial function, and the outcome of chronic exposure is maladaptive plasticity. The negative consequences of coexposure to HIV and drug abuse are determined by numerous factors including genetics, sex, age, and multidrug exposure. Glia and some neurons are generated throughout life, and their progenitors appear to be targets of HIV and opiates/psychostimulants. The chronic nature of HIV and drug abuse appears to result in sustained alterations in the maturation and fate of neural progenitors, which may affect the balance of glial populations within multiple brain regions.

1. INTRODUCTION

For several decades, drug abuse has been recognized as a significant risk factor for acquiring HIV infection and has been suggested to worsen some aspects of HIV infection within the brain. However, the enormity of the problem has been emphasized by a sustained nearly two-decade effort of the National Institute on Drug Abuse (NIDA) to study and treat this problem. As initially affirmed by Dr. Alan I. Leshner, former director of NIDA, "Drug abuse and HIV are truly interlinked epidemics" (Andres et al., 2011; Biber, Neumann, Inoue, & Boddeke, 2007; Swan, 1997). Despite widespread acceptance of this concept and impressive new gains in our understanding of this problem, there is a realization that the interplay between HIV and drug abuse is more complex than initially surmised. Many authoritative reviews describing the effects of drug abuse on neuro-acquired immunodeficiency syndrome (neuroAIDS) have been published, and our goal here is not to repeat past dialogue. During the past decade, there has been considerable new information—especially with respect to increasing evidence that glia are fundamental sites of convergent drug abuse–HIV interactions and the emerging realization of the influence of host genetic factors on the severity of drug abuse and neuroAIDS comorbid interactions. This review will highlight recent findings, emphasizing the role of glia and genetic factors in shaping the interactions of opiates and psychostimulants with neuroAIDS. Reviews on the effects of cannabinoids and ethanol on HIV neuropathogenesis are provided elsewhere within this issue, and not repeated here.

1.1. Opiates and HIV—Preclinical and clinical findings

In this review, we will use the term "opiate" to refer to products or derivatives that can be found naturally in the opium poppy, *Papaver somniferum*. Opiates include heroin, morphine, and codeine, whereas "opioids" refer to the endogenous system of related peptides and receptors initially revealed through the actions of opiate drugs.

Chronic opiate abuse alone is sufficient to promote neurodegenerative changes in the central nervous system (CNS). Clinical evidence from a cohort of preferential opiate abusers in Edinburgh, UK, demonstrates that chronic abuse accelerates Alzheimer's disease-like pathology in HIV-negative individuals (Anthony et al., 2010; Ramage et al., 2005). Chronic preferential opiate abuse caused an accumulation of hyperphosphorylated tau-positive neuropil threads in the frontal and temporal cortex, and in the locus coeruleus compared to similarly aged-control subjects (Anthony et al., 2010). Increases in hyperphosphorylated tau were accompanied by increases in glycogen synthase kinase 3β and cyclin-dependent kinase-5 levels, as well as microgliosis—indicative of accelerated aging, and signaling events associated with Alzheimer's disease prematurely (Anthony et al., 2010).

There are compelling reasons to investigate opioid and HIV interactions and their role in a more severe and/or accelerated neuropathogenesis. In a seminal precombination antiretroviral therapy (cART) era investigation, the presence of multinucleated giant cells and HIV p24 reactivity in the CNS was found more frequently in preferential opioid drug users (25 of 45; 56%) than in nondrug-abusing men (6 of 35; 17%) with AIDS (Bell, Brettle, Chiswick, & Simmonds, 1998). Chronic opiate exposure has been reported to increase the progression to HIV encephalitis (HIVE) in pre-cART era reports (Bell et al., 2002, 1998; Nath et al., 2000) and also worsens neuropathology in cART-treated patients (Smith, Simmonds, & Bell,

2014). Even in cART-treated patients, chronic opiates aggravate CNS inflammation (Anthony, Arango, Stephens, Simmonds, & Bell, 2008; Anthony, Ramage, Carnie, Simmonds, & Bell, 2005) and worsen HIV-associated neurocognitive disorders (HAND) symptomatology—including deficits in verbal and working memory and increased peripheral neuropathy (Byrd et al., 2011; Byrd, Murray, Safdieh, & Morgello, 2012; Cohen, 2009; Meyer et al., 2013; Robinson-Papp et al., 2012). A recent study of a cART naïve population of injecting drug users who preferentially abuse heroin in Indonesia showed consistent reductions in CD4 counts compared to non-drug abusers (Meijerink et al., 2014). Another recent report suggests that concurrent exposure can selectively increase the severity of some features of HIVE (Smith et al., 2014). Despite increasing clinical evidence that chronic opiate exposure can worsen neuroAIDS, it remains unknown how opioids interact with individual HIV expressed gene products to affect subclasses of neurons, astroglia, and microglia (Hauser, Fitting, Dever, Podhaizer, & Knapp, 2012; Nath et al., 2002).

Some clinical studies are inconsistent with the findings cited above and report minimal or no neurocognitive differences between HIV $\pm$ opiate abuse (Donahoe & Vlahov, 1998; Royal et al., 1991), suggesting that there may be critical genetic (Kreek, Bart, Lilly, LaForge, & Nielsen, 2005; Proudnikov et al., 2012), pharmacokinetic (Eap, Buclin, & Baumann, 2002), pharmacodynamic, sex (Becker & Hu, 2008; Hahn et al., 2014; Zubieta et al., 2002), and/or possible age-dependent differences among opiate abusers that can influence outcomes (discussed later). Moreover, the timing of opiate coexposure in relation to the onset of HIV infection or vice versa may have a marked influence on outcome (Fitting et al., 2012). It is hoped that advances in the understanding of disease mechanisms, experimental models, and methodology will reveal opiate–HIV interactions with increasing clarity.

Preclinical studies more regularly suggest that chronic opiate exposure plays a fundamental role in the pathogenesis of HIV in the CNS (Hauser et al., 2012). Briefly, morphine can exacerbate HIV-1 toxicity through separate actions in neurons (Bruce-Keller et al., 2008; Gurwell et al., 2001; Hu, Sheng, Lokensgard, & Peterson, 2005), including human neurons (Hu et al., 2005; Turchan-Cholewo et al., 2006), and in glia. Many of the harmful neurotoxic effects are mediated through μ opioid receptors (MOR) (Hauser et al., 2012; Zou et al., 2011) and differ depending on the CNS cell type involved. Glial targets can include MOR-expressing astroglia (El-Hage et al., 2013; Hauser et al., 2007, 2012; Zou et al., 2011), microglia

(El-Hage et al., 2013; Sorrell & Hauser, 2014; Suzuki et al., 2011; Turchan-Cholewo et al., 2008, 2009; Zou et al., 2011), oligodendroglia (Hahn et al., 2014; Hauser et al., 2009, 2012), and neuronal/glial precursors (Buch et al., 2007; Hahn, Podhaizer, Hauser, & Knapp, 2012; Hahn et al., 2010; Khurdayan et al., 2004). Details of opiate and HIV interactions in neurons and glia have been reviewed previously (Abt & Meucci, 2011; Banerjee et al., 2011; Dutta & Roy, 2012; Festa & Meucci, 2012; Hauser, El-Hage, et al., 2005; Hauser et al., 2006, 2007, 2012; Reddy, Pilakka-Kanthikeel, Saxena, Saiyed, & Nair, 2012).

1.2. Psychostimulants and HIV

Numerous authoritative reviews exist on the generalized effects of chronic psychostimulant (such as methamphetamine, MDMA) abuse on brain pathology (e.g., Buch et al., 2012; Buttner, 2011; Cadet, Bisagno, & Milroy, 2014; Cadet & Krasnova, 2009; Clark, Wiley, & Bradberry, 2013; Halpin, Collins, & Yamamoto, 2014). Evidence suggests that psychostimulants can act directly on both neurons and glia to disrupt CNS function and promote the injury of vulnerable subpopulations of neurons in the CNS. The vulnerable subpopulations include dopaminergic, noradrenergic, and serotonergic neurons. Minimal attention has been given to stimulants besides methamphetamine, such as 3,4-methylenedioxymethamphetamine (MDMA or ecstasy), ketamine, or γ-hydroxybutyrate (GHB) (Colfax & Guzman, 2006), other than evidence that club drug use increases risky sexual behavior associated with increasing HIV transmission (Zuckerman & Boyer, 2012).

Chronic methamphetamine abuse and disruptions to dopaminergic function are especially deleterious to HIV-infected individuals (Nath, 2010; Nath et al., 2000; Theodore, Cass, Nath, & Maragos, 2007). Methamphetamine addiction results in marked structural pathology in the brain (Berman, O'Neill, Fears, Bartzokis, & London, 2008) and increases the probability of neuropsychological deficiencies in HIV-infected individuals (Rippeth et al., 2004). Many comprehensive reviews exist on the mechanisms underlying the neurotoxic effects of methamphetamine alone (Berman et al., 2008; Cadet et al., 2014; Cadet & Krasnova, 2009; Cadet, Krasnova, Jayanthi, & Lyles, 2007; Cisneros & Ghorpade, 2012; Clark et al., 2013; Coller & Hutchinson, 2012; Ferris, Mactutus, & Booze, 2008; Kaushal & Matsumoto, 2011; Krasnova & Cadet, 2009; Nath, 2010; Quinton & Yamamoto, 2006; Theodore et al., 2007). Briefly, methamphetamine has both direct and indirect toxic effects on neurons. Indirect neurotoxicity is mediated by disrupting glia and/or targets such as the blood–brain barrier (BBB) or the immune system. Chronic methamphetamine exposure can be accompanied by neurodegeneration. Dopaminergic neurons and presynaptic terminals are particularly vulnerable (Czub et al., 2001; Flora et al., 2003; Theodore, Cass, & Maragos, 2006a).

Methamphetamine selectively injures neurons in specific brain regions such as the basal ganglia (Seiden & Ricaurte, 1987), although the precise sequence of direct and indirect events leading to neuronal injury is not fully understood. Acute exposure to methamphetamine induces rapid release of dopamine from presynaptic terminals, while chronic dopamine use results in lasting decreases in striatal dopamine and serotonin and their metabolites (Cass, 1997; Kogan, Nichols, & Gibb, 1976; Wagner et al., 1980) that are accompanied by the destruction of dopaminergic presynaptic terminals in the caudate nucleus (Brunswick, Benmansour, Tejani-Butt, & Hauptmann, 1992; Halpin et al., 2014; Krasnova & Cadet, 2009; Nakayama, Koyama, & Yamashita, 1993; Nickell, Siripurapu, Vartak, Crooks, & Dwoskin, 2014; Pereira et al., 2012; Wagner et al., 1980). Methamphetamine inhibits the function of the vesicular monoamine transporter 2 (VMAT2) (Hanson, Rau, & Fleckenstein, 2004; Hanson, Sandoval, Riddle, & Fleckenstein, 2004) and the dopamine transporter (DAT) (Cadet & Krasnova, 2009; Halpin et al., 2014; Krasnova & Cadet, 2009; Nickell et al., 2014; Theodore, Cass, & Maragos, 2006b; Volkow et al., 2001). VMAT2 transports dopamine (and other monoamines) from the cytosol into presynaptic vesicles (Cartier et al., 2010; Qi, Miller, & Voit, 2008), while DAT sequesters released dopamine from the synaptic cleft into the presynaptic terminal. A transmembrane pH gradient is necessary for vesicular uptake (Sulzer & Rayport, 1990). Methamphetamine, a weak base, limits acidification of the presynaptic cytoplasm and therefore can disrupt vesicular transport. By targeting VMAT2, methamphetamine elevates levels of dopamine within the presynaptic terminal. Cumulative increases in dopamine and oxidative byproducts further reduce levels of VMAT2, creating a destructive positive feedback cycle. In humans, many (but not all) markers of dopamine nerve terminals, such as dopamine itself, tyrosine hydroxylase, and DAT, are decreased in brains of psychostimulant abusers (Wilson et al., 1996; Wilson & Kish, 1996). These effects may be enduring (McCann et al., 1998).

Methamphetamine-related disruption of intracellular redox potential, nitrogen metabolism, and pH exerts the greatest burden on vesicular trafficking at presynaptic terminals, which include losses in DAT (Brown et al., 2002; Fleckenstein, Metzger, Wilkins, Gibb, & Hanson, 1997;

Hanson, Rau, et al., 2004) and VMAT2 function (Larsen, Fon, Hastings, Edwards, & Sulzer, 2002; Miller, Gainetdinov, Levey, & Caron, 1999). The selective presynaptic harm may be worsened by excessive peripheral ammonia caused by concurrent methamphetamine-induced hepatotoxicity (Halpin & Yamamoto, 2012).

Dopamine and the accumulation of other amines within the presynaptic cytoplasm can activate trace amine-associated receptor 1 (TAAR1) (Borowsky et al., 2001; Bunzow et al., 2001; Reese, Bunzow, Arttamangkul, Sonders, & Grandy, 2007). TAAR1 is localized within the presynaptic membranes of monoaminergic neurons and located intracellularly suggesting segregation to internalized vesicles (Revel et al., 2011; Xie & Miller, 2009). TAAR1 reportedly can dimerize with dopamine D2 receptors (Espinoza et al., 2011) and TAAR1 activation phosphorylates DAT, resulting in increased dopamine efflux and eventually DAT internalization (Miller, 2011). TAAR1 likely mediates key aspects of aberrant function following methamphetamine (Xie & Miller, 2009) and perhaps cocaine (Revel et al., 2011) exposure.

Methamphetamine and cocaine exposure can result in excessive glutamate within the extracellular space (ECS) at synaptic and extrasynaptic sites (Cadet et al., 2007; Davidson et al., 2007; Kaushal & Matsumoto, 2011; Miyatake, Narita, Shibasaki, Nakamura, & Suzuki, 2005; Pereira et al., 2012; Quinton & Yamamoto, 2006). Glutamate overflow, especially at extrasynaptic sites (Hardingham, Fukunaga, & Bading, 2002; Sattler, Xiong, Lu, MacDonald, & Tymianski, 2000), can induce excitotoxic neuronal injury and even death through overactivation of extrasynaptic GluN2B NMDA receptors (Ivanov et al., 2006; Liu et al., 2007). The NMDA receptor antagonists MK-801 or dextromethorphan can attenuate methamphetamine neurotoxicity (Thomas & Kuhn, 2005), suggesting that microglial activation and dopamine terminal losses may be intimately linked to excitotoxic glutamatergic transmission and imbalances in synaptic and extrasynaptic glutamate (Beardsley & Hauser, 2014).

Chronic HIV-1 or simian immunodeficiency virus (SIV) infection depletes dopamine and reduces the number of dopamine terminals in the basal ganglia (Czub et al., 2001; Maragos et al., 2002; Nath et al., 2000; Reyes, Faraldi, Senseng, Flowers, & Fariello, 1991; Wang et al., 2004), which appears to be worsened by methamphetamine exposure (Cass, Harned, Peters, Nath, & Maragos, 2003; Czub et al., 2001; Maragos et al., 2002; Nath et al., 2000; Scheller et al., 2005) or dopamine agonists (Czub et al., 2004, 2001; Nath, Maragos, Avison, Schmitt, & Berger, 2001).

Dopaminergic neurons are especially vulnerable to Tat and/or gp120-induced insults (Hu, Sheng, Lokensgard, Peterson, & Rock, 2009). When methamphetamine is administered after intrastriatal HIV-1 Tat injection, it acts synergistically to diminish levels of dopamine and dopamine metabolites (Maragos et al., 2002). Dopamine D2 receptor levels are also decreased as indicated by reductions in raclopride binding (Maragos et al., 2002). Interestingly, HIV-1 Tat may directly inhibit VMAT2 function in the CNS (Theodore, Cass, Dwoskin, & Maragos, 2012). More recently, HIV-1 Tat has been demonstrated to interact directly with the DAT (Midde, Gomez, & Zhu, 2012; Midde et al., 2013; Zhu, Ananthan, Mactutus, & Booze, 2011; Zhu, Mactutus, Wallace, & Booze, 2009). Thus, HIV-1 Tat disrupts some of the same molecular targets as methamphetamine albeit through independent mechanisms, which is likely to contribute to the devastating neurological and psychiatric consequences of methamphetamine and HIV comorbidity.

Brain aging has been demonstrated to be accelerated in cocaine abusers (Ersche, Jones, Williams, Robbins, & Bullmore, 2013). Since psychostimulants alone cause marked pathogenesis, it is not surprising that in combination with HIV, the pathological consequences to the brain can be severe (Cadet et al., 2014). Many authoritative reviews exist on cocaine neurotoxicity (Bowers, Chen, & Bonci, 2010; Cadet et al., 2014; Clark et al., 2013; Ferris et al., 2008; Kousik, Napier, & Carvey, 2012; Narayanan, Mesangeau, Poupaert, & McCurdy, 2011; Nath, 2010; Pierce & Wolf, 2013). Accordingly, only some of the more recent findings regarding the actions of cocaine will be discussed.

The DAT is thought to be a major site of cocaine action. By inhibiting DAT function, dopamine accumulates in the synaptic cleft, overactivating dopamine receptors postsynaptically and increasing its rewarding properties. Restricting DAT function alone as demonstrated in DAT knockout mice, however, is insufficient to block cocaine conditioned-place preference (Sora et al., 1998). When dopamine and serotonin transporters are both deleted, mice no longer self-administer or show cocaine place preference (Sora et al., 2001). These findings and others suggest that dopamine and serotonin, as well as norepinephrine, act in concert to contribute to cocaine's addictive properties (Hall et al., 2004, 2009; Sora et al., 2001). Overall, cocaine appears to primarily act by inhibiting presynaptic DATs, but also hinders serotonin and norepinephrine transporters (Hall et al., 2009, 2004; Kreek, Bart, et al., 2005; Kreek, Nielsen, Butelman, & LaForge, 2005; Sora et al., 2001), and may also secondarily dysregulate inhibitory GABAergic function (Cameron & Williams, 1994). Cocaine also modulates the endogenous opioid system, especially MOR, κ opioid receptors (KOR), and preprodynorphin (Kreek, Bart, et al., 2005).

1.3. The central role of glia

Although neuronal interconnections form the synaptic circuitry that underlie behavior (Hebb, 1949; Yuste & Bonhoeffer, 2001), astroglia and microglia provide essential structural, trophic, and metabolic support necessary for maintaining synaptic integrity and function. Importantly, in HIV, glia are both targets and effectors in the progression of disease. Not only do glia harbor the virus and release inflammatory mediators, but they may also be functionally compromised and/or killed in the process. Unlike neurogenesis, which is restricted to limited brain regions and neuronal types, glia continue to be generated throughout ontogeny (discussed below). This includes microglia and all classes of macroglia (including astroglia, oligodendroglia, and ependymal cells). The ability of glial populations to respond in a dynamic fashion to both HIV infection/exposure and drugs of abuse prompts our glia-centric viewpoint, which provides extensive insight into the pathogenesis of neuroAIDS and into the molecular and cellular mechanisms by which drug abuse affects the progression of the disease.

2. MICROGLIA

In the CNS, the principal cell types that are productively infected are perivascular macrophages and brain-resident microglia. Astrocytes can also become infected, and this is particularly evident *in vitro*; however, the production of new virions by astrocytes is greatly restricted compared to microglia (discussed below). Although the virus itself can infect and replicate in microglia, much of the subsequent spiraling inflammation, synaptodendritic injury, and neurotoxicity arise from the response of bystander microglia and astroglia that are not necessarily infected.

2.1. Microglia as innate immune effectors

A seminal study by Ginhoux et al. (2010) reports that microglia originate from a mesenchymal (incipient myeloid) precursor in the murine yolk sac on embryonic day 8.25–8.5, seed the incipient CNS on embryonic day 9.25–9.5, and remain in the CNS throughout life. Despite considerable overlap with myeloid-lineage cells ("myeloid" refers to the bone marrow) such as monocytes and macrophages, which also originate from a common progenitor in the yolk sac, microglia are an ontogenically distinct population—since they never inhabit the bone marrow (Saijo & Glass, 2011). This finding concurs with evidence that microglia and perivascular CNS macrophages can differ functionally and phenotypically—especially in their response to HIV (see below).

Akin to monocyte-derived macrophages (MDMs), microglia possess a wide variety of "pattern" or "pathogen" recognition receptors (PRRs) related to innate immune function. PRRs include Toll-like receptors (TLRs), nucleotide-binding oligomerization domain receptors (NOD-like receptors or NLRs), Mac-1, CD14 (Nadeau & Rivest, 2000), and a wide variety of scavenger receptors, including those that recognize and remove low-density lipoproteins (Coraci et al., 2002; Husemann, Loike, Kodama, & Silverstein, 2001), and receptors for advanced glycation end-products (RAGE) (Farina, Aloisi, & Meinl, 2007). Microglia can express major histocompatibility complex-I (MHC-I) and MHC-II complexes that allow them to contribute to adaptive immunity by processing both intracellular and extracellular foreign proteins for presentation as antigens to T-lymphocytes.

2.2. Microglia and HIV

Despite substantial overlap, microglia and perivascular CNS macrophages can differ functionally and phenotypically in their response to HIV (Fischer-Smith et al., 2001; Guillemin & Brew, 2004) or SIV (Williams et al., 2001). Given that microglia reside and replicate within the brain throughout life (Ginhoux et al., 2010; Saijo & Glass, 2011), the virus must enter the brain before microglia can become infected. Perivascular macrophages are the major cell by which HIV (Fischer-Smith et al., 2001) or SIV (Williams et al., 2001) seeds the CNS. Perivascular macrophages originate from HIV-infected and -uninfected MDMs, which are thought to have some ability to traffic bidirectionally between the blood and perivascular sites within the CNS parenchyma (Gonzalez-Scarano & Martin-Garcia, 2005). Although infected MDMs are the major source of virions initially seeding CNS microglia, once the infection is established in microglia (Cosenza, Zhao, Si, & Lee, 2002), current evidence suggests that resident microglia become a sustained site of both active and latent infection (Fischer-Smith & Rappaport, 2005; Garden, 2002; Gonzalez-Scarano &

Martin-Garcia, 2005; Kaul, Garden, & Lipton, 2001; Kramer-Hammerle, Rothenaigner, Wolff, Bell, & Brack-Werner, 2005; Persidsky & Gendelman, 2003), and that HIV evolves independently in distinct CNS cell types (Schnell, Joseph, Spudich, Price, & Swanstrom, 2011). The exchange of HIV between MDMs and microglia makes the brain a reservoir of latent infection (Fischer-Smith & Rappaport, 2005; Gonzalez-Scarano & Martin-Garcia, 2005; Joseph et al., 2014; Kaul et al., 2001; Kramer-Hammerle, Rothenaigner, et al., 2005; Persidsky & Gendelman, 2003; Schnell et al., 2011). This is thought to be particularly important in the post-cART era in which antiretroviral drugs have more limited access to the CNS parenchyma (where microglia reside) because of the BBB.

Infected macrophages and microglia produce "virotoxins" (Nath & Geiger, 1998), i.e., viral protein products such as Tat and gp120, Vpr, and others, as well as "cellular toxins" including extracellular reactive oxygen species (ROS), reactive nitrogen species (RNS), and numerous cytokines, including TNF-α, IL-1β, IFN-γ, and IL-6, and chemokines (Fiala et al., 1997; Kraft-Terry, Buch, Fox, & Gendelman, 2009; Persidsky, Buttini, Limoges, Bock, & Gendelman, 1997; Seilhean et al., 1997). Both virotoxins and cellular toxins can independently interact in unique ways with opiates and psychostimulants and will be the topic of the discussions that follow. In addition, cellular toxins [e.g., quinolinic acid (Guillemin, Kerr, & Brew, 2005) or the neurotoxic amine, Ntox (Giulian et al., 1996)], as well as excess glutamate (Gupta et al., 2010), released by macrophages/microglia also have toxic bystander effects on neighboring neurons (Alirezaei, Kiosses, Flynn, Brady, & Fox, 2008; Eugenin et al., 2006; Kiebala & Maggirwar, 2011; Langford & Masliah, 2001; Mayne, Holden, Nath, & Geiger, 2000, Saha & Pahan, 2003; Schuenke & Gelman, 2003; Turchan-Cholewo et al., 2009; Yi, Lee, Liu, Freedman, & Collman, 2004). The actions of macrophages and microglia in the context of HIV have been extensively reviewed (Gonzalez-Scarano & Martin-Garcia, 2005; Kaul et al., 2001; Masliah et al., 1997).

Microglia can express a wide variety of neurotransmitter receptors, including AMPA and NMDA receptors, which presumably allows microglia to coordinate and synchronize their responses with neuronal function (Eggen, Raj, Hanisch, & Boddeke, 2013; Gras et al., 2012; Hanisch & Kettenmann, 2007; Pocock & Kettenmann, 2007; Prada, Furlan, Matteoli, & Verderio, 2013). Excitotoxic levels of glutamate can trigger inflammatory responses by microglia, including the release of proinflammatory cytokines and ROS

(Hagino et al., 2004; Noda, Nakanishi, Nabekura, & Akaike, 2000). HIV-1 virions and gp120 have been shown to increase levels of extracellular glutamate by direct effects on uptake mechanisms in astroglia (Wang et al., 2003), and the increased glutamate likely drives further microglial reactivity.

2.2.1 Opioid and HIV actions in microglia

Microglia display more pronounced glial reactivity than astroglia in HIV-infected opiate abusers (Anderson et al., 2003). HIV-1 gp120 and/or Tat released from infected glia (microglia and astrocytes) (1) trigger cytokine release, (2) initiate inflammatory lipid production (Bandaru, Patel, Ewaleifoh, & Haughey, 2011), (3) destabilize intracellular ion homeostasis, and (4) increase extracellular glutamate (Podhaizer et al., 2012; Wang et al., 2003; Zou et al., 2011) and extracellular adenosine triphosphate (ATP) levels (Sorrell & Hauser, 2014) (Fig. 9.1) by bystander glia. Bystander neurons are directly and indirectly damaged (Ellis, Langford, & Masliah, 2007; Gonzalez-Scarano & Martin-Garcia, 2005; Kaul et al., 2001; Masliah et al., 1997; Mattson, Haughey, & Nath, 2005). Opiates can affect all aspects of the above processes, including (1) MDM trafficking across the BBB; (2) viral replication in MDMs, microglia, and astroglia (largely in *in vitro* studies); (3) the production of proinflammatory cytokines and chemokines; and (4) losses in extracellular ion homeostasis. The effects of opiates in MDMs and microglia have been comprehensively reviewed elsewhere (Banerjee et al., 2011; Chao, Hu, & Peterson, 1996; Dutta & Roy, 2012; Hauser et al., 2007, 2012; Reddy et al., 2012; Regan, Dave, Datta, & Khalili, 2012; Rock & Peterson, 2006). Accordingly, only a few recent findings will be briefly considered in this review.

Acute exposure to opiate drugs such as morphine (El-Hage et al., 2013) or methadone (Li et al., 2002) tends to increase HIV replication by infected microglia. However, depending on the duration and timing of exposure, morphine can increase, act in a neutral manner, or inhibit HIV expression (Peterson, Gekker, Hu, Cabral, & Lokensgard, 2004). Moreover, selective MOR agonists such as endomorphin-1, but not DAMGO or morphine (Peterson et al., 1999), can increase HIV-1 replication in infected microglia—suggesting the involvement of a nontraditional MOR variant in HIV replication (Peterson et al., 1999) or suggesting that "biased agonism" (Hauser et al., 2012) may be operative. We have recently found that specific subsets of MOR splice variants, including MOR-1X and MOR-1K, were differentially expressed by human astrocytes, but not expressed at detectable levels in microglia (Dever et al., 2014; Dever,

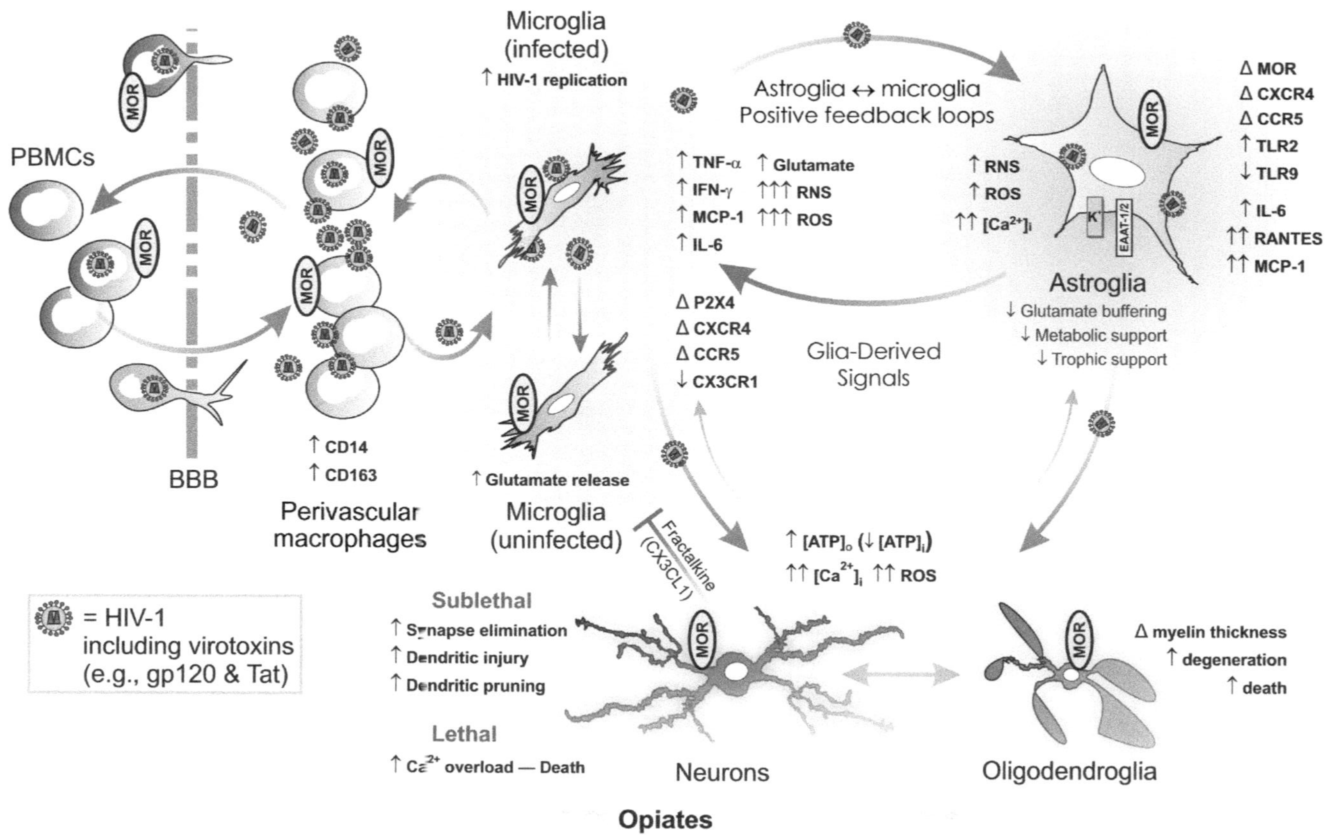

Figure 9.1 See legend on next page.

Xu, Fitting, Knapp, & Hauser, 2012). Moreover, the expression of each MOR variant may be differentially regulated by HIV and in a cell-specific manner (Dever et al., 2014, 2012). Thus, microglia express a subset of MOR variants each of which may respond uniquely to morphine and/or HIV (Dever et al., 2014, 2012). Collectively, the findings indicate that the effects of MOR activation on HIV replication and the response of microglia to HIV (discussed below) are complex and may differ significantly depending on context.

Acute exposure of microglia to HIV-1 Tat increases glutamate release via the x_c^- cystine–glutamate antiporter (Gupta et al., 2010). Tat-dependent increases in extracellular glutamate were attenuated by inhibitors of p38, p42/44 MAPK, or NADPH oxidase or by inhibiting the x_c^- cystine–glutamate antiporter directly (Gupta et al., 2010). Interestingly, morphine coexposure with Tat can significantly increase glutamate release from

Figure 9.1 Opiate drugs exacerbate HIV-1 neuropathogenesis through direct actions on glia—especially microglia and astroglia. Microglia are likely infected through interactions with infiltrating, perivascular macrophages and propagate the bulk of HIV infection in the CNS. HIV-1 also infects astroglia, but to a far lesser extent, and perhaps without production of new virus. Infection results in the production of reactive oxygen and nitrogen species (ROS and RNS, respectively), proinflammatory cytokines, and the release of HIV-1 proteins such as gp120 and Tat. All of these promote inflammation and cytotoxicity in bystander neurons and glia. Opiate abuse alone can cause premature Alzheimer-like changes (Anthony et al., 2010) and morphine by itself can enhance neurotoxicity *in vitro* (Zou et al., 2011); however, opiates appear to potentiate many of the pathophysiological effects of HIV in the central nervous system of infected individuals. Multiple neuronal and glial types express μ-opioid receptors (MOR). Many of the neurodegenerative effects of opioid–HIV interactions are the result of direct actions on microglia and astroglia, which then lead to a positive feedback cycle of inflammatory/cytotoxic signaling between HIV-1-infected microglia and astroglia. Abbreviations: α-chemokine "C–X–C" receptor 4 (CXCR4); altered or changed (Δ); β-chemokine "C–C" receptor 5 (CCR5); blood–brain barrier (BBB); decreased (↓); fractalkine (CX3CL1); fractalkine receptor (CX3CR1); increased (↑); interferon-γ (IFN-γ); interleukin-6 (IL-6); intracellular Ca^{2+} concentration ($[Ca^{2+}]_i$); intracellular sodium concentration ($[Na^+]_i$); monocyte chemoattractant protein-1 (MCP-1 [or CCL2]); peripheral blood mononuclear cells (PBMCs); regulated upon activation, normal T-cell expressed, and secreted (RANTES [or CCL5]); Toll-like receptor (TLR). Fractalkine released by neurons (and astroglia) can be neuroprotective by limiting the neurotoxic actions of microglia (blue "T"; gray in the print version); red (gray in the print version) arrows suggest proinflammatory/cytotoxic interactions. *Modified and reprinted from Hauser et al. (2012) an "open access article distributed under the terms of the Creative Commons Attribution License (http://creativecommons.org/licenses/by/2.5/), which permits unrestrictive use, distribution, and reproduction in any medium, provided the original work is properly cited."*

microglia above maximal levels of secretion seen with Tat alone (Gupta et al., 2010). Although excitatory amino acid transporters-1 (EAAT1 or glutamate/aspartate transporter (GLAST)) and (EAAT2 (or "glial" glutamate transporter 1 (GLT-1)), respectively) are minimally expressed by resting microglia and thought to be primarily expressed by astrocytes, recent evidence suggests that both transporters are inducible in microglia with immune activation (Gras et al., 2012). Because the function of EAAT1/2 is reduced markedly by morphine by itself (Zou et al., 2011), or in combination with Tat (Zou et al., 2011) or gp120 (Podhaizer et al., 2012) in astrocytes, future studies examining the potential contributions of microglial EAAT1/2 to HIV protein ± morphine-induced increases in extracellular glutamate are warranted.

Chronic opiates disrupt glial function, which is especially problematic in microglia where normal cellular functions have been hijacked by the virus (Hauser et al., 2012). HIV-1 alone causes increases in extracellular glutamate, ROS, and RNS by overactivating microglia (Gendelman et al., 1997; Gonzalez-Scarano & Martin-Garcia, 2005; Kaul et al., 2001; Kaul & Lipton, 2005; Li, Li, Steiner, & Nath, 2009; Nath, 1999). Opiates have been shown to modulate (typically, but not always, worsening) all of these neuroinflammatory events (Banerjee et al., 2011; Chao, Hu, et al., 1996; Dutta & Roy, 2012; Hauser et al., 2007, 2012; Reddy et al., 2012; Regan et al., 2012; Rock & Peterson, 2006). However, the nature of opiate–HIV interactions in microglia is complicated and depends on a variety of factors that are incompletely understood. For example, we find that isolated microglia display transient increases in cytokine and ROS production in response to acute morphine and HIV-1 Tat coexposure that are quite robust (Turchan-Cholewo et al., 2009). However, after 24 h of sustained exposure to morphine and Tat, the inflammatory response of isolated microglia has faded to levels below that seen with Tat alone and this is not due to increased microglial death (Turchan-Cholewo et al., 2009). Alternatively, when microglia are cocultured with astroglia (Hauser et al., 2012; Podhaizer et al., 2012; Zou et al., 2011), or when glial inflammation and/or neuronal injury is examined *in vivo* (Bruce-Keller et al., 2008; El-Hage, Bruce-Keller, Knapp, & Hauser, 2008; Fitting, Xu, et al., 2010), prolonged morphine and HIV-1 Tat coexposure result in neuroinflammation and/or neuronal injury that is evident for as long as 10 days. Although the sustained microglial activation is presumed to be driven through reverberating inflammatory signaling between MOR-expressing astroglia and microglia (Hauser et al., 2012; Podhaizer et al.,

2012; Suzuki et al., 2011; Zou et al., 2011), additional study is needed to confirm this notion and to identify the mechanisms involved.

MDMs can display phenotypic heterogeneity in their expression of PRRs and in response to regional differences in the extracellular milieu (Kigerl et al., 2009). CD163 and CD16 coexpressing MDMs appear to be preferentially involved in HIV (Ancuta, Wang, & Gabuzda, 2006; Fischer-Smith, Tedaldi, & Rappaport, 2008) or SIV (Borda et al., 2008) replication and AIDS progression. Akin to MDMs, microglia can also display considerable functional heterogeneity (Carson et al., 2007; Hanisch, 2013; Hanisch & Kettenmann, 2007; Saijo & Glass, 2011; Scheffel et al., 2012) and a variety of intermediate states of activation (Colton, 2009). The phenotypic heterogeneity extends to opioid receptors and endogenous opioid peptides, since both macrophages and microglia can variably express MOR, δ, and κ opioid receptors (Chao, Gekker, et al., 1996; Chao, Hu, et al., 1996; Gekker et al., 2004; Peterson, Molitor, & Chao, 1998; Sheng, Hu, Lokensgard, & Peterson, 2003; Turchan-Cholewo et al., 2008).

Emerging evidence indicates that microglia contribute to synaptic plasticity and the stability of synaptodendritic structure during maturation and in response to CNS disorders in adults (Antonucci et al., 2012; Paolicelli et al., 2011; Ransohoff & Stevens, 2011; Tremblay, Lowery, & Majewska, 2010; Wake, Moorhouse, Jinno, Kohsaka, & Nabekura, 2009). Opiate drugs such as morphine and methadone can directly trigger the retraction of dendritic spines in cerebral cortical neurons (Liao, Grigoriants, Loh, & Law, 2007; Liao, Grigoriants, Wang, et al., 2007; Liao, Lin, Law, & Loh, 2005) and affect the plasticity of adult neurons (Liao, Grigoriants, Loh, et al., 2007; Liao, Grigoriants, Wang, et al., 2007; Liao et al., 2005; Robinson & Kolb, 1999, 2004). Morphine's actions at MOR trigger decreases in NeuroD phosphorylation that impede glutamatergic signals originating from AMPA receptors (Liao et al., 2005). Subsequent increases in MOR-driven, dynamin-dependent receptor internalization retract spines (Liao, Grigoriants, Loh, et al., 2007; Liao, Grigoriants, Wang, et al., 2007; Liao et al., 2005). In striatal medium spiny neurons, the relationship is less clear since only a subset of medium spiny neurons express MOR, despite evidence indicating that opiate-induced spine losses are consistent among all medium spiny neurons (Fitting, Xu, et al., 2010). While morphine may disrupt the excitotoxic response by decreasing NeuroD phosphorylation, and restricting glutamatergic transmission through neuroprotective AMPA and NMDA receptor subtypes (Liao, Grigoriants, Loh, et al., 2007; Liao,

Grigoriants, Wang, et al., 2007; Liao et al., 2005), neurons in the striatum are less likely to be directly affected than in the cerebral cortex since only a subset of medium spiny neurons express MOR. MOR-expressing microglia and astroglia appear to contribute to the interactive neurotoxicity of morphine and Tat in the striatum (Sorrell & Hauser, 2014; Zou et al., 2011). In addition, e.g., GABA and fractalkine may serve as "off" signals—switching off overactive microglia (Beardsley & Hauser, 2014), and opiates can possibly modify these signals (Bagley et al., 2011; Krebs, Gauchy, Desban, Glowinski, & Kemel, 1994; McQuiston, 2007; Steiner & Gerfen, 1998; Suzuki et al., 2011; You et al., 1996). Thus, the innate immune and neurotransmitter signals disrupted by opiates and HIV strategically converge and are integrated into a unique "neuroimmune" logic by microglia.

2.2.2 Psychostimulant and HIV actions in microglia

2.2.2.1 Methamphetamine and HIV

Methamphetamine enhances HIV-1 replication in microglia (Liang et al., 2008). In addition to direct effects on viral replication, combined HIV Tat or gp120 and methamphetamine induce oxidative stress and free radical production in the CNS, which likely originates from reactive microglia (Banerjee, Zhang, Manda, Banks, & Ercal, 2010).

Psychostimulants including methamphetamine, cocaine, and ecstasy have all been suggested to activate the innate immune system (Clark et al., 2013). Immune activation may be an essential component of neurobiological adaption in alcohol and cocaine addiction (Crews, Zou, & Qin, 2011). Neuronal damage-associated molecular patterns (DAMPs) can directly activate or under pathophysiological conditions overactivate microglia (Biber et al., 2007; Block, Zecca, & Hong, 2007). DAMPs are released from stressed or injured cells (Bianchi, 2007; Srikrishna & Freeze, 2009) and trigger innate immune activation. Multiple classes of PRRs appear to be triggered through drug and alcohol abuse (Beardsley & Hauser, 2014; Crews, Zou, et al., 2011; Yakovleva, Bazov, Watanabe, Hauser, & Bakalkin, 2011). Methamphetamine also reportedly affects neuronal subpopulations through the activation of TAAR1 (Bunzow et al., 2001; Reese et al., 2007; Xie & Miller, 2009). TAAR1 is coexpressed on dopamine D2 receptor and DAT-positive neurons in the striatum, and TAAR1 activation results in increases in cAMP levels (Espinoza et al., 2011; Xie & Miller, 2009).

Alternatively, dopamine and norepinephrine, which accumulate in the synaptic cleft following psychostimulant exposure, are thought to be able to elicit responses in microglia (Farber, Pannasch, & Kettenmann, 2005), while $GABA_B$ receptor activation reduces lipopolysaccharide (LPS)-induced IL-6 and IL-12 p40 release (Kuhn et al., 2004). Minocycline preferentially blocks macrophage/microglial activation (IL-1β and IL-6 are attenuated, but not TNF-α), but fails to mitigate striatal dopaminergic neurotoxicity because minocycline does not attenuate methamphetamine-induced increases in TNF-α (Sriram, Miller, & O'Callaghan, 2006). Thus, as noted earlier in the context of opiate abuse, the innate immune and neurotransmitter signals disrupted by psychostimulants converge in microglia, which attempt to integrate the diverse input into a coordinated and measured response.

HIV-1 activates microglia directly causing increases in both viral and cellular toxins as outlined earlier. Tat alone appears to mediates many of the glial proliferative and cytokine/chemokine-secreting effects of HIV-1. Methamphetamine exacerbates the neurotoxic effects of HIV-1 through enhanced cytokine production and microglial activation (Theodore et al., 2006a, 2006b, 2007) (Fig. 9.2). Dopamine losses evident in the SIV model can be prevented by inhibiting macrophage/microglial activity (Scheller et al., 2005). As with methamphetamine, HIV-1 Tat causes a rapid increase in cortical neuronal excitability that is exacerbated by cocaine (Napier, Chen, Kashanchi, & Hu, 2014). However, in the case of cocaine, despite considerable overlap, some of the events triggering neuroinflammation and microglial activation differ. For example, TAAR1 is activated by methamphetamine and MDMA, but may play a less central role in cocaine's actions (Bunzow et al., 2001).

Considerable evidence points toward sigma-1 receptors as a molecular target of cocaine actions (Matsumoto, Nguyen, Kaushal, & Robson, 2014; Su, Hayashi, Maurice, Buch, & Ruoho, 2010). Sigma-1 receptors are widely expressed throughout the cells of the nervous system and elsewhere and can contribute to a variety of pathological processes (Maurice & Su, 2009; Su et al., 2010). Cocaine increases chemokine (C–C motif) receptor 5 (CCR5) and CXCR4 HIV coreceptor expression *in vitro* and *in vivo*, while transiently increasing viral replication in human peripheral blood mononuclear cells (PBMCs) in humanized severe combined immunodeficiency (SCID) mice (Roth, Whittaker, Choi, Tashkin, & Baldwin, 2005). In these studies, the sigma-1 receptor antagonist, BD1047, attenuated the effects of cocaine on HIV replication in HIV-infected humanized SCID mice, suggesting that the sigma-1 receptor is a molecular target of cocaine's actions

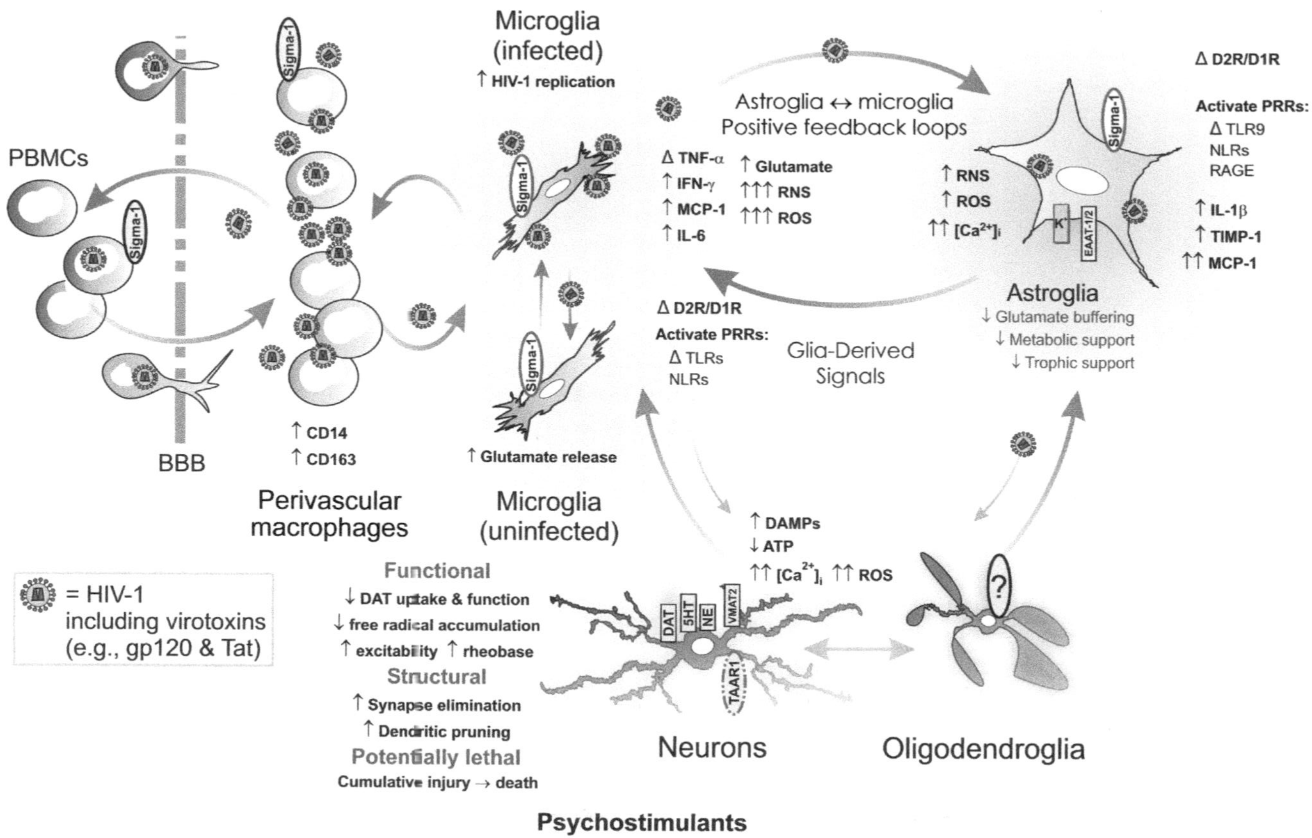

Figure 9.2 See legend on next page.

(Roth et al., 2005). Peterson and colleagues similarly demonstrated using BD1047 that sigma-1 receptor blockade prevented cocaine-induced HIV replication in microglia (Gekker et al., 2006). These investigators additionally showed the TGF-β inhibitor, SB-431542 (Inman et al., 2002), or immunoneutralizing TGF-β1 antibodies, to be effective in negating the cocaine-induced increases in HIV-1 expression (Gekker et al., 2006). Cocaine treatment accelerates monocyte extravasation across the endothelium of the BBB through a monocyte chemoattractant protein-1 (MCP-1 or CCL2) - dependent mechanism that is initiated via sigma-1 receptors (Yao et al., 2011, 2010). Understanding the regulation of MCP-1 expression and functional role of sigma-1 receptors following cocaine exposure should provide novel insight into the basic mechanisms by which cocaine augments the severity of neuroAIDS (Yao et al., 2010).

Figure 9.2 Psychostimulants can increase synaptic damage through direct actions on neurons and glia, including both microglia and astroglia. Psychostimulants block dopamine, serotonin (5HT), and norepinephrine (NE) transport resulting in excessive accumulations of these neurotransmitters in the synaptic cleft. Dopaminergic neurons are particularly vulnerable to methamphetamine, which disrupts dopamine transporter (DAT) and vesicular monoamine transporter 1 (VMAT2) function and can damage presynaptic terminals of neurons. Synaptic injury is accompanied by the production of reactive oxygen species (ROS) and reactive nitrogen species (RNS), and the production of damage-associated molecular patterns (DAMPs) that trigger activation of pattern recognition receptors (PRRs), including Toll-like receptor 9 (TLR9), nucleotide-binding oligomerization domain-like receptors (NLRs), and other PRRs (e.g., receptor for advanced glycation end-products or RAGE) expressed by microglia and astroglia. Importantly, psychostimulants (especially methamphetamine) appear to activate neurons directly through the disruption of monoaminergic transporters and VMAT2 mentioned above and through the activation of trace amine-associated receptor 1 (TAAR1). Psychostimulants also disrupt glial function directly by increasing intracellular ROS and likely Ca^{2+} concentrations ($[Ca^{2+}]_i$), NF-κB transcriptional activity, and by activating sigma-1-receptors (sigma-1; red, dashed-line outline; gray in the print version), especially in the case of cocaine, and enzyme systems driving oxidative and nitrosative stress especially in microglia (and other cell types). Increases in NF-κB transcriptional activity result in increased microglial, and to a lesser extent astroglial, production of tumor necrosis factor-α (TNF-α), interferon-γ (IFN-γ), interleukin-1β (IL-1β), and various other cytokines, as well as tissue inhibitor of metalloproteinase-1 (TIMP-1). Psychostimulants also obstruct the buffering of extracellular glutamate by inhibiting excitatory amino acid transporters-1/2 (EAAT1/2) and the conversion of glutamate to glutamine by inhibiting glutamine synthetase, as well as by limiting glucose metabolism in astrocytes. Collectively, neuronal injury and intensified glial activation promote positive microglial–astroglial, and neuronal–glial feedback that cause spiraling increases in neuroinflammation and neuronal injury. If unrestrained, the cumulative insults result in lasting neurodegenerative changes. *Modified and reprinted from Beardsley and Hauser (2014). Copyright 2014, with permission from Elsevier.*

3. ASTROGLIA

3.1. Critical functions in neuronal support and gliotransmission

Astrocytes form a close association with neurons and are strategically positioned to provide structural support and to maintain metabolic, trophic, and functional processes including synaptic transmission that were previously thought to be regulated by neurons themselves (Araque, Parpura, Sanzgiri, & Haydon, 1999; Haydon & Carmignoto, 2006; Parpura, Basarsky, Liu, Jeftinija, & Haydon, 1994; Volterra & Meldolesi, 2005). Astrocytes also express nearly every class of neurotransmitter receptor, which permits them to coordinate their response precisely with neurons (Beardsley & Hauser, 2014; Zhang & Barres, 2010). The "tripartite synapse" refers to the intimate association between astrocytes and pre- and postsynaptic interconnections (Araque et al., 1999; Parpura et al., 1994; Perea, Navarrete, & Araque, 2009). Gliotransmission refers to the selective uptake and/or release of specific neurotransmitters by astroglia through vesicular (Montana, Malarkey, Verderio, Matteoli, & Parpura, 2006), extracellular membrane microvesicles (Verderio, 2013; Verderio et al., 2012), and/or nanotubes (Verderio, 2013). Gliotransmission significantly modulates neurotransmission (De Pittà et al., 2012; Tewari & Parpura, 2013).

In addition to modulating synaptic transmission, astrocytes regulate extrasynaptic transmission within the ECS of the CNS (Sykova & Nicholson, 2008). This includes regulating "intersynaptic cross talk" or the movement of excess neurotransmitters, including glutamate, between synapses, as well as the management of ion homeostasis (Sykova, 2005; Vargova, Jendelova, Chvatal, & Sykova, 2001) and of the movement of water within the ECS (Amiry-Moghaddam et al., 2003; King, Kozono, & Agre, 2004). The coordinated movement of ions and water within the ECS ultimately regulates tissue volume, including brain swelling during specific pathological conditions (Anderova et al., 2011; Sykova, 2005; Zamecnik et al., 2012). Aquaporin-4 (AQP-4) channels expressed largely by astrocytes are critical for regulating the volume of water within the ECS (Amiry-Moghaddam et al., 2003; King et al., 2004). Because HIV-induced deficits in astroglial glutamate and perhaps K^+ buffering capacity, as well as AQP-4 function, appear to be independently disrupted by HIV and substance abuse (Berman et al., 2006; Cisneros & Ghorpade, 2012; Hauser et al., 2012; Kalivas & Volkow, 2011; Knackstedt,

Melendez, & Kalivas, 2010; Li et al., 2006; Podhaizer et al., 2012; Wang et al., 2003; Zou et al., 2011), extrasynaptic transmission within the ECS is likely to be highly compromised in HIV-infected substance abusers.

3.2. Innate immune effectors

Astroglia are particularly adept at sensing metabolic instability in neurons and in the surrounding microenvironment. They are critical for interpreting and translating intercellular communication between neurons and microglia—especially during pathological situations (Maragakis & Rothstein, 2006; Molofsky et al., 2012; Verkhratsky, Rodriguez, & Parpura, 2013). Astrocytes can express a variety of types of PRRs against PAMPs and DAMPs, including TLRs, NODs, complement receptors [CR1, CR2, C3aR, C5aR (Gasque, Dean, McGreal, VanBeek, & Morgan, 2000)], mannose receptor (Liu et al., 2004), and RAGE (Beardsley & Hauser, 2014; Farina et al., 2007; Husemann & Silverstein, 2001). Thus, astroglia can act as transducers—both sensing neuronal injury and conveying information about neuron injury to microglia. Emerging evidence also suggests that astrocytes may express MHC-II following injury or stress (Jensen, Massie, & De Keyser, 2013).

3.3. Astroglia and HIV

Astroglia (Brack-Werner, 1999; Gorry, Purcell, Howard, & McPhee, 1998; Gorry et al., 2003; Kramer-Hammerle, Hahn, et al., 2005; Kramer-Hammerle, Rothenaigner, et al., 2005) and perhaps also pericytes (Nakagawa, Castro, & Toborek, 2012) are the only resident CNS cells besides microglia that can become infected. Unlike microglia, astroglia tend not to display productive infection; rather, they harbor latent infection that can be reactivated from latency by specific proinflammatory cytokines such as TNF-α, GM-CSF, or IFN-γ in SIV_{mac251}-infected astrocytes (Carroll-Anzinger, Kumar, Adarichev, Kashanchi, & Al-Harthi, 2007; Guillemin et al., 2000; Narasipura et al., 2012) or specific, class I histone deacetylase (HDAC) inhibitors (Narasipura, Kim, & Al-Harthi, 2014). Latent HIV is dormant, meaning that viral RNA has been reverse transcribed and integrated into the host DNA, but that new virions are not being produced. A number of hypotheses have been put forth in an attempt to explain fundamental differences in the regulation of HIV infectivity by astroglia. The activation of NF-κB was proposed to play a less central role in driving viral production by astrocytes than in microglia (Conant, Atwood, Traub,

Tornatore, & Major, 1994). Subsequent studies also suggested differences in Rev–astroglial RNA helicase DDX1 interactions (Fang et al., 2005). In addition, a specific class I HDACs and a lysine-specific histone methyltransferase, SU(VAR)3–9, demonstrated in an U87MG astroglial cell line (Narasipura et al., 2014) were proposed as uniquely restricting HIV transcription in astroglia. Overexpression of *nef* (an early, regulatory gene), but not *gag* (a late structural gene), is seen in approximately 20% of astrocytes in infected individuals (Saito et al., 1994). Accordingly, Nef serves as a marker and perhaps contributing factor of restricted HIV infection in astroglia (Saito et al., 1994).

Besides being cellular sites of latent infection (Bencheikh, Bentsman, Sarkissian, Canki, & Volsky, 1999; Brack-Werner, 1999; Canki et al., 1997; Kramer-Hammerle, Rothenaigner, et al., 2005), astroglia respond robustly following exposure to HIV proteins (Tat and gp120) or intact virions (Li, Bentsman, Potash, & Volsky, 2007; Wang et al., 2003) by releasing proinflammatory cytokines. Following exposure to viral products, astroglia release toxic and inflammatory cellular products (e.g., glutamate, ROS or cytokines such as TNF-α, IFN-γ, and IL-6) creating pathophysiological conditions that are detrimental for neurons (Bell, 1998; Garden, 2002; Genis et al., 1992; Kaul et al., 2001; Nath, Conant, Chen, Scott, & Major, 1999; Persidsky & Gendelman, 2003).

The nature of the inflammatory response can differ among individual astrocytes (Fitting, Zou, et al., 2010; Zhang & Barres, 2010) as well as among microglia (Carson et al., 2007; Hanisch, 2013; Scheffel et al., 2012). Astrocytes are quite heterogeneous in the expression of a wide variety of phenotypic characteristics (Emsley & Macklis, 2006), including many receptor classes, and/or in their response to cues within the local microenvironment of the CNS (Fitting, Zou, et al., 2010; Shao & McCarthy, 1994; Shao, Porter, & McCarthy, 1994; Zhang & Barres, 2010). Astroglial heterogeneity has been historically attributed to the unique environmental milieu imparted by neighboring neurons. Conversely, astrocytes generated along specific spatiotemporal domains within the ventricular zone retain unique phenotypic characteristics throughout life (Tsai et al., 2012). Domain-specific astroglial variants have recently been shown to specify synaptic identity and regulate the ability of neurons to regenerate (Tsai et al., 2012).

Opiates and psychostimulants destabilize astroglial function directly. The destabilization usurps the ability of astrocytes to support neurons metabolically and trophically, while disrupting gliotransmission. Despite some attempts at neuroprotection, the net consequences of exposing astroglia to opiates or psychostimulants are that they are less likely to aid neurons or to mitigate a reactive microglial response to HIV infection (Hauser et al., 2007).

3.4. Effects of opiates and HIV in astroglia

Histopathological studies demonstrate that astrocytes display fewer reactive changes than microglia in a chronic opiate abusing cohort of HIV-infected patients (Anderson et al., 2003). Nevertheless, astroglial function is markedly affected and astroglia are able to transduce and amplify signals from HIV-infected or -uninfected perivascular macrophages and microglia—even in the absence of substance abuse coexposure (Hauser et al., 2007). The release of proinflammatory cytokines and chemokines (e.g., MCP-1, MCP-5, and RANTES) can recruit macrophages/microglia, and these newly arriving cells likely contribute to neurotoxicity (El-Hage et al., 2005). The consequences of opiate abuse and HIV interactions in astroglia have been exhaustively reviewed previously (Banerjee et al., 2011; Dutta & Roy, 2012; Hauser, Aldrich, et al., 2005; Hauser et al., 2007, 2012; Peterson et al., 1998; Reddy et al., 2012). Accordingly, we will only highlight key aspects of the interactions in the paragraphs that follow.

The high degree of phenotypic heterogeneity and plasticity that occurs among individual astrocytes in the expression of HIV coreceptors (Podhaizer et al., 2012) or PRRs such as TLRs (El-Hage, Podhaizer, Sturgill, & Hauser, 2011), as well as in response to HIV proteins Tat and gp120 (Fitting, Zou, et al., 2010), are often overlooked. Moreover, the prevalence of HIV infection appears to increase in immature astroglia (Lawrence et al., 2004; Messam & Major, 2000; Tornatore, Chandra, Berger, & Major, 1994; Tornatore, Meyers, Atwood, Conant, & Major, 1994; Tornatore, Nath, Amemiya, & Major, 1991), suggesting developmentally regulated differences in susceptibility to HIV-1 by astrocytes. With respect to opioids, astrocytes can express MOR, δ (DOR), and KOR (Curtis, Faull, & Eriksson, 2007; Eriksson, Hansson, & Rönnbäck, 1992; Eriksson, Nilsson, Wagberg, Hansson, & Rönnbäck, 1993; Gurwell et al., 1996; Hauser & Mangoura, 1998; Hauser et al., 1996; Ruzicka et al., 1995; Stiene-Martin & Hauser, 1991; Stiene-Martin, Zhou, & Hauser, 1998; Stiene-Martin et al., 2001; Turchan-Cholewo et al., 2008). Moreover, astrocytes can express endogenous opioid peptides associated with the preproenkephalin gene (Hauser, Osborne, Stiene-Martin, & Melner, 1990; Shinoda, Marini, Cosi, & Schwartz, 1989; Spruce, Curtis,

Wilkin, & Glover, 1990) and preproenkephalin is upregulated by cytokines including IL-1β and interferon-γ (Low & Melner, 1990; Ruzicka & Akil, 1997). Unlike neurons, which typically produce and release fully processed enkephalin pentapeptides with a high affinity for DOR, astroglia tend to release larger, intact, or partially processed proenkephalin peptides that can have high affinity at DOR or KOR. Overall, the wide expression of MOR, DOR, and KOR by astroglia makes them a significant target for both endogenous and exogenous opioids/opiates.

By disrupting astrocyte function, opiate drug abuse likely subverts their ability to maintain a homeostatic balance of ions and neurochemicals within the ECS, which promotes neuronal injury and death. Morphine can modify cytokine and chemokine production by astroglia (Avdoshina, Biggio, Palchik, Campbell, & Mocchetti, 2010; El-Hage, Bruce-Keller, Knapp, et al., 2008; El-Hage, Bruce-Keller, Yakovleva, et al., 2008; El-Hage et al., 2005; Mahajan, Aalinkeel, et al., 2005; Mahajan, Schwartz, et al., 2005; Mahajan, Schwartz, Shanahan, Chawda, & Nair, 2002; Sawaya, Deshmane, Mukerjee, Fan, & Khalili, 2009). Opiates short-circuit the ability of astroglia to protect neurons from HIV (Hauser, El-Hage, et al., 2005; Hauser et al., 2007; reviewed in Hauser et al., 2012). Opiates can intrinsically affect the expression of the glutamate transporters EAAT1 and EAAT2 (Mao, Sung, Ji, & Lim, 2002; Ozawa, Nakagawa, Shige, Minami, & Satoh, 2001). In the presence of HIV-1 Tat, opiates exacerbate the deleterious effects of the disease on intracellular signaling and $[Ca^{2+}]_i$ homeostasis (El-Hage, Bruce-Keller, Yakovleva, et al., 2008; El-Hage et al., 2005), which further reduces the ability to buffer extracellular glutamate (Zou et al., 2011). The failure to buffer extracellular glutamate lowers the threshold for excitotoxicity in neurons (Podhaizer et al., 2012; Zou et al., 2011). Furthermore, opiate exposure alone can increase ROS (Podhaizer et al., 2012; Zou et al., 2011) and some proinflammatory cytokines (Mahajan, Schwartz, et al., 2005; Mahajan et al., 2002) in sufficient amounts to potentially be directly neurotoxic (Zou et al., 2011).

HIV-1 Tat is a potent activator of NF-κB (Conant, Ma, Nath, & Major, 1996; El-Hage, Bruce-Keller, Yakovleva, et al., 2008), resulting in the release of a large number of cytokines and chemokines by astroglia (Conant et al., 1998; El-Hage, Bruce-Keller, Yakovleva, et al., 2008; El-Hage, Wu, Wang, et al., 2006; El-Hage et al., 2005; Kutsch, Oh, Nath, & Benveniste, 2000). Besides destabilizing glutamate and triggering inflammation, Tat shares a Cys-Cys-Phe motif found in β-chemokine sequences such as CCL5 (Albini et al., 1998) that endows it with chemotactic properties. HIV-1 Tat also destabilizes Ca^{2+} in astroglia (El-Hage et al., 2005) by mechanisms involving IP_3-dependent release (Kumar, Manna, Dhawan, & Aggarwal, 1998). Increased $[Ca^{2+}]_i$ dysregulates nuclear–cytoplasmic trafficking of NF-κB subunits (El-Hage, Bruce-Keller, Yakovleva, et al., 2008) and leads to release of CCL2, CCL5, IL-6, and TNF-α. Morphine exacerbates this cycle (El-Hage, Bruce-Keller, Yakovleva, et al., 2008; El-Hage et al., 2005; El-Hage, Wu, Ambati, et al., 2006; El-Hage, Wu, Wang, et al., 2006), presumably by augmenting Tat-induced increases $[Ca^{2+}]_i$. MOR can couple to Gβγ (Bonacci et al., 2006; Mathews, Smrcka, & Bidlack, 2008), $G_{q/11}$-α (Hauser et al., 1996), and/or G_sα via MOR-1K splice variants (Dever et al., 2014) in astroglia resulting in cellular excitation. Opiate and HIV-induced increases in astroglial-derived cytokines in turn enhance microglial recruitment and activation (El-Hage, Wu, Wang, et al., 2006). Morphine's unique actions in HIV-1-exposed astrocytes, in particular, appear to drive escalating, intercellular feedback loops with microglia and perivascular macrophages that increase and sustain inflammation (El-Hage, Wu, Wang, et al., 2006; Hauser et al., 2007). We have proposed that, unlike other HIV-1-infected organs, which also can harbor MOR-expressing macrophages, the brain is unique because of the inflated response of astroglia to opiates (Hauser et al., 2007, 2012). As a partial test of this assertion, we recently tested whether drugs with selective anti-inflammatory activity in glia could attenuate the deleterious effects of HIV and opiate exposure. We found that ibudilast (also known as AV411 or MN-166) or an analogue lacking phosphodiesterase activity (AV1013), both of which preferentially suppress glial inflammation, attenuates HIV-1 ± morphine-dependent increases in HIV-1 replication and in HIV-1 Tat ± morphine-induced cytokine release and neurotoxicity *in vitro* (El-Hage et al., 2014).

This concept is supported by findings that HIV Tat ± morphine-induced death of medium spiny neurons is largely mediated via MOR-expressing glia (Zou et al., 2011), including astroglia (El-Hage, Bruce-Keller, Knapp, et al., 2008; El-Hage, Bruce-Keller, Yakovleva, et al., 2008; El-Hage et al., 2005; El-Hage, Wu, Wang, et al., 2006) and microglia (Bokhari et al., 2009; Gupta et al., 2010; Turchan-Cholewo et al., 2008, 2009). Alternatively, the extent to which synaptodendritic culling is similarly driven by glia has not yet been established. Although glia undoubtedly play a significant role, as noted earlier, there is some evidence that morphine can converge with HIV Tat to cause spine retraction through direct actions on the dendrites of cerebral cortical neurons (Liao, Grigoriants, Loh, et al., 2007; Liao, Grigoriants, Wang, et al., 2007; Liao et al., 2005). Additionally, since morphine can excite dopaminergic neurons projecting from the ventral tegmental area (VTA) to striatal neurons by hyperpolarizing inhibitory GABAergic interneurons in the VTA (Johnson & North, 1992), it is likely that HIV-1 and opiate-related alterations in synaptic organization are affected by a complex interplay of events.

3.5. Effects of psychostimulants and HIV in astroglia

A number of reviews on the effects of psychomotor stimulants by themselves (Beardsley & Hauser, 2014; Cadet et al., 2014; Cadet & Krasnova, 2009; Clark et al., 2013) and in the context of HIV (Buch et al., 2012; Cisneros & Ghorpade, 2012; Hauser et al., 2007; Nath, 2010) on astrocytes are available. Briefly, methamphetamine and cocaine are thought to affect astrocyte function through a variety of indirect and direct actions. The profile of inflammatory cytokines released by astrocytes in response to methamphetamine notably includes the release of TNF-α, IL-1β, IL-6, and the chemokine MCP-1, as well as intercellular adhesion molecule-1 (ICAM-1) (Clark et al., 2013; Flora et al., 2002; Goncalves et al., 2008; Nakajima et al., 2004; Theodore et al., 2006a). The astroglial response to cocaine is more limited than the response to methamphetamine. Cocaine can increase the expression of TNF-α, IL-1β, and IL-6 transcripts, while downregulating the anti-inflammatory cytokine IL-10 (Clark et al., 2013). These cytokines fuel inflammatory cascades and the release of chemokines such as MCP-1, which recruit macrophages and activate microglia within the CNS (Yao et al., 2010). Glial inflammation is proposed to be an essential step in the maladaptive neuroplasticity accompanying addiction (Crews, Zou, et al., 2011; Clark et al., 2013; Frank, Watkins, & Maier, 2011). Excessive or sustained high levels of inflammation result in neuronal injury and potentially neuronal death (Buttner, 2011; Jayanthi et al., 2005; Krasnova & Cadet, 2009).

Gliotransmission has been reported to be necessary for reinstatement of cocaine-seeking behavior (Turner, Ecke, Briand, Haydon, & Blendy, 2013). Mice expressing a dominant-negative soluble *N*-ethylmaleimide-sensitive factor attachment protein receptor (SNARE) variant driven by a GFAP-dependent promoter were used to selectively disrupt gliotransmission (Turner et al., 2013). However, in these studies, the transmitter substances released during gliotransmission are uncertain. Gliotransmission can involve the vesicular release of excitatory transmitters, including glutamate, serine, and ATP (D'Ascenzo et al., 2007; Martineau, 2013; Parpura & Zorec, 2010;

Pascual et al., 2005; Santello, Cali, & Bezzi, 2012; Van Horn, Sild, & Ruthazer, 2013). Of these, glutamate appears to contribute to drug-seeking behavior and other aspects of cocaine addiction (Beardsley & Hauser, 2014).

In combination with methamphetamine, HIV Tat exacerbates the disruption of EAAT-2 and perhaps EAAT-1 in astroglia (Cisneros & Ghorpade, 2012), and may additionally increase the release of glutamate from injured presynaptic terminals. Interestingly, ceftriaxone, which upregulates EAAT2 expression in astroglia, protects neurons against Tat or gp120-induced injury (Rumbaugh, Li, Rothstein, & Nath, 2007). Collectively, methamphetamine and HIV appear to dysregulate the buffering of extracellular glutamate by astrocytes, which contributes to excitotoxic injury in neurons. Propentofylline, which is thought to affect multiple molecular targets in both astroglia and microglia (Sweitzer & De Leo, 2011; Sweitzer, Schubert, & De Leo, 2001), impairs reinstatement to cocaine through an EAAT-2-related mechanism (Reissner et al., 2014). Furthermore, anti-inflammatory drugs with preferential actions in glia such as minocycline and/or ibudilast can limit key aspects of methamphetamine's locomotor behaviors and/or reinforcing properties (Snider, Hendrick, & Beardsley, 2013; Snider, Vunck, van den Oord, et al., 2012) or aspects of cocaine sensitization (Chen & Manev, 2011; Chen, Uz, & Manev, 2009).

Cocaine affects BBB permeability and increases MDM trafficking across the barrier (Fiala et al., 1998; Gan et al., 1999; Zhang et al., 1998). Key aspects of the higher rates of CNS infection (Fiala et al., 2005) and encephalitis (Clark et al., 2013) caused by cocaine are fueled by sigma 1 receptor-induced increases in MCP-1 derived from glia, and especially astrocytes, recruiting new MDMs into the CNS (Yao et al., 2010). Moreover, sigma 1 receptor-dependent increases in the expression of activated-leukocyte cell adhesion molecule by endothelial cells increases diapedesis of MDMs and the recruitment of perivascular macrophages (Yao et al., 2011). Methamphetamine increases astroglial shedding of matrix metalloproteinases (MMPs), especially MMP-1 and MMP-2, which can disrupt the BBB (Conant et al., 2004). Besides MMP-1 and MMP-2, HIV-1 Tat can increase MMP-5, which may reduce long-term potentiation in the hippocampus (Conant et al., 2010). MMP-5 can cleave GluN1 NMDAR subunits (Szklarczyk et al., 2008) and AQP-4.

AQP-4 expressed by astroglia (Rash, Yasumura, Hudson, Agre, & Nielsen, 1998) is essential for moving water from the ECS through astrocytes and across the BBB into the vasculature (King et al., 2004; Tait, Saadoun, Bell, & Papadopoulos, 2008). AQP-4 is intimately linked to astroglial function (King et al., 2004). AQP-4 levels are increased in HIV-associated dementia (HAD); however, it is uncertain whether this is a maladaptive or a compensatory response to counteract the effects of chronic inflammation (St Hillaire et al., 2005). Interestingly, the effects of cocaine are attenuated in AQP-4-null mice, leading to speculation that AQP-4 regulates cocaine reinforcement and dependence by alternating dopamine and glutamate release associated with drug reward (Li et al., 2006). Although AQP-4 expression *per se* appears to be unaffected by cocaine exposure (Narayana et al., 2014), AQP-4-null mice display attenuated locomotor and reward responses to cocaine, suggesting its involvement in the neurobiological actions of cocaine (Li et al., 2006). By virtue of their critical role in regulating the volume of water within the ECS (Amiry-Moghaddam et al., 2003; King et al., 2004), including pathological brain swelling (Anderova et al., 2011; Sykova, 2005; Zamecnik et al., 2012), AQP-4 channels are likely to be important in regulating HIV and psychostimulant interactions.

4. GENETIC FACTORS THAT MODULATE HIV-1 INFECTIVITY AND NEUROPATHOGENESIS

4.1. Introduction

There are huge differences in the susceptibility of individuals to addiction or to acquiring HIV. Emerging evidence indicates that different gene polymorphisms may underlie these differences, and may also contribute to differences in the response to cART among individuals. Gene profiling differences have also suggested that HAND-defining disorders may represent fundamentally different disease courses, and not a continuum of a single pathophysiological process (Gelman et al., 2012).

Polymorphisms in the genes associated with HIV-1 coreceptors and/or their endogenous ligands can markedly influence AIDS progression (Carrington, Dean, Martin, & O'Brien, 1999; Smith et al., 1997; Winkler et al., 1998). CCR5 in particular (Carrington et al., 1999), and mutations thereof, e.g., CCR5Δ32 (Boven, van der Bruggen, van Asbeck, Marx, & Nottet, 1999; Huang et al., 1996), as well as mutations in key cytokines, e.g., IL-10 (Shin et al., 2000) or TNF-α (Quasney et al., 2001), or other genes linked to specific neurodegenerative disorders such as those encoding apolipoprotein ε4 (ApoE4) (Verghese, Castellano, & Holtzman, 2011) may have a marked influence on neuroAIDS outcome measures (Shapshak, Duncan, Minagar, et al., 2004; Shapshak et al., 2011). Mutations in other human gene products, such as the specific apolipoprotein B mRNA-editing enzyme catalytic polypeptide-like 3G encoded by the *APOBEC3G* gene (Kim et al., 2010), an innate viral restriction factor that inhibits the production of HIV (Bishop, Holmes, Sheehy, & Malim, 2004; Mangeat et al., 2003; Shindo et al., 2003), may have deleterious consequences. In this section, we cite and discuss several examples in which human genetic variability is beginning to uncover essential sites of drug abuse and neuroAIDS interplay.

CCR5 plays a critical role in HIV infection as a coreceptor for entry of CCR5-preferring strains that appear largely responsible for initial infection (Berger, Murphy, & Farber, 1999; Moore, Kitchen, Pugach, & Zack, 2004). The importance of CCR5 in HIV infectivity is supported by evidence from a wide variety of approaches. CCR5 levels in general, and in brain MDMs and microglia, have been correlated with the severity of HIV neurologic disease (An, Osuntokun, Groves, & Scaravilli, 2001; Vallat et al., 1998). Individuals homozygous for the CCR5Δ32 mutation resist infection by HIV (Boven et al., 1999; Huang et al., 1996). CCR5 blockade employing the antagonist maraviroc to inhibit viral entry has marked clinic efficacy (MacArthur & Novak, 2008). Experimental reductions in CCR5 levels using gene silencing strategies have been successful in reducing HIV infectivity experimentally (Anderson & Akkina, 2007; Lee et al., 1999).

The importance of HIV coreceptors in viral infectivity and pathogenesis should not be underestimated. Polymorphisms of CCR5, such as CCR5Δ32 (Boven et al., 1999; Huang et al., 1996) or CCR2 (Smith et al., 1997), can confer significant protection against HIV progression, while other mutations can worsen disease progression. Two studies have demonstrated that dual single nucleotide polymorphisms (SNPs) in the *RANTES* gene promoter (−471 and −96) reduce HIV disease advancement (Gonzalez et al., 2001; McDermott et al., 2000), while only one of these studies found an effect on transmission risk (McDermott et al., 2000). The RANTES-28G mutation increases RANTES transcript levels and is associated with increased protection against the clinical progression of HIV infection (Liu et al., 1999). The apparent eradication of HIV in a patient receiving a transplant of hematopoietic stem cells harboring a mutation (CCR5Δ32) in CCR5 has highlighted the insights that might be gained into mechanisms of HIV infectivity/pathogenesis by studying the role of genetic variability (Allers et al., 2011; Hutter et al., 2009).

Although polymorphisms of CXCR4 also exist, these are far less frequently identified than CCR5 polymorphisms. A likely reason may be that the deletion of CXCR4 or its cognate ligand SDF-1/CXCL12 is lethal (Ma et al., 1998; Zou, Kottmann, Kuroda, Taniuchi, & Littman, 1998).

Combined mutations in multiple HIV coreceptors and/or in the cognate ligands of these coreceptors can interact to confer more or less protection against HIV infectivity or the subsequent pathogenesis (Gelman et al., 2012; Levine, Sinsheimer, Bilder, Shapshak, & Singer, 2012; Shapshak, Duncan, Torres-Munoz, et al., 2004). Lastly, "elite suppressors" or "controllers" is the term given a subset of individuals who maintain plasma HIV copy numbers below 50 copies/mL (Han et al., 2008). In sum, gene variations in β-chemokines and their receptors can have marked influences on the clinical course HIV infection. There is considerable debate regarding the extent to which combinations of protective/nonprotective allelic variants contribute to the subset of patients who are elite suppressors and are able to intrinsically suppress HIV replication (Baker, Block, Rothchild, & Walker, 2009; Miura et al., 2008).

Levels of MCP-1 and its cognate receptor, CCR2, are increased with HIV infection and coincide with neurological impairment (Cinque et al., 1998; Sozzani et al., 1997). As noted earlier, MCP-1 is released by HIV-exposed MDMs, microglia, and astrocytes (Conant et al., 1998; Nath et al., 1999). MCP-1 released by resident glia has been proposed as a key event in recruiting MDMs into the brain, an event that is exacerbated by opiates and psychostimulants (Berman et al., 2008; Eugenin, Dyer, Calderon, & Berman, 2005; Eugenin et al., 2006; Fiala et al., 2005; Hauser et al., 2007; Yao et al., 2010; Zhang et al., 1998). MCP-1 release from glia, especially astroglia, is significantly affected by drug abuse and HIV interactions (Hauser et al., 2007). Morphine exacerbates the release of chemokines, especially RANTES and MCP-1, from HIV Tat-exposed astrocytes (El-Hage et al., 2005; El-Hage, Wu, Wang, et al., 2006) and microglia (Turchan-Cholewo et al., 2009) in a time- and concentration-dependent manner, while cocaine accelerates monocyte extravasation across the BBB endothelium through a MCP-1-dependent mechanism that is absent in CCR2 knockout mice (Yao et al., 2010). A number of mutations in the *CCR2* and *CCL2* genes have been demonstrated to affect aspects of HIV/AIDS. The CCR2 V64I allele is associated with a more rapid onset of neurocognitive impairment, but even after adjusting for estimated time of seroconversion, there is no correlation with increased viral loads in cerebrospinal fluid (CSF) or in plasma, of HIV-infected subjects (Singh et al., 2004). Interestingly, the CCR2 V64I allele affords some protection against progression to AIDS, especially during early phases of the disease process (Ioannidis et al., 2001; Mulherin et al., 2003), perhaps at the expense of CNS function. An MCP-1 (CCL2)-2578G/A promoter polymorphism was shown to enhance protein production and was associated with significantly reduced risk of acquiring HIV infection (Gonzalez et al., 2002). However, once infected, patients with this genotype showed faster disease progression and enhanced risk for HAD, presumably due to enhanced infiltration of infected monocytes (Gonzalez et al., 2002). *CCR2* gene polymorphisms can act in a cooperative manner with other genes to affect HIV pathogenesis, including *CCR5* (Rigato et al., 2008). By contrast, no connections were found between CCR2 V64I or CCR5Δ32 mutations and HIV infectivity in "preferential" opiate abusers in northeastern India (Sarkar et al., 2010). However, as is common with most drug abusing cohorts, complicated individual abuse patterns and polydrug use confound the interpretation of the findings: 59% of these subjects abused spasmo-proxyvon, which contains the synthetic opiates dextropropoxyphene or propoxyphene and acetaminophen (Mahanta, Borkakoty, Das, & Chelleng, 2009), 54% abused heroin, and 15% abused "brown sugar" (partially purified heroin) (Sarkar et al., 2010).

TNF-α is important in triggering subsequent proinflammatory cascades such that any abnormalities in the regulation of TNF-α responsiveness are likely to have resounding consequences for the CNS (Bradley, 2008; McCoy & Tansey, 2008). Accordingly, it is perhaps not surprising that polymorphisms in the TNF-α promoter are associated with higher incidence of HAND (Quasney et al., 2001). Notably, both classes of abused drugs, opiates and psychostimulants, can increase the release of TNF-α from HIV or virotoxin-exposed glia (Fiala et al., 2005, 1997; Flora et al., 2002; Gendelman et al., 1997; Sawaya et al., 2009; Sriram et al., 2006; Zhang et al., 1998). Alternatively, it has been argued that while TNF-α appears to serve as a marker for HIV progression, interferon-γ may play a more central role as a causal factor in the development of the disease based on genetic analyses of polymorphisms of both genes in the same patient population (Shapshak, Duncan, Minagar, et al., 2004).

4.1.1 Mitochondrial genetics

Within the CHARTER cohort, several mtDNA SNP haplotypes are associated with marked protection from peripheral neuropathies (African haplogroup L1c and European haplogroup J) (Holzinger et al., 2012). mtDNA polymorphisms have also been linked to bipolar disorder and may be associated with an increased risk for neurodegenerative/neurocognitive disorders (Chinnery et al., 2001; Lin & Beal, 2006), especially those associated with aging (Kato, 2001). Dopaminergic neurons may be particularly susceptible to mtDNA damage (Bender et al., 2008). Interestingly, methamphetamine increases mtDNA damage (Bachmann et al., 2009), and CNS damage caused by *in utero* methamphetamine exposure can be rescued by increasing DNA repair through the enhancement of oxoguanine glycosylase 1 activity (Wong, McCallum, Jeng, & Wells, 2008).

4.2. Gene variation in opiate drug abuse and HIV interactions

Given the importance of host genetic variability in determining and revealing fundamental mechanisms underlying HIV infectivity and pathogenesis, might host genetics also reveal basic processes underlying the interactions between substance abuse and HIV? An examination of specific human gene polymorphisms, especially genes for drug receptors (Bond et al., 1998; Kreek, Bart, et al., 2005; Kreek et al., 2012; Kreek, Nielsen, et al., 2005), enzymes affecting drug metabolism (Meyer & Zanger, 1997), and/or neurochemical systems thought to underlie addiction (Kreek, Bart, et al., 2005; Kreek, Nielsen, et al., 2005; Lachman et al., 1996; Levine et al., 2012; Li et al., 2004; Nebert, McKinnon, & Puga, 1996), has identified significant correlative relationships between gene polymorphisms and substance abuse (Crystal et al., 2012; Jacobs, Murray, Byrd, Hurd, & Morgello, 2013; Kreek, Bart, et al., 2005; Kreek, Nielsen, et al., 2005; Manini, Jacobs, Vlahov, & Hurd, 2013; Yuferov, Levran, Proudnikov, Nielsen, & Kreek, 2010) (see Table 9.1). Because addiction is principally a CNS disorder with neurobehavioral/neuropsychiatric underpinnings (Leshner, 1997; Volkow, Wang, Fowler, & Tomasi, 2012), examining substance abuse–HIV interactions in the brain seems a logical direction. The potential importance of studying human gene polymorphisms as an approach to identify novel factors and mechanisms underlying drug abuse and neuroAIDS pathogenesis cannot be overemphasized.

In a sample population of 1031 women, polymorphisms in *OPRM1*, the gene encoding MOR, were associated with the severity of HIV infection or the response to cART (Proudnikov et al., 2012). These investigators found both negative and positive correlations with HIV severity in a small subset of *OPRM1* polymorphisms, while most variants displayed no association with HIV progression. Interestingly, although a subset of the patients sampled undoubtedly were substance abusers, many were not—suggesting that MOR receptor is inextricably linked to processes influencing HIV pathogenesis irrespective of opiate exposure. Although the mode of action is unclear, MOR activation can alter the expression of HIV-1 chemokine

Table 9.1 Examples of gene variants contributing to substance abuse that also correlate with the severity of HIV disease progression

Neurochemical system	Gene	Variant (polymorphism/isoform)	Outcome(s) measured	Finding/interpretation	Selected reference(s)
Dopamine					
	Drd1	*rs265975*	Drug dependence	Altered opiate/cocaine dependence	Jacobs et al. (2013)
	Drd2	*multiple*	Drug dependence	Altered opiate/cocaine dependence	Jacobs et al. (2013)
	Drd3	*rs6280TC*	Cognition	Increased cognitive impairment	Gupta et al. (2011)
	Drd4		Impulsivity	Altered inhibitory control	Congdon, Lesch, and Canli (2008)
	SLC6A3 (DAT1)			Altered inhibitory control	Congdon et al. (2008)
	SLC6A3 (DAT1)	*3′-UTR 40 bp*	HAND	No relationship to cognitive impairment	Levine et al. (2012)
	MAO1				
	MAO2				
	COMT	*rs4680 (val158met)*	HIV serostatus	Significant correlation	Sundermann et al. (2013)
	COMT	*rs4680 (val158met)*	HAND	No relationship to cognitive impairment	Levine et al. (2012)
	BDNF	*rs6265 (val66met)*	HAND	No relationship to cognitive impairment; dopamine neurotrophin	Levine et al. (2012)

Opioid	*OPRM1*		HIV progression response to cART	Specific polymorphisms associated with HIV progression and response to treatment	Proudnikov et al. (2012)
	OPRM1	MOR-1K MOR-1	HAND and HIVE	↑MOR-1K with HAND and HIVE; MOR-1 unaltered	Dever et al. (2014)
	OPRM1	MOR-1	Susceptibility to drug [opiate?] overdose		Manini et al. (2013)
	OPRM1	*A C17T*	Drug use		Crystal et al. (2012)
	OPRM1	MOR-1	Drug use	MOR-1 isoform unaltered	Crystal et al. (2012)
	OPRK1		HIV risk	Specific polymorphisms associated with HIV progression and response to treatment	Proudnikov et al. (2013)
	OPKR1	KOR-1		↑*OPRK1* in anterior cingulate; homeostatic attempt to decrease inflammation in exchange for motor impairment	Yuferov, Butelman, Ho, Morgello, and Kreek (2014)
	PDYN			↓*PDYN* in anterior cingulate; homeostatic attempt to decrease inflammation in exchange for motor impairment	Yuferov et al. (2014)

coreceptors involved in HIV entry, and MOR can undergo heterologous desensitization with CXCR4 (Burbassi, Aloyo, Simansky, & Meucci, 2008; Finley et al., 2008; Patel et al., 2006; Pitcher et al., 2014; Steele, Henderson, & Rogers, 2003; Szabo et al., 2002) or CCR5 (Chen et al., 2004; Happel, Steele, Finley, Kutzler, & Rogers, 2008; Song et al., 2011). Lastly, non-opioid genes may influence opiate drug and HIV interactions. For example, the presence of an apolipoprotein ε4 (ApoE4) allele increases the likelihood of neurotoxicity in response to combined morphine and HIV-1 Tat exposure in isolated human neurons *in vitro* (Turchan-Cholewo et al., 2006). The mechanisms responsible for opiate interactions with ApoE4 are unclear. Possessing an ApoE4 allele and increased ApoE4 levels in CSF was associated with an increased probability of neurocognitive impairment in HIV-infected patients in one study (Andres et al., 2011), while another failed to find a linkage between the expression of an ApoE4 allele and cognitive deficits in HIV-infected subjects (Morgan et al., 2013).

A potentially important role of *OPRM1* splice variants in neuroAIDS is indicated by our findings, showing quantitative differences in specific human MOR splice variant expression levels with HIVE and/or neurocognitive status (Dever et al., 2014, 2012). In addition, MOR-1 (exon 1), MOR-1A, MOR-1X, and MOR-1K splice variants appear to differ across CNS cell types (Dever et al., 2014, 2012). Evolving evidence suggests that there are significant functional differences among MOR splice variants (Dever et al., 2012; Lu, Xu, Xu, Pasternak, & Pan, 2014; Majumdar et al., 2011; Pan et al., 2005; Xu et al., 2014). Thus far, the findings are limited to neural cells isolated from a relative small number of individuals and from imprecisely defined brain regions. However, if our initial findings remain supported, this would be highly significant—indicating the existence of quantitative and functional differences in MOR subtypes among cell types.

A novel, truncated 6-transmembrane spanning (6TM) MOR-1K splice variant has been described, which, unlike canonical 7-transmembrane spanning (7TM) MOR isoforms, that appears to preferentially couple to $G_{\alpha}s$, increases $[Ca^{2+}]_i$ and nitric oxide, and causes cellular excitation (Gris et al., 2010; Shabalina et al., 2009). Interestingly, HIVE and perhaps cognitive impairment correlate with increased MOR-1K (Dever et al., 2014), but not with MOR-1A or MOR-1X splice variants (Dever et al., 2012), in patient samples obtained through the National NeuroAIDS Tissue Consortium.

Global RANTES/CCL5 knockout reduced microgliosis and was neuroprotective in mice concurrently treated with morphine and HIV-1 Tat protein (El-Hage, Bruce-Keller, Knapp, et al., 2008). This insinuates that CCL5-to-CCR5 signaling increases neuroinflammation even in non-infectious models, suggesting that in addition to viral entry, CCR5 blockade may be inherently neuroprotective (El-Hage, Bruce-Keller, Knapp, et al., 2008). Preclinical studies in the SIV model demonstrating maraviroc protection against inflammatory markers in neuroAIDS lend support to this assertion (Kelly et al., 2013). Moreover, findings from other investigators suggest that CCR5 blockade may have advantages in other aspects of disease management besides preventing viral entry (Hunt et al., 2013). Restricting CCR5 function, either genetically or by maraviroc, eliminates synergistic morphine and Tat-related neurotoxicity in glial-neuronal cocultures (unpublished results).

The above findings and work of others demonstrating heterologous, bidirectional cross-desensitization (Chen, Geller, Rogers, & Adler, 2007; Happel et al., 2008; Rogers & Peterson, 2003; Song et al., 2011; Steele et al., 2003) or direct molecular interactions (Chen et al., 2004; Suzuki, Chuang, Yau, Doi, & Chuang, 2002) between CCR5 and MOR, are guiding the development of novel therapeutics that selectively target CCR5–MOR and HIV interactions. Using translationally relevant, infectious models, we have started to screen novel bivalent ligands comprised of linked CCR5 and opioid receptor antagonists, which selectively target putative CCR5–MOR heterodimers and found one of these selectively restricts HIV entry/infectivity in astrocytes but not microglia (El-Hage et al., 2013; Yuan et al., 2013, 2012) (Fig. 9.3).

5. NEURAL/GLIAL PROGENITORS AND HIV

Neural progenitor cells (NPCs) are the undifferentiated precursors of both neurons and macroglia (astroglia and oligodendroglia); infection of NPCs might either lead to (occasional reports of) viral expression in more differentiated derivatives or directly influence their development and/or survival. NPCs may also be targets of HIV through interactions with HIV-1 proteins or through the influence of extracellular changes that occur as a consequence of microglial or astroglial infection (e.g., enhanced levels of glutamate, upregulation of inflammatory signals) (Buch et al., 2007; Hahn et al., 2012, 2010; Krathwohl & Kaiser, 2004; Lee, Wang, et al., 2011; Mishra, Taneja, Malik, Khalique, & Seth, 2010; Okamoto et al., 2007; Peng et al., 2011, 2008). During development, both neurons and macroglia derive from NPCs in the subventricular zone (SVZ) lining the central canal

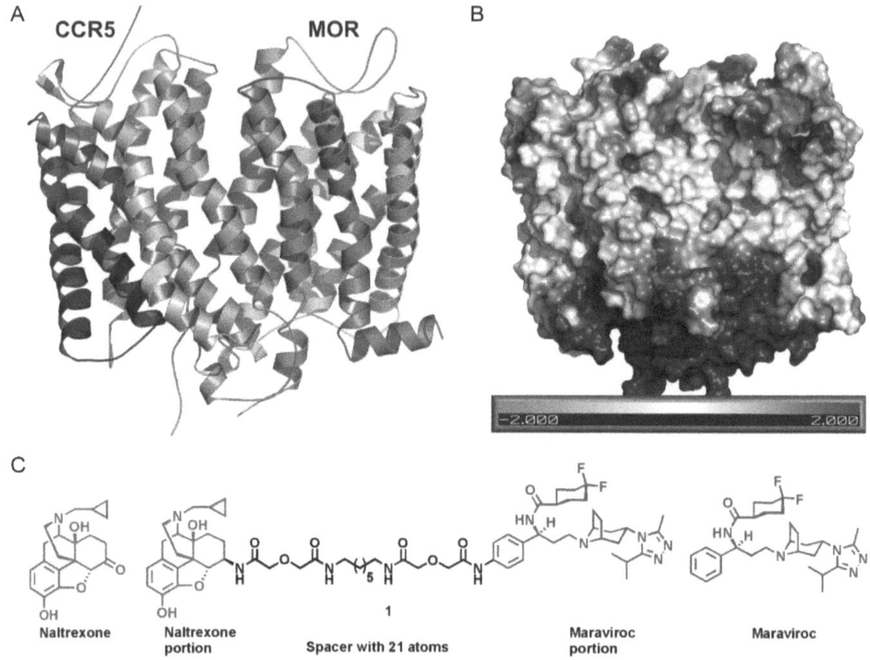

Figure 9.3 Computer-generated model of a MOR–CCR5 dimer (A). The helical portions colored in blue (black in the print version) and green (gray in the print version) represent CCR5, while the red (dark gray in the print version) and yellow (light gray in the print version) helices represent MOR (A). Each ribbon was given an arbitrary color in order to distinguish individual helices from one another (A). Representation of the Poisson–Boltzmann electrostatic potentials at the surface of the heterodimer using the APBS plugin by PYMOL (El-Hage et al., 2013) (B). Acidic residues are shown in red (light gray in the print version) (−2 *kBT/e*); basic residues are shown in blue (dark gray in the print version) (+2 *kBT/e*); white represents uncharged residues (B). The model predicts that a majority of the interactions between the two receptors are hydrophobic (B). Chemical structure of a bivalent ligand that binds both MOR and CCR5 receptors concurrently (C) (for complete description, see El-Hage et al., 2013). *Reprinted with permission from Lippincott Williams and Wilkins/Wolters Kluwer Health: AIDS (El-Hage et al., 2013). Copyright 2013.*

of the developing CNS. While most neurons in both rodents and humans are formed prior to birth, production of astroglia and oligodendroglia continues postnatally (Chan, Lorke, Tiu, & Yew, 2002; Geha et al., 2010; Lee, Mayer-Proschel, & Rao, 2000; Skoff, 1990; Skoff & Knapp, 1991). In humans, for example, oligodendrocyte formation and myelination continue well into the late teenage years, to accommodate CNS growth and maturation. While NPCs are obviously critical for CNS development, they are also present and functional in the adult CNS, although their characteristics, localization, and cell-specific markers are somewhat different from NPCs in the developing brain. In adults, *neuro*genesis is normally quite limited, occurring only in specific regions. In humans and other primates, these regions include the subgranular zone of the hippocampal dentate gyrus (DG) and the SVZ of the lateral ventricles (aka, ependymal/subependymal zones) (Doetsch, Garcia-Verdugo, & Alvarez-Buylla, 1997; Eriksson et al., 1998; Kaplan & Hinds, 1977; Kornack & Rakic, 1999). In vertebrates, including rodents, that are heavily dependent upon olfactory cues, neurogenesis within the adult SVZ results in a pool of cells that continually enters the "rostral migratory stream," a migratory route specialized for the delivery of newly formed neurons to the olfactory bulb (Doetsch et al., 1997). A structure analogous to the rostral migratory stream remains to be validated in adult humans, although the pathway exists during development (Bhardwaj et al., 2006; Pencea, Bingaman, Freedman, & Luskin, 2001; Sanai et al., 2011; Wang, Liu, et al., 2011). Unlike neurons, glia are normally formed throughout life and glial progenitors undergo a constant, slow turnover throughout the adult CNS parenchyma (Imamoto, Paterson, & Leblond, 1978; Kornack & Rakic, 1999, 2001; Messier, Leblond, & Smart, 1958; Sturrock, 1979). Whereas neonatal NPCs proliferate routinely and robustly, adult NPCs in the SVZ and DG are mostly quiescent. They display somewhat different cell markers and often respond to milieu signals or interact with surrounding cells differently than NPCs in young tissues. In response to exercise, injury/perturbation, inflammation, or other stimuli, adult NPCs can become more active, generating a subset of highly proliferative progenitors that can form neurons and/or glia that integrate into surrounding tissue (Kernie & Parent, 2010; Thored et al., 2006; Wang, Plane, et al., 2011). Whether the enhanced turnover of adult NPCs after injury involves "active" adult NPCs or transit-amplifying cells that derive from "quiescent" NPCs, or both, it is a somewhat different process from formation of cell populations in development, with different effectors and results (Burns, Murphy, Danzer, & Kuan, 2009; Rola et al., 2006; Romanko et al., 2004). Deficits in NPC populations appear to contribute to diseases like Parkinson's, Huntington's, and Alzheimer's, which involve specific neuron types (Crews, Patrick, et al., 2011; Curtis et al., 2007, 2003), and to more global injuries such as stroke/ischemia and epilepsy (Curtis et al., 2007; Ohira et al., 2010; Parent et al., 1997; Thored et al., 2006). Increased NPC proliferation is often assumed to indicate beneficial plasticity. However, proliferation must renew NPCs while also producing new neurons or glia (asymmetric division), and cell production must be balanced to tissue requirements. In Huntington's disease, for example, there is a loss of striatal neurons even though proliferation and SVZ size are significantly increased (Curtis et al., 2007; Kazanis, 2009) because the ratio of GFAP$^+$ glia (type B cells) produced is too high (Curtis, Waldvogel, Synek, & Faull, 2005). Increased proliferation may also be offset by death or aberrant migration, as in some epilepsy models (Parent et al., 1997).

The effects of HIV and/or drugs of abuse on adult hippocampal neurogenesis have been recently reviewed (Eisch & Harburg, 2006; Venkatesan, Nath, Ming, & Song, 2007). Accordingly, the discussion here will include other brain regions and gliogenesis. It will also include developmental studies, because of relevance to HIV acquired in the perinatal and adolescent periods. Since NPCs are uniquely positioned as the forerunners of CNS neurons and macroglia, the potential effects of NPC infection or dysregulation by HIV are extensive. For example, the evidence that nestin$^+$ or Sox-2$^+$ human NPCs can be infected by HIV (Hahn et al., 2012; Lawrence et al., 2004; Rothenaigner et al., 2007; Schwartz & Major, 2006) under conditions or in numbers that are disease-relevant remains controversial. In numerous experimental paradigms, HIV or HIV proteins have been shown to alter the proliferative and survival characteristics of NPCs. These include rodent (Buch et al., 2007; Hahn et al., 2012; Khurdayan et al., 2004) and human NPCs (Hahn et al., 2012; Krathwohl & Kaiser, 2004; Mishra et al., 2010) exposed in culture or *ex vivo* settings, NPCs within the brains of gp120 and Tat transgenic rodent HIV models (Avraham et al., 2014; Hahn et al., 2012; Lee, Chen, et al., 2008; Lee, Wang, et al., 2011; Okamoto et al., 2007), and in HIV postmortem brains (Krathwohl & Kaiser, 2004). In general, whether the models have examined NPCs from/in adult or developing systems, HIV or HIV proteins, have been reported to depress NPC proliferation. This was also the case in human hippocampal slice cultures exposed to either X4 HIV coat proteins or the CSF from HIV patients with dementia (Krathwohl & Kaiser, 2004). Alternatively, NPCs showed enhanced proliferation when exposed to medium from infected macrophages costimulated with LPS (Peng et al., 2008). Survival is variably reported as being unchanged by Tat, gp120, or HIV (Hahn et al., 2012; Lee, Chen, et al., 2008; Lee, Wang, et al., 2011; Malik, Saha, & Seth, 2014; Mishra et al., 2010) or reduced by Tat or gp120 (Avraham et al., 2014; Buch et al., 2007), feline immunodeficiency virus (van Marle, Antony, Silva, Sullivan, & Power, 2005), or by infection with a viral vector expressing HIV envelope (van Marle et al., 2005). The different outcomes likely reflect varied susceptibility among stages of differentiation (Brazel, Nunez, Yang, & Levison, 2005; Levison, Rothstein, Brazel, Young, &

Albrecht, 2000). Changes in these parameters have the potential to alter the balance of mature cells that derive from the NPCs.

Exposure to HIV, HIV proteins, or milieu changes that occur in the HIV-infected brain might also redirect the lineage of undifferentiated cells. Normally, NPCs undergo a series of self-renewing divisions prior to an asynchronous division that results in one daughter cell that is committed to a neuron or glial cell fate/lineage. HIV infection or viral protein exposure appears to skew NPC fate toward production of glia/astroglia at the expense of neurons and/or oligodendrocytes (Hahn et al., 2012, 2010; Peng et al., 2011). Since viral production reportedly increases as infected NPCs differentiate into astroglia (Lawrence et al., 2004), cell lineage choices, i.e., the formation of astroglia at the expense of other NPC derivatives, may increase CNS consequences of NPC infection, including the potential for more cellular sites harboring latent virus in the CNS. This may be especially relevant in pediatric/adolescent patients given their larger and more mitotically active NPC population. Changes in the milieu can push more NPCs toward a particular lineage, for example, increasing glia or astroglia at the expense of neurons, which may explain the enhanced numbers of astroglia that we have seen in adult mice exposed to HIV-1 Tat for 3 months (Hahn et al., 2014). Increased astrogliogenesis also appears to be directly triggered by HIV-infected microglia both *in vitro* and in an infected SCID mouse model via a STAT3-dependent mechanism (Peng et al., 2011). While lineage redirection has not been directly demonstrated in the context of HIV, it has been well documented during development and in some diseases (Bithell, Finch, Hornby, & Williams, 2008; Lu et al., 2011; Ninkovic & Gotz, 2013; Sabo, Kilpatrick, & Cate, 2009).

HIV is a situation of reverberating inflammation. It is perhaps not surprising that HIV infection or viral protein exposure limits NPC proliferation, since brain inflammation is well known to impair neurogenesis (Ekdahl, Claasen, Bonde, Kokaia, & Lindvall, 2003; Lu et al., 2011; Monje, Toda, & Palmer, 2003; Moreno-Lopez et al., 2004; Whitney, Eidem, Peng, Huang, & Zheng, 2009). However, the relationship between inflammation and NPC function is nuanced. Endogenous ROS and nitric oxide may actually be necessary for NPC proliferation (Yoneyama, Kawada, Gotoh, Shiba, & Ogita, 2010), and the inflammatory milieu can inhibit, stimulate, and influence the general direction of cell lineage (neuronal vs. glial), depending on the characteristics of the inflammation and the phenotype of the microglia involved in the inflammatory response (Butovsky et al., 2006).

5.1. Opiate and opiate–HIV interactions on progenitors and cell populations

Morphine, which is a legal but regulated analgesic, is also an active metabolite of heroin (diacetylmorphine) and is used in many studies as a surrogate for heroin exposure/abuse. NPCs (nestin$^+$, Sox2$^+$; also GFAP$^+$ in adult) and their early derivatives (e.g., DCX$^+$/β-III tubulin$^+$ neurons; CD44$^+$/vimentin$^+$ young astrocytes; Nkx2.2$^+$/A2B5$^+$/O_4^+/Olig1$^+$/Olig2$^+$ young oligodendrocytes) express MOR, DOR, and KOR (Buch et al., 2007; Hahn et al., 2010; Tripathi, Khurshid, Kumar, & Iyengar, 2008). Furthermore, diverse agonists and antagonists for MOR, KOR, and DOR (e.g., heroin, morphine, β-endorphin, naltrindole, [D-Ala^2, D-Leu^5]-enkephalin (DADLE), β-funaltrexamine (β-FNA), naltrexone) (Eisch, Barrot, Schad, Self, & Nestler, 2000; Holmes & Galea, 2002; Koehl et al., 2008; Mandyam, Norris, & Eisch, 2004; Persson, Thorlin, Bull, & Eriksson, 2003; Persson, Thorlin, Bull, Zarnegar, et al., 2003; Tsai, Lee, Hayashi, Freed, & Su, 2010) affect their proliferation and other behaviors. To date, most opiate studies have focused on *adult* hippocampal neurogenesis, and a number have included physical exercise as a parameter, which promotes neurogenesis partly via β-endorphin effects (Persson, Thorlin, Bull, & Eriksson, 2003). Interestingly, the reduced NPCs and neurogenesis seen in transgenic HIV-1 gp120 mice are normalized by exercise, although the role of endogenous opiates in this effect was not assessed (Lee et al., 2013). When exercise is removed from consideration, morphine reduces adult neurogenesis (Arguello et al., 2008; Eisch et al., 2000; Kolodziej, Stumm, Becker, & Hollt, 2008; Mandyam et al., 2004), even after controlling for glucocorticoid levels (Eisch et al., 2000), and MOR-null mice show a transiently enhanced postinjury neurogenesis in hippocampus (Kolodziej et al., 2008). Important new evidence shows that opiates used in addiction therapy, such as buprenorphine, may also reduce NPC formation in either the perinatal (Wu et al., 2014) or adult (Pettit, Desroches, & Bennett, 2012) CNS, disturbing the normal balance of glial populations through actions at MOR and/or the nociceptin/orphanin FQ receptors (Eschenroeder, Vestal-Laborde, Sanchez, Robinson, & Sato-Bigbee, 2012).

NPC proliferation is reduced by exposure to either HIV/HIV proteins or opiates independently, suggesting the potential for interactive, comorbid effects of combined HIV and opiate exposure. Morphine interactively increases CNS inflammation in most HIV models (Bokhari et al., 2011;

Bruce-Keller et al., 2008; Dever et al., 2014; El-Hage, Bruce-Keller, Knapp, et al., 2008; El-Hage, Bruce-Keller, Yakovleva, et al., 2008; Perez-Casanova, Husain, Noel, Rivera-Amill, & Kumar, 2008; Turchan-Cholewo et al., 2009), so it might be expected to exacerbate the effects of HIV on NPCs through generalized inflammation. However, as NPCs and their progeny express opiate receptors, there may be interactive effects that are independent of inflammation. Several studies have now shown that morphine exacerbates acute effects of HIV and Tat *in vitro* and in the developing brain. Our *in vitro* work has shown that combined exposure to morphine and Tat, or morphine and HIV, further depressed proliferation of murine and human NPCs and decreased NPC pools (Hahn et al., 2012). Morphine exposure in perinatal HIV-1 Tat-transgenic mice had a remarkably similar effect *in vivo* (Hahn et al., 2012). In human NPCs, this effect was shown to reflect a delay at the G1 phase of the cell cycle involving increased extracellular-signal regulated kinase-1/2 (ERK1/2) activation and concomitant increases in p21 and p53 (Malik et al., 2014). The ramifications of reduced numbers of NPCs, or an increased astroglial population at the expense of neurons and/or oligodendroglia, are likely to significantly impact cognitive and motor function. It remains unclear whether the chronic exposure to opiates that occurs in patients or drug abusers, either during development (neonates, adolescents) or in adult brains, can permanently alter CNS cell populations. The very critical question of whether the nature of opiate exposure might yield fundamentally different outcomes has also not been addressed. For example, do opiates administered for pain yield a different outcome than opiates that have been self-administered (resulting in dependence and addiction)? Studies that compare outcomes of contemporaneous exposure to opiates and HIV/HIV proteins versus the more "real-life" situation where drug exposure occurs prior to HIV infection are also lacking.

5.2. Psychostimulant–HIV interactions on progenitors and cell populations

Among the psychostimulants, both cocaine and amphetamine/methamphetamine exposure can influence NPC behaviors. Similar to opiates, in most studies, cocaine exposure appears to inhibit the proliferation of human (Hu et al., 2006) and rodent (Lee et al., 2008) NPCs *in vitro* as well as in various perinatal and adult rodent models *in vivo* (Dominguez-Escriba et al., 2006; Garcia-Fuster, Perez, Clinton, Watson, & Akil, 2010; Lee et al., 2008; Yao, Duan, Yang, & Buch, 2012). A number of mechanisms for these effects have been put forward, including downregulation of cyclin A related to ER stress induced by cocaine metabolites (Lee et al., 2008), and changes in cytoskeletal-associated genes (Lee et al., 2009). In one study, exposure to cocaine during critical prenatal periods reduced normal numbers and disturbed radial migration of GABAergic and glutamatergic neurons in the neocortex of developing rats (Lee, Chen, et al., 2011), presumably in part through effects on NPCs. In the DG of adult rats, both 8- and 24-day exposure to cocaine decreased NPC proliferation, but did not affect survival or newly generated cells or their morphology, dendritic arborization, or localization in the areas examined (Dominguez-Escriba et al., 2006). These disparate findings are no doubt partly due to the inherently different properties of NPCs in developing and adult systems. Cocaine also reduced the motility of human NPCs *in vitro* through inhibition of CXCL12-to-CXCR4 signaling (Hu et al., 2006). Reduced motility is associated with downregulation of a transcription factor (SOX2) that supports progenitor phenotype, and the early differentiation of young neurons. Prenatal cocaine exposure also appears to have lasting, sex-specific effects on resident NPCs in adult brains (Patel, Booze, & Mactutus, 2012). In a genetic study with relevance to addicted individuals, cocaine had quite different effects on NPCs in two lines of rats bred to express different propensities for cocaine "abuse." NPC proliferation was suppressed by chronic cocaine exposure in rats with low responses to novelty, but in rats with high novelty responses (enhanced sensitization/psychomotor response to cocaine), NPC proliferation was normal, as was the ratio of neurons-to-glia generated, although there was less survival of newly formed neurons (Garcia-Fuster et al., 2010).

The preponderance of studies shows that methamphetamine or amphetamine exposure also reduces NPC proliferation and formation of new glia and/or neurons, although a few disparate outcomes may provide insight into important variables. The proliferation of both adult hippocampal and striatal progenitor cells *in vivo* is frequently assessed by 5-bromo-2′-deoxyuridine (BrdU) incorporation (a thymidine analogue incorporated during DNA synthesis) or Ki67 antigenicity (a nuclear protein associated with cell division) (Scholzen & Gerdes, 2000) in species including rats, mice, and gerbils (Mandyam, Wee, Eisch, Richardson, & Koob, 2007; Mao & Wang, 2001; Teuchert-Noodt, Dawirs, & Hildebrandt, 2000; Yuan, Quiocho, Kim, Wee, & Mandyam, 2011). Similar effects are seen *in vitro*, where survival is also variably reduced (Bento, Baptista, Malva, Silva, & Agasse, 2011; Tian, Murrin, & Zheng, 2009; Venkatesan et al., 2011). These effects have been correlated to enhanced nitration and modification of function of key metabolic proteins (Venkatesan et al., 2011). One report suggests that a single dose of methamphetamine at 14 days of age can result in reduced NPC proliferation in the hippocampus of adults (Hildebrandt, Teuchert-Noodt, & Dawirs, 1999). Another set of studies suggests that exposure to a neurotoxic dose of methamphetamine may temporarily increase the activity of quiescent NPCs in the adult striatum, by effects mediated through dopamine D2 (but not D1) and neurokinin 1 receptors (Tulloch, Ghazaryan, Mexhitaj, Ordonez, & Angulo, 2011). These cells, most of which have glial phenotypes, contribute to an increase in the size of the striatum, but the increased glial population is not maintained in the absence of methamphetamine (Tulloch et al., 2014, 2011). The importance of the timing of drug exposure is highlighted in a study comparing self-administration in intermittent (biweekly) versus daily exposure (Mandyam et al., 2008). Increased NPC proliferation was reported with intermittent exposure to methamphetamine, while daily exposure had the opposite effect. Interestingly, intermittent exposure also resulted in the formation of more immature neurons, although there was no net neuron gain due to offsetting effects on later stages of differentiation. Daily exposure decreased all aspects of cytogenesis. Other extended access self-administration models also showed reduced NPC proliferation and decreased neuron and/or glial formation (Mandyam et al., 2007; Yuan et al., 2011).

There has been very little exploration of psychostimulant–HIV interactions on NPCs. In both a rat hippocampal NPC cell line and transgenic mice, HIV-1 Tat and cocaine reduced NPC proliferation without affecting NPC survival. These individual effects and a modest tendency toward interaction were reversed by platelet-derived growth factor-BB (PDGF-BB) through a mechanism involving both the transient receptor potential cation channel-C1 (TRPC1) and ERK/CREB and mammalian target of rapamycin (mTOR) activation (Yao et al., 2012).

6. CONCLUSIONS

Although considerable progress has been made during the last decade toward identifying molecular and cellular sites of HIV and drug abuse convergence in the CNS, and several new therapeutic strategies have emerged that may prove beneficial, much work remains to be done. The high degree of variability in response among individuals to HIV alone reveals a highly complex chemistry between host and viral genetics, a complexity that is undoubtedly more convoluted by drug abuse-neuroAIDS comorbidity.

Moreover, many of the events underlying drug abuse–HIV neuropathological interactions occur only within specific cell types and/or at specific times during ontogeny, which adds greatly to the intricacy of the problem. A full appreciation of the mechanisms underlying drug abuse and neuroAIDS interactions will likely require an understanding of the interrelationship of gene networks, rather than individual genes.

ACKNOWLEDGMENTS

This work was supported by Grants K02 DA027374, R01 DA018633, and R01 DA024461 from the NIH, National Institute on Drug Abuse. The authors wish to thank Dr. MaryPeace McRae for helpful thoughts and comments in preparing the manuscript.

Conflict of Interest Statement: The authors have no conflicts of interest to declare.

REFERENCES

Abt, A. C., & Meucci, O. (2011). Regulation of neuronal ferritin heavy chain, a new player in opiate-induced chemokine dysfunction. *Journal of Neuroimmune Pharmacology*, *6*(4), 466–476.

Albini, A., Ferrini, S., Benelli, R., Sforzini, S., Giunciuglio, D., Aluigi, M. G., et al. (1998). HIV-1 Tat protein mimicry of chemokines. *Proceedings of the National Academy of Sciences of the United States of America*, *95*(22), 13153–13158. Retrieved from, http://www.ncbi.nlm.nih.gov/cgi-bin/Entrez/referer?http://www.pnas.org/cgi/content/full/95/22/13153.

Alirezaei, M., Kiosses, W. B., Flynn, C. T., Brady, N. R., & Fox, H. S. (2008). Disruption of neuronal autophagy by infected microglia results in neurodegeneration. *PLoS One*, *3*(8), e2906, Retrieved from PM:18682838.

Allers, K., Hutter, G., Hofmann, J., Loddenkemper, C., Rieger, K., Thiel, E., et al. (2011). Evidence for the cure of HIV infection by CCR5Delta32/Delta32 stem cell transplantation. *Blood*, *117*(10), 2791–2799. http://dx.doi.org/10.1182/blood-2010-09-309591, Retrieved from PM:21148083.

Amiry-Moghaddam, M., Williamson, A., Palomba, M., Eid, T., de Lanerolle, N. C., Nagelhus, E. A., et al. (2003). Delayed K^+ clearance associated with aquaporin-4 mislocalization: Phenotypic defects in brains of α-syntrophin-null mice. *Proceedings of the National Academy of Sciences of the United States of America*, *100*(23), 13615–13620. http://dx.doi.org/10.1073/pnas.2336064100, Retrieved from PM: 14597704.

An, S. F., Osuntokun, O., Groves, M., & Scaravilli, F. (2001). Expression of CCR-5/CXCR-4 in spinal cord of patients with AIDS. *Acta Neuropathologica (Berlin)*, *102*(2), 175–180, Retrieved from PM:11563633.

Ancuta, P., Wang, J., & Gabuzda, D. (2006). CD16+ monocytes produce IL-6, CCL2, and matrix metalloproteinase-9 upon interaction with CX3CL1-expressing endothelial cells. *Journal of Leukocyte Biology*, *80*(5), 1156–1164, Retrieved from PM:17056766.

Anderova, M., Vorisek, I., Pivonkova, H., Benesova, J., Vargova, L., Cicanic, M., et al. (2011). Cell death/proliferation and alterations in glial morphology contribute to changes in diffusivity in the rat hippocampus after hypoxia-ischemia. *Journal of Cerebral Blood Flow and Metabolism*, *31*(3), 894–907. http://dx.doi.org/10.1038/jcbfm.2010.168, Retrieved from PM:20877389.

Anderson, J., & Akkina, R. (2007). Complete knockdown of CCR5 by lentiviral vector-expressed siRNAs and protection of transgenic macrophages against HIV-1 infection.

Gene Therapy, *14*(17), 1287–1297. http://dx.doi.org/10.1038/sj.gt.3302958, Retrieved from PM:17597795.

Anderson, C. E., Tomlinson, G. S., Pauly, B., Brannan, F. W., Chiswick, A., Brack-Werner, R., et al. (2003). Relationship of Nef-positive and GFAP-reactive astrocytes to drug use in early and late HIV infection. *Neuropathology and Applied Neurobiology*, *29*(4), 378–388.

Andres, M. A., Feger, U., Nath, A., Munsaka, S., Jiang, C. S., & Chang, L. (2011). APOE epsilon 4 allele and CSF APOE on cognition in HIV-infected subjects. *Journal of Neuroimmune Pharmacology*, *6*(3), 389–398, Retrieved from PM:21184197.

Anthony, I. C., Arango, J. C., Stephens, B., Simmonds, P., & Bell, J. E. (2008). The effects of illicit drugs on the HIV infected brain. *Frontiers in Bioscience*, *13*, 1294–1307.

Anthony, I. C., Norrby, K. E., Dingwall, T., Carnie, F. W., Millar, T., Arango, J. C., et al. (2010). Predisposition to accelerated Alzheimer-related changes in the brains of human immunodeficiency virus negative opiate abusers. *Brain*, *133*(Pt 12), 3685–3698, Retrieved from PM:21126996.

Anthony, I. C., Ramage, S. N., Carnie, F. W., Simmonds, P., & Bell, J. E. (2005). Does drug abuse alter microglial phenotype and cell turnover in the context of advancing HIV infection? *Neuropathology and Applied Neurobiology*, *31*(3), 325–338.

Antonucci, F., Turola, E., Riganti, L., Caleo, M., Gabrielli, M., Perrotta, C., et al. (2012). Microvesicles released from microglia stimulate synaptic activity via enhanced sphingolipid metabolism. *The EMBO Journal*, *31*(5), 1231–1240. http://dx.doi.org/10.1038/emboj.2011.489, Retrieved from PM:22246184.

Araque, A., Parpura, V., Sanzgiri, R. P., & Haydon, P. G. (1999). Tripartite synapses: Glia, the unacknowledged partner. *Trends in Neurosciences*, *22*(5), 208–215.

Arguello, A. A., Harburg, G. C., Schonborn, J. R., Mandyam, C. D., Yamaguchi, M., & Eisch, A. J. (2008). Time course of morphine's effects on adult hippocampal subgranular zone reveals preferential inhibition of cells in S phase of the cell cycle and a subpopulation of immature neurons. *Neuroscience*, *157*(1), 70–79. http://dx.doi.org/10.1016/j.neuroscience.2008.08.064, Retrieved from PM:18832014.

Avdoshina, V., Biggio, F., Palchik, G., Campbell, L. A., & Mocchetti, I. (2010). Morphine induces the release of CCL5 from astrocytes: Potential neuroprotective mechanism against the HIV protein gp120. *Glia*, *58*(13), 1630–1639.

Avraham, H. K., Jiang, S., Fu, Y., Rockenstein, E., Makriyannis, A., Zvonok, A., et al. (2014). The cannabinoid CB(2) receptor agonist AM1241 enhances neurogenesis in GFAP/Gp120 transgenic mice displaying deficits in neurogenesis. *British Journal of Pharmacology*, *171*(2), 468–479. http://dx.doi.org/10.1111/bph.12478, Retrieved from PM:24148086.

Bachmann, R. F., Wang, Y., Yuan, P., Zhou, R., Li, X., Alesci, S., et al. (2009). Common effects of lithium and valproate on mitochondrial functions: Protection against methamphetamine-induced mitochondrial damage. *The International Journal of Neuropsychopharmacology*, *12*(6), 805–822. http://dx.doi.org/10.1017/S1461145708009802, Retrieved from PM:19149911.

Bagley, E. E., Hacker, J., Chefer, V. I., Mallet, C., McNally, G. P., Chieng, B. C., et al. (2011). Drug-induced GABA transporter currents enhance GABA release to induce opioid withdrawal behaviors. *Nature Neuroscience*, *14*(12), 1548–1554, Retrieved from PM:22037500.

Baker, B. M., Block, B. L., Rothchild, A. C., & Walker, B. D. (2009). Elite control of HIV infection: Implications for vaccine design. *Expert Opinion on Biological Therapy*, *9*(1), 55–69. http://dx.doi.org/10.1517/14712590802571928, Retrieved from PM:19063693.

Bandaru, V. V., Patel, N., Ewaleifoh, O., & Haughey, N. J. (2011). A failure to normalize biochemical and metabolic insults during morphine withdrawal disrupts synaptic repair in mice transgenic for HIV-gp120. *Journal of Neuroimmune Pharmacology*, *6*(4), 640–649, Retrieved from PM:21748284.

Banerjee, A., Strazza, M., Wigdahl, B., Pirrone, V., Meucci, O., & Nonnemacher, M. R. (2011). Role of mu-opioids as cofactors in human immunodeficiency virus type 1 disease progression and neuropathogenesis. *Journal of Neurovirology*, *17*(4), 291–302.

Banerjee, A., Zhang, X., Manda, K. R., Banks, W. A., & Ercal, N. (2010). HIV proteins (gp120 and Tat) and methamphetamine in oxidative stress-induced damage in the brain: Potential role of the thiol antioxidant N-acetylcysteine amide. *Free Radical Biology and Medicine*, *48*(10), 1388–1398. http://dx.doi.org/10.1016/j.freeradbiomed.2010.02.023, Retrieved from PM:20188164.

Beardsley, P. M., & Hauser, K. F. (2014). Glial modulators as potential treatments of psychostimulant abuse. *Advances in Pharmacology*, *69*, 1–69. http://dx.doi.org/10.1016/B978-0-12-420118-7.00001-9, Retrieved from PM:24484974.

Becker, J. B., & Hu, M. (2008). Sex differences in drug abuse. *Frontiers in Neuroendocrinology*, *29*(1), 36–47. http://dx.doi.org/10.1016/j.yfrne.2007.07.003, Retrieved from PM:17904621.

Bell, J. E. (1998). The neuropathology of adult HIV infection. *Revue Neurologique (Paris)*, *154*(12), 816–829.

Bell, J. E., Arango, J. C., Robertson, R., Brettle, R. P., Leen, C., & Simmonds, P. (2002). HIV and drug misuse in the Edinburgh cohort. *Journal of Acquired Immune Deficiency Syndromes*, *31*(Suppl. 2), S35–S42.

Bell, J. E., Brettle, R. P., Chiswick, A., & Simmonds, P. (1998). HIV encephalitis, proviral load and dementia in drug users and homosexuals with AIDS. Effect of neocortical involvement. *Brain*, *121*(Pt 11), 2043–2052.

Bencheikh, M., Bentsman, G., Sarkissian, N., Canki, M., & Volsky, D. J. (1999). Replication of different clones of human immunodeficiency virus type 1 in primary fetal human astrocytes: Enhancement of viral gene expression by Nef. *Journal of Neurovirology*, *5*(2), 115–124.

Bender, A., Schwarzkopf, R. M., McMillan, A., Krishnan, K. J., Rieder, G., Neumann, M., et al. (2008). Dopaminergic midbrain neurons are the prime target for mitochondrial DNA deletions. *Journal of Neurology*, *255*(8), 1231–1235.

Bento, A. R., Baptista, S., Malva, J. O., Silva, A. P., & Agasse, F. (2011). Methamphetamine exerts toxic effects on subventricular zone stem/progenitor cells and inhibits neuronal differentiation. *Rejuvenation Research*, *14*(2), 205–214. http://dx.doi.org/10.1089/rej.2010.1109, Retrieved from PM:21453012.

Berger, E. A., Murphy, P. M., & Farber, J. M. (1999). Chemokine receptors as HIV-1 coreceptors: Roles in viral entry, tropism, and disease. *Annual Review of Immunology*, *17*, 657–700.

Berman, J. W., Carson, M. J., Chang, L., Cox, B. M., Fox, H. S., Gonzalez, R. G., et al. (2006). NeuroAIDS, drug abuse, and inflammation: Building collaborative research activities. *Journal of Neuroimmune Pharmacology*, *1*(4), 351–399.

Berman, S., O'Neill, J., Fears, S., Bartzokis, G., & London, E. D. (2008). Abuse of amphetamines and structural abnormalities in the brain. *Annals of the New York Academy of Sciences*, *1141*, 195–220. http://dx.doi.org/10.1196/annals.1441.031, Retrieved from PM:18991959.

Bhardwaj, R. D., Curtis, M. A., Spalding, K. L., Buchholz, B. A., Fink, D., Bjork-Eriksson, T., et al. (2006). Neocortical neurogenesis in humans is restricted to development. *Proceedings of the National Academy of Sciences of the United States of America*, *103*(33), 12564–12568. http://dx.doi.org/10.1073/pnas.0605177103, Retrieved from PM:16901981.

Bianchi, M. E. (2007). DAMPs, PAMPs and alarmins: all we need to know about danger. *Journal of Leukocyte Biology*, *81*(1), 1–5. Retrieved from PM:17.

Biber, K., Neumann, H., Inoue, K., & Boddeke, H. W. (2007). Neuronal 'On' and 'Off' signals control microglia. *Trends in Neurosciences*, *30*(11), 596–602.

Bishop, K. N., Holmes, R. K., Sheehy, A. M., & Malim, M. H. (2004). APOBEC-mediated editing of viral RNA. *Science, 305*(5684), 645. http://dx.doi.org/10.1126/science.1100658, Retrieved from PM:15286366.

Bithell, A., Finch, S. E., Hornby, M. F., & Williams, B. P. (2008). Fibroblast growth factor 2 maintains the neurogenic capacity of embryonic neural progenitor cells in vitro but changes their neuronal subtype specification. *Stem Cells, 26*(6), 1565–1574. http://dx.doi.org/10.1634/stemcells.2007-0832, Retrieved from PM:18339769.

Block, M. L., Zecca, L., & Hong, J. S. (2007). Microglia-mediated neurotoxicity: Uncovering the molecular mechanisms. *Nature Reviews Neuroscience, 8*(1), 57–69, Retrieved from PM:17180163.

Bokhari, S. M., Hegde, R., Callen, S., Yao, H., Adany, I., Li, Q., et al. (2011). Morphine potentiates neuropathogenesis of SIV infection in Rhesus macaques. *Journal of Neuroimmune Pharmacology, 6*(4), 626–639, Retrieved from PM:21431470.

Bokhari, S. M., Yao, H., Bethel-Brown, C., Fuwang, P., Williams, R., Dhillon, N. K., et al. (2009). Morphine enhances Tat-induced activation in murine microglia. *Journal of Neurovirology, 15*(3), 219–228. http://dx.doi.org/10.1080/13550280902913628, Retrieved from PM:19462331.

Bonacci, T. M., Mathews, J. L., Yuan, C., Lehmann, D. M., Malik, S., Wu, D., et al. (2006). Differential targeting of Gbetagamma-subunit signaling with small molecules. *Science, 312*(5772), 443–446, Retrieved from PM:16627746.

Bond, C., LaForge, K. S., Tian, M., Melia, D., Zhang, S., Borg, L., et al. (1998). Single-nucleotide polymorphism in the human mu opioid receptor gene alters β-endorphin binding and activity: Possible implications for opiate addiction. *Proceedings of the National Academy of Sciences of the United States of America, 95*(16), 9608–9613, Retrieved from PM:9689128.

Borda, J. T., Alvarez, X., Mohan, M., Hasegawa, A., Bernardino, A., Jean, S., et al. (2008). CD163, a marker of perivascular macrophages, is up-regulated by microglia in simian immunodeficiency virus encephalitis after haptoglobin-hemoglobin complex stimulation and is suggestive of breakdown of the blood–brain barrier. *The American Journal of Pathology, 172*(3), 725–737, Retrieved from PM:18276779.

Borowsky, B., Adham, N., Jones, K. A., Raddatz, R., Artymyshyn, R., Ogozalek, K. L., et al. (2001). Trace amines: Identification of a family of mammalian G protein-coupled receptors. *Proceedings of the National Academy of Sciences of the United States of America, 98*(16), 8966–8971. http://dx.doi.org/10.1073/pnas.151105198, Retrieved from PM:11459929.

Boven, L. A., van der Bruggen, T., van Asbeck, B. S., Marx, J. J., & Nottet, H. S. (1999). Potential role of CCR5 polymorphism in the development of AIDS dementia complex. *FEMS Immunology and Medical Microbiology, 26*(3–4), 243–247. http://dx.doi.org/10.1111/j.1574-695X.1999.tb01395.x, Retrieved from PM:10575135.

Bowers, M. S., Chen, B. T., & Bonci, A. (2010). AMPA receptor synaptic plasticity induced by psychostimulants: The past, present, and therapeutic future. *Neuron, 67*(1), 11–24, Retrieved from PM:20624588.

Brack-Werner, R. (1999). Astrocytes: HIV cellular reservoirs and important participants in neuropathogenesis. *AIDS, 13*(1), 1–22.

Bradley, J. R. (2008). TNF-mediated inflammatory disease. *The Journal of Pathology, 214*(2), 149–160. http://dx.doi.org/10.1002/path.2287, Retrieved from PM:18161752.

Brazel, C. Y., Nunez, J. L., Yang, Z., & Levison, S. W. (2005). Glutamate enhances survival and proliferation of neural progenitors derived from the subventricular zone. *Neuroscience, 131*(1), 55–65. http://dx.doi.org/10.1016/j.neuroscience.2004.10.038, Retrieved from PM:15680691.

Brown, J. M., Riddle, E. L., Sandoval, V., Weston, R. K., Hanson, J. E., Crosby, M. J., et al. (2002). A single methamphetamine administration rapidly decreases vesicular dopamine uptake. *The Journal of Pharmacology and Experimental Therapeutics, 302*(2), 497–501.

Bruce-Keller, A. J., Turchan-Cholewo, J., Smart, E. J., Guerin, T., Chauhan, A., Reid, R., et al. (2008). Morphine causes rapid increases in glial activation and neuronal injury in the striatum of inducible HIV-1 Tat transgenic mice. *Glia*, *56*, 1414–1427.

Brunswick, D. J., Benmansour, S., Tejani-Butt, S. M., & Hauptmann, M. (1992). Effects of high-dose methamphetamine on monoamine uptake sites in rat brain measured by quantitative autoradiography. *Synapse*, *11*(4), 287–293, Retrieved from PM:1502685.

Buch, S. K., Khurdayan, V. K., Lutz, S. E., Knapp, P. E., El-Hage, N., & Hauser, K. F. (2007). Glial-restricted precursors: Patterns of expression of opioid receptors and relationship to HIV-1 Tat and morphine susceptibility in vitro. *Neuroscience*, *146*, 1546–1554.

Buch, S., Yao, H., Guo, M., Mori, T., Mathias-Costa, B., Singh, V., et al. (2012). Cocaine and HIV-1 interplay in CNS: Cellular and molecular mechanisms. *Current HIV Research*, *10*(5), 425–428. http://dx.doi.org/10.2174/157016212802138823, Retrieved from PM:22591366.

Bunzow, J. R., Sonders, M. S., Arttamangkul, S., Harrison, L. M., Zhang, G., Quigley, D. I., et al. (2001). Amphetamine, 3,4-methylenedioxymethamphetamine, lysergic acid diethylamide, and metabolites of the catecholamine neurotransmitters are agonists of a rat trace amine receptor. *Molecular Pharmacology*, *60*(6), 1181–1188, Retrieved from PM:11723224.

Burbassi, S., Aloyo, V. J., Simansky, K. J., & Meucci, O. (2008). GTPγS incorporation in the rat brain: A study on μ-opioid receptors and CXCR4. *Journal of Neuroimmune Pharmacology*, *3*(1), 26–34, Retrieved from PM:18247130.

Burns, K. A., Murphy, B., Danzer, S. C., & Kuan, C. Y. (2009). Developmental and post-injury cortical gliogenesis: A genetic fate-mapping study with Nestin-CreER mice. *Glia*, *57*(10), 1115–1129. http://dx.doi.org/10.1002/glia.20835, Retrieved from PM:19115384.

Butovsky, O., Ziv, Y., Schwartz, A., Landa, G., Talpalar, A. E., Pluchino, S., et al. (2006). Microglia activated by IL-4 or IFN-γ differentially induce neurogenesis and oligodendrogenesis from adult stem/progenitor cells. *Molecular and Cellular Neurosciences*, *31*(1), 149–160. http://dx.doi.org/10.1016/j.mcn.2005.10.006, Retrieved from PM:16297637.

Buttner, A. (2011). The neuropathology of drug abuse. *Neuropathology and Applied Neurobiology*, *37*(2), 118–134, Retrieved from PM:20946118.

Byrd, D. A., Fellows, R. P., Morgello, S., Franklin, D., Heaton, R. K., Deutsch, R., et al. (2011). Neurocognitive impact of substance use in HIV infection. *Journal of Acquired Immune Deficiency Syndromes*, *58*(2), 154–162.

Byrd, D., Murray, J., Safdieh, G., & Morgello, S. (2012). Impact of opiate addiction on neuroinflammation in HIV. *Journal of Neurovirology*, *18*(5), 364–373. http://dx.doi.org/10.1007/s13365-012-0118-x, Retrieved from PM:22797933.

Cadet, J. L., Bisagno, V., & Milroy, C. M. (2014). Neuropathology of substance use disorders. *Acta Neuropathologica (Berlin)*, *127*(1), 91–107. http://dx.doi.org/10.1007/s00401-013-1221-7, Retrieved from PM:24292887.

Cadet, J. L., & Krasnova, I. N. (2009). Molecular bases of methamphetamine-induced neurodegeneration. *International Review of Neurobiology*, *88*, 101–119. http://dx.doi.org/10.1016/S0074-7742(09)88005-7, Retrieved from PM:19897076.

Cadet, J. L., Krasnova, I. N., Jayanthi, S., & Lyles, J. (2007). Neurotoxicity of substituted amphetamines: Molecular and cellular mechanisms. *Neurotoxicity Research*, *11*(3–4), 183–202, Retrieved from PM:17449459.

Cameron, D. L., & Williams, J. T. (1994). Cocaine inhibits GABA release in the VTA through endogenous 5-HT. *Journal of Neuroscience*, *14*(11 Pt 1), 6763–6767, Retrieved from PM:7965077.

Canki, M., Potash, M. J., Bentsman, G., Chao, W., Flynn, T., Heinemann, M., et al. (1997). Isolation and long-term culture of primary ocular human immunodeficiency virus type 1 isolates in primary astrocytes. *Journal of Neurovirology*, *3*(1), 10–15.

Carrington, M., Dean, M., Martin, M. P., & O'Brien, S. J. (1999). Genetics of HIV-1 infection: Chemokine receptor CCR5 polymorphism and its consequences. *Human Molecular Genetics*, *8*(10), 1939–1945.

Carroll-Anzinger, D., Kumar, A., Adarichev, V., Kashanchi, F., & Al-Harthi, L. (2007). Human immunodeficiency virus-restricted replication in astrocytes and the ability of gamma interferon to modulate this restriction are regulated by a downstream effector of the Wnt signaling pathway. *Journal of Virology*, *81*(11), 5864–5871, Retrieved from PM:17392368.

Carson, M. J., Bilousova, T. V., Puntambekar, S. S., Melchior, B., Doose, J. M., & Ethell, I. M. (2007). A rose by any other name? The potential consequences of microglial heterogeneity during CNS health and disease. *Neurotherapeutics*, *4*(4), 571–579, Retrieved from PM:17920538.

Cartier, E. A., Parra, L. A., Baust, T. B., Quiroz, M., Salazar, G., Faundez, V., et al. (2010). A biochemical and functional protein complex involving dopamine synthesis and transport into synaptic vesicles. *The Journal of Biological Chemistry*, *285*(3), 1957–1966. http://dx.doi.org/10.1074/jbc.M109.054510, Retrieved from PM:19903816.

Cass, W. A. (1997). Decreases in evoked overflow of dopamine in rat striatum after neurotoxic doses of methamphetamine. *The Journal of Pharmacology and Experimental Therapeutics*, *280*(1), 105–113, Retrieved from PM:8996187.

Cass, W. A., Harned, M. E., Peters, L. E., Nath, A., & Maragos, W. F. (2003). HIV-1 protein Tat potentiation of methamphetamine-induced decreases in evoked overflow of dopamine in the striatum of the rat. *Brain Research*, *984*(1–2), 133–142, Retrieved from PM:12932847.

Chan, W. Y., Lorke, D. E., Tiu, S. C., & Yew, D. T. (2002). Proliferation and apoptosis in the developing human neocortex. *The Anatomical Record*, *267*(4), 261–276. http://dx.doi.org/10.1002/ar.10100, Retrieved from PM:12124904.

Chao, C. C., Gekker, G., Hu, S., Sheng, W. S., Shark, K. B., Bu, D. F., et al. (1996). κ opioid receptors in human microglia downregulate human immunodeficiency virus 1 expression. *Proceedings of the National Academy of Sciences of the United States of America*, *93*(15), 8051–8056.

Chao, C. C., Hu, S., & Peterson, P. K. (1996). Glia: The not so innocent bystanders. *Journal of Neurovirology*, *2*(4), 234–239, Retrieved from PM:8799214.

Chen, X., Geller, E. B., Rogers, T. J., & Adler, M. W. (2007). Rapid heterologous desensitization of antinociceptive activity between mu or delta opioid receptors and chemokine receptors in rats. *Drug and Alcohol Dependence*, *88*(1), 36–41. http://dx.doi.org/10.1016/j.drugalcdep.2006.09.010, Retrieved from PM:17049756.

Chen, C., Li, J., Bot, G., Szabo, I., Rogers, T. J., & Liu-Chen, L. Y. (2004). Heterodimerization and cross-desensitization between the mu-opioid receptor and the chemokine CCR5 receptor. *European Journal of Pharmacology*, *483*(2–3), 175–186.

Chen, H., & Manev, H. (2011). Effects of minocycline on cocaine sensitization and phosphorylation of GluR1 receptors in 5-lipoxygenase deficient mice. *Neuropharmacology*, *60*(7–8), 1058–1063. http://dx.doi.org/10.1016/j.neuropharm.2010.09.006, Retrieved from PM:20868701.

Chen, H., Uz, T., & Manev, H. (2009). Minocycline affects cocaine sensitization in mice. *Neuroscience Letters*, *452*(3), 258–261. http://dx.doi.org/10.1016/j.neulet.2009.01.078, Retrieved from PM:19348734.

Chinnery, P. F., Taylor, G. A., Howell, N., Brown, D. T., Parsons, T. J., & Turnbull, D. M. (2001). Point mutations of the mtDNA control region in normal and neurodegenerative human brains. *American Journal of Human Genetics*, *68*(2), 529–532.

Cinque, P., Vago, L., Mengozzi, M., Torri, V., Ceresa, D., Vicenzi, E., et al. (1998). Elevated cerebrospinal fluid levels of monocyte chemotactic protein-1 correlate with HIV-1 encephalitis and local viral replication. *AIDS*, *12*(11), 1327–1332.

Cisneros, I. E., & Ghorpade, A. (2012). HIV-1, methamphetamine and astrocyte glutamate regulation: Combined excitotoxic implications for neuro-AIDS. *Current HIV Research, 10*(5), 392–406. http://dx.doi.org/10.2174/157016212802138832, Retrieved from PM:22591363.

Clark, K. H., Wiley, C. A., & Bradberry, C. W. (2013). Psychostimulant abuse and neuroinflammation: Emerging evidence of their interconnection. *Neurotoxicity Research, 23*(2), 174–188. http://dx.doi.org/10.1007/s12640-012-9334-7, Retrieved from PM:22714667.

Cohen, R. A. (2009). The changing face of HIV-associated cognitive and neuropsychiatric disturbance. In R. H. Paul, N. C. Sacktor, V. Valcour, & K. T. Tashima (Eds.), *Current clinical neurology. HIV and the brain: New challenges in the modern era* (pp. 133–186). New York: Humana Press Inc.

Colfax, G., & Guzman, R. (2006). Club drugs and HIV infection: A review. *Clinical Infectious Diseases, 42*(10), 1463–1469. http://dx.doi.org/10.1086/503259, Retrieved from PM:16619161.

Coller, J. K., & Hutchinson, M. R. (2012). Implications of central immune signaling caused by drugs of abuse: Mechanisms, mediators and new therapeutic approaches for prediction and treatment of drug dependence. *Pharmacology & Therapeutics, 134*(2), 219–245.

Colton, C. A. (2009). Heterogeneity of microglial activation in the innate immune response in the brain. *Journal of Neuroimmune Pharmacology, 4*(4), 399–418, Retrieved from PM:19655259.

Conant, K., Atwood, W. J., Traub, R., Tornatore, C., & Major, E. O. (1994). An increase in p50/p65 NF-kB binding to the HIV-1 LTR is not sufficient to increase viral expression in the primary human astrocyte. *Virology, 205*, 586–590.

Conant, K., Garzino-Demo, A., Nath, A., McArthur, J. C., Halliday, W., Power, C., et al. (1998). Induction of monocyte chemoattractant protein-1 in HIV-1 Tat-stimulated astrocytes and elevation in AIDS dementia. *Proceedings of the National Academy of Sciences of the United States of America, 95*(6), 3117–3121.

Conant, K., Ma, M., Nath, A., & Major, E. O. (1996). Extracellular human immunodeficiency virus type 1 Tat protein is associated with an increase in both NF-kappa B binding and protein kinase C activity in primary human astrocytes. *Journal of Virology, 70*(3), 1384–1389.

Conant, K., St, H. C., Anderson, C., Galey, D., Wang, J., & Nath, A. (2004). Human immunodeficiency virus type 1 Tat and methamphetamine affect the release and activation of matrix-degrading proteinases. *Journal of Neurovirology, 10*(1), 21–28, Retrieved from PM:14982725.

Conant, K., Wang, Y., Szklarczyk, A., Dudak, A., Mattson, M. P., & Lim, S. T. (2010). Matrix metalloproteinase-dependent shedding of intercellular adhesion molecule-5 occurs with long-term potentiation. *Neuroscience, 166*(2), 508–521. http://dx.doi.org/10.1016/j.neuroscience.2009.12.061, Retrieved from PM:20045450.

Congdon, E., Lesch, K. P., & Canli, T. (2008). Analysis of DRD4 and DAT polymorphisms and behavioral inhibition in healthy adults: Implications for impulsivity. *American Journal of Medical Genetics. Part B, Neuropsychiatric Genetics, 147B*(1), 27–32. http://dx.doi.org/10.1002/ajmg.b.30557, Retrieved from PM:17525955.

Coraci, I. S., Husemann, J., Berman, J. W., Hulette, C., Dufour, J. H., Campanella, G. K., et al. (2002). CD36, a class B scavenger receptor, is expressed on microglia in Alzheimer's disease brains and can mediate production of reactive oxygen species in response to beta-amyloid fibrils. *The American Journal of Pathology, 160*(1), 101–112.

Cosenza, M. A., Zhao, M. L., Si, Q., & Lee, S. C. (2002). Human brain parenchymal microglia express CD14 and CD45 and are productively infected by HIV-1 in HIV-1 encephalitis. *Brain Pathology, 12*(4), 442–455, Retrieved from PM:12408230.

Crews, L., Patrick, C., Adame, A., Rockenstein, E., & Masliah, E. (2011). Modulation of aberrant CDK5 signaling rescues impaired neurogenesis in models of Alzheimer's disease. *Cell Death and Disease*, *2*, e120. http://dx.doi.org/10.1038/cddis.2011.2, cddis20112 [pii]. Retrieved from PM:21368891.

Crews, F. T., Zou, J., & Qin, L. (2011). Induction of innate immune genes in brain create the neurobiology of addiction. *Brain, Behavior, and Immunity*, *25*(Suppl. 1), S4–S12. http://dx.doi.org/10.1016/j.bbi.2011.03.003, Retrieved from PM:21402143.

Crystal, H. A., Hamon, S., Randesi, M., Cook, J., Anastos, K., Lazar, J., et al. (2012). A C17T polymorphism in the mu opiate receptor is associated with quantitative measures of drug use in African American women. *Addiction Biology*, *17*(1), 181–191 Retrieved from PM:21070507.

Curtis, M. A., Faull, R. L., & Eriksson, P. S. (2007). The effect of neurodegenerative diseases on the subventricular zone. *Nature Reviews. Neuroscience*, *8*(9), 712–723. http://dx.doi.org/10.1038/nrn2216, Retrieved from PM:17704813.

Curtis, M. A., Penney, E. B., Pearson, A. G., van Roon-Mom, W. M., Butterworth, N. J., Dragunow, M., et al. (2003). Increased cell proliferation and neurogenesis in the adult human Huntington's disease brain. *Proceedings of the National Academy of Sciences of the United States of America*, *100*(15), 9023–9027. http://dx.doi.org/10.1073/pnas.1532244100, Retrieved from PM:12853570.

Curtis, M. A., Waldvogel, H. J., Synek, B., & Faull, R. L. (2005). A histochemical and immunohistochemical analysis of the subependymal layer in the normal and Huntington's disease brain. *Journal of Chemical Neuroanatomy*, *30*(1), 55–66, Retrieved from PM:16108100.

Czub, S., Czub, M., Koutsilieri, E., Sopper, S., Villinger, F., Muller, J. G., et al. (2004). Modulation of simian immunodeficiency virus neuropathology by dopaminergic drugs. *Acta Neuropathologica (Berlin)*, *107*(3), 216–226, Retrieved from PM:14712399.

Czub, S., Koutsilieri, E., Sopper, S., Czub, M., Stahl-Hennig, C., Muller, J. G., et al. (2001). Enhancement of central nervous system pathology in early simian immunodeficiency virus infection by dopaminergic drugs. *Acta Neuropathologica (Berlin)*, *101*(2), 85–91.

D'Ascenzo, M., Fellin, T., Terunuma, M., Revilla-Sanchez, R., Meaney, D. F., Auberson, Y. P., et al. (2007). mGluR5 stimulates gliotransmission in the nucleus accumbens. *Proceedings of the National Academy of Sciences of the United States of America*, *104*(6), 1995–2000, Retrieved from PM:17259307.

Davidson, C., Chen, Q., Zhang, X., Xiong, X., Lazarus, C., Lee, T. H., et al. (2007). Deprenyl treatment attenuates long-term pre- and post-synaptic changes evoked by chronic methamphetamine. *European Journal of Pharmacology*, *573*(1–3), 100–110, Retrieved from PM:17651730.

De Pittà, M., Volman, V., Berry, H., Parpura, V., Volterra, A., & Ben-Jacob, E. (2012). Computational quest for understanding the role of astrocyte signaling in synaptic transmission and plasticity. *Frontiers in Computational Neuroscience*, *6*, 98. http://dx.doi.org/10.3389/fncom.2012.00098, Retrieved from PM:23267326.

Dever, S. M., Costin, B. N., Xu, R., El-Hage, N., Balinang, J., Samoshkin, A., et al. (2014). Differential expression of the alternatively spliced *OPRM1* isoform μ-opioid receptor-1K in HIV-infected individuals. *AIDS*, *28*(1), 19–30.

Dever, S. M., Xu, R., Fitting, S., Knapp, P. E., & Hauser, K. F. (2012). Differential expression and HIV-1 regulation of μ-opioid receptor splice variants across human central nervous system cell types. *Journal of Neurovirology*, *18*(3), 181–190. Retrieved from, http://www.springerlink.com/content/b540406322297676-z, Retrieved from PM:22528479.

Doetsch, F., Garcia-Verdugo, J. M., & Alvarez-Buylla, A. (1997). Cellular composition and three-dimensional organization of the subventricular germinal zone in the adult mammalian brain. *Journal of Neuroscience*, *17*(13), 5046–5061, Retrieved from PM:9185542.

Dominguez-Escriba, L., Hernandez-Rabaza, V., Soriano-Navarro, M., Barcia, J. A., Romero, F. J., Garcia-Verdugo, J. M., et al. (2006). Chronic cocaine exposure impairs progenitor proliferation but spares survival and maturation of neural precursors in adult rat dentate gyrus. *The European Journal of Neuroscience*, *24*(2), 586–594. http://dx.doi.org/10.1111/j.1460-9568.2006.04924.x, Retrieved from PM:16903860.

Donahoe, R. M., & Vlahov, D. (1998). Opiates as potential cofactors in progression of HIV-1 infections to AIDS. *Journal of Neuroimmunology*, *83*(1–2), 77–87.

Dutta, R., & Roy, S. (2012). Mechanism(s) involved in opioid drug abuse modulation of HAND. *Current HIV Research*, *10*(5), 469–477. http://dx.doi.org/10.2174/157016212802138805, Retrieved from PM:22591371.

Eap, C. B., Buclin, T., & Baumann, P. (2002). Interindividual variability of the clinical pharmacokinetics of methadone: Implications for the treatment of opioid dependence. *Clinical Pharmacokinetics*, *41*(14), 1153–1193. http://dx.doi.org/10.2165/00003088-200241140-00003, Retrieved from PM:12405865.

Eggen, B. J., Raj, D., Hanisch, U. K., & Boddeke, H. W. (2013). Microglial phenotype and adaptation. *Journal of Neuroimmune Pharmacology*, *8*(4), 807–823. http://dx.doi.org/10.1007/s11481-013-9490-4, Retrieved from PM:23881706.

Eisch, A. J., Barrot, M., Schad, C. A., Self, D. W., & Nestler, E. J. (2000). Opiates inhibit neurogenesis in the adult rat hippocampus. *Proceedings of the National Academy of Sciences of the United States of America*, *97*(13), 7579–7584.

Eisch, A. J., & Harburg, G. C. (2006). Opiates, psychostimulants, and adult hippocampal neurogenesis: Insights for addiction and stem cell biology. *Hippocampus*, *16*(3), 271–286, Retrieved from PM:16411230.

Ekdahl, C. T., Claasen, J. H., Bonde, S., Kokaia, Z., & Lindvall, O. (2003). Inflammation is detrimental for neurogenesis in adult brain. *Proceedings of the National Academy of Sciences of the United States of America*, *100*(23), 13632–13637. http://dx.doi.org/10.1073/pnas.2234031100, Retrieved from PM:14581618.

El-Hage, N., Bruce-Keller, A. J., Knapp, P. E., & Hauser, K. F. (2008). CCL5/RANTES gene deletion attenuates opioid-induced increases in glial CCL2/MCP-1 immunoreactivity and activation in HIV-1 Tat exposed mice. *Journal of Neuroimmune Pharmacology*, *3*(4), 275–285.

El-Hage, N., Bruce-Keller, A. J., Yakovleva, T., Bakalkin, G., Knapp, P. E., & Hauser, K. F. (2008). Morphine exacerbates HIV-1 Tat-induced cytokine production in astrocytes through convergent effects on $[Ca^{2+}]_i$, NF-κB trafficking and transcription. *PLoS One*, *3*(12), e4093.

El-Hage, N., Dever, S. M., Podhaizer, E. M., Arnatt, C. K., Zhang, Y., & Hauser, K. F. (2013). A novel bivalent HIV-1 entry inhibitor reveals fundamental differences in CCR5-μ-opioid receptor interactions between human astroglia and microglia. *AIDS*, *27*(14), 2181–2190, Retrieved from PM:23751259.

El-Hage, N., Gurwell, J. A., Singh, I. N., Knapp, P. E., Nath, A., & Hauser, K. F. (2005). Synergistic increases in intracellular Ca^{2+}, and the release of MCP-1, RANTES, and IL-6 by astrocytes treated with opiates and HIV-1 Tat. *Glia*, *50*(2), 91–106.

El-Hage, N., Podhaizer, E. M., Sturgill, J., & Hauser, K. F. (2011). Toll-like receptor expression and activation in astroglia: Differential regulation by HIV-1 Tat, gp120, and morphine. *Immunological Investigations*, *40*(5), 498–522, Retrieved from PM:21425908.

El-Hage, N., Rodriguez, M., Podhaizer, E. M., Zou, S., Dever, S. M., Snider, S. E., et al. (2014). Ibudilast (AV411), and its AV1013 analog, reduce HIV-1 replication and neuronal death induced by HIV-1 and morphine. *AIDS*. http://dx.doi.org/10.1097/QAD.0000000000000291, Retrieved from PM:24732776.

El-Hage, N., Wu, G., Ambati, J., Bruce-Keller, A. J., Knapp, P. E., & Hauser, K. F. (2006). CCR2 mediates increases in glial activation caused by exposure to HIV-1 Tat and opiates. *Journal of Neuroimmunology*, *178*(1–2), 9–16.

El-Hage, N., Wu, G., Wang, J., Ambati, J., Knapp, P. E., Reed, J. L., et al. (2006). HIV-1 Tat and opiate-induced changes in astrocytes promote chemotaxis of microglia through the expression of MCP-1 and alternative chemokines. *Glia, 53*(2), 132–146.

Ellis, R., Langford, D., & Masliah, E. (2007). HIV and antiretroviral therapy in the brain: Neuronal injury and repair. *Nature Reviews. Neuroscience, 8*(1), 33–44, Retrieved from PM:17180161.

Emsley, J. G., & Macklis, J. D. (2006). Astroglial heterogeneity closely reflects the neuronal-defined anatomy of the adult murine CNS. *Neuron Glia Biology, 2*(3), 175–186, Retrieved from PM:17356684.

Eriksson, P. S., Hansson, E., & Rönnbäck, L. (1992). δ and kappa opiate receptors in primary astroglial cultures. Part II: Receptor sets in cultures from various brain regions and interactions with β-receptor activated cyclic AMP. *Neurochemical Research, 17*, 545–551.

Eriksson, P. S., Nilsson, M., Wagberg, M., Hansson, E., & Rönnbäck, L. (1993). Kappa-opioid receptors on astrocytes stimulate L-type Ca^{2+} channels. *Neuroscience, 54*(2), 401–407.

Eriksson, P. S., Perfilieva, E., Bjork-Eriksson, T., Alborn, A. M., Nordborg, C., Peterson, D. A., et al. (1998). Neurogenesis in the adult human hippocampus. *Nature Medicine, 4*(11), 1313–1317.

Ersche, K. D., Jones, P. S., Williams, G. B., Robbins, T. W., & Bullmore, E. T. (2013). Cocaine dependence: A fast-track for brain ageing? *Molecular Psychiatry, 18*(2), 134–135. http://dx.doi.org/10.1038/mp.2012.31, Retrieved from PM:22525488.

Eschenroeder, A. C., Vestal-Laborde, A. A., Sanchez, E. S., Robinson, S. E., & Sato-Bigbee, C. (2012). Oligodendrocyte responses to buprenorphine uncover novel and opposing roles of mu-opioid- and nociceptin/orphanin FQ receptors in cell development: Implications for drug addiction treatment during pregnancy. *Glia, 60*(1), 125–136. http://dx.doi.org/10.1002/glia.21253, Retrieved from PM:22002899.

Espinoza, S., Salahpour, A., Masri, B., Sotnikova, T. D., Messa, M., Barak, L. S., et al. (2011). Functional interaction between trace amine-associated receptor 1 and dopamine D2 receptor. *Molecular Pharmacology, 80*(3), 416–425. http://dx.doi.org/10.1124/mol.111.073304, Retrieved from PM:21670104.

Eugenin, E. A., Dyer, G., Calderon, T. M., & Berman, J. W. (2005). HIV-1 tat protein induces a migratory phenotype in human fetal microglia by a CCL2 (MCP-1)-dependent mechanism: Possible role in NeuroAIDS. *Glia, 49*(4), 501–510.

Eugenin, E. A., Osiecki, K., Lopez, L., Goldstein, H., Calderon, T. M., & Berman, J. W. (2006). CCL2/monocyte chemoattractant protein-1 mediates enhanced transmigration of human immunodeficiency virus (HIV)-infected leukocytes across the blood-brain barrier: A potential mechanism of HIV-CNS invasion and NeuroAIDS. *Journal of Neuroscience, 26*(4), 1098–1106.

Fang, J., Acheampong, E., Dave, R., Wang, F., Mukhtar, M., & Pomerantz, R. J. (2005). The RNA helicase DDX1 is involved in restricted HIV-1 Rev function in human astrocytes. *Virology, 336*(2), 299–307.

Farber, K., Pannasch, U., & Kettenmann, H. (2005). Dopamine and noradrenaline control distinct functions in rodent microglial cells. *Molecular and Cellular Neurosciences, 29*(1), 128–138.

Farina, C., Aloisi, F., & Meinl, E. (2007). Astrocytes are active players in cerebral innate immunity. *Trends in Immunology, 28*(3), 138–145, Retrieved from PM:17276138.

Ferris, M. J., Mactutus, C. F., & Booze, R. M. (2008). Neurotoxic profiles of HIV, psychostimulant drugs of abuse, and their concerted effect on the brain: Current status of dopamine system vulnerability in NeuroAIDS. *Neuroscience and Biobehavioral Reviews, 32*(5), 883–909.

Festa, L., & Meucci, O. (2012). Effects of opiates and HIV proteins on neurons: The role of ferritin heavy chain and a potential for synergism. *Current HIV Research, 10*(5), 453–462. http://dx.doi.org/10.2174/157016212802138751, Retrieved from PM:22591369.

Fiala, M., Eshleman, A. J., Cashman, J., Lin, J., Lossinsky, A. S., Suarez, V., et al. (2005). Cocaine increases human immunodeficiency virus type 1 neuroinvasion through remodeling brain microvascular endothelial cells. *Journal of Neurovirology*, *11*(3), 281–291 Retrieved from PM:16036808.

Fiala, M., Gan, X. H., Zhang, L., House, S. D., Newton, T., Graves, M. C., et al. (1998). Cocaine enhances monocyte migration across the blood-brain barrier. Cocaine's connection to AIDS dementia and vasculitis? *Advances in Experimental Medicine and Biology*, *437*, 199–205.

Fiala, M., Looney, D. J., Stins, M., Way, D. D., Zhang, L., Gan, X., et al. (1997). TNF-alpha opens a paracellular route for HIV-1 invasion across the blood-brain barrier. *Molecular Medicine*, *3*(8), 553–564.

Finley, M. J., Chen, X., Bardi, G., Davey, P., Geller, E. B., Zhang, L., et al. (2008). Bidirectional heterologous desensitization between the major HIV-1 co-receptor CXCR4 and the kappa-opioid receptor. *Journal of Neuroimmunology*, *197*(2), 114–123, Retrieved from PM:18533278.

Fischer-Smith, T., Croul, S., Sverstiuk, A. E., Capini, C., L'Heureux, D., Regulier, E. G., et al. (2001). CNS invasion by CD14+/CD16+ peripheral blood-derived monocytes in HIV dementia: Perivascular accumulation and reservoir of HIV infection. *Journal of Neurovirology*, 7(6), 528–541. http://dx.doi.org/10.1080/135502801753248114, Retrieved from PM:11704885.

Fischer-Smith, T., & Rappaport, J. (2005). Evolving paradigms in the pathogenesis of HIV-1-associated dementia. *Expert Reviews in Molecular Medicine*, 7(27), 1–26.

Fischer-Smith, T., Tedaldi, E. M., & Rappaport, J. (2008). CD163/CD16 coexpression by circulating monocytes/macrophages in HIV: Potential biomarkers for HIV infection and AIDS progression. *AIDS Research and Human Retroviruses*, *24*(3), 417–421. http://dx.doi.org/10.1089/aid.2007.0193, Retrieved from PM:18373432.

Fitting, S., Scoggins, K. L., Xu, R., Dever, S. M., Knapp, P. E., Dewey, W. L., et al. (2012). Morphine efficacy is altered in conditional HIV-1 Tat transgenic mice. *European Journal of Pharmacology*, *689*(1–3), 96–103.

Fitting, S., Xu, R., Bull, C. M., Buch, S. K., El-Hage, N., Nath, A., et al. (2010). Interactive comorbidity between opioid drug abuse and HIV-1 Tat: Chronic exposure augments spine loss and sublethal dendritic pathology in striatal neurons. *The American Journal of Pathology*, *177*(3), 1397–1410.

Fitting, S., Zou, S., Chen, W., Vo, P., Hauser, K. F., & Knapp, P. E. (2010). Regional heterogeneity and diversity in cytokine and chemokine production by astroglia: Differential responses to HIV-1 Tat, gp120 and morphine revealed by multiplex analysis. *Journal of Proteome Research*, *9*(4), 1795–1804.

Fleckenstein, A. E., Metzger, R. R., Wilkins, D. G., Gibb, J. W., & Hanson, G. R. (1997). Rapid and reversible effects of methamphetamine on dopamine transporters. *The Journal of Pharmacology and Experimental Therapeutics*, *282*(2), 834–838, Retrieved from PM:9262348.

Flora, G., Lee, Y. W., Nath, A., Hennig, B., Maragos, W., & Toborek, M. (2003). Methamphetamine potentiates HIV-1 Tat protein-mediated activation of redox-sensitive pathways in discrete regions of the brain. *Experimental Neurology*, *179*(1), 60–70.

Flora, G., Lee, Y. W., Nath, A., Maragos, W., Hennig, B., & Toborek, M. (2002). Methamphetamine-induced TNF-alpha gene expression and activation of AP-1 in discrete regions of mouse brain: Potential role of reactive oxygen intermediates and lipid peroxidation. *Neuromolecular Medicine*, *2*(1), 71–85.

Frank, M. G., Watkins, L. R., & Maier, S. F. (2011). Stress- and glucocorticoid-induced priming of neuroinflammatory responses: Potential mechanisms of stress-induced vulnerability to drugs of abuse. *Brain, Behavior, and Immunity*, *25*(Suppl. 1), S21–S28. http://dx.doi.org/10.1016/j.bbi.2011.01.005, Retrieved from PM:21256955.

Gan, X., Zhang, L., Berger, O., Stins, M. F., Way, D., Taub, D. D., et al. (1999). Cocaine enhances brain endothelial adhesion molecules and leukocyte migration. *Clinical Immunology, 91*(1), 68–76.

Garcia-Fuster, M. J., Perez, J. A., Clinton, S. M., Watson, S. J., & Akil, H. (2010). Impact of cocaine on adult hippocampal neurogenesis in an animal model of differential propensity to drug abuse. *The European Journal of Neuroscience, 31*(1), 79–89, Retrieved from PM:20104651.

Garden, G. A. (2002). Microglia in human immunodeficiency virus-associated neurodegeneration. *Glia, 40*(2), 240–251.

Gasque, P., Dean, Y. D., McGreal, E. P., VanBeek, J., & Morgan, B. P. (2000). Complement components of the innate immune system in health and disease in the CNS. *Immunopharmacology, 49*(1–2), 171–186, Retrieved from PM:10904116.

Geha, S., Pallud, J., Junier, M. P., Devaux, B., Leonard, N., Chassoux, F., et al. (2010). NG2+/Olig2+ cells are the major cycle-related cell population of the adult human normal brain. *Brain Pathology, 20*(2), 399–411. http://dx.doi.org/10.1111/j.1750-3639.2009.00295.x, Retrieved from PM:19486010.

Gekker, G., Hu, S., Sheng, W. S., Rock, R. B., Lokensgard, J. R., & Peterson, P. K. (2006). Cocaine-induced HIV-1 expression in microglia involves sigma-1 receptors and transforming growth factor-beta1. *6*(6), 1029–1033. http://dx.doi.org/10.1016/j.intimp.2005.12.005, Retrieved from PM:16644490.

Gekker, G., Hu, S., Wentland, M. P., Bidlack, J. M., Lokensgard, J. R., & Peterson, P. K. (2004). Kappa-opioid receptor ligands inhibit cocaine-induced HIV-1 expression in microglial cells. *Journal of Pharmacology and Experimental Therapeutics, 309*(2), 600–606.

Gelman, B. B., Chen, T., Lisinicchia, J. G., Soukup, V. M., Carmical, J. R., Starkey, J. M., et al. (2012). The national NeuroAIDS tissue consortium brain gene array: Two types of HIV-associated neurocognitive impairment. *PLoS One*, 7(9), e46178. http://dx.doi.org/10.1371/journal.pone.0046178, Retrieved from PM:23049970.

Gendelman, H. E., Persidsky, Y., Ghorpade, A., Limoges, J., Stins, M., Fiala, M., et al. (1997). The neuropathogenesis of the AIDS dementia complex. *AIDS, 11*(Suppl. A), 35–45.

Genis, P., Jett, M., Bernton, E. W., Boyle, T., Gelbard, H. A., Dzenko, K., et al. (1992). Cytokines and arachidonic metabolites produced during human immunodeficiency virus (HIV)-infected macrophage-astroglia interactions: Implications for the neuropathogenesis of HIV disease. *The Journal of Experimental Medicine, 176*(6), 1703–1718.

Ginhoux, F., Greter, M., Leboeuf, M., Nandi, S., See, P., Gokhan, S., et al. (2010). Fate mapping analysis reveals that adult microglia derive from primitive macrophages. *Science, 330*(6005), 841–845.

Giulian, D., Yu, J., Li, X., Tom, D., Li, J., Wendt, E., et al. (1996). Study of receptor-mediated neurotoxins released by HIV-1-infected mononuclear phagocytes found in human brain. *Journal of Neuroscience, 16*(10), 3139–3153, Retrieved from PM:8627353.

Goncalves, J., Martins, T., Ferreira, R., Milhazes, N., Borges, F., Ribeiro, C. F., et al. (2008). Methamphetamine-induced early increase of IL-6 and TNF-α mRNA expression in the mouse brain. *Annals of the New York Academy of Sciences, 1139*, 103–111. http://dx.doi.org/10.1196/annals.1432.043, Retrieved from PM:18991854.

Gonzalez, E., Dhanda, R., Bamshad, M., Mummidi, S., Geevarghese, R., Catano, G., et al. (2001). Global survey of genetic variation in CCR5, RANTES, and MIP-1α: Impact on the epidemiology of the HIV-1 pandemic. *Proceedings of the National Academy of Sciences of the United States of America, 98*(9), 5199–5204. http://dx.doi.org/10.1073/pnas.091056898, Retrieved from PM:11320252.

Gonzalez, E., Rovin, B. H., Sen, L., Cooke, G., Dhanda, R., Mummidi, S., et al. (2002). HIV-1 infection and AIDS dementia are influenced by a mutant MCP-1 allele linked to increased monocyte infiltration of tissues and MCP-1 levels. *Proceedings of the National Academy of Sciences of the United States of America, 99*(21), 13795–13800.

Gonzalez-Scarano, F., & Martin-Garcia, J. (2005). The neuropathogenesis of AIDS. *Nature Reviews. Immunology, 5*(1), 69–81.

Gorry, P. R., Ong, C., Thorpe, J., Bannwarth, S., Thompson, K. A., Gatignol, A., et al. (2003). Astrocyte infection by HIV-1: Mechanisms of restricted virus replication, and role in the pathogenesis of HIV-1-associated dementia. *Current HIV Research, 1*(4), 463–473.

Gorry, P., Purcell, D., Howard, J., & McPhee, D. (1998). Restricted HIV-1 infection of human astrocytes: Potential role of nef in the regulation of virus replication. *Journal of Neurovirology, 4*(4), 377–386.

Gras, G., Samah, B., Hubert, A., Leone, C., Porcheray, F., & Rimaniol, A. C. (2012). EAAT expression by macrophages and microglia: Still more questions than answers. *Amino Acids, 42*(1), 221–229, Retrieved from PM:21373769.

Gris, P., Gauthier, J., Cheng, P., Gibson, D. G., Gris, D., Laur, O., et al. (2010). A novel alternatively spliced isoform of the mu-opioid receptor: Functional antagonism. *Molecular Pain, 6*(33), Retrieved from PM:20525224.

Guillemin, G. J., & Brew, B. J. (2004). Microglia, macrophages, perivascular macrophages, and pericytes: A review of function and identification. *Journal of Leukocyte Biology, 75*(3), 388–397. http://dx.doi.org/10.1189/jlb.0303114, Retrieved from PM:14612429.

Guillemin, G., Croitoru, J., Le Grand, R. L., Franck-Duchenne, M., Dormont, D., & Boussin, F. D. (2000). Simian immunodeficiency virus mac251 infection of astrocytes. *Journal of Neurovirology, 6*(3), 173–186, Retrieved from PM:10878708.

Guillemin, G. J., Kerr, S. J., & Brew, B. J. (2005). Involvement of quinolinic acid in AIDS dementia complex. *Neurotoxicity Research*, 7(1–2), 103–123.

Gupta, S., Bousman, C. A., Chana, G., Cherner, M., Heaton, R. K., Deutsch, R., et al. (2011). Dopamine receptor D3 genetic polymorphism (rs6280TC) is associated with rates of cognitive impairment in methamphetamine-dependent men with HIV: Preliminary findings. *Journal of Neurovirology, 17*(3), 239–247. http://dx.doi.org/10.1007/s13365-011-0028-3, Retrieved from PM:21491142.

Gupta, S., Knight, A. G., Gupta, S., Knapp, P. E., Hauser, K. F., Keller, J. N., et al. (2010). HIV-Tat elicits microglial glutamate release: Role of NAPDH oxidase and the cystine-glutamate antiporter. *Neuroscience Letters, 485*(3), 233–236, Retrieved from PM:20849923.

Gurwell, J. A., Duncan, M. J., Maderspach, K., Stiene-Martin, A., Elde, R. P., & Hauser, K. F. (1996). κ-Opioid receptor expression defines a phenotypically distinct subpopulation of astroglia: Relationship to Ca^{2+} mobilization, development, and the antiproliferative effect of opioids. *Brain Research, 737*(1–2), 175–187.

Gurwell, J. A., Nath, A., Sun, Q., Zhang, J., Martin, K. M., Chen, Y., et al. (2001). Synergistic neurotoxicity of opioids and human immunodeficiency virus-1 Tat protein in striatal neurons in vitro. *Neuroscience, 102*(3), 555–563.

Hagino, Y., Kariura, Y., Manago, Y., Amano, T., Wang, B., Sekiguchi, M., et al. (2004). Heterogeneity and potentiation of AMPA type of glutamate receptors in rat cultured microglia. *Glia, 47*(1), 68–77.

Hahn, Y. K., Podhaizer, E. M., Farris, S. P., Miles, M. F., Hauser, K. F., & Knapp, P. E. (2014). Effects of chronic HIV-1 Tat exposure in the CNS: Heightened vulnerability of males versus females to changes in cell numbers, synaptic integrity, and behavior. *Brain Structure and Function*. http://dx.doi.org/10.1007/s00429-013-0676-6, Epub. Retrieved from PM:24352707.

Hahn, Y. K., Podhaizer, E. M., Hauser, K. F., & Knapp, P. E. (2012). HIV-1 alters neural and glial progenitor cell dynamics in the central nervous system: Coordinated response to opiates during maturation. *Glia, 60*(12), 1871–1887, Retrieved from PM:22865725.

Hahn, Y. K., Vo, P., Fitting, S., Block, M. L., Hauser, K. F., & Knapp, P. E. (2010). Beta-chemokine production by neural and glial progenitor cells is enhanced by HIV-1 Tat: Effects on microglial migration. *Journal of Neurochemistry, 114*(1), 97–109, Retrieved from PM:20403075.

Hall, F. S., Li, X. F., Randall-Thompson, J., Sora, I., Murphy, D. L., Lesch, K. P., et al. (2009). Cocaine-conditioned locomotion in dopamine transporter, norepinephrine transporter and 5-HT transporter knockout mice. *Neuroscience, 162*(4), 870–880. http://dx.doi.org/10.1016/j.neuroscience.2009.05.058, Retrieved from PM:19482066.

Hall, F. S., Sora, I., Drgonova, J., Li, X. F., Goeb, M., & Uhl, G. R. (2004). Molecular mechanisms underlying the rewarding effects of cocaine. *Annals of the New York Academy of Sciences, 1025*, 47–56. http://dx.doi.org/10.1196/annals.1316.006, Retrieved from PM:15542699.

Halpin, L. E., Collins, S. A., & Yamamoto, B. K. (2014). Neurotoxicity of methamphetamine and 3,4-methylenedioxymethamphetamine. *Life Sciences, 97*(1), 37–44. http://dx.doi.org/10.1016/j.lfs.2013.07.014, Retrieved from PM:23892199.

Halpin, L. E., & Yamamoto, B. K. (2012). Peripheral ammonia as a mediator of methamphetamine neurotoxicity. *Journal of Neuroscience, 32*(38), 13155–13163. http://dx.doi.org/10.1523/JNEUROSCI.2530-12.2012, Retrieved from PM:22993432.

Han, Y., Lai, J., Barditch-Crovo, P., Gallant, J. E., Williams, T. M., Siliciano, R. F., et al. (2008). The role of protective HCP5 and HLA-C associated polymorphisms in the control of HIV-1 replication in a subset of elite suppressors. *AIDS, 22*(4), 541–544. http://dx.doi.org/10.1097/QAD.0b013e3282f470e4, Retrieved from PM:18301071.

Hanisch, U. K. (2013). Functional diversity of microglia - how heterogeneous are they to begin with? *Frontiers in Cellular Neuroscience*, 7, 65. http://dx.doi.org/10.3389/fncel.2013.00065, Retrieved from PM:23717262.

Hanisch, U. K., & Kettenmann, H. (2007). Microglia: Active sensor and versatile effector cells in the normal and pathologic brain. *Nature Neuroscience, 10*(11), 1387–1394, Retrieved from PM:17965659.

Hanson, G. R., Rau, K. S., & Fleckenstein, A. E. (2004). The methamphetamine experience: A NIDA partnership. *Neuropharmacology, 47*(Suppl. 1), 92–100.

Hanson, G. R., Sandoval, V., Riddle, E., & Fleckenstein, A. E. (2004). Psychostimulants and vesicle trafficking: A novel mechanism and therapeutic implications. *Annals of the New York Academy of Sciences, 1025*, 146–150.

Happel, C., Steele, A. D., Finley, M. J., Kutzler, M. A., & Rogers, T. J. (2008). DAMGO-induced expression of chemokines and chemokine receptors: The role of TGF-beta1. *Journal of Leukocyte Biology, 83*(4), 956–963.

Hardingham, G. E., Fukunaga, Y., & Bading, H. (2002). Extrasynaptic NMDARs oppose synaptic NMDARs by triggering CREB shut-off and cell death pathways. *Nature Neuroscience, 5*(5), 405–414. http://dx.doi.org/10.1038/nn835, Retrieved from PM:11953750.

Hauser, K. F., Aldrich, J. V., Anderson, K. J., Bakalkin, G., Christie, M. J., Hall, E. D., et al. (2005). Pathobiology of dynorphins in trauma and disease. *Frontiers in Bioscience, 10*, 216–235.

Hauser, K. F., El-Hage, N., Buch, S., Berger, J. R., Tyor, W. R., Nath, A., et al. (2005). Molecular targets of opiate drug abuse in neuroAIDS. *Neurotoxicity Research, 8*(1–2), 63–80.

Hauser, K. F., El-Hage, N., Buch, S., Tyor, W. R., Nath, A., Bruce-Keller, A. J., et al. (2006). Impact of opiate-HIV-1 interactions on neurotoxic signaling. *Journal of Neuroimmune Pharmacology, 1*, 98–105.

Hauser, K. F., El-Hage, N., Stiene-Martin, A., Maragos, W. F., Nath, A., Persidsky, Y., et al. (2007). HIV-1 neuropathogenesis: Glial mechanisms revealed through substance abuse. *Journal of Neurochemistry, 100*(3), 567–586.

Hauser, K. F., Fitting, S., Dever, S. M., Podhaizer, E. M., & Knapp, P. E. (2012). Opiate drug use and the pathophysiology of neuroAIDS. *Current HIV Research*, *10*(5), 435–452, Retrieved from PM:22591368.

Hauser, K. F., Hahn, Y. K., Adjan, V. V., Zou, S., Buch, S. K., Nath, A., et al. (2009). HIV-1 Tat and morphine have interactive effects on oligodendrocyte survival and morphology. *Glia*, *57*(2), 194–206, Retrieved from PM:18756534.

Hauser, K. F., & Mangoura, D. (1998). Diversity of the endogenous opioid system in development: Novel signal transduction translates multiple extracellular signals into neural cell growth and differentiation. *Perspectives on Developmental Neurobiology*, *5*(4), 437–449.

Hauser, K. F., Osborne, J. G., Stiene-Martin, A., & Melner, M. H. (1990). Cellular localization of proenkephalin mRNA and enkephalin peptide products in cultured astrocytes. *Brain Research*, *522*, 347–353.

Hauser, K. F., Stiene-Martin, A., Mattson, M. P., Elde, R. P., Ryan, S. E., & Godleske, C. C. (1996). μ-Opioid receptor-induced Ca^{2+} mobilization and astroglial development: Morphine inhibits DNA synthesis and stimulates cellular hypertrophy through a Ca^{2+}-dependent mechanism. *Brain Research*, *720*(1–2), 191–203.

Haydon, P. G., & Carmignoto, G. (2006). Astrocyte control of synaptic transmission and neurovascular coupling. *Physiological Reviews*, *86*(3), 1009–1031, Retrieved from PM:16816144.

Hebb, D. O. (1949). *The organization of behavior*. New York: Wiley.

Hildebrandt, K., Teuchert-Noodt, G., & Dawirs, R. R. (1999). A single neonatal dose of methamphetamine suppresses dentate granule cell proliferation in adult gerbils which is restored to control values by acute doses of haloperidol. *Journal of Neural Transmission*, *106*(5–6), 549–558.

Holmes, M. M., & Galea, L. A. (2002). Defensive behavior and hippocampal cell proliferation: Differential modulation by naltrexone during stress. *Behavioral Neuroscience*, *116*(1), 160–168, Retrieved from PM:11898802.

Holzinger, E. R., Hulgan, T., Ellis, R. J., Samuels, D. C., Ritchie, M. D., Haas, D. W., et al. (2012). Mitochondrial DNA variation and HIV-associated sensory neuropathy in CHARTER. *Journal of Neurovirology*, *18*(6), 511–520. http://dx.doi.org/10.1007/s13365-012-0133-y, Retrieved from PM:23073667.

Hu, S., Cheeran, M. C., Sheng, W. S., Ni, H. T., Lokensgard, J. R., & Peterson, P. K. (2006). Cocaine alters proliferation, migration, and differentiation of human fetal brain-derived neural precursor cells. *The Journal of Pharmacology and Experimental Therapeutics*, *318*(3), 1280–1286. http://dx.doi.org/10.1124/jpet.106.103853, Retrieved from PM:16766721.

Hu, S., Sheng, W. S., Lokensgard, J. R., & Peterson, P. K. (2005). Morphine potentiates HIV-1 gp120-induced neuronal apoptosis. *The Journal of Infectious Diseases*, *191*(6), 886–889.

Hu, S., Sheng, W. S., Lokensgard, J. R., Peterson, P. K., & Rock, R. B. (2009). Preferential sensitivity of human dopaminergic neurons to gp120-induced oxidative damage. *Journal of Neurovirology*, *15*(5–6), 401–410, Retrieved from PM:20175694.

Huang, Y., Paxton, W. A., Wolinsky, S. M., Neumann, A. U., Zhang, L., He, T., et al. (1996). The role of a mutant CCR5 allele in HIV-1 transmission and disease progression. *Nature Medicine*, *2*(11), 1240–1243, Retrieved from PM:8898752.

Hunt, P. W., Shulman, N., Hayes, T. L., Dahl, V., Somsouk, M., Funderburg, N. T., et al. (2013). The immunologic effects of maraviroc intensification in treated HIV-infected individuals with incomplete CD4+ T cell recovery: A randomized trial. *Blood*, *121*(23), 4635–4646, Retrieved from PM:23589670.

Husemann, J., Loike, J. D., Kodama, T., & Silverstein, S. C. (2001). Scavenger receptor class B type I (SR-BI) mediates adhesion of neonatal murine microglia to fibrillar

beta-amyloid. *Journal of Neuroimmunology, 114*(1–2), 142–150. http://dx.doi.org/10.1016/S0165-5728(01)00239-9, Retrieved from PM:11240025.
Husemann, J., & Silverstein, S. C. (2001). Expression of scavenger receptor class B, type I, by astrocytes and vascular smooth muscle cells in normal adult mouse and human brain and in Alzheimer's disease brain. *The American Journal of Pathology, 158*(3), 825–832. http://dx.doi.org/10.1016/S0002-9440(10)64030-8, Retrieved from PM:11238031.
Hutter, G., Nowak, D., Mossner, M., Ganepola, S., Mussig, A., Allers, K., et al. (2009). Long-term control of HIV by CCR5 Delta32/Delta32 stem-cell transplantation. *The New England Journal of Medicine, 360*(7), 692–698. http://dx.doi.org/10.1056/NEJMoa0802905, Retrieved from PM:19213682.
Imamoto, K., Paterson, J. A., & Leblond, C. P. (1978). Radioautographic investigation of gliogenesis in the corpus callosum of young rats. I. Sequential changes in oligodendrocytes. *The Journal of Comparative Neurology, 180*(1), 115–117, Retrieved from PM:649784.
Inman, G. J., Nicolas, F. J., Callahan, J. F., Harling, J. D., Gaster, L. M., Reith, A. D., et al. (2002). SB-431542 is a potent and specific inhibitor of transforming growth factor-beta superfamily type I activin receptor-like kinase (ALK) receptors ALK4, ALK5, and ALK7. *Molecular Pharmacology, 62*(1), 65–74, Retrieved from PM:12065756.
Ioannidis, J. P., Rosenberg, P. S., Goedert, J. J., Ashton, L. J., Benfield, T. L., Buchbinder, S. P., et al. (2001). Effects of CCR5-Δ32, CCR2-64I, and SDF-1 3'A alleles on HIV-1 disease progression: An international meta-analysis of individual-patient data. *Annals of Internal Medicine, 135*(9), 782–795. http://dx.doi.org/10.7326/0003-4819-135-9-200111060-00008, Retrieved from PM:11694103.
Ivanov, A., Pellegrino, C., Rama, S., Dumalska, I., Salyha, Y., Ben-Ari, Y., et al. (2006). Opposing role of synaptic and extrasynaptic NMDA receptors in regulation of the extracellular signal-regulated kinases (ERK) activity in cultured rat hippocampal neurons. *The Journal of Physiology, 572*(Pt 3), 789–798.
Jacobs, M., Murray, J., Byrd, D. A., Hurd, Y. L., & Morgello, S. (2013). HIV-related cognitive impairment shows bi-directional association with dopamine receptor DRD1 and DRD2 polymorphisms in substance-dependent and substance-independent populations. *Journal of Neurovirology, 19*(5), 495–504. http://dx.doi.org/10.1007/s13365-013-0204-8, Retrieved from PM:24078558.
Jayanthi, S., Deng, X., Ladenheim, B., McCoy, M. T., Cluster, A., Cai, N. S., et al. (2005). Calcineurin/NFAT-induced up-regulation of the Fas ligand/Fas death pathway is involved in methamphetamine-induced neuronal apoptosis. *Proceedings of the National Academy of Sciences of the United States of America, 102*(3), 868–873.
Jensen, C. J., Massie, A., & De Keyser, J. (2013). Immune players in the CNS: The astrocyte. *Journal of Neuroimmune Pharmacology, 8*(4), 824–839. http://dx.doi.org/10.1007/s11481-013-9480-6, Retrieved from PM:23821340.
Johnson, S. W., & North, R. A. (1992). Opioids excite dopamine neurons by hyperpolarization of local interneurons. *Journal of Neuroscience, 12*, 483–488.
Joseph, S. B., Arrildt, K. T., Swanstrom, A. E., Schnell, G., Lee, B., Hoxie, J. A., et al. (2014). Quantification of entry phenotypes of macrophage-tropic HIV-1 across a wide range of CD4 densities. *Journal of Virology, 88*(4), 1858–1869.
Kalivas, P. W., & Volkow, N. D. (2011). New medications for drug addiction hiding in glutamatergic neuroplasticity. *Molecular Psychiatry, 16*(10), 974–986.
Kaplan, M. S., & Hinds, J. W. (1977). Neurogenesis in the adult rat: Electron microscopic analysis of light radioautographs. *Science, 197*(4308), 1092–1094, Retrieved from PM:887941.
Kato, T. (2001). The other, forgotten genome: Mitochondrial DNA and mental disorders. *Molecular Psychiatry, 6*(6), 625–633. http://dx.doi.org/10.1038/sj.mp.4000926, Retrieved from PM:11673790.

Kaul, M., Garden, G. A., & Lipton, S. A. (2001). Pathways to neuronal injury and apoptosis in HIV-associated dementia. *Nature, 410*(6831), 988–994.

Kaul, M., & Lipton, S. A. (2005). Experimental and potential future therapeutic approaches for HIV-1 associated dementia targeting receptors for chemokines, glutamate and erythropoietin. *Neurotoxicity Research, 8*(1–2), 167–186.

Kaushal, N., & Matsumoto, R. R. (2011). Role of sigma receptors in methamphetamine-induced neurotoxicity. *Current Neuropharmacology, 9*(1), 54–57.

Kazanis, I. (2009). The subependymal zone neurogenic niche: A beating heart in the centre of the brain: How plastic is adult neurogenesis? Opportunities for therapy and questions to be addressed. *Brain, 132*(Pt 11), 2909–2921. http://dx.doi.org/10.1093/brain/awp237, Retrieved from PM:19773354.

Kelly, K. M., Beck, S. E., Pate, K. A., Queen, S. E., Dorsey, J. L., Adams, R. J., et al. (2013). Neuroprotective maraviroc monotherapy in simian immunodeficiency virus-infected macaques: Reduced replicating and latent SIV in the brain. *AIDS*. http://dx.doi.org/10.1097/QAD.0000000000000074, Retrieved from PM:24051706.

Kernie, S. G., & Parent, J. M. (2010). Forebrain neurogenesis after focal Ischemic and traumatic brain injury. *Neurobiology of Disease, 37*(2), 267–274. http://dx.doi.org/10.1016/j.nbd.2009.11.002, Retrieved from PM:19909815.

Khurdayan, V. K., Buch, S., El-Hage, N., Lutz, S. E., Goebel, S. M., Singh, I. N., et al. (2004). Preferential vulnerability of astroglia and glial precursors to combined opioid and HIV-1 Tat exposure *in vitro*. *The European Journal of Neuroscience, 19*(12), 3171–3182.

Kiebala, M., & Maggirwar, S. B. (2011). Ibudilast, a pharmacologic phosphodiesterase inhibitor, prevents human immunodeficiency virus-1 Tat-mediated activation of microglial cells. *PLoS One, 6*(4), e18633, Retrieved from PM:21494611.

Kigerl, K. A., Gensel, J. C., Ankeny, D. P., Alexander, J. K., Donnelly, D. J., & Popovich, P. G. (2009). Identification of two distinct macrophage subsets with divergent effects causing either neurotoxicity or regeneration in the injured mouse spinal cord. *Journal of Neuroscience, 29*(43), 13435–13444. http://dx.doi.org/10.1523/JNEUROSCI.3257-09.2009, Retrieved from PM:19864556.

Kim, E. Y., Bhattacharya, T., Kunstman, K., Swantek, P., Koning, F. A., Malim, M. H., et al. (2010). Human APOBEC3G-mediated editing can promote HIV-1 sequence diversification and accelerate adaptation to selective pressure. *Journal of Virology, 84*(19), 10402–10405. http://dx.doi.org/10.1128/JVI.01223-10, Retrieved from PM:20660203.

King, L. S., Kozono, D., & Agre, P. (2004). From structure to disease: The evolving tale of aquaporin biology. *Nature Reviews. Molecular Cell Biology, 5*(9), 687–698. http://dx.doi.org/10.1038/nrm1469, Retrieved from PM:15340377.

Knackstedt, L. A., Melendez, R. I., & Kalivas, P. W. (2010). Ceftriaxone restores glutamate homeostasis and prevents relapse to cocaine seeking. *Biological Psychiatry, 67*(1), 81–84. http://dx.doi.org/10.1016/j.biopsych.2009.07.018, Retrieved from PM:19717140.

Koehl, M., Meerlo, P., Gonzales, D., Rontal, A., Turek, F. W., & Abrous, D. N. (2008). Exercise-induced promotion of hippocampal cell proliferation requires beta-endorphin. *FASEB Journal, 22*(7), 2253–2262. http://dx.doi.org/10.1096/fj.07-099101, Retrieved from PM:18263701.

Kogan, F. J., Nichols, W. K., & Gibb, J. W. (1976). Influence of methamphetamine on nigral and striatal tyrosine hydroxylase activity and on striatal dopamine levels. *European Journal of Pharmacology, 36*(2), 363–371.

Kolodziej, A., Stumm, R., Becker, A., & Hollt, V. (2008). Endogenous opioids inhibit ischemia-induced generation of immature hippocampal neurons via the mu-opioid receptor. *The European Journal of Neuroscience, 27*(6), 1311–1319. http://dx.doi.org/10.1111/j.1460-9568.2008.06111.x, Retrieved from PM:18331339.

Kornack, D. R., & Rakic, P. (1999). Continuation of neurogenesis in the hippocampus of the adult macaque monkey. *Proceedings of the National Academy of Sciences of the United States of America, 96*(10), 5768–5773, Retrieved from PM:10318959.

Kornack, D. R., & Rakic, P. (2001). Cell proliferation without neurogenesis in adult primate neocortex. *Science*, *294*(5549), 2127–2130. http://dx.doi.org/10.1126/science.1065467, Retrieved from PM:11739948.

Kousik, S. M., Napier, T. C., & Carvey, P. M. (2012). The effects of psychostimulant drugs on blood brain barrier function and neuroinflammation. *Frontiers in Pharmacology*, *3*, 121. http://dx.doi.org/10.3389/fphar.2012.00121, Retrieved from PM:22754527.

Kraft-Terry, S. D., Buch, S. J., Fox, H. S., & Gendelman, H. E. (2009). A coat of many colors: Neuroimmune crosstalk in human immunodeficiency virus infection. *Neuron*, *64*(1), 133–145. http://dx.doi.org/10.1016/j.neuron.2009.09.042, Retrieved from PM:19840555.

Kramer-Hammerle, S., Hahn, A., Brack-Werner, R., & Werner, T. (2005). Elucidating effects of long-term expression of HIV-1 Nef on astrocytes by microarray, promoter, and literature analyses. *Gene*, *358*(31–8), 31–38.

Kramer-Hammerle, S., Rothenaigner, I., Wolff, H., Bell, J. E., & Brack-Werner, R. (2005). Cells of the central nervous system as targets and reservoirs of the human immunodeficiency virus. *Virus Research*, *111*(2), 194–213.

Krasnova, I. N., & Cadet, J. L. (2009). Methamphetamine toxicity and messengers of death. *Brain Research Reviews*, *60*(2), 379–407. http://dx.doi.org/10.1016/j.brainresrev.2009.03.002, Retrieved from PM:19328213.

Krathwohl, M. D., & Kaiser, J. L. (2004). HIV-1 promotes quiescence in human neural progenitor cells. *The Journal of Infectious Diseases*, *190*(2), 216–226.

Krebs, M. O., Gauchy, C., Desban, M., Glowinski, J., & Kemel, M. L. (1994). Role of dynorphin and GABA in the inhibitory regulation of NMDA-induced dopamine release in striosome- and matrix-enriched areas of the rat striatum. *Journal of Neuroscience*, *14*(4), 2435–2443.

Kreek, M. J., Bart, G., Lilly, C., LaForge, K. S., & Nielsen, D. A. (2005). Pharmacogenetics and human molecular genetics of opiate and cocaine addictions and their treatments. *Pharmacological Reviews*, *57*(1), 1–26. http://dx.doi.org/10.1124/pr.57.1.1, Retrieved from PM:15734726.

Kreek, M. J., Levran, O., Reed, B., Schlussman, S. D., Zhou, Y., & Butelman, E. R. (2012). Opiate addiction and cocaine addiction: Underlying molecular neurobiology and genetics. *The Journal of Clinical Investigation*, *122*(10), 3387–3393. http://dx.doi.org/10.1172/JCI60390, Retrieved from PM:23023708.

Kreek, M. J., Nielsen, D. A., Butelman, E. R., & LaForge, K. S. (2005). Genetic influences on impulsivity, risk taking, stress responsivity and vulnerability to drug abuse and addiction. *Nature Neuroscience*, *8*(11), 1450–1457, Retrieved from PM:16251987.

Kuhn, S. A., van Landeghem, F. K., Zacharias, R., Farber, K., Rappert, A., Pavlovic, S., et al. (2004). Microglia express GABA(B) receptors to modulate interleukin release. *Molecular and Cellular Neurosciences*, *25*(2), 312–322. http://dx.doi.org/10.1016/j.mcn.2003.10.023, Retrieved from PM:15019947.

Kumar, A., Manna, S. K., Dhawan, S., & Aggarwal, B. B. (1998). HIV-Tat protein activates c-Jun N-terminal kinase and activator protein- 1. *Journal of Immunology*, *161*(2), 776–781.

Kutsch, O., Oh, J., Nath, A., & Benveniste, E. N. (2000). Induction of the chemokines interleukin-8 and IP-10 by human immunodeficiency virus type 1 tat in astrocytes. *Journal of Virology*, *74*(19), 9214–9221.

Lachman, H. M., Papolos, D. F., Saito, T., Yu, Y. M., Szumlanski, C. L., & Weinshilboum, R. M. (1996). Human catechol-O-methyltransferase pharmacogenetics: Description of a functional polymorphism and its potential application to neuropsychiatric disorders. *Pharmacogenetics*, *6*(3), 243–250, Retrieved from PM:8807664.

Langford, D., & Masliah, E. (2001). Crosstalk between components of the blood brain barrier and cells of the CNS in microglial activation in AIDS. *Brain Pathology*, *11*(3), 306–312.

Larsen, K. E., Fon, E. A., Hastings, T. G., Edwards, R. H., & Sulzer, D. (2002). Methamphetamine-induced degeneration of dopaminergic neurons involves autophagy and upregulation of dopamine synthesis. *Journal of Neuroscience, 22*(20), 8951–8960, Retrieved from PM:12388602.

Lawrence, D. M., Durham, L. C., Schwartz, L., Seth, P., Maric, D., & Major, E. O. (2004). Human immunodeficiency virus type 1 infection of human brain-derived progenitor cells. *Journal of Virology, 78*(14), 7319–7328.

Lee, M. H., Amin, N. D., Venkatesan, A., Wang, T., Tyagi, R., Pant, H. C., et al. (2013). Impaired neurogenesis and neurite outgrowth in an HIV-gp120 transgenic model is reversed by exercise via BDNF production and Cdk5 regulation. *Journal of Neurovirology, 19*(5), 418–431. http://dx.doi.org/10.1007/s13365-013-0194-6, Retrieved from PM:23982957.

Lee, C. T., Chen, J., Hayashi, T., Tsai, S. Y., Sanchez, J. F., Errico, S. L., et al. (2008). A mechanism for the inhibition of neural progenitor cell proliferation by cocaine. *PLoS Medicine, 5*(6), e117. http://dx.doi.org/10.1371/journal.pmed.0050117, Retrieved from PM:18593214.

Lee, C. T., Chen, J., Worden, L. T., & Freed, W. J. (2011). Cocaine causes deficits in radial migration and alters the distribution of glutamate and GABA neurons in the developing rat cerebral cortex. *Synapse, 65*(1), 21–34. http://dx.doi.org/10.1002/syn.20814, Retrieved from PM:20506319.

Lee, C. T., Lehrmann, E., Hayashi, T., Amable, R., Tsai, S. Y., Chen, J., et al. (2009). Gene expression profiling reveals distinct cocaine-responsive genes in human fetal CNS cell types. *Journal of Addiction Medicine, 3*(4), 218–226. http://dx.doi.org/10.1097/ADM.0b013e318199d863, Retrieved from PM:20948987.

Lee, J. C., Mayer-Proschel, M., & Rao, M. S. (2000). Gliogenesis in the central nervous system. *Glia, 30*(2), 105–121.

Lee, B., Sharron, M., Blanpain, C., Doranz, B. J., Vakili, J., Setoh, P., et al. (1999). Epitope mapping of CCR5 reveals multiple conformational states and distinct but overlapping structures involved in chemokine and coreceptor function. *The Journal of Biological Chemistry, 274*(14), 9617–9626 Retrieved from PM:10092648.

Lee, M. H., Wang, T., Jang, M. H., Steiner, J., Haughey, N., Ming, G. L., et al. (2011). Rescue of adult hippocampal neurogenesis in a mouse model of HIV neurologic disease. *Neurobiology of Disease, 41*(3), 678–687. http://dx.doi.org/10.1016/j.nbd.2010.12.002, Retrieved from PM:21146610.

Leshner, A. I. (1997). Addiction is a brain disease, and it matters. *Science, 278*(5335), 45–47.

Levine, A. J., Sinsheimer, J. S., Bilder, R., Shapshak, P., & Singer, E. J. (2012). Functional polymorphisms in dopamine-related genes: Effect on neurocognitive functioning in HIV+ adults. *Journal of Clinical and Experimental Neuropsychology, 34*(1), 78–91. http://dx.doi.org/10.1080/13803395.2011.623118, Retrieved from PM:22082040.

Levison, S. W., Rothstein, R. P., Brazel, C. Y., Young, G. M., & Albrecht, P. J. (2000). Selective apoptosis within the rat subependymal zone: A plausible mechanism for determining which lineages develop from neural stem cells. *Developmental Neuroscience, 22*(1–2), 106–115. http://dx.doi.org/10.1159/000017432, Retrieved from PM:10657703.

Li, J., Bentsman, G., Potash, M. J., & Volsky, D. J. (2007). Human immunodeficiency virus type 1 efficiently binds to human fetal astrocytes and induces neuroinflammatory responses independent of infection. *BMC Neuroscience, 8*, 31, Retrieved from PM:17498309.

Li, T., Chen, C. K., Hu, X., Ball, D., Lin, S. K., Chen, W., et al. (2004). Association analysis of the DRD4 and COMT genes in methamphetamine abuse. *American Journal of Medical Genetics. Part B, Neuropsychiatric Genetics, 129B*(1), 120–124. http://dx.doi.org/10.1002/ajmg.b.30024, Retrieved from PM:15274053.

Li, Z., Gao, L., Liu, Q., Cao, C., Sun, X. L., Ding, J. H., et al. (2006). Aquaporin-4 knockout regulated cocaine-induced behavior and neurochemical changes in mice. *Neuroscience Letters, 403*(3), 294–298. http://dx.doi.org/10.1016/j.neulet.2006.05.004, Retrieved from PM:16797122.

Li, W., Li, G., Steiner, J., & Nath, A. (2009). Role of Tat protein in HIV neuropathogenesis. *Neurotoxicity Research, 16*(3), 205–220, Retrieved from PM:19526283.

Li, Y., Wang, X., Tian, S., Guo, C. J., Douglas, S. D., & Ho, W. Z. (2002). Methadone enhances human immunodeficiency virus infection of human immune cells. *The Journal of Infectious Diseases, 185*(1), 118–122, Retrieved from PM:11756991.

Liang, H., Wang, X., Chen, H., Song, L., Ye, L., Wang, S. H., et al. (2008). Methamphetamine enhances HIV infection of macrophages. *The American Journal of Pathology, 172*(6), 1617–1624. http://dx.doi.org/10.2353/ajpath.2008.070971, Retrieved from PM:18458095.

Liao, D., Grigoriants, O. O., Loh, H. H., & Law, P. Y. (2007). Agonist-dependent postsynaptic effects of opioids on miniature excitatory postsynaptic currents in cultured hippocampal neurons. *Journal of Neurophysiology, 97*(2), 1485–1494.

Liao, D., Grigoriants, O. O., Wang, W., Wiens, K., Loh, H. H., & Law, P. Y. (2007). Distinct effects of individual opioids on the morphology of spines depend upon the internalization of mu opioid receptors. *Molecular and Cellular Neurosciences, 35*(3), 456–469.

Liao, D., Lin, H., Law, P. Y., & Loh, H. H. (2005). Mu-opioid receptors modulate the stability of dendritic spines. *Proceedings of the National Academy of Sciences of the United States of America, 102*(5), 1725–1730.

Lin, M. T., & Beal, M. F. (2006). Mitochondrial dysfunction and oxidative stress in neurodegenerative diseases. *Nature, 443*(7113), 787–795, Retrieved from PM:17051205.

Liu, H., Chao, D., Nakayama, E. E., Taguchi, H., Goto, M., Xin, X., et al. (1999). Polymorphism in RANTES chemokine promoter affects HIV-1 disease progression. *Proceedings of the National Academy of Sciences of the United States of America, 96*(8), 4581–4585, Retrieved from PM:10200305.

Liu, Y., Liu, H., Kim, B. O., Gattone, V. H., Li, J., Nath, A., et al. (2004). CD4-independent infection of astrocytes by human immunodeficiency virus type 1: Requirement for the human mannose receptor. *Journal of Virology, 78*(8), 4120–4133.

Liu, Y., Wong, T. P., Aarts, M., Rooyakkers, A., Liu, L., Lai, T. W., et al. (2007). NMDA receptor subunits have differential roles in mediating excitotoxic neuronal death both in vitro and in vivo. *Journal of Neuroscience, 27*(11), 2846–2857.

Low, K. G., & Melner, M. H. (1990). Expression of high levels of proenkephalin in an isolated glial cell type: Inhibition by gamma-interferon. *Annals of the New York Academy of Sciences, 594*, 475–478.

Lu, J., Esposito, G., Scuderi, C., Steardo, L., Delli-Bovi, L. C., Hecht, J. L., et al. (2011). S100B and APP promote a gliocentric shift and impaired neurogenesis in Down syndrome neural progenitors. *PLoS One, 6*(7), e22126. http://dx.doi.org/10.1371/journal.pone.0022126, Retrieved from PM:21779383.

Lu, Z., Xu, J., Xu, M., Pasternak, G. W., & Pan, Y. X. (2014). Morphine regulates expression of mu-opioid receptor MOR-1A, an intron-retention carboxyl terminal splice variant of the mu-opioid receptor (*OPRM1*) gene via miR-103/miR-107. *Molecular Pharmacology, 85*(2), 368–380. http://dx.doi.org/10.1124/mol.113.089292, Retrieved from PM:24302561.

Ma, Q., Jones, D., Borghesani, P. R., Segal, R. A., Nagasawa, T., Kishimoto, T., et al. (1998). Impaired B-lymphopoiesis, myelopoiesis, and derailed cerebellar neuron migration in CXCR4- and SDF-1-deficient mice. *Proceedings of the National Academy of Sciences of the United States of America, 95*(16), 9448–9453.

MacArthur, R. D., & Novak, R. M. (2008). Reviews of anti-infective agents: Maraviroc: The first of a new class of antiretroviral agents. *Clinical Infectious Diseases, 47*(2), 236–241. http://dx.doi.org/10.1086/589289, Retrieved from PM:18532888.

Mahajan, S. D., Aalinkeel, R., Reynolds, J. L., Nair, B. B., Fernandez, S. F., Schwartz, S. A., et al. (2005). Morphine exacerbates HIV-1 viral protein gp120 induced modulation of chemokine gene expression in U373 astrocytoma cells. *Current HIV Research*, *3*(3), 277–288.

Mahajan, S. D., Schwartz, S. A., Aalinkeel, R., Chawda, R. P., Sykes, D. E., & Nair, M. P. (2005). Morphine modulates chemokine gene regulation in normal human astrocytes. *Clinical Immunology*, *115*(3), 323–332.

Mahajan, S. D., Schwartz, S. A., Shanahan, T. C., Chawda, R. P., & Nair, M. P. (2002). Morphine regulates gene expression of alpha- and beta-chemokines and their receptors on astroglial cells via the opioid mu receptor. *Journal of Immunology*, *169*(7), 3589–3599.

Mahanta, J., Borkakoty, B., Das, H. K., & Chelleng, P. K. (2009). The risk of HIV and HCV infections among injection drug users in northeast India. *AIDS Care*, *21*(11), 1420–1424. http://dx.doi.org/10.1080/09540120902862584, Retrieved from PM:20024719.

Majumdar, S., Grinnell, S., Le Rouzic, V., Burgman, M., Polikar, L., Ansonoff, M., et al. (2011). Truncated G protein-coupled mu opioid receptor MOR-1 splice variants are targets for highly potent opioid analgesics lacking side effects. *Proceedings of the National Academy of Sciences of the United States of America*, *108*(49), 19778–19783, Retrieved from PM:22106286.

Malik, S., Saha, R., & Seth, P. (2014). Involvement of extracellular signal-regulated kinase (ERK1/2)-p53-p21 axis in mediating neural stem/progenitor cell cycle arrest in co-morbid HIV-drug abuse exposure. *Journal of Neuroimmune Pharmacology*, *9*(3), 340–353. http://dx.doi.org/10.1007/s11481-014-9523-7, Retrieved from PM:24469921.

Mandyam, C. D., Norris, R. D., & Eisch, A. J. (2004). Chronic morphine induces premature mitosis of proliferating cells in the adult mouse subgranular zone. *Journal of Neuroscience Research*, *76*(6), 783–794.

Mandyam, C. D., Wee, S., Crawford, E. F., Eisch, A. J., Richardson, H. N., & Koob, G. F. (2008). Varied access to intravenous methamphetamine self-administration differentially alters adult hippocampal neurogenesis. *Biological Psychiatry*, *64*(11), 958–965. http://dx.doi.org/10.1016/j.biopsych.2008.04.010, Retrieved from PM:18490002.

Mandyam, C. D., Wee, S., Eisch, A. J., Richardson, H. N., & Koob, G. F. (2007). Methamphetamine self-administration and voluntary exercise have opposing effects on medial prefrontal cortex gliogenesis. *Journal of Neuroscience*, *27*(42), 11442–11450. http://dx.doi.org/10.1523/JNEUROSCI.2505-07.2007, Retrieved from PM:17942739.

Mangeat, B., Turelli, P., Caron, G., Friedli, M., Perrin, L., & Trono, D. (2003). Broad antiretroviral defence by human APOBEC3G through lethal editing of nascent reverse transcripts. *Nature*, *424*(6944), 99–103. http://dx.doi.org/10.1038/nature01709, Retrieved from PM:12808466.

Manini, A. F., Jacobs, M. M., Vlahov, D., & Hurd, Y. L. (2013). Opioid receptor polymorphism A118G associated with clinical severity in a drug overdose population. *Journal of Medical Toxicology*, *9*(2), 148–154. http://dx.doi.org/10.1007/s13181-012-0286-3, Retrieved from PM:23318993.

Mao, J., Sung, B., Ji, R. R., & Lim, G. (2002). Chronic morphine induces downregulation of spinal glutamate transporters: Implications in morphine tolerance and abnormal pain sensitivity. *Journal of Neuroscience*, *22*(18), 8312–8323.

Mao, L., & Wang, J. Q. (2001). Gliogenesis in the striatum of the adult rat: Alteration in neural progenitor population after psychostimulant exposure. *Developmental Brain Research*, *130*(1), 41–51. http://dx.doi.org/10.1016/S0165-3806(01)00195-X, Retrieved from PM:11557092.

Maragakis, N. J., & Rothstein, J. D. (2006). Mechanisms of disease: Astrocytes in neurodegenerative disease. *Nature Clinical Practice. Neurology*, *2*(12), 679–689. http://dx.doi.org/10.1038/ncpneuro0355, Retrieved from PM:17117171.

Maragos, W. F., Young, K. L., Turchan, J. T., Guseva, M., Pauly, J. R., Nath, A., et al. (2002). Human immunodeficiency virus-1 Tat protein and methamphetamine interact synergistically to impair striatal dopaminergic function. *Journal of Neurochemistry*, *83*(4), 955–963.

Martineau, M. (2013). Gliotransmission: Focus on exocytotic release of L-glutamate and D-serine from astrocytes. *Biochemical Society Transactions*, *41*(6), 1557–1561. http://dx.doi.org/10.1042/BST20130195, Retrieved from PM:24256254.

Masliah, E., Heaton, R. K., Marcotte, T. D., Ellis, R. J., Wiley, C. A., Mallory, M., et al. (1997). Dendritic injury is a pathological substrate for human immunodeficiency virus-related cognitive disorders. HNRC Group. The HIV Neurobehavioral Research Center. *Annals of Neurology*, *42*(6), 963–972.

Mathews, J. L., Smrcka, A. V., & Bidlack, J. M. (2008). A novel Gbetagamma-subunit inhibitor selectively modulates mu-opioid-dependent antinociception and attenuates acute morphine-induced antinociceptive tolerance and dependence. *Journal of Neuroscience*, *28*(47), 12183–12189, Retrieved from PM:19020012.

Matsumoto, R. R., Nguyen, L., Kaushal, N., & Robson, M. J. (2014). Sigma (sigma) receptors as potential therapeutic targets to mitigate psychostimulant effects. *Advances in Pharmacology*, *69*, 323–386. http://dx.doi.org/10.1016/B978-0-12-420118-7.00009-3, Retrieved from PM:24484982.

Mattson, M. P., Haughey, N. J., & Nath, A. (2005). Cell death in HIV dementia. *Cell Death and Differentiation*, *12*(Suppl 1), 893–904.

Maurice, T., & Su, T. P. (2009). The pharmacology of sigma-1 receptors. *Pharmacology & Therapeutics*, *124*(2), 195–206. http://dx.doi.org/10.1016/j.pharmthera.2009.07.001, Retrieved from PM:19619582.

Mayne, M., Holden, C. P., Nath, A., & Geiger, J. D. (2000). Release of calcium from inositol 1,4,5-trisphosphate receptor-regulated stores by HIV-1 Tat regulates TNF-alpha production in human macrophages. *Journal of Immunology*, *164*(12), 6538–6542.

McCann, U. D., Wong, D. F., Yokoi, F., Villemagne, V., Dannals, R. F., & Ricaurte, G. A. (1998). Reduced striatal dopamine transporter density in abstinent methamphetamine and methcathinone users: Evidence from positron emission tomography studies with [11C]WIN-35,428. *Journal of Neuroscience*, *18*(20), 8417–8422, Retrieved from PM:9763484.

McCoy, M. K., & Tansey, M. G. (2008). TNF signaling inhibition in the CNS: Implications for normal brain function and neurodegenerative disease. *Journal of Neuroinflammation*, *5*, 45. http://dx.doi.org/10.1186/1742-2094-5-45, Retrieved from PM:18925972.

McDermott, D. H., Beecroft, M. J., Kleeberger, C. A., Al-Sharif, F. M., Ollier, W. E., Zimmerman, P. A., et al. (2000). Chemokine RANTES promoter polymorphism affects risk of both HIV infection and disease progression in the Multicenter AIDS Cohort Study. *AIDS*, *14*(17), 2671–2678.

McQuiston, A. R. (2007). Effects of mu-opioid receptor modulation on GABAB receptor synaptic function in hippocampal CA1. *Journal of Neurophysiology*, *97*(3), 2301–2311, Retrieved from PM:17215502.

Meijerink, H., Wisaksana, R., Iskandar, S., den, H. M., van der Ven, A. J., Alisjahbana, B., et al. (2014). Injecting drug use is associated with a more rapid CD4 cell decline among treatment naive HIV-positive patients in Indonesia. *Journal of the International AIDS Society*. *17*(1). http://dx.doi.org/10.7448/IAS.17.1.18844 (PMC), Retrieved from PM:24388495.

Messam, C. A., & Major, E. O. (2000). Stages of restricted HIV-1 infection in astrocyte cultures derived from human fetal brain tissue. *Journal of Neurovirology*, *6*(Suppl. 1), S90–S94, Retrieved from PM:10871771.

Messier, B., Leblond, C. P., & Smart, I. (1958). Presence of DNA synthesis and mitosis in the brain of young adult mice. *Experimental Cell Research*, *14*(1), 224–226. http://dx.doi.org/10.1016/0014-4827(58)90235-0, Retrieved from PM:13512326.

Meyer, V. J., Rubin, L. H., Martin, E., Weber, K. M., Cohen, M. H., Golub, E. T., et al. (2013). HIV and recent illicit drug use interact to affect verbal memory in women. *Journal of Acquired Immune Deficiency Syndromes*, *63*(1), 67–76, Retrieved from PM:23392462.

Meyer, U. A., & Zanger, U. M. (1997). Molecular mechanisms of genetic polymorphisms of drug metabolism. *Annual Review of Pharmacology and Toxicology, 37*, 269–296. http://dx.doi.org/10.1146/annurev.pharmtox.37.1.269, Retrieved from PM:9131254.

Midde, N. M., Gomez, A. M., & Zhu, J. (2012). HIV-1 Tat protein decreases dopamine transporter cell surface expression and vesicular monoamine transporter-2 function in rat striatal synaptosomes. *Journal of Neuroimmune Pharmacology*, 7(3), 629–639. http://dx.doi.org/10.1007/s11481-012-9369-9, Retrieved from PM:22570010.

Midde, N. M., Huang, X., Gomez, A. M., Booze, R. M., Zhan, C. G., & Zhu, J. (2013). Mutation of tyrosine 470 of human dopamine transporter is critical for HIV-1 Tat-induced inhibition of dopamine transport and transporter conformational transitions. *Journal of Neuroimmune Pharmacology, 8*(4), 975–987. http://dx.doi.org/10.1007/s11481-013-9464-6, Retrieved from PM:23645138.

Miller, G. M. (2011). The emerging role of trace amine-associated receptor 1 in the functional regulation of monoamine transporters and dopaminergic activity. *Journal of Neurochemistry, 116*(2), 164–176. http://dx.doi.org/10.1111/j.1471-4159.2010.07109.x, Retrieved from PM:21073468.

Miller, G. W., Gainetdinov, R. R., Levey, A. I., & Caron, M. G. (1999). Dopamine transporters and neuronal injury. *Trends in Pharmacological Sciences, 20*(10), 424–429.

Mishra, M., Taneja, M., Malik, S., Khalique, H., & Seth, P. (2010). Human immunodeficiency virus type 1 Tat modulates proliferation and differentiation of human neural precursor cells: Implication in NeuroAIDS. *Journal of Neurovirology, 16*(5), 355–367. http://dx.doi.org/10.3109/13550284.2010.513028, Retrieved from PM:20839920.

Miura, T., Brockman, M. A., Brumme, C. J., Brumme, Z. L., Carlson, J. M., Pereyra, F., et al. (2008). Genetic characterization of human immunodeficiency virus type 1 in elite controllers: Lack of gross genetic defects or common amino acid changes. *Journal of Virology, 82*(17), 8422–8430. http://dx.doi.org/10.1128/JVI.00535-08, Retrieved from PM:18562530.

Miyatake, M., Narita, M., Shibasaki, M., Nakamura, A., & Suzuki, T. (2005). Glutamatergic neurotransmission and protein kinase C play a role in neuron-glia communication during the development of methamphetamine-induced psychological dependence. *The European Journal of Neuroscience, 22*(6), 1476–1488.

Molofsky, A. V., Krencik, R., Ullian, E. M., Tsai, H. H., Deneen, B., Richardson, W. D., et al. (2012). Astrocytes and disease: A neurodevelopmental perspective. *Genes and Development, 26*(9), 891–907.

Monje, M. L., Toda, H., & Palmer, T. D. (2003). Inflammatory blockade restores adult hippocampal neurogenesis. *Science, 302*(5651), 1760–1765.

Montana, V., Malarkey, E. B., Verderio, C., Matteoli, M., & Parpura, V. (2006). Vesicular transmitter release from astrocytes. *Glia, 54*(7), 700–715. http://dx.doi.org/10.1002/glia.20367, Retrieved from PM:17006898.

Moore, J. P., Kitchen, S. G., Pugach, P., & Zack, J. A. (2004). The CCR5 and CXCR4 coreceptors—Central to understanding the transmission and pathogenesis of human immunodeficiency virus type 1 infection. *AIDS Research and Human Retroviruses, 20*(1), 111–126. http://dx.doi.org/10.1089/088922204322749567, Retrieved from PM:15000703.

Moreno-Lopez, B., Romero-Grimaldi, C., Noval, J. A., Murillo-Carretero, M., Matarredona, E. R., & Estrada, C. (2004). Nitric oxide is a physiological inhibitor of neurogenesis in the adult mouse subventricular zone and olfactory bulb. *Journal of Neuroscience, 24*(1), 85–95. http://dx.doi.org/10.1523/JNEUROSCI.1574-03.2004, Retrieved from PM:14715941.

Morgan, E. E., Woods, S. P., Letendre, S. L., Franklin, D. R., Bloss, C., Goate, A., et al. (2013). Apolipoprotein E4 genotype does not increase risk of HIV-associated neurocognitive disorders. *Journal of Neurovirology*, *19*(2), 150–156. http://dx.doi.org/10.1007/s13365-013-0152-3, Retrieved from PM:23408335.

Mulherin, S. A., O'Brien, T. R., Ioannidis, J. P., Goedert, J. J., Buchbinder, S. P., Coutinho, R. A., et al. (2003). Effects of CCR5-Delta32 and CCR2-64I alleles on HIV-1 disease progression: The protection varies with duration of infection. *AIDS*, *17*(3), 377–387. http://dx.doi.org/10.1097/01.aids.0000050783.28043.3e, Retrieved from PM:12556692.

Nadeau, S., & Rivest, S. (2000). Role of microglial-derived tumor necrosis factor in mediating CD14 transcription and nuclear factor kappa B activity in the brain during endotoxemia. *Journal of Neuroscience*, *20*(9), 3456–3468, Retrieved from PM:10777809.

Nakagawa, S., Castro, V., & Toborek, M. (2012). Infection of human pericytes by HIV-1 disrupts the integrity of the blood-brain barrier. *Journal of Cellular and Molecular Medicine*, *16*(12), 2950–2957. http://dx.doi.org/10.1111/j.1582-4934.2012.01622.x, Retrieved from PM:22947176.

Nakajima, A., Yamada, K., Nagai, T., Uchiyama, T., Miyamoto, Y., Mamiya, T., et al. (2004). Role of tumor necrosis factor-alpha in methamphetamine-induced drug dependence and neurotoxicity. *Journal of Neuroscience*, *24*(9), 2212–2225, Retrieved from PM:14999072.

Nakayama, M., Koyama, T., & Yamashita, I. (1993). Long-lasting decrease in dopamine uptake sites following repeated administration of methamphetamine in the rat striatum. *Brain Research*, *601*(1–2), 209–212, Retrieved from PM:8431767.

Napier, T. C., Chen, L., Kashanchi, F., & Hu, X. T. (2014). Repeated cocaine treatment enhances HIV-1 Tat-induced cortical excitability via over-activation of L-type calcium channels. *Journal of Neuroimmune Pharmacology*, *9*(3), 354–368. http://dx.doi.org/10.1007/s11481-014-9524-6, Retrieved from PM:24567038.

Narasipura, S. D., Henderson, L. J., Fu, S. W., Chen, L., Kashanchi, F., & Al-Harthi, L. (2012). Role of beta-catenin and TCF/LEF family members in transcriptional activity of HIV in astrocytes. *Journal of Virology*, *86*(4), 1911–1921.

Narasipura, S. D., Kim, S., & Al-Harthi, L. (2014). Epigenetic regulation of HIV-1 latency in astrocytes. *Journal of Virology*, *88*(5), 3031–3038. http://dx.doi.org/10.1128/JVI.03333-13, Retrieved from PM:24352441.

Narayana, P. A., Herrera, J. J., Bockhorst, K. H., Esparza-Coss, E., Xia, Y., Steinberg, J. L., et al. (2014). Chronic cocaine administration causes extensive white matter damage in brain: Diffusion tensor imaging and immunohistochemistry studies. *Psychiatry Research*, *221*(3), 220–230. http://dx.doi.org/10.1016/j.pscychresns.2014.01.005, Retrieved from PM:24507117.

Narayanan, S., Mesangeau, C., Poupaert, J. H., & McCurdy, C. R. (2011). Sigma receptors and cocaine abuse. *Current Topics in Medicinal Chemistry*, *11*(9), 1128–1150. http://dx.doi.org/10.2174/156802611795371323, Retrieved from PM:21050176.

Nath, A. (1999). Pathobiology of human immunodeficiency virus dementia. *Seminars in Neurology*, *19*(2), 113–127.

Nath, A. (2010). Human immunodeficiency virus-associated neurocognitive disorder: Pathophysiology in relation to drug addiction. *Annals of the New York Academy of Sciences*, *1187*, 122–128, Retrieved from PM:20201849.

Nath, A., Conant, K., Chen, P., Scott, C., & Major, E. O. (1999). Transient exposure to HIV-1 Tat protein results in cytokine production in macrophages and astrocytes: A hit and run phenomenon. *The Journal of Biological Chemistry*, *274*, 17098–17102.

Nath, A., & Geiger, J. (1998). Neurobiological aspects of human immunodeficiency virus infection: Neurotoxic mechanisms. *Progress in Neurobiology*, *54*(1), 19–33.

Nath, A., Hauser, K. F., Wojna, V., Booze, R. M., Maragos, W., Prendergast, M., et al. (2002). Molecular basis for interactions of HIV and drugs of abuse. *Journal of Acquired Immune Deficiency Syndromes, 31*(Suppl. 2), S62–S69.

Nath, A., Jones, M., Maragos, W., Booze, R., Mactutus, C., Bell, J., et al. (2000). Neurotoxicity and dysfunction of dopamine systems associated with AIDS dementia. *Journal of Psychopharmacology, 14*(3), 222–227.

Nath, A., Maragos, W. F., Avison, M. J., Schmitt, F. A., & Berger, J. R. (2001). Acceleration of HIV dementia with methamphetamine and cocaine. *Journal of Neurovirology*, 7(1), 66–71.

Nebert, D. W., McKinnon, R. A., & Puga, A. (1996). Human drug-metabolizing enzyme polymorphisms: Effects on risk of toxicity and cancer. *DNA and Cell Biology, 15*(4), 273–280, Retrieved from PM:8639263.

Nickell, J. R., Siripurapu, K. B., Vartak, A., Crooks, P. A., & Dwoskin, L. P. (2014). The vesicular monoamine transporter-2: An important pharmacological target for the discovery of novel therapeutics to treat methamphetamine abuse. *Advances in Pharmacology, 69*, 71–106. http://dx.doi.org/10.1016/B978-0-12-420118-7.00002-0, Retrieved from PM:24484975.

Ninkovic, J., & Gotz, M. (2013). Fate specification in the adult brain-lessons for eliciting neurogenesis from glial cells. *Bioessays, 35*(3), 242–252. http://dx.doi.org/10.1002/bies.201200108, Retrieved from PM:23335359.

Noda, M., Nakanishi, H., Nabekura, J., & Akaike, N. (2000). AMPA-kainate subtypes of glutamate receptor in rat cerebral microglia. *Journal of Neuroscience, 20*(1), 251–258.

Ohira, K., Furuta, T., Hioki, H., Nakamura, K. C., Kuramoto, E., Tanaka, Y., et al. (2010). Ischemia-induced neurogenesis of neocortical layer 1 progenitor cells. *Nature Neuroscience, 13*(2), 173–179. http://dx.doi.org/10.1038/nn.2473, Retrieved from PM:20037576.

Okamoto, S., Kang, Y. J., Brechtel, C. W., Siviglia, E., Russo, R., Clemente, A., et al. (2007). HIV/gp120 decreases adult neural progenitor cell proliferation via checkpoint kinase-mediated cell-cycle withdrawal and G1 arrest. *Cell Stem Cell, 1*(2), 230–236. http://dx.doi.org/10.1016/j.stem.2007.07.010, Retrieved from PM:18371353.

Ozawa, T., Nakagawa, T., Shige, K., Minami, M., & Satoh, M. (2001). Changes in the expression of glial glutamate transporters in the rat brain accompanied with morphine dependence and naloxone-precipitated withdrawal. *Brain Research, 905*(1–2), 254–258.

Pan, Y. X., Xu, J., Bolan, E., Moskowitz, H. S., Xu, M., & Pasternak, G. W. (2005). Identification of four novel exon 5 splice variants of the mouse mu-opioid receptor gene: Functional consequences of C-terminal splicing. *Molecular Pharmacology, 68*(3), 866–875, Retrieved from PM:15939800.

Paolicelli, R. C., Bolasco, G., Pagani, F., Maggi, L., Scianni, M., Panzanelli, P., et al. (2011). Synaptic pruning by microglia is necessary for normal brain development. *Science, 333*(6048), 1456–1458.

Parent, J. M., Yu, T. W., Leibowitz, R. T., Geschwind, D. H., Sloviter, R. S., & Lowenstein, D. H. (1997). Dentate granule cell neurogenesis is increased by seizures and contributes to aberrant network reorganization in the adult rat hippocampus. *Journal of Neuroscience, 17*(10), 3727–3738, Retrieved from PM:9133393.

Parpura, V., Basarsky, T. A., Liu, F., Jeftinija, S., & Haydon, P. G. (1994). Glutamate-mediated astrocyte-neuron signalling. *Nature, 369*, 744–747.

Parpura, V., & Zorec, R. (2010). Gliotransmission: Exocytotic release from astrocytes. *Brain Research Reviews, 63*(1–2), 83–92. http://dx.doi.org/10.1016/j.brainresrev.2009.11.008, Retrieved from PM:19948188.

Pascual, O., Casper, K. B., Kubera, C., Zhang, J., Revilla-Sanchez, R., Sul, J. Y., et al. (2005). Astrocytic purinergic signaling coordinates synaptic networks. *Science, 310*(5745), 113–116, Retrieved from PM:16210541.

Patel, D. A., Booze, R. M., & Mactutus, C. F. (2012). Prenatal cocaine exposure alters progenitor cell markers in the subventricular zone of the adult rat brain. *International Journal of Developmental Neuroscience*, *30*(1), 1–9. http://dx.doi.org/10.1016/j.ijdevneu.2011.11.001, Retrieved from PM:22119286.

Patel, J. P., Sengupta, R., Bardi, G., Khan, M. Z., Mullen-Przeworski, A., & Meucci, O. (2006). Modulation of neuronal CXCR4 by the μ-opioid agonist DAMGO. *Journal of Neurovirology*, *12*(6), 492–500.

Pencea, V., Bingaman, K. D., Freedman, L. J., & Luskin, M. B. (2001). Neurogenesis in the subventricular zone and rostral migratory stream of the neonatal and adult primate forebrain. *Experimental Neurology*, *172*(1), 1–16. http://dx.doi.org/10.1006/exnr.2001.7768, Retrieved from PM:11681836.

Peng, H., Sun, L., Jia, B., Lan, X., Zhu, B., Wu, Y., et al. (2011). HIV-1-infected and immune-activated macrophages induce astrocytic differentiation of human cortical neural progenitor cells via the STAT3 pathway. *PLoS One*, *6*(5), e19439. http://dx.doi.org/10.1371/journal.pone.0019439, Retrieved from PM:21637744.

Peng, H., Whitney, N., Wu, Y., Tian, C., Dou, H., Zhou, Y., et al. (2008). HIV-1-infected and/or immune-activated macrophage-secreted TNF-alpha affects human fetal cortical neural progenitor cell proliferation and differentiation. *Glia*, *56*(8), 903–916. http://dx.doi.org/10.1002/glia.20665, Retrieved from PM:18383342.

Perea, G., Navarrete, M., & Araque, A. (2009). Tripartite synapses: Astrocytes process and control synaptic information. *Trends in Neurosciences*, *32*(8), 421–431. http://dx.doi.org/10.1016/j.tins.2009.05.001, Retrieved from PM:19615761.

Pereira, F. C., Cunha-Oliveira, T., Viana, S. D., Travassos, A. S., Nunes, S., Silva, C., et al. (2012). Disruption of striatal glutamatergic/GABAergic homeostasis following acute methamphetamine in mice. *Neurotoxicology and Teratology*, *34*(5), 522–529.

Perez-Casanova, A., Husain, K., Noel, R. J., Jr., Rivera-Amill, V., & Kumar, A. (2008). Interaction of SIV/SHIV infection and morphine on plasma oxidant/antioxidant balance in macaque. *Molecular and Cellular Biochemistry*, *308*(1–2), 169–175, Retrieved from PM:17934700.

Persidsky, Y., Buttini, M., Limoges, J., Bock, P., & Gendelman, H. E. (1997). An analysis of HIV-1-associated inflammatory products in brain tissue of humans and SCID mice with HIV-1 encephalitis. *Journal of Neurovirology*, *3*(6), 401–416.

Persidsky, Y., & Gendelman, H. E. (2003). Mononuclear phagocyte immunity and the neuropathogenesis of HIV-1 infection. *Journal of Leukocyte Biology*, *74*(5), 691–701.

Persson, A. I., Thorlin, T., Bull, C., & Eriksson, P. S. (2003). Opioid-induced proliferation through the MAPK pathway in cultures of adult hippocampal progenitors. *Molecular and Cellular Neurosciences*, *23*(3), 360–372.

Persson, A. I., Thorlin, T., Bull, C., Zarnegar, P., Ekman, R., Terenius, L., et al. (2003). Mu- and delta-opioid receptor antagonists decrease proliferation and increase neurogenesis in cultures of rat adult hippocampal progenitors. *The European Journal of Neuroscience*, *17*(6), 1159–1172.

Peterson, P. K., Gekker, G., Hu, S., Cabral, G., & Lokensgard, J. R. (2004). Cannabinoids and morphine differentially affect HIV-1 expression in CD4(+) lymphocyte and microglial cell cultures. *Journal of Neuroimmunology*, *147*(1–2), 123–126.

Peterson, P. K., Gekker, G., Hu, S., Lokensgard, J., Portoghese, P. S., & Chao, C. C. (1999). Endomorphin-1 potentiates HIV-1 expression in human brain cell cultures: Implication of an atypical mu-opioid receptor. *Neuropharmacology*, *38*(2), 273–278.

Peterson, P. K., Molitor, T. W., & Chao, C. C. (1998). The opioid-cytokine connection. *Journal of Neuroimmunology*, *83*(1–2), 63–69.

Pettit, A. S., Desroches, R., & Bennett, S. A. (2012). The opiate analgesic buprenorphine decreases proliferation of adult hippocampal neuroblasts and increases survival of their progeny. *Neuroscience, 200*, 211–222. http://dx.doi.org/10.1016/j.neuroscience.2011.10.039, Retrieved from PM:22079577.

Pierce, R. C., & Wolf, M. E. (2013). Psychostimulant-induced neuroadaptations in nucleus accumbens AMPA receptor transmission. *Cold Spring Harbor Perspectives in Medicine, 3*(2), a012021, Retrieved from PM:23232118.

Pitcher, J., Abt, A., Myers, J., Han, R., Snyder, M., Graziano, A., et al. (2014). Neuronal ferritin heavy chain and drug abuse affect HIV-associated cognitive dysfunction. *The Journal of Clinical Investigation, 124*(2), 656–669. http://dx.doi.org/10.1172/JCI70090, Retrieved from PM:24401274.

Pocock, J. M., & Kettenmann, H. (2007). Neurotransmitter receptors on microglia. *Trends in Neurosciences, 30*(10), 527–535.

Podhaizer, E. M., Zou, S., Fitting, S., Samano, K. L., El-Hage, N., Knapp, P. E., et al. (2012). Morphine and gp120 toxic interactions in striatal neurons are dependent on HIV-1 strain. *Journal of Neuroimmune Pharmacology*, 7(4), 877–891. Retrieved from, http://www.springerlink.com/content/6020qx40u36uk8lw/export-citation/.

Prada, I., Furlan, R., Matteoli, M., & Verderio, C. (2013). Classical and unconventional pathways of vesicular release in microglia. *Glia, 61*(7), 1003–1017. http://dx.doi.org/10.1002/glia.22497, Retrieved from PM:23625857.

Proudnikov, D., Randesi, M., Levran, O., Yuferov, V., Crystal, H., Ho, A., et al. (2013). Polymorphisms of the kappa opioid receptor and prodynorphin genes: HIV risk and HIV natural history. *Journal of Acquired Immune Deficiency Syndromes, 63*(1), 17–26. http://dx.doi.org/10.1097/QAI.0b013e318285cd0c, Retrieved from PM:23392455.

Proudnikov, D., Randesi, M., Orna, L., Crystal, H., Dorn, M., Ott, J., et al. (2012). Association of polymorphisms of the mu opioid receptor gene with the severity of HIV infection and response to HIV treatment. *The Journal of Infectious Diseases, 205*(11), 1745–1756, Retrieved from PM:22457278.

Qi, Z., Miller, G. W., & Voit, E. O. (2008). Computational systems analysis of dopamine metabolism. *PLoS One, 3*(6), e2444. http://dx.doi.org/10.1371/journal.pone.0002444, Retrieved from PM:18568086.

Quasney, M. W., Zhang, Q., Sargent, S., Mynatt, M., Glass, J., & McArthur, J. (2001). Increased frequency of the tumor necrosis factor-alpha-308 A allele in adults with human immunodeficiency virus dementia. *Annals of Neurology, 50*(2), 157–162, Retrieved from PM:11506397.

Quinton, M. S., & Yamamoto, B. K. (2006). Causes and consequences of methamphetamine and MDMA toxicity. *The AAPS Journal, 8*(2), E337–E347. http://dx.doi.org/10.1208/aapsj080238, Retrieved from PM:16796384.

Ramage, S. N., Anthony, I. C., Carnie, F. W., Busuttil, A., Robertson, R., & Bell, J. E. (2005). Hyperphosphorylated tau and amyloid precursor protein deposition is increased in the brains of young drug abusers. *Neuropathology and Applied Neurobiology, 31*(4), 439–448, Retrieved from PM:16008828.

Ransohoff, R. M., & Stevens, B. (2011). Neuroscience. How many cell types does it take to wire a brain? *Science, 333*(6048), 1391–1392.

Rash, J. E., Yasumura, T., Hudson, C. S., Agre, P., & Nielsen, S. (1998). Direct immunogold labeling of aquaporin-4 in square arrays of astrocyte and ependymocyte plasma membranes in rat brain and spinal cord. *Proceedings of the National Academy of Sciences of the United States of America, 95*(20), 11981–11986, Retrieved from PM:9751776.

Reddy, P. V., Pilakka-Kanthikeel, S., Saxena, S. K., Saiyed, Z., & Nair, M. P. (2012). Interactive effects of morphine on HIV infection: Role in HIV-associated neurocognitive disorder. *AIDS Research and Treatment, 2012*, 953678. http://dx.doi.org/10.1155/2012/953678, Retrieved from PM:22666564.

Reese, E. A., Bunzow, J. R., Arttamangkul, S., Sonders, M. S., & Grandy, D. K. (2007). Trace amine-associated receptor 1 displays species-dependent stereoselectivity for isomers of methamphetamine, amphetamine, and para-hydroxyamphetamine. *The Journal of Pharmacology and Experimental Therapeutics*, *321*(1), 178–186. http://dx.doi.org/10.1124/jpet.106.115402, Retrieved from PM:17218486.

Regan, P. M., Dave, R. S., Datta, P. K., & Khalili, K. (2012). Epigenetics of micro-opioid receptors: Intersection with HIV-1 infection of the central nervous system. *Journal of Cellular Physiology*, *227*(7), 2832–2841.

Reissner, K. J., Brown, R. M., Spencer, S., Tran, P. K., Thomas, C. A., & Kalivas, P. W. (2014). Chronic administration of the methylxanthine propentofylline impairs reinstatement to cocaine by a GLT-1-dependent mechanism. *Neuropsychopharmacology*, *39*(2), 499–506. http://dx.doi.org/10.1038/npp.2013.223, Retrieved from PM:23985782.

Revel, F. G., Moreau, J. L., Gainetdinov, R. R., Bradaia, A., Sotnikova, T. D., Mory, R., et al. (2011). TAAR1 activation modulates monoaminergic neurotransmission, preventing hyperdopaminergic and hypoglutamatergic activity. *Proceedings of the National Academy of Sciences of the United States of America*, *108*(20), 8485–8490. http://dx.doi.org/10.1073/pnas.1103029108, Retrieved from PM:21525407.

Reyes, M. G., Faraldi, F., Senseng, C. S., Flowers, C., & Fariello, R. (1991). Nigral degeneration in acquired immune deficiency syndrome (AIDS). *Acta Neuropathologica (Berlin)*, *82*(1), 39–44.

Rigato, P. O., Hong, M. A., Casseb, J., Ueda, M., de Castro, I., Benard, G., et al. (2008). Better CD4+ T cell recovery in Brazilian HIV-infected individuals under HAART due to cumulative carriage of SDF-1-3'A, CCR2-V64I, CCR5-D32 and CCR5-promoter 59029A/G polymorphisms. *Current HIV Research*, *6*(5), 466–473, Retrieved from PM:18855658.

Rippeth, J. D., Heaton, R. K., Carey, C. L., Marcotte, T. D., Moore, D. J., Gonzalez, R., et al. (2004). Methamphetamine dependence increases risk of neuropsychological impairment in HIV infected persons. *Journal of the International Neuropsychological Society*, *10*(1), 1–14. http://dx.doi.org/10.1017/S1355617704101021, Retrieved from PM:14751002.

Robinson, T. E., & Kolb, B. (1999). Morphine alters the structure of neurons in the nucleus accumbens and neocortex of rats. *Synapse*, *33*(2), 160–162.

Robinson, T. E., & Kolb, B. (2004). Structural plasticity associated with exposure to drugs of abuse. *Neuropharmacology*, *47*(Suppl. 1), 33–46, Retrieved from PM:15464124.

Robinson-Papp, J., Gelman, B. B., Grant, I., Singer, E., Gensler, G., & Morgello, S. (2012). Substance abuse increases the risk of neuropathy in an HIV-infected cohort. *Muscle & Nerve*, *45*(4), 471–476.

Rock, R. B., & Peterson, P. K. (2006). Microglia as a pharmacological target in infectious and inflammatory diseases of the brain. *Journal of Neuroimmune Pharmacology*, *1*(2), 117–126, Retrieved from PM:18040778.

Rogers, T. J., & Peterson, P. K. (2003). Opioid G protein-coupled receptors: Signals at the crossroads of inflammation. *Trends in Immunology*, *24*(3), 116–121.

Rola, R., Mizumatsu, S., Otsuka, S., Morhardt, D. R., Noble-Haeusslein, L. J., Fishman, K., et al. (2006). Alterations in hippocampal neurogenesis following traumatic brain injury in mice. *Experimental Neurology*, *202*(1), 189–199. http://dx.doi.org/10.1016/j.expneurol.2006.05.034, Retrieved from PM:16876159.

Romanko, M. J., Rola, R., Fike, J. R., Szele, F. G., Dizon, M. L., Felling, R. J., et al. (2004). Roles of the mammalian subventricular zone in cell replacement after brain injury. *Progress in Neurobiology*, *74*(2), 77–99. http://dx.doi.org/10.1016/j.pneurobio.2004.07.001, Retrieved from PM:15518954.

Roth, M. D., Whittaker, K. M., Choi, R., Tashkin, D. P., & Baldwin, G. C. (2005). Cocaine and sigma-1 receptors modulate HIV infection, chemokine receptors, and the HPA axis in the huPBL-SCID model. *Journal of Leukocyte Biology*, *78*(6), 1198–1203, Retrieved from PM:16204638.

Rothenaigner, I., Kramer, S., Ziegler, M., Wolff, H., Kleinschmidt, A., & Brack-Werner, R. (2007). Long-term HIV-1 infection of neural progenitor populations. *AIDS*, *21*(17), 2271–2281. http://dx.doi.org/10.1097/QAD.0b013e3282f12f27, Retrieved from PM:18090275.

Royal, W., III, Updike, M., Selnes, O. A., Proctor, T. V., Nance-Sproson, L., Solomon, L., et al. (1991). HIV-1 infection and nervous system abnormalities among a cohort of intravenous drug users. *Neurology*, *41*(12), 1905–1910, Retrieved from PM:1745346.

Rumbaugh, J. A., Li, G., Rothstein, J., & Nath, A. (2007). Ceftriaxone protects against the neurotoxicity of human immunodeficiency virus proteins. *Journal of Neurovirology*, *13*(2), 168–172. http://dx.doi.org/10.1080/13550280601178218, Retrieved from PM:17505985.

Ruzicka, B. B., & Akil, H. (1997). The interleukin-1beta-mediated regulation of proenkephalin and opioid receptor messenger RNA in primary astrocyte-enriched cultures. *Neuroscience*, *79*(2), 517–524.

Ruzicka, B. B., Fox, C. A., Thompson, R. C., Meng, F., Watson, S. J., & Akil, H. (1995). Primary astroglial cultures derived from several rat brain regions differentially express μ, δ and kappa opioid receptor mRNA. *Molecular Brain Research*, *34*, 209–220.

Sabo, J. K., Kilpatrick, T. J., & Cate, H. S. (2009). Effects of bone morphogenic proteins on neural precursor cells and regulation during central nervous system injury. *Neurosignals*, *17*(4), 255–264. http://dx.doi.org/10.1159/000231892, Retrieved from PM:19816062.

Saha, R. N., & Pahan, K. (2003). Tumor necrosis factor-alpha at the crossroads of neuronal life and death during HIV-associated dementia. *Journal of Neurochemistry*, *86*(5), 1057–1071.

Saijo, K., & Glass, C. K. (2011). Microglial cell origin and phenotypes in health and disease. *Nature Reviews. Immunology*, *11*(11), 775–787. http://dx.doi.org/10.1038/nri3086, Retrieved from PM:22025055.

Saito, Y., Sharer, L. R., Epstein, L. G., Michaels, J., Mintz, M., Louder, M., et al. (1994). Overexpression of nef as a marker for restricted HIV-1 infection of astrocytes in postmortem pediatric central nervous tissues. *Neurology*, *44*(3 Pt 1), 474–481.

Sanai, N., Nguyen, T., Ihrie, R. A., Mirzadeh, Z., Tsai, H. H., Wong, M., et al. (2011). Corridors of migrating neurons in the human brain and their decline during infancy. *Nature*, *478*(7369), 382–386. http://dx.doi.org/10.1038/nature10487, Retrieved from PM:21964341.

Santello, M., Cali, C., & Bezzi, P. (2012). Gliotransmission and the tripartite synapse. *Advances in Experimental Medicine and Biology*, *970*, 307–331. http://dx.doi.org/10.1007/978-3-7091-0932-8_14, Retrieved from PM:22351062.

Sarkar, K., Das, S. S., Pal, R., Bal, B., Madhusudan, P., & Chakraborti, S. (2010). HIV infection and host genetic mutation among injecting drug-users of northeastern states of India. *Journal of Health, Population, and Nutrition*, *28*(2), 130–136, Retrieved from PM:20411675.

Sattler, R., Xiong, Z., Lu, W. Y., MacDonald, J. F., & Tymianski, M. (2000). Distinct roles of synaptic and extrasynaptic NMDA receptors in excitotoxicity. *Journal of Neuroscience*, *20*(1), 22–33, Retrieved from PM:10627577.

Sawaya, B. E., Deshmane, S. L., Mukerjee, R., Fan, S., & Khalili, K. (2009). TNF alpha production in morphine-treated human neural cells is NF-κB-dependent. *Journal of Neuroimmune Pharmacology*, *4*(1), 140–149, Retrieved from PM:19023660.

Scheffel, J., Regen, T., Van, R. D., Seifert, S., Ribes, S., Nau, R., et al. (2012). Toll-like receptor activation reveals developmental reorganization and unmasks responder subsets of microglia. *Glia*, *60*(12), 1930–1943. http://dx.doi.org/10.1002/glia.22409, Retrieved from PM:22911652.

Scheller, C., Sopper, S., Jenuwein, M., Neuen-Jacob, E., Tatschner, T., Grunblatt, E., et al. (2005). Early impairment in dopaminergic neurotransmission in brains of SIV-infected rhesus monkeys due to microglia activation. *Journal of Neurochemistry*, *95*(2), 377–387.

Schnell, G., Joseph, S., Spudich, S., Price, R. W., & Swanstrom, R. (2011). HIV-1 replication in the central nervous system occurs in two distinct cell types. *PLoS Pathogens*, 7(10), e1002286.

Scholzen, T., & Gerdes, J. (2000). The Ki-67 protein: From the known and the unknown. *Journal of Cellular Physiology*, *182*(3), 311–322. http://dx.doi.org/10.1002/(SICI)1097-4652(200003)182:3<311::AID-JCP1>3.0.CO;2-9, Retrieved from PM:10653597.

Schuenke, K., & Gelman, B. B. (2003). Human microglial cell isolation from adult autopsy brain: Brain pH, regional variation, and infection with human immunodeficiency virus type 1. *Journal of Neurovirology*, *9*(3), 346–357, Retrieved from PM:12775418.

Schwartz, L., & Major, E. O. (2006). Neural progenitors and HIV-1-associated central nervous system disease in adults and children. *Current HIV Research*, *4*(3), 319–327, Retrieved from PM:16842084.

Seiden, L. S., & Ricaurte, G. A. (1987). Neurotoxicity of methamphetamine and related drugs. In H. Meltzer (Ed.), *Psychopharmacology: A third generation of progress* (pp. 359–366). New York: Raven.

Seilhean, D., Kobayashi, K., He, Y., Uchihara, T., Rosenblum, O., Katlama, C., et al. (1997). Tumor necrosis factor-alpha, microglia and astrocytes in AIDS dementia complex. *Acta Neuropathologica (Berlin)*, *93*(5), 508–517.

Shabalina, S. A., Zaykin, D. V., Gris, P., Ogurtsov, A. Y., Gauthier, J., Shibata, K., et al. (2009). Expansion of the human mu-opioid receptor gene architecture: Novel functional variants. *Human Molecular Genetics*, *18*(6), 1037–1051, Retrieved from PM:19103668.

Shao, Y., & McCarthy, K. D. (1994). Plasticity of astrocytes. *Glia*, *11*, 147–155.

Shao, Y., Porter, J. T., & McCarthy, K. D. (1994). Neuroligand receptor heterogeneity among astroglia. *Perspectives on Developmental Neurobiology*, *2*(3), 205–215, Retrieved from PM:7850353.

Shapshak, P., Duncan, R., Minagar, A., Rodriguez de la Vega, P., Stewart, R. V., & Goodkin, K. (2004). Elevated expression of IFN-gamma in the HIV-1 infected brain. *Frontiers in Bioscience*, *9*, 1073–1081.

Shapshak, P., Duncan, R., Torres-Munoz, J. E., Duran, E. M., Minagar, A., & Petito, C. K. (2004). Analytic approaches to differential gene expression in AIDS versus control brains. *Frontiers in Bioscience*, *9*, 2935–2946, Retrieved from PM:15353327.

Shapshak, P., Kangueane, P., Fujimura, R. K., Commins, D., Chiappelli, F., Singer, E., et al. (2011). Editorial neuroAIDS review. *AIDS*, *25*(2), 123–141. http://dx.doi.org/10.1097/QAD.0b013e328340fd42, Retrieved from PM:21076277.

Sheng, W. S., Hu, S., Lokensgard, J. R., & Peterson, P. K. (2003). U50,488 inhibits HIV-1 Tat-induced monocyte chemoattractant protein-1 (CCL2) production by human astrocytes. *Biochemical Pharmacology*, *65*(1), 9–14.

Shin, H. D., Winkler, C., Stephens, J. C., Bream, J., Young, H., Goedert, J. J., et al. (2000). Genetic restriction of HIV-1 pathogenesis to AIDS by promoter alleles of IL10. *Proceedings of the National Academy of Sciences of the United States of America*, *97*(26), 14467–14472. http://dx.doi.org/10.1073/pnas.97.26.14467, Retrieved from PM:11121048.

Shindo, K., Takaori-Kondo, A., Kobayashi, M., Abudu, A., Fukunaga, K., & Uchiyama, T. (2003). The enzymatic activity of CEM15/Apobec-3G is essential for the regulation of the infectivity of HIV-1 virion but not a sole determinant of its antiviral activity. *The Journal of Biological Chemistry*, *278*(45), 44412–44416. http://dx.doi.org/10.1074/jbc.C300376200, Retrieved from PM:12970355.

Shinoda, H., Marini, A. M., Cosi, C., & Schwartz, J. P. (1989). Brain region and gene specificity of neuropeptide gene expression in cultured astrocytes. *Science*, *245*, 415–417.

Singh, K. K., Ellis, R. J., Marquie-Beck, J., Letendre, S., Heaton, R. K., Grant, I., et al. (2004). CCR2 polymorphisms affect neuropsychological impairment in HIV-1-infected adults. *Journal of Neuroimmunology*, *157*(1–2), 185–192.

Skoff, R. P. (1990). Gliogenesis in the rat optic nerve: Astrocytes are generated in a single wave before oligodendrocytes. *Developmental Biology*, *139*, 149–168.

Skoff, R. P., & Knapp, P. E. (1991). Division of astroblasts and oligodendroblasts in postnatal rodent brain: Evidence for separate astrocyte and oligodendrocyte lineages. *Glia*, *4*, 165–174.

Smith, M. W., Dean, M., Carrington, M., Winkler, C., Huttley, G. A., Lomb, D. A., et al. (1997). Contrasting genetic influence of CCR2 and CCR5 variants on HIV-1 infection and disease progression. Hemophilia growth and development study (HGDS), multicenter AIDS cohort study (MACS), multicenter hemophilia cohort study (MHCS), San Francisco city cohort (SFCC), ALIVE study. *Science*, *277*(5328), 959–965, Retrieved from PM:9252328.

Smith, D. B., Simmonds, P., & Bell, J. E. (2014). Brain viral burden, neuroinflammation and neurodegeneration in HAART-treated HIV positive injecting drug users. *Journal of Neurovirology*, *20*(1), 28–38. http://dx.doi.org/10.1007/s13365-013-0225-3, Retrieved from PM:24420447.

Snider, S. E., Hendrick, E. S., & Beardsley, P. M. (2013). Glial cell modulators attenuate methamphetamine self-administration in the rat. *European Journal of Pharmacology*, *701* (1-3), 124–130. http://dx.doi.org/10.1016/j.ejphar.2013.01.016, Retrieved from PM:23375937.

Snider, S. E., Vunck, S. A., van den Oord, E. J., Adkins, D. E., McClay, J. L., & Beardsley, P. M. (2012). The glial cell modulators, ibudilast and its amino analog, AV1013, attenuate methamphetamine locomotor activity and its sensitization in mice. *European Journal of Pharmacology*, *679*(1–3), 75–80, Retrieved from PM:22306241.

Song, C., Rahim, R. T., Davey, P. C., Bednar, F., Bardi, G., Zhang, L., et al. (2011). Protein kinase Czeta mediates micro-opioid receptor-induced cross-desensitization of chemokine receptor CCR5. *Journal of Biological Chemistry*, *286*(23), 20354–20365, Retrieved from PM:21454526.

Sora, I., Hall, F. S., Andrews, A. M., Itokawa, M., Li, X. F., Wei, H. B., et al. (2001). Molecular mechanisms of cocaine reward: Combined dopamine and serotonin transporter knockouts eliminate cocaine place preference. *Proceedings of the National Academy of Sciences of the United States of America*, *98*(9), 5300–5305. http://dx.doi.org/10.1073/pnas.091039298, Retrieved from PM:11320258.

Sora, I., Wichems, C., Takahashi, N., Li, X. F., Zeng, Z., Revay, R., et al. (1998). Cocaine reward models: Conditioned place preference can be established in dopamine- and in serotonin-transporter knockout mice. *Proceedings of the National Academy of Sciences of the United States of America*, *95*(13), 7699–7704, Retrieved from PM:9636213.

Sorrell, M. E., & Hauser, K. F. (2014). Ligand-gated purinergic receptors regulate HIV-1 Tat and morphine related neurotoxicity in primary mouse striatal neuron-glia co-cultures. *Journal of Neuroimmune Pharmacology*, *9*(2), 233–244. http://dx.doi.org/10.1007/s11481-013-9507-z, Retrieved from PM:24158495.

Sozzani, S., Introna, M., Bernasconi, S., Polentarutti, N., Cinque, P., Poli, G., et al. (1997). MCP-1 and CCR2 in HIV infection: Regulation of agonist and receptor expression. *Journal of Leukocyte Biology*, *62*(1), 30–33.

Spruce, B. A., Curtis, R., Wilkin, G. P., & Glover, D. M. (1990). A neuropeptide precursor in cerebellum: Proenkephalin exists in subpopulations of both neurons and astrocytes. *The EMBO Journal*, *9*, 1787–1795.

Srikrishna, G., & Freeze, H. H. (2009). Endogenous damage-associated molecular pattern molecules at the crossroads of inflammation and cancer. *Neoplasia*, *11*(7), 615–628, Retrieved from PM:19568407.

Sriram, K., Miller, D. B., & O'Callaghan, J. P. (2006). Minocycline attenuates microglial activation but fails to mitigate striatal dopaminergic neurotoxicity: Role of tumor necrosis factor-alpha. *Journal of Neurochemistry*, *96*(3), 706–718.

St Hillaire, C., Vargas, D., Pardo, C. A., Gincel, D., Mann, J., Rothstein, J. D., et al. (2005). Aquaporin 4 is increased in association with human immunodeficiency virus dementia:

Implications for disease pathogenesis. *Journal of Neurovirology, 11*(6), 535–543. http://dx.doi.org/10.1080/13550280500385203, Retrieved from PM:16338747.

Steele, A. D., Henderson, E. E., & Rogers, T. J. (2003). Mu-opioid modulation of HIV-1 coreceptor expression and HIV-1 replication. *Virology, 309*(1), 99–107.

Steiner, H., & Gerfen, C. R. (1998). Role of dynorphin and enkephalin in the regulation of striatal output pathways and behavior. *Experimental Brain Research, 123*(1–2), 60–76.

Stiene-Martin, A., & Hauser, K. F. (1991). Glial growth is regulated by agonists selective for multiple opioid receptor types in vitro. *Journal of Neuroscience Research, 29*, 538–548.

Stiene-Martin, A., Knapp, P. E., Martin, K. M., Gurwell, J. A., Ryan, S., Thornton, S. R., et al. (2001). Opioid system diversity in developing neurons, astroglia, and oligodendroglia in the subventricular zone and striatum: Impact on gliogenesis *in vivo*. *Glia, 36*(1), 78–88.

Stiene-Martin, A., Zhou, R., & Hauser, K. F. (1998). Regional, developmental, and cell cycle-dependent differences in μ, δ, and κ-opioid receptor expression among cultured mouse astrocytes. *Glia, 22*(3), 249–259.

Sturrock, R. R. (1979). A quantitative lifespan study of changes in cell number, cell division and cell death in various regions of the mouse forebrain. *Neuropathology and Applied Neurobiology, 5*(6), 433–456.

Su, T. P., Hayashi, T., Maurice, T., Buch, S., & Ruoho, A. E. (2010). The sigma-1 receptor chaperone as an inter-organelle signaling modulator. *Trends in Pharmacological Sciences, 31*(12), 557–566.

Sulzer, D., & Rayport, S. (1990). Amphetamine and other psychostimulants reduce pH gradients in midbrain dopaminergic neurons and chromaffin granules: A mechanism of action. *Neuron, 5*(6), 797–808. http://dx.doi.org/10.1016/0896-6273(90)90339-H, Retrieved from PM:2268433.

Sundermann, E. E., Bishop, J. R., Rubin, L. H., Aouizerat, B., Wilson, T. E., Weber, K. M., et al. (2013). HIV serostatus differs by catechol-O-methyltransferase Val158Met genotype. *AIDS, 27*(11), 1779–1782. http://dx.doi.org/10.1097/QAD.0b013e328361c6a1, Retrieved from PM:23807274.

Suzuki, S., Chuang, L. F., Yau, P., Doi, R. H., & Chuang, R. Y. (2002). Interactions of opioid and chemokine receptors: Oligomerization of mu, kappa, and delta with CCR5 on immune cells. *Experimental Cell Research, 280*(2), 192–200.

Suzuki, M., El-Hage, N., Zou, S., Hahn, Y. K., Sorrell, M. E., Sturgill, J. L., et al. (2011). Fractalkine/CX3CL1 protects striatal neurons from synergistic morphine and HIV-1 Tat-induced dendritic losses and death. *Molecular Neurodegeneration, 6*(1), 78, Retrieved from PM:22093090.

Swan, N. (1997). CDC report highlights link between drug abuse and spread of HIV. *National Institute on Drug Abuse Notes (AIDS Research), 12*(2), March/April, Retrieved from http://archives.drugabuse.gov/NIDA_Notes/NNVol12N2/CDCReports.html.

Sweitzer, S., & De Leo, J. (2011). Propentofylline: Glial modulation, neuroprotection, and alleviation of chronic pain. *Handbook of Experimental Pharmacology, 200*, 235–250. http://dx.doi.org/10.1007/978-3-642-13443-2_8, Retrieved from PM:20859798.

Sweitzer, S. M., Schubert, P., & De Leo, J. A. (2001). Propentofylline, a glial modulating agent, exhibits antiallodynic properties in a rat model of neuropathic pain. *The Journal of Pharmacology and Experimental Therapeutics, 297*(3), 1210–1217, Retrieved from PM:11356948.

Sykova, E. (2005). Glia and volume transmission during physiological and pathological states. *Journal of Neural Transmission, 112*, 137–147.

Sykova, E., & Nicholson, C. (2008). Diffusion in brain extracellular space. *Physiological Reviews, 88*(4), 1277–1340. http://dx.doi.org/10.1152/physrev.00027.2007, Retrieved from PM:18923183.

Szabo, I., Chen, X. H., Xin, L., Adler, M. W., Howard, O. M., Oppenheim, J. J., et al. (2002). Heterologous desensitization of opioid receptors by chemokines inhibits

chemotaxis and enhances the perception of pain. *Proceedings of the National Academy of Sciences of the United States of America*, *99*(16), 10276–10281.
Szklarczyk, A., Ewaleifoh, O., Beique, J. C., Wang, Y., Knorr, D., Haughey, N., et al. (2008). MMP-7 cleaves the NR1 NMDA receptor subunit and modifies NMDA receptor function. *FASEB Journal*, *22*(11), 3757–3767. http://dx.doi.org/10.1096/fj.07-101402, Retrieved from PM:18644839.
Tait, M. J., Saadoun, S., Bell, B. A., & Papadopoulos, M. C. (2008). Water movements in the brain: Role of aquaporins. *Trends in Neurosciences*, *31*(1), 37–43. http://dx.doi.org/10.1016/j.tins.2007.11.003, Retrieved from PM:18054802.
Teuchert-Noodt, G., Dawirs, R. R., & Hildebrandt, K. (2000). Adult treatment with methamphetamine transiently decreases dentate granule cell proliferation in the gerbil hippocampus. *Journal of Neural Transmission*, *107*(2), 133–143.
Tewari, S., & Parpura, V. (2013). A possible role of astrocytes in contextual memory retrieval: An analysis obtained using a quantitative framework. *Frontiers in Computational Neuroscience*, 7, 145. http://dx.doi.org/10.3389/fncom.2013.00145, Retrieved from PM:24204341.
Theodore, S., Cass, W. A., Dwoskin, L. P., & Maragos, W. F. (2012). HIV-1 protein Tat inhibits vesicular monoamine transporter-2 activity in rat striatum. *Synapse*, *66*(8), 755–757. http://dx.doi.org/10.1002/syn.21564, Retrieved from PM:22517264.
Theodore, S., Cass, W. A., & Maragos, W. F. (2006a). Involvement of cytokines in human immunodeficiency virus-1 protein Tat and methamphetamine interactions in the striatum. *Experimental Neurology*, *199*(2), 490–498.
Theodore, S., Cass, W. A., & Maragos, W. F. (2006b). Methamphetamine and human immunodeficiency virus protein Tat synergize to destroy dopaminergic terminals in the rat striatum. *Neuroscience*, *137*(3), 925–935, Retrieved from PM:16338084.
Theodore, S., Cass, W. A., Nath, A., & Maragos, W. F. (2007). Progress in understanding basal ganglia dysfunction as a common target for methamphetamine abuse and HIV-1 neurodegeneration. *Current HIV Research*, *5*(3), 301–313, Retrieved from PM:17504172.
Thomas, D. M., & Kuhn, D. M. (2005). MK-801 and dextromethorphan block microglial activation and protect against methamphetamine-induced neurotoxicity. *Brain Research*, *1050*(1–2), 190–198.
Thored, P., Arvidsson, A., Cacci, E., Ahlenius, H., Kallur, T., Darsalia, V., et al. (2006). Persistent production of neurons from adult brain stem cells during recovery after stroke. *Stem Cells*, *24*(3), 739–747. http://dx.doi.org/10.1634/stemcells.2005-0281, Retrieved from PM:16210404.
Tian, C., Murrin, L. C., & Zheng, J. C. (2009). Mitochondrial fragmentation is involved in methamphetamine-induced cell death in rat hippocampal neural progenitor cells. *PLoS One*, *4*(5), e5546. http://dx.doi.org/10.1371/journal.pone.0005546, Retrieved from PM:19436752.
Tornatore, C., Chandra, R., Berger, J. R., & Major, E. O. (1994). HIV-1 infection of subcortical astrocytes in the pediatric central nervous system. *Neurology*, *44*(3 Pt 1), 481–487.
Tornatore, C., Meyers, K., Atwood, W., Conant, K., & Major, E. (1994). Temporal patterns of human immunodeficiency virus type 1 transcripts in human fetal astrocytes. *Journal of Virology*, *68*, 93–102.
Tornatore, C., Nath, A., Amemiya, K., & Major, E. O. (1991). Persistent human immunodeficiency virus type 1 infection in human fetal glial cells reactivated by T-cell factor(s) or by the cytokines tumor necrosis factor alpha and interleukin-1 beta. *Journal of Virology*, *65*(11), 6094–6100.
Tremblay, M. E., Lowery, R. L., & Majewska, A. K. (2010). Microglial interactions with synapses are modulated by visual experience. *PLoS Biology*, *8*(11), e1000527, Retrieved from PM:21072242.

Tripathi, A., Khurshid, N., Kumar, P., & Iyengar, S. (2008). Expression of delta- and mu-opioid receptors in the ventricular and subventricular zones of the developing human neocortex. *Neuroscience Research*, *61*(3), 257–270, Retrieved from PM:18455254.

Tsai, S. Y., Lee, C. T., Hayashi, T., Freed, W. J., & Su, T. P. (2010). Delta opioid peptide DADLE and naltrexone cause cell cycle arrest and differentiation in a CNS neural progenitor cell line. *Synapse*, *64*(4), 267–273. http://dx.doi.org/10.1002/syn.20727, Retrieved from PM:19953654.

Tsai, H. H., Li, H., Fuentealba, L. C., Molofsky, A. V., Taveira-Marques, R., Zhuang, H., et al. (2012). Regional astrocyte allocation regulates CNS synaptogenesis and repair. *Science*, *337*(6092), 358–362. http://dx.doi.org/10.1126/science.1222381, Retrieved from PM:22745251.

Tulloch, I. K., Afanador, L., Baker, L., Ordonez, D., Payne, H., Mexhitaj, I., et al. (2014). Methamphetamine induces low levels of neurogenesis in striatal neuron subpopulations and differential motor performance. *Neurotoxicity Research*, *26*(2), 115–129. http://dx.doi.org/10.1007/s12640-014-9456-1, Retrieved from PM:24549503.

Tulloch, I., Ghazaryan, N., Mexhitaj, I., Ordonez, D., & Angulo, J. A. (2011). Role of neurokinin-1 and dopamine receptors on the striatal methamphetamine-induced proliferation of new cells in mice. *Brain Research*, *1399*, 33–39. http://dx.doi.org/10.1016/j.brainres.2011.05.017, Retrieved from PM:21652034.

Turchan-Cholewo, J., Dimayuga, F. O., Ding, Q., Keller, J. N., Hauser, K. F., Knapp, P. E., et al. (2008). Cell-specific actions of HIV-Tat and morphine on opioid receptor expression in glia. *Journal of Neuroscience Research*, *86*(9), 2100–2110.

Turchan-Cholewo, J., Dimayuga, F. O., Gupta, S., Keller, J. N., Knapp, P. E., Hauser, K. F., et al. (2009). Morphine and HIV-Tat increase microglial-free radical production and oxidative stress: Possible role in cytokine regulation. *Journal of Neurochemistry*, *108*, 202–215.

Turchan-Cholewo, J., Liu, Y., Gartner, S., Reid, R., Jie, C., Peng, X., et al. (2006). Increased vulnerability of ApoE4 neurons to HIV proteins and opiates: Protection by diosgenin and L-deprenyl. *Neurobiology of Disease*, *23*(1), 109–119, Retrieved from PM:16697650.

Turner, J. R., Ecke, L. E., Briand, L. A., Haydon, P. G., & Blendy, J. A. (2013). Cocaine-related behaviors in mice with deficient gliotransmission. *Psychopharmacology*, *226*(1), 167–176. http://dx.doi.org/10.1007/s00213-012-2897-4, Retrieved from PM:23104263.

Vallat, A. V., De Girolami, U., He, J., Mhashilkar, A., Marasco, W., Shi, B., et al. (1998). Localization of HIV-1 co-receptors CCR5 and CXCR4 in the brain of children with AIDS. *The American Journal of Pathology*, *152*(1), 167–178.

Van Horn, M. R., Sild, M., & Ruthazer, E. S. (2013). D-serine as a gliotransmitter and its roles in brain development and disease. *Frontiers in Cellular Neuroscience*, 7, 39. http://dx.doi.org/10.3389/fncel.2013.00039, Retrieved from PM:23630460.

van Marle, G., Antony, J. M., Silva, C., Sullivan, A., & Power, C. (2005). Aberrant cortical neurogenesis in a pediatric neuroAIDS model: Neurotrophic effects of growth hormone. *AIDS*, *19*(16), 1781–1791.

Vargova, L., Jendelova, P., Chvatal, A., & Sykova, E. (2001). Glutamate, NMDA, and AMPA induced changes in extracellular space volume and tortuosity in the rat spinal cord. *Journal of Cerebral Blood Flow and Metabolism*, *21*(9), 1077–1089.

Venkatesan, A., Nath, A., Ming, G. L., & Song, H. (2007). Adult hippocampal neurogenesis: Regulation by HIV and drugs of abuse. *Cellular and Molecular Life Sciences*, *64*(16), 2120–2132.

Venkatesan, A., Uzasci, L., Chen, Z., Rajbhandari, L., Anderson, C., Lee, M. H., et al. (2011). Impairment of adult hippocampal neural progenitor proliferation by methamphetamine: Role for nitrotyrosination. *Molecular Brain*, *4*, 28. http://dx.doi.org/10.1186/1756-6606-4-28, Retrieved from PM:21708025.

Verderio, C. (2013). Extracellular membrane microvesicles and nanotubes in the brain: Understanding their nature, their function in cell-to-cell communication, their role in transcellular spreading of pathological agents and their therapeutic potential. *Frontiers in Physiology*, *4*, 163. http://dx.doi.org/10.3389/fphys.2013.00163, Retrieved from PM:23847543.

Verderio, C., Muzio, L., Turola, E., Bergami, A., Novellino, L., Ruffini, F., et al. (2012). Myeloid microvesicles are a marker and therapeutic target for neuroinflammation. *Annals of Neurology*, *72*(4), 610–624. http://dx.doi.org/10.1002/ana.23627, Retrieved from PM:23109155.

Verghese, P. B., Castellano, J. M., & Holtzman, D. M. (2011). Apolipoprotein E in Alzheimer's disease and other neurological disorders. *Lancet Neurology*, *10*(3), 241–252. http://dx.doi.org/10.1016/S1474-4422(10)70325-2, Retrieved from PM:21349439.

Verkhratsky, A., Rodriguez, J. J., & Parpura, V. (2013). Astroglia in neurological diseases. *Future Neurology*, *8*(2), 149–158. http://dx.doi.org/10.2217/fnl.12.90, Retrieved from PM:23658503.

Volkow, N. D., Chang, L., Wang, G. J., Fowler, J. S., Franceschi, D., Sedler, M., et al. (2001). Loss of dopamine transporters in methamphetamine abusers recovers with protracted abstinence. *Journal of Neuroscience*, *21*(23), 9414–9418, Retrieved from PM:11717374.

Volkow, N. D., Wang, G. J., Fowler, J. S., & Tomasi, D. (2012). Addiction circuitry in the human brain. *Annual Review of Pharmacology and Toxicology*, *52*, 321–336. http://dx.doi.org/10.1146/annurev-pharmtox-010611-134625, Retrieved from PM:21961707.

Volterra, A., & Meldolesi, J. (2005). Astrocytes, from brain glue to communication elements: The revolution continues. *Nature Reviews. Neuroscience*, *6*(8), 626–640.

Wagner, G. C., Ricaurte, G. A., Seiden, L. S., Schuster, C. R., Miller, R. J., & Westley, J. (1980). Long-lasting depletions of striatal dopamine and loss of dopamine uptake sites following repeated administration of methamphetamine. *Brain Research*, *181*(1), 151–160.

Wake, H., Moorhouse, A. J., Jinno, S., Kohsaka, S., & Nabekura, J. (2009). Resting microglia directly monitor the functional state of synapses in vivo and determine the fate of ischemic terminals. *Journal of Neuroscience*, *29*(13), 3974–3980.

Wang, G. J., Chang, L., Volkow, N. D., Telang, F., Logan, J., Ernst, T., et al. (2004). Decreased brain dopaminergic transporters in HIV-associated dementia patients. *Brain*, *127*(Pt 11), 2452–2458, Retrieved from PM:15319273.

Wang, C., Liu, F., Liu, Y. Y., Zhao, C. H., You, Y., Wang, L., et al. (2011). Identification and characterization of neuroblasts in the subventricular zone and rostral migratory stream of the adult human brain. *Cell Research*, *21*(11), 1534–1550. http://dx.doi.org/10.1038/cr.2011.83, Retrieved from PM:21577236.

Wang, Z., Pekarskaya, O., Bencheikh, M., Chao, W., Gelbard, H. A., Ghorpade, A., et al. (2003). Reduced expression of glutamate transporter EAAT2 and impaired glutamate transport in human primary astrocytes exposed to HIV-1 or gp120. *Virology*, *312*(1), 60–73.

Wang, Y. Z., Plane, J. M., Jiang, P., Zhou, C. J., & Deng, W. (2011). Concise review: Quiescent and active states of endogenous adult neural stem cells: Identification and characterization. *Stem Cells*, *29*(6), 907–912. http://dx.doi.org/10.1002/stem.644, Retrieved from PM:21557389.

Whitney, N. P., Eidem, T. M., Peng, H., Huang, Y., & Zheng, J. C. (2009). Inflammation mediates varying effects in neurogenesis: Relevance to the pathogenesis of brain injury and neurodegenerative disorders. *Journal of Neurochemistry*, *108*(6), 1343–1359. http://dx.doi.org/10.1111/j.1471-4159.2009.05886.x, Retrieved from PM:19154336.

Williams, K. C., Corey, S., Westmoreland, S. V., Pauley, D., Knight, H., deBakker, C., et al. (2001). Perivascular macrophages are the primary cell type productively infected by simian immunodeficiency virus in the brains of macaques: Implications for the neuropathogenesis of AIDS. *The Journal of Experimental Medicine*, *193*(8), 905–915, Retrieved from PM:11304551.

Wilson, J. M., Kalasinsky, K. S., Levey, A. I., Bergeron, C., Reiber, G., Anthony, R. M., et al. (1996). Striatal dopamine nerve terminal markers in human, chronic methamphetamine users. *Nature Medicine*, *2*(6), 699–703, Retrieved from PM:8640565.

Wilson, J. M., & Kish, S. J. (1996). The vesicular monoamine transporter, in contrast to the dopamine transporter, is not altered by chronic cocaine self-administration in the rat. *Journal of Neuroscience*, *16*(10), 3507–3510.

Winkler, C., Modi, W., Smith, M. W., Nelson, G. W., Wu, X., Carrington, M., et al. (1998). Genetic restriction of AIDS pathogenesis by an SDF-1 chemokine gene variant. ALIVE Study, Hemophilia Growth and Development Study (HGDS), multicenter AIDS Cohort Study (MACS), Multicenter Hemophilia Cohort Study (MHCS), San Francisco City Cohort (SFCC). *Science*, *279*(5349), 389–393, Retrieved from PM:9430590.

Wong, A. W., McCallum, G. P., Jeng, W., & Wells, P. G. (2008). Oxoguanine glycosylase 1 protects against methamphetamine-enhanced fetal brain oxidative DNA damage and neurodevelopmental deficits. *Journal of Neuroscience*, *28*(36), 9047–9054. http://dx.doi.org/10.1523/JNEUROSCI.2557-08.2008, Retrieved from PM:18768699.

Wu, C. C., Hung, C. J., Shen, C. H., Chen, W. Y., Chang, C. Y., Pan, H. C., et al. (2014). Prenatal buprenorphine exposure decreases neurogenesis in rats. *Toxicology Letters*, *225*(1), 92–101. http://dx.doi.org/10.1016/j.toxlet.2013.12.001, Retrieved from PM:24321744.

Xie, Z., & Miller, G. M. (2009). A receptor mechanism for methamphetamine action in dopamine transporter regulation in brain. *The Journal of Pharmacology and Experimental Therapeutics*, *330*(1), 316–325. http://dx.doi.org/10.1124/jpet.109.153775, Retrieved from PM:19364908.

Xu, J., Xu, M., Bolan, E., Gilbert, A. K., Pasternak, G. W., & Pan, Y. X. (2014). Isolating and characterizing three alternatively spliced mu opioid receptor variants: mMOR-1A, mMOR-1O, and mMOR-1P. *Synapse*, *68*(4), 144–152. http://dx.doi.org/10.1002/syn.21727, Retrieved from PM:24375714.

Yakovleva, T., Bazov, I., Watanabe, H., Hauser, K. F., & Bakalkin, G. (2011). Transcriptional control of maladaptive and protective responses in alcoholics: A role of the NF-κB system. *Brain, Behavior, and Immunity*, *25*(Suppl. 1), S29–S38. http://dx.doi.org/10.1016/j.bbi.2010.12.019, Retrieved from PM:21195164.

Yao, H., Duan, M., Yang, L., & Buch, S. (2012). Platelet-derived growth factor-BB restores human immunodeficiency virus Tat-cocaine-mediated impairment of neurogenesis: Role of TRPC1 channels. *Journal of Neuroscience*, *32*(29), 9835–9847. http://dx.doi.org/10.1523/JNEUROSCI.0638-12.2012, Retrieved from PM:22815499.

Yao, H., Kim, K., Duan, M., Hayashi, T., Guo, M., Morgello, S., et al. (2011). Cocaine hijacks sigma1 receptor to initiate induction of activated leukocyte cell adhesion molecule: Implication for increased monocyte adhesion and migration in the CNS. *Journal of Neuroscience*, *31*(16), 5942–5955. http://dx.doi.org/10.1523/JNEUROSCI.5618-10.2011, Retrieved from PM:21508219.

Yao, H., Yang, Y., Kim, K. J., Bethel-Brown, C., Gong, N., Funa, K., et al. (2010). Molecular mechanisms involving sigma receptor-mediated induction of MCP-1: Implication for increased monocyte transmigration. *Blood*, *115*(23), 4951–4962. http://dx.doi.org/10.1182/blood-2010-01-266221, Retrieved from PM:20354174.

Yi, Y., Lee, C., Liu, Q. H., Freedman, B. D., & Collman, R. G. (2004). Chemokine receptor utilization and macrophage signaling by human immunodeficiency virus type 1 gp120: Implications for neuropathogenesis. *Journal of Neurovirology*, *10*(Suppl. 1), 91–96.

Yoneyama, M., Kawada, K., Gotoh, Y., Shiba, T., & Ogita, K. (2010). Endogenous reactive oxygen species are essential for proliferation of neural stem/progenitor cells. *Neurochemistry International*, *56*(6–7), 740–746. http://dx.doi.org/10.1016/j.neuint.2009.11.018, Retrieved from PM:19958807.

You, Z. B., Herrera-Marschitz, M., Nylander, I., Goiny, M., Kehr, J., Ungerstedt, U., et al. (1996). Effect of morphine on dynorphin B and GABA release in the basal ganglia of rats. *Brain Research*, *710*(1–2), 241–248.

Yuan, Y., Arnatt, C. K., El-Hage, N., Dever, S. M., Jacob, J. C., Selley, D. E., et al. (2013). A bivalent ligand targeting the putative mu opioid receptor and chemokine receptor CCR5 heterodimers: Binding affinity versus functional activities. *Medicinal Chemistry Communications*, *4*(5), 847–851, Retrieved from PM:23682308.

Yuan, Y., Arnatt, C. K., Li, G., Haney, K. M., Ding, D., Jacob, J. C., et al. (2012). Design and synthesis of a bivalent ligand to explore the putative heterodimerization of the mu opioid receptor and the chemokine receptor CCR5. *Organic & Biomolecular Chemistry*, *10*(13), 2633–2646, Retrieved from PM:22354464.

Yuan, C. J., Quiocho, J. M., Kim, A., Wee, S., & Mandyam, C. D. (2011). Extended access methamphetamine decreases immature neurons in the hippocampus which results from loss and altered development of neural progenitors without altered dynamics of the S-phase of the cell cycle. *Pharmacology, Biochemistry, and Behavior*, *100*(1), 98–108. http://dx.doi.org/10.1016/j.pbb.2011.08.004, Retrieved from PM:21855565.

Yuferov, V., Butelman, E. R., Ho, A., Morgello, S., & Kreek, M. J. (2014). Neurocognitive and neuroinflammatory correlates of PDYN and OPRK1 mRNA expression in the anterior cingulate in postmortem brain of HIV-infected subjects. *Journal of Neuroinflammation*, *11*(1), 5. http://dx.doi.org/10.1186/1742-2094-11-5, Retrieved from PM:24405578.

Yuferov, V., Levran, O., Proudnikov, D., Nielsen, D. A., & Kreek, M. J. (2010). Search for genetic markers and functional variants involved in the development of opiate and cocaine addiction and treatment. *Annals of the New York Academy of Sciences*, *1187*, 184–207, Retrieved from PM:20201854.

Yuste, R., & Bonhoeffer, T. (2001). Morphological changes in dendritic spines associated with long-term synaptic plasticity. *Annual Review of Neuroscience*, *24*(1071–89), 1071–1089.

Zamecnik, J., Homola, A., Cicanic, M., Kuncova, K., Marusic, P., Krsek, P., et al. (2012). The extracellular matrix and diffusion barriers in focal cortical dysplasias. *The European Journal of Neuroscience*, *36*(1), 2017–2024, Retrieved from PM:22536791.

Zhang, Y., & Barres, B. A. (2010). Astrocyte heterogeneity: An underappreciated topic in neurobiology. *Current Opinion in Neurobiology*, *20*(5), 588–594, Retrieved from PM:20655735.

Zhang, L., Looney, D., Taub, D., Chang, S. L., Way, D., Witte, M. H., et al. (1998). Cocaine opens the blood-brain barrier to HIV-1 invasion. *Journal of Neurovirology*, *4*(6), 619–626.

Zhu, J., Ananthan, S., Mactutus, C. F., & Booze, R. M. (2011). Recombinant human immunodeficiency virus-1 transactivator of transcription1-86 allosterically modulates dopamine transporter activity. *Synapse*, *65*(11), 1251–1254. http://dx.doi.org/10.1002/syn.20949, Retrieved from PM:21538554.

Zhu, J., Mactutus, C. F., Wallace, D. R., & Booze, R. M. (2009). HIV-1 Tat protein-induced rapid and reversible decrease in [3H]dopamine uptake: Dissociation of [3H] dopamine uptake and [3H]2beta-carbomethoxy-3-beta-(4-fluorophenyl)tropane (WIN 35,428) binding in rat striatal synaptosomes. *The Journal of Pharmacology and Experimental Therapeutics*, *329*(3), 1071–1083. http://dx.doi.org/10.1124/jpet.108.150144, Retrieved from PM:19325033.

Zou, S., Fitting, S., Hahn, Y. K., Welch, S. P., El-Hage, N., Hauser, K. F., et al. (2011). Morphine potentiates neurodegenerative effects of HIV-1 Tat through actions at μ-opioid receptor-expressing glia. *Brain*, *134*(12), 3613–3628. Retrieved from, http://brain.oxfordjournals.org.proxy.library.vcu.edu/content/early/2011/11/16/brain.awr281.abstract.

Zou, Y. R., Kottmann, A. H., Kuroda, M., Taniuchi, I., & Littman, D. R. (1998). Function of the chemokine receptor CXCR4 in haematopoiesis and in cerebellar development. *Nature*, *393*(6685), 595–599.

Zubieta, J. K., Smith, Y. R., Bueller, J. A., Xu, Y., Kilbourn, M. R., Jewett, D. M., et al. (2002). mu-Opioid receptor-mediated antinociceptive responses differ in men and women. *Journal of Neuroscience*, *22*(12), 5100–5107, Retrieved from PM:12077205.
Zuckerman, M. D., & Boyer, E. W. (2012). HIV and club drugs in emerging adulthood. *Current Opinion in Pediatrics*, *24*(2), 219–224. http://dx.doi.org/10.1097/MOP.0b013e32834faa9b, Retrieved from PM:22227785.

CHAPTER TEN

Neuroimmune Basis of Alcoholic Brain Damage

Fulton T. Crews[1], Ryan P. Vetreno
Bowles Center for Alcohol Studies, University of North Carolina at Chapel Hill, Chapel Hill, North Carolina, USA
[1]Corresponding author: e-mail address: ftcrews@med.unc.edu

Contents

Abstract

Alcohol-induced brain damage likely contributes to the dysfunctional poor decisions associated with alcohol dependence. Human alcoholics have a global loss of brain volume that is most severe in the frontal cortex. Neuroimmune gene induction by binge drinking increases neurodegeneration through increased oxidative stress, particularly NADPH oxidase-induced oxidative stress. In addition, HMGB1-TLR4 and innate immune NF-κB target genes are increased leading to persistent and sensitized neuroimmune responses to ethanol and other agents that release HMGB1 or directly stimulate TLR receptors and/or NMDA receptors. Neuroimmune signaling and glutamate excitotoxicity are linked to alcoholic neurodegeneration. Models of adolescent alcohol abuse lead to significant frontal cortical degeneration and show the most severe loss of hippocampal neurogenesis. Adolescence is a period of high risk for ethanol-induced neurodegeneration and alterations in brain structure, gene expression, and maturation of adult phenotypes. Together, these findings support the hypothesis that adolescence is a period of risk for persistent and long-lasting increases in brain neuroimmune gene expression that promote persistent and long-term increases in alcohol consumption, neuroimmune gene induction, and neurodegeneration that we find associated with alcohol use disorders.

International Review of Neurobiology, Volume 118
ISSN 0074-7742
http://dx.doi.org/10.1016/B978-0-12-801284-0.00010-5

1. INTRODUCTION

A characteristic of alcohol use disorders is the consumption of large quantities of alcohol despite the knowledge that problems occur during drinking. Approximately 10% of the population drinks 75% of the alcohol consumed in the United States (Li, 2008). In addition to the increased risk of developing a variety of health problems, heavy drinking patterns also increase the risk of developing an addiction to alcohol. Chronic repetitive use of alcohol results in persistent changes in brain and behavior that are characterized by diminished behavioral control, difficulty avoiding negative consequences, increased preoccupation with drinking alcohol, and mounting craving and limbic negative affect. Alcohol dependence and addiction involve difficulty in moderating or stopping drinking even though it has become a problem. The prefrontal cortex (PFC) and limbic system structures are particularly vulnerable to the neurotoxic effects of alcohol, and are critically involved in decision-making, motivation, planning, goal setting, and impulse inhibition. A variety of preclinical and clinical studies suggest that chronic repetitive alcohol consumption persistently changes neurobiology thereby increasing risky decision-making, impulsivity, and anxiety that further drive repeated cycles of binge drinking, disrupting cognitive-limbic circuits through progressive neuroimmune activation and loss of control over alcohol use. Impulsivity, which is regulated by the PFC, increases during intoxication and alcoholism (Potenza & De Wit, 2010). Further, loss of attention, poor decision-making, and increased impulsivity are found in individuals with lesions of the PFC consistent with PFC degeneration contributing to alcoholism and other drug dependence (Bechara, Damasio, Damasio, & Anderson, 1994; Bechara et al., 2005). These studies suggest that alcoholism is at least in part related to heavy alcohol consumption-induced neurodegeneration and other alterations in brain and behavior contributing to the chronic relapsing disorder characterized by repeated use of alcohol and/or other drugs despite problems related to use of the substance (O'Brien, 2008).

Alcoholism is a chronic relapsing disorder that can include extended periods of abstinence followed by relapse to heavy drinking. Relapse rapidly returns to excessive drinking suggesting long-term permanent alterations in neurobiology that can be controlled, but remain persistently altered. A key component of the risks of addiction and relapse are subtle changes in cognitive flexibility, risky decision-making, and mounting alcohol reminders (e.g., cue-induced limbic anxiety and impulsivity). The chronic relapsing nature of alcoholism suggests persistent alterations in neurobiology that continuously increase risk of relapse to heavy drinking. Successful abstinence is most often achieved through complete abstinence and active avoidance of places and situations involving alcohol and associated reminders of alcohol that drive craving and relapse.

Brain neuroimmune signaling is activated in models of binge drinking and neurodegeneration. Neurons and glia (both astrocytes and microglia) contribute to the release and responses to signaling molecules first discovered within the immune system. Neuroimmune factors in this review are defined as proteins such as cytokines, Toll-like receptors (TLRs), and HMGB1 that are known to be peripheral immune signaling molecules that were recently discovered to function as brain signaling proteins and protein receptors. Immune signaling pathways activated by infection are known to remain sensitized for life through adaptive immunity. Similarly, neuroimmune signaling in brain, which consists of innate immune signals, persists for long periods that could contribute to long-lasting changes in neurobiology. Neuroimmune signaling increases alcohol drinking, risky decision-making, and blunts ability to change (e.g., loss of behavior flexibility) in alcohol-treated animals. Postmortem human alcoholic brain studies find neuroimmune gene expression correlates with lifetime alcohol consumption (Crews, Qin, Sheedy, Vetreno, & Zou, 2013) consistent with persistent neuroimmune signaling repeatedly increased by alcohol consumption and signaling contributing to the chronic relapsing nature of alcoholism (Crews et al., 2013; see Table 10.1).

2. ALCOHOL-INDUCED NEURODEGENERATION AND ALCOHOLISM

Human studies find that active, recent, and frequent heavy drinking behaviors are the best indicators of alcoholic brain damage (Shelton & Parsons, 1987; Sullivan & Pfefferbaum, 2005). Over a 5-year interval, the degree of excessive drinking in alcoholics corresponded with gray matter loss, particularly in the frontal lobes (Pfefferbaum et al., 1995). Frontal cortical choline-containing compounds measured by MRI were decreased in alcoholics with significant correlations between alcohol consumption in the last 90 days and decreases in anterior cingulated cortex and frontal white matter (Ende et al., 2006). Although studies suggest a recovery of some brain structure volumes in abstinence (Drake et al., 1995; O'Neill,

Table 10.1 Studies linking neuroimmune activation to alcoholism

Marker	Brain region
Rodent	
LPS increases ethanol drinking	Blednov et al. (2011)
CD14 KO reduces ethanol drinking	Blednov et al. (2012)
AIE-induced cognitive inflexibility correlates with neuroimmune expression	Vetreno and Crews (2012)
Amygdalar TLR4 siRNA administration reduces ethanol responding in dependent rats	Liu et al. (2011)
Age of drinking onset in humans correlates with neuroimmune expression	Vetreno et al. (2013)
TLR4 KO blocks ethanol-induced dopamine release	Alfonso-Loeches et al. (2010)
Human	
Neuroimmune expression correlates with lifetime alcohol consumption	Crews (2012)
Neuroimmune expression correlates with age of drinking onset	Vetreno et al. (2013)

Cardenas, & Meyerhoff, 2001), there is also evidence of greater vulnerability to damage during relapse (Pfefferbaum et al., 1995). Neuropathology studies have found a 22% reduction in the number of neurons in the superior frontal cortex of alcoholics (Harper, Kril, & Daly, 1987), and reduced brain weight of alcoholics relative to controls that was correlated with the rate and amount of lifetime alcohol consumption (Harding, Halliday, Ng, Harper, & Kril, 1996). Brain imaging studies by Fein et al. (2002) and Fein, Di Sclafani, and Finn (2010) found cortical gray matter volumes in alcohol dependent individuals were negatively associated with age and lifetime duration of alcohol use. Individuals with less severe alcohol abuse histories showed reduced whole brain, prefrontal, and parietal cortical gray matter volumes compared with nonalcoholic control participants. Other studies focused on cortical thickness find abstinent alcoholics have reduced whole brain cortical thickness. Decreases in cortical thickness were largest in frontal lobes and were related to the severity of alcohol abuse (Fortier et al., 2011). Investigations of recently detoxified individuals with chronic alcoholism have consistently shown significant impairments in the cognitive domains of executive function, nonverbal memory, visuospatial function, and gait and balance (Gansler et al., 2000; Oscar-Berman & Marinkovic, 2007, e.g., Parsons, 1993). Deficits in these functions suggest dysfunction in cerebellum, posterior parietal lobe, and most significantly, frontal lobes (specifically, orbitofrontal and PFC) (Deshmukh, Rosenbloom, Pfefferbaum, & Sullivan, 2002; Sullivan, Rosenbloom, & Pfefferbaum, 2000). Multiple studies find the frontal lobes are the most insulted region in the alcoholic brain (Kubota et al., 2001; Sullivan & Pfefferbaum, 2005). The frontal lobes regulate complex cognitive skills such as working memory, temporal ordering, discrimination and reversal learning that underlie judgment, attention, risk taking, motivation, mood, and wanting. Thus, the impaired judgment, blunted affect, poor insight, social withdrawal, reduced motivation, distractibility, and attention- and impulse-control deficits associated with alcohol use disorders are consistent with neurodegeneration, particularly within the cortex and more specifically the frontal cortex (Oscar-Berman & Hutner, 1993; Shelton & Parsons, 1987; Sullivan & Pfefferbaum, 2005; Sullivan et al., 2000).

The Crews laboratory and others using preclinical binge drinking models in rats find neurodegeneration in multiple cortical areas that extend from frontal to perirhinal, piriform, and entorhinal cortex (Collins, Zou, & Neafsey, 1998; Corso, Mostafa, Collins, & Neafsey, 1998; Crews et al., 2004; Hamelink, Hampson, Wink, Eiden, & Eskay, 2005; Obernier, Bouldin, & Crews, 2002). The cortical areas damaged in binge drinking rat models of neurodegeneration overlap with those found to be thinner in alcoholics (see Fig. 10.1). Binge alcohol-induced brain damage in adult rats is found in multiple cortical regions, namely agranular insular cortex, anterior piriform cortex, perirhinal cortex, and entorhinal cortex as well as the hippocampal dentate gyrus, particularly ventral dentate gyrus which is close to the amygdala, analogous to human temporal lobe where hippocampus is adjacent to the amygdala. In these regions, dark cell degeneration, a necrotic form of cell death with shrunken soma, is the predominant form of neuronal death (Obernier, Bouldin, et al., 2002). Studies in humans find cortical thinning in alcoholics in multiple cortical areas that overlap with the rat binge neurodegeneration studies (see Fig. 10.1), with several regions showing greater thinning with severity of alcoholism (Fortier et al., 2011). Genetics is implicated in the increased risk for the development of alcohol dependence, and genetically bred high alcohol preferring P rats are more sensitive to binge drinking-induced cortical neurodegeneration than non-preferring low alcohol drinking rats. Interestingly, humans who begin

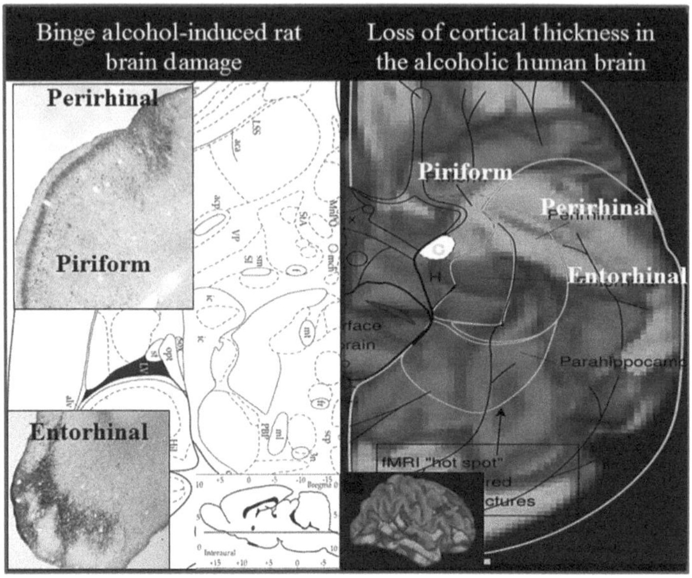

Figure 10.1 Comparison of human alcoholic cortical thinning with rat binge alcohol treatment-induced neurodegeneration. Shown are horizontal sections through the ventral rat brain and the ventral surface of the human brain. *Left*: Rat brain anatomical section in the lower right indicates location of ventral horizontal section. The left micrographs shows brain sections with silver cell death stain (black areas) showing neurodegeneration from a binge drinking model that includes the association cortical areas piriform and perirhinal cortex as well as entorhinal cortex (Crews et al., 2004; Obernier, Bouldin, et al., 2002). *Right*: Human alcoholic ventral cortical thinning is shown as a significance map of group differences in cortical thickness in abstinent alcoholics as compared to nonalcoholic controls with areas showing brain regions with significant cortical thinning in alcoholics highlighted in blue (light blue (light gray in the print version) $p < 0.01$, dark blue (dark gray in the print version) $p < 0.05$). Note: perirhinal, entorhinal, and piriform cortex are areas with cortical thinning in humans and areas of binge drinking neurodegeneration in rats, suggesting binge drinking human alcoholics are damaging association cortical areas. Right: *Adapted from Fortier et al. (2011).*

drinking in the early teenage years are more likely to have an alcohol use disorder in their lifetime (Grant & Dawson, 1997) and binge drinking models of neurodegeneration comparing adolescent and adult rats find significantly more frontal cortical neurodegeneration in adolescent animals (Coleman, Liu, Oguz, Styner, & Crews, 2014). Adolescence has been suggested to be a critical period of risk for developing alcoholism since the frontal cortex and executive functions develop during adolescence, and the developing brain is uniquely sensitive to alcohol (Crews, He, & Hodge, 2007). Binge drinking-induced brain damage is associated with a loss of behavioral flexibility (Obernier, White, Swartzwelder, & Crews, 2002). Binge drinking models in rats find no persistent deficits in spatial learning on the Morris water maze (Obernier, White, et al., 2002) or Barnes Maze (Vetreno & Crews, 2012) but show deficits in reversal learning, a task requiring animals to change what was learned and find a new site. Reversal learning deficits are consistent with frontal cortical dysfunction, particularly orbital frontal cortex (Schoenbaum, Roesch, Stalnaker, & Takahashi, 2009). Consistent with adolescence being a critical period of risk, studies find that adolescent binge drinking models in both rats (Vetreno & Crews, 2012) and mice (Coleman, He, Lee, Styner, & Crews, 2011) lead to reversal learning deficits that persists into adulthood. Further, they find persistent changes in brain structure, neurotransmitter, and other gene expression (Coleman et al., 2011, 2014; Vetreno & Crews, 2012; Vetreno, Qin, & Crews, 2013). The preclinical finding that the adolescent brain is particularly sensitive to frontal cortical damage and that adolescent age as well as genetic factors increases sensitivity to alcohol-induced brain damage that is related to a persistent loss of behavioral flexibility is consistent with human studies suggesting neurodegeneration contributes to the neurobiology of alcoholism.

3. LOSS OF NEUROGENESIS COULD CONTRIBUTE TO ALCOHOLIC NEURODEGENERATION

New neurons from neural stem cells are continuously produced in at least two regions of the normal adult brain, the subventricular zone of the lateral ventricles and the hippocampal dentate gyrus (Altman & Das, 1965; Alvarez-Buylla & Garcia-Verdugo, 2002). The Crews laboratory and many others have found that both acute and chronic alcohol exposure reduce hippocampal neurogenesis (He, Nixon, Shetty, & Crews, 2005; Herrera et al., 2003; Jang, Shin, Jung, et al., 2002; Jang, Shin, Kim, & Kim, 2002; Nixon & Crews, 2002). Multiple markers covering all stages of neurogenesis, specifically proliferation, neuronal maturation, and survival, have been examined in the 4-day binge model and find all are decreased (Crews, Bechara, et al., 2006; He et al., 2005; Nixon & Crews, 2002, 2004). Further, ethanol treatment during adult neurogenesis blunts the growth of the progenitor's dendritic arbor (He et al., 2005). Adolescents have high levels of neurogenesis in hippocampus and are more sensitive to acute alcohol inhibition of neurogenesis (Crews, Mdzinarishvili, Kim, He, & Nixon, 2006). Interestingly, adult hippocampal neurogenesis is resilient, recovering over a 30-day period from the 4-day binge alcohol model (Nixon & Crews, 2004) and a 7-week chronic prolonged relapsing model of alcohol dependence, whereas the ventricular neurogenesis is persistently reduced (Hansson et al., 2010). In contrast, in models of underage drinking (e.g., adolescent intermittent ethanol (AIE) treatment), hippocampal neurogenesis is persistently inhibited into adulthood (Broadwater, Liu, Crews, & Spear, 2013; Ehlers, Liu, Wills, & Crews, 2013). AIE exposure reductions in neurogenesis were found to be associated with more "disinhibitory" behavior in the open field conflict test at 2 and 8 weeks following termination of vapor exposure (Ehlers, Liu, et al., 2013). Similarly, AIE exposure of rats found reduced hippocampal volumes assessed using MRI consistent with those found in alcoholism (Ehlers, Oguz, Budin, Wills, & Crews, 2013). Loss of neurogenesis and/or gliogenesis likely contributes to alcoholic neurobiology and the reductions in brain volume found in human alcoholics (Crews & Nixon, 2009).

4. MONOCYTES AND INNATE IMMUNE GENES

Studies on the mechanisms of alcoholic neurodegeneration and loss of neurogenesis led the Crews laboratory to the discovery that innate immune gene induction is involved in alcoholic neurodegeneration. Innate immune genes are associated with rapid monocyte responses to infections (e.g., the acute phase response) that include increases in multiple cytokines as well as increases in their cellular receptors that together amplify expression of a large number of genes through kinase signaling pathways that converge on nuclear factor kappa-light-chain-enhancer of activated B cells (NF-κ B) and activator protein-1 (AP-1) transcription factors. NF-κB and AP-1 transcription factors promote expression of innate immune cytokines, including tumor necrosis factor-alpha (TNFα), interleukin-1β (IL-1β), monocyte chemotactic protein-1 (MCP-1), and IL-6 as well as TLRs and cytokine receptors (see Fig. 10.2). In addition, innate immune proteases and oxidases are induced, particularly cyclooxygenase and nicotinamide adenine dinucleotide phosphate (NADPH) oxidase, as well as major histocompatibility (MHC) signaling molecules, such as beta-2 microglobulin. NF-κB transcription of proinflammatory genes is amplified within and across cells by induction of TLR and cytokine receptors, many belonging to the IL-1β receptor family, which induce innate immune gene expression. Amplification of innate immune gene induction across cells and tissues can cause pathology, with one example being sepsis. Sepsis and systemic

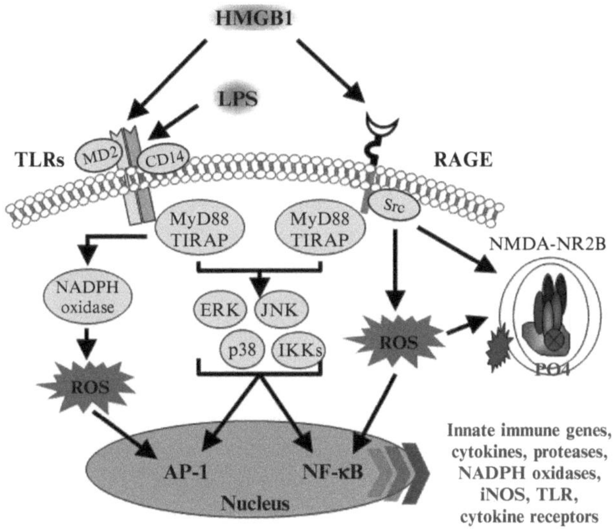

Figure 10.2 HMGB1-TLR/RAGE signaling leads to activation of NF-κB transcription of innate immune genes and alterations in NMDA receptors. A simplified schematic of the TLR and RAGE signaling cascades. Stimulation of TLRs with HMGB1 and other inflammagens leads to the generation of ROS and downstream activation of NF-κB. Similarly, HMGB1 activation of the RAGE receptor leads to downstream activation of NF-κB and induction of ROS. Nuclear translocation of NF-κB leads to the secretion of proinflammatory gene expression, neuroimmune induction, and cell death. Innate immune receptor stimulation also leads to activation of glutamatergic *N*-methyl-D-aspartate (NMDA) receptors (Iori et al., 2013; Maroso et al., 2010), which increases Ca^{2+} flux triggering further induction of neuroimmune genes. AP-1, activator protein-1; CD14, cluster of differentiation 14; ERK, extracellular signal-regulated kinase; HMGB1, high-mobility group box-1; IKK, inhibitor of nuclear factor kappa-B; JNK, c-Jun N-terminal kinases; LPS, lipopolysaccharide; MyD88, myeloid differentiation primary response gene 88; NADPH oxidase, nicotinamide adenine dinucleotide phosphate-oxidase; NF-κB, nuclear factor kappa-light-chain-enhancer of activated B cells; RAGE, receptor for advanced glycation end products; ROS: reactive oxygen species; Src, Proto-oncogene tyrosine-protein kinase; TIRAP, Toll/Interleukin-1 receptor domain-containing adaptor protein; TLRs, Toll-like receptors.

inflammatory response syndrome involve a "cytokine storm," a potentially fatal innate immune reaction consisting of positive feedback loops between cytokines, and immune and tissue cells, resulting in highly elevated levels of blood cytokines, multiorgan failure, and death (Osterholm, 2005). Sepsis is modeled with high doses of bacterial endotoxin, lipopolysaccharide (LPS), or other polypathogen infections that activate an acute phase-like response. Multiple cytokines increase in blood, with both TNFα and IL-1β increasing in blood during the first several hours after infection and then subsiding.

Interestingly, HMGB1, a cytokine-like protein that can activate TLR4 receptors and potentiate cytokine responses, rises in blood about 16 h after infection and remains elevated for several days (Wang, Yang, Czura, Sama, & Tracey, 2001). In mouse models, sepsis-induced death occurs days after infection, which is associated with HMGB1, and is blocked by HMGB1 antibody treatment. Survivors of sepsis show prolonged increases in serum HMGB1 and cognitive deficits that can be prevented with HMGB1 antibody treatment (Chavan et al., 2012). In humans, about half of those released from the hospital after surviving a cytokine storm-sepsis insult die within 5 years (Quartin, Schein, Kett, & Peduzzi, 1997). Thus, innate immune responses can be long lasting and can induce pathology long after initial activation. Although most studies support innate immune signaling converging upon NF-κB transcription of proinflammatory cytokines, proteases, and oxidases, the precise mechanisms that regulate individual cell or cytokine activation, and how different tissues and cells contribute *in vivo* to amplification of specific innate immune genes is poorly understood.

Monocytes are the primary innate immune cell studied in acute proinflammatory responses as well as signaling adaptive immune cells through MHC molecules creating a persistent sensitization to the pathogen (e.g., antibodies that mediate immunization). In addition to responding to pathogens, monocytes and the innate immune system respond to tissue damage, cell death, and degeneration. The innate immune systems response to tissue damage includes the activation of monocyte proinflammatory signaling, termed M1 monocytes or M1 microglia, the brain-specific monocyte. Microglia and other monocyte-like cells consistently express multiple cytokine and TLRs that activate microglia–monocyte-like cells inducing innate immune genes, particularly proinflammatory cytokines, proteases, and oxidases that contribute to the breakdown, processing, and removal of damaged cells and tissue. In addition, monocyte innate immune responses to tissue damage have a delayed response that initiates wound healing trophic signaling, associated with M2 trophic monocyte phenotypes that appear critical for healing. Although poorly understood, both monocytes and brain microglia have M1 and M2, proinflammatory M1, and trophic M2 phenotypes (Colton, 2009; Michelucci, Heurtaux, Grandbarbe, Morga, & Heuschling, 2009). Although monocyte M1 to M2 phenotypes are poorly understood, monocyte proinflammatory activation is clearly linked to NF-κB transcription of multiple innate immune genes. Activation of monocyte NF-κB transcription by pathogens and tissue damage shares common pathways (see Fig. 10.2). Toll-like receptor 4 (TLR4) is a receptor that responds to bacterial endotoxin (e.g., LPS), and HMGB1, a ubiquitously expressed nuclear protein released during necrotic cell death, leading to proinflammatory responses to tissue damage. Proinflammatory gene induction is also amplified by cytokine receptor activated release of HMGB1 that further contributes to innate immune gene induction. The role of these innate immune signaling molecules is well characterized within the immune system, but only recently they have been discovered to contribute to brain signaling. Recent studies find MHC molecules contribute to brain development (Huh et al., 2000), to most neurodegenerative diseases (Gage, 2002; Glass, Saijo, Winner, Marchetto, & Gage, 2010), and alcohol and drug dependence (Crews, 2012). Neuroimmune signaling has not been extensively studied and most knowledge is based on the assumption that monocyte responses represent microglial and brain innate immune responses.

The immune system is not normally active in the healthy brain. The healthy normal brain does not contain antibodies and has only one immune-like cell, the microglia. Neurons, astrocytes, and all other brain cells are formed from ectoderm; whereas, microglia migrates from mesoderm to brain at a specific time during fetal development (Ginhoux et al., 2010). In the healthy brain, ramified or "resting" microglia equal neurons in number and contribute to the integration of sensory systems and overall survey of the brain milieu (Raivich, 2005). Healthy brain microglia are poorly understood, but clearly participate in overall brain health. Microglia, along with astrocytes, modulate important metabolic, trophic, and synaptic functions in addition to responding to brain damage induced neuroimmune responses (Farina, Aloisi, & Meinl, 2007; Streit, Mrak, & Griffin, 2004). Although poorly understood, microglia respond to endogenous or exogenous insults with distinct morphological changes in shape (see Fig. 10.3) as well as marked alterations in gene expression, including proinflammatory innate immune response genes (Graeber, 2010). However, it is sometimes unclear whether microglia are responding to a brain insult or causing the brain insult. Microglial signaling involves both neuroimmune signals and neurotransmitter signals. For example, acetylcholine, an important neurotransmitter involved in multiple brain functions including cognition, inhibits peripheral monocyte and brain microglial proinflammatory activation and is anti-inflammatory. Our laboratory found an increase in expression of the microglial marker, Iba-1 in the brains of alcoholic individuals (Fig. 10.3 He & Crews, 2007), suggesting that microglia contribute to the neurobiology of alcoholism. Microglia in postmortem human alcoholic

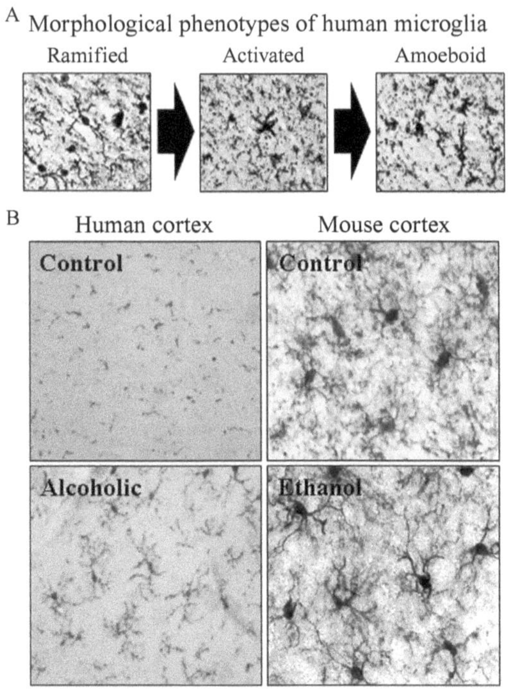

Figure 10.3 Increased expression of microglial markers in postmortem alcoholic and mouse brain. (A) Representative photomicrographs depicting the characteristic stages of microglial activation. Ramified or "resting" microglia are characterized by long, highly ramified processes with comparatively small cell bodies. Activated microglia are characterized by swollen, truncated processes, and enlarged cell bodies. Amoeboid or "phagocytic" microglia are characterized by large, amoeba-like cell body with no or few small processes (Kreutzberg, 1996; Raivich et al., 1999). (B) *Left*, Photomicrographs depicting microglial activation in postmortem human alcoholic brain tissue (He & Crews, 2008). *Right*: Representative figures depicting microglial activation in the mouse cortex following ethanol treatment. Male C57BL/6 mice were treated with saline or ethanol (5 g/kg, i.g.) for 10 days and were sacrificed 1 h later (Qin et al., 2008).

brain and chronic alcohol-treated mouse and rat brain show increased MHC gene expression, but not bushy or phagocytic activation profiles (see Fig. 10.3). Thus, microglia increase MHC and likely TLR4 receptor expression during chronic ethanol treatment but are not activated to bushy or phagocytic phenotypes associated with marked brain damage. Thus, microglia are the only immune cells in healthy brain and are integrated into brain responding to both neurotransmitters and neuroimmune signals as well as contributing to chronic alcohol-induced responses.

5. ALCOHOL, NEUROIMMUNE SIGNALING, AND NEURODEGENERATION

Chronic binge drinking models have repeatedly found that ethanol treatment increases the expression of a variety of neuroimmune genes in brain (see Table 10.2 and Fig. 10.4). Cyclooxygenase 2 (COX2) induction by chronic ethanol treatment in brain was found to persist in multiple cortical and limbic brain regions long after physical signs of withdrawal had subsided (Knapp & Crews, 1999). Guerri's laboratory has found ethanol induction of COX2 is blocked in transgenic mice-lacking TLR4 receptors (Alfonso-Loeches, Pascual-Lucas, Blanco, Sanchez-Vera, & Guerri, 2010). Zou & Crews (2006), using HEC brain slice cultures found that ethanol treatment increased NF-κB binding to DNA probes modeling gene promoter regions and decreased cyclic AMP-responsive element binding protein (CREB) binding to DNA probes modeling CREB gene promoter DNA. The CREB family of transcription factors are activated by phosphorylation and promote neuronal survival, protecting neurons from excitotoxicity and apoptosis through regulating the transcription of prosurvival factors (Lonze & Ginty, 2002; Mantamadiotis et al., 2002). Conversely, NF-κB is a transcriptional factor that is known widely for its ubiquitous roles in inflammatory and immune responses (O'Neill & Kaltschmidt, 1997). As described earlier, NF-κB and CREB have different target genes. For example, neuropeptide Y (NPY) and brain-derived neurotrophic factor (BDNF) are CREB target genes involved in promoting neuronal growth and resilience to insults including protection against excitotoxicity and neuronal death (Lonze & Ginty, 2002). Excitation of neurons increases synaptic plasticity-related to CREB and induction of synaptic proteins and BDNF, whereas excessive excitation triggers excitation of extrasynaptic *N*-methyl-D-aspartate (NMDA) receptors and excitotoxicity, either rapid or delayed neuronal death, which is associated with reduced CREB (Hardingham & Bading, 2010). Levels of CREB-DNA binding and phosphorylated CREB, as well as the target gene BDNF, are decreased in the rat frontal cortex following a 24-h withdrawal from chronic ethanol exposure (Pandey, Roy, & Mittal, 2001; Pandey, Zhang, Mittal, & Nayyar, 1999). In addition, NPY levels are reduced in the cortex following ethanol treatment, an effect that was complemented by reduced phosphorylated CREB (Bison & Crews, 2003). A reciprocal relationship between NF-κB and CREB transcription sensitizes neurons to excitotoxicity (Zou & Crews,

Table 10.2 Neuroimmune markers in the human alcoholic postmortem brain

Marker	Brain region	Effect	Method	Citations
RAGE	Orbitofrontal cortex	↑	IHC, RTPCR, Western blot	Vetreno et al. (2013)
TLR2	Orbitofrontal cortex	↑	IHC, Western blot	Crews et al. (2013)
TLR3	Orbitofrontal cortex	↑	IHC, Western blot	Crews et al. (2013)
TLR4	Orbitofrontal cortex	↑	IHC, Western blot	Crews et al. (2013)
HMGB1	Orbitofrontal cortex	↑	IHC, Western blot	Crews et al. (2013)
IL-1β	Hippocampus	↑	IHC	Zou and Crews (2012)
NALP1	Hippocampus	↑	IHC	Zou and Crews (2012)
$gp91^{phox}$ (NOX2)	Orbitofrontal cortex	↑	IHC	Qin and Crews (2012a, 2012b)
MCP-1	Ventral tegmental area Substantia nigra Hippocampus Amygdala	↑ ↑ ↑ ↑	ELISA	He and Crews (2008)
Iba-1	Cingulate cortex Ventral tegmental area Midbrain Amygdala	↑ *ns* *ns* *ns*	IHC	He and Crews (2008)
$GluT_5$	Cingulate cortex Ventral tegmental area Midbrain Amygdala	↑ ↑ ↑ *ns*	IHC	He and Crews (2008)
Active NF-κB	Prefrontal cortex	↓	EMSA binding activity	Okvist et al. (2007)

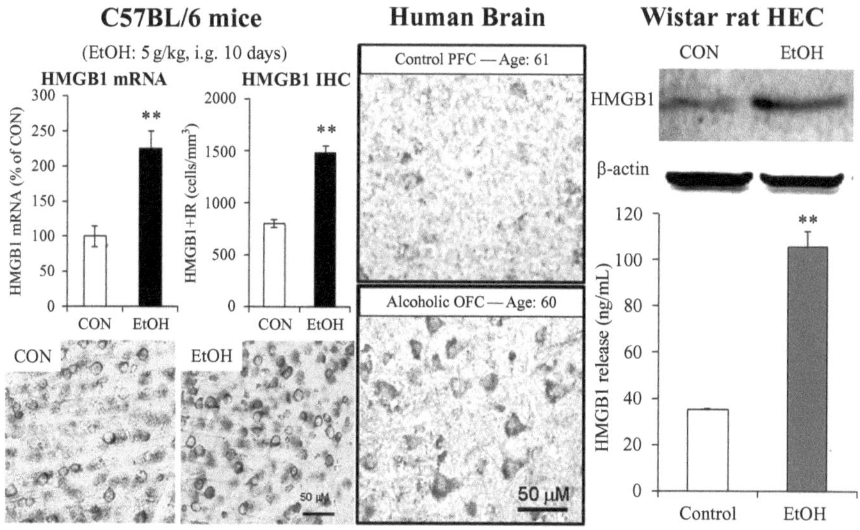

Figure 10.4 Alcohol increases high-mobility group box 1 (HMGB1) expression in mouse brain, human brain, and releases HMGB1 from rat brain slices. *Left*: Chronic ethanol treatment of mice for 10 days increases expression of HMGB1 and TLR receptor mRNA and protein. *Middle*: Postmortem human alcoholic orbitofrontal cortex has significantly more HMGB1-immunoreactive cells than age-matched controls. *Right*: Ethanol causes the release of HMGB1 into the media from hippocampal-entorhinal cortex (HEC) slice culture. *Top*: Representative Western blot of HMGB1 protein in medium following control (CON) and ethanol (EtOH) treatment. *Bottom*: ELISA assessment of HMGB1 protein in medium in control and EtOH-treated HEC slice cultures. $^{**}p<0.01$, relative to corresponding control group. *Adapted from (Crews et al., 2013).*

2006). This reciprocal relationship appears to be due to kinases, such as protein kinase A, which activates CREB transcription and inhibits NF-κB activation. However, the reciprocal relationship may represent neuronal-glial signaling as CREB is principally neuronal and NF-κB activation of proinflammatory genes primarily glial consistent with signals that crossing cells reducing CREB in neurons while activating NF-κB in glia. Thus, ethanol can directly increase NF-κB transcription of proinflammatory genes in brain as well as decreasing trophic protective factor transcription decreasing resilience to insults.

There are multiple mechanisms of neuroimmune gene induction by ethanol. We recently discovered that one mechanism involves alcohol release of HMGB1 (a TLR4 agonist) that increases NF-κB transcription of proinflammatory cytokines (Crews et al., 2013; Zou & Crews, 2014). A number of transmitters and neuroimmune signaling receptors as well as neuronal excitability increase release of HMGB1 (Maroso et al., 2010). HMGB1, also known as amphoterin (Huttunen & Rauvala, 2004), has multiple signaling mechanisms in brain that impact astrocytes, microglia, neurogenesis-neurite growth, and excitability in adjacent neurons. HMGB1 released by neuronal activity stimulates TLR4 receptors inducing IL-1β and increasing phosphorylation of the NR2B subunit of the NMDA receptor increasing excitation to seizures (Maroso et al., 2010; Vezzani, Maroso, Balosso, Sanchez, & Bartfai, 2011). Actively released HMGB1 is acetylated. We found that ethanol increases HMGB1 acetylation and histone deacetylases in brain slice cultures. Acetyl-HMGB1 is increased in cytosolic fractions, likely to vesicles, and is increased progressively in the media consistent with neuronal release (Zou & Crews, 2014). The importance of ethanol release of HMGB1 and activation of TLR4 receptors became apparent due to the elegant experiments of Consuelo Guerri's laboratory establishing TLR4 receptors as critical to ethanol-induced neurodegeneration and behavioral pathology. Using TLR4 knock-out transgenic cells and mice, her group discovered that chronic ethanol-induced neurodegeneration and induction of proinflammatory gene expression is markedly blunted by knockout of TLR4 receptors (Alfonso-Loeches et al., 2010; Blanco, Valles, Pascual, & Guerri, 2005; Fernandez-Lizarbe, Pascual, & Guerri, 2009; Pascual, Balino, Alfonso-Loeches, Aragon, & Guerri, 2011; Valles, Blanco, Pascual, & Guerri, 2004). Guerri's laboratory has also shown that ethanol treatment induces neuroimmune genes in microglia and astrocyte primary cultures as well as *in vivo* in mice that is dependent upon the expression of TLR4 receptors. TLR4 receptors are constitutively expressed on microglia, making microglia a key component of drug-induced neuroimmune activation (Alfonso-Loeches & Guerri, 2011; Schwarz & Bilbo, 2013). More recent studies by Guerri's laboratory have found that TLR4 receptors are integral to ethanol-induced dopamine release (Alfonso-Loeches & Guerri, 2011), damage to white matter (Alfonso-Loeches et al., 2012), and other pathologies associated with chronic ethanol-induced changes in brain (Pascual et al., 2011). In culture, ethanol treatment increases innate immune gene expression in a time-dependent fashion mimicking responses to LPS or IL-1β administration, although ethanol induces a much smaller response (Crews et al., 2013). *In vivo*, ethanol induces neuroimmune genes in the brains of wild type mice, but not TLR4 transgenic mice (Alfonso-Loeches et al., 2010). These studies support the hypothesis that TLR4 signaling is critical to many of the effects of alcohol on the brain. The critical role of TLR4 is somewhat surprising since multiple proinflammatory receptors converge upon common signaling through NF-κB. Also, Vetreno et al. (2013) found that chronic intermittent treatment of adolescent rats led to persistent increases in expression of RAGE in brain, a receptor also stimulated by HMGB1 (see Fig. 10.2). It is not understood why TLR4 signaling appears to contribute significantly to ethanol responses since TLR and cytokine receptors generally are within the IL1-receptor family and share kinase cascades in monocytes and microglia that all converge upon NF-κB. These findings suggest that the TLR4 receptors on neurons or other cells in brain may have some unique properties that differ from NF-κB activation by TNFα, IL-1β, and other cytokine receptors known to induce NF-κB transcription of proinflammatory cytokines. One explanation may be related to amplification of overlapping signaling, particularly IL-1β and HMGB1. IL-1β can be induced by HMGB1-TLR signaling, including formation of the inflammasome, a unique intracellular multiprotein organelle containing specific proteases and NLRP proteins. We found increased expression of IL-1β and NLRP-inflammasome proteins within alcoholic human hippocampus (see Fig. 10.5) consistent with neuroimmune gene activation contributing to ethanol-induced loss of neurogenesis (Zou & Crews, 2014). Although the mechanisms remain complicated, together these studies suggest that HMGB1-TLR4, and perhaps RAGE signaling, involves multiple brain cells type, with neuronal-glial neuroimmune signaling and microglial–astrocyte activation contributing to alcohol-induced brain damage.

Although chronic alcohol treatment increases proinflammatory gene expression in brain through activation of TLR4 receptors, this is confounded by alcohol inhibition of TLR4 signaling in monocytes and possibly other cells. Time-dependent acute and chronic opposing effects of ethanol confound many studies (Crews, Bechara, et al., 2006; Crews, Zou, & Qin, 2011; Szabo & Mandrekar, 2009). Ethanol suppresses the innate immune response to LPS, a TLR4 agonist, in both *in vivo* and *in vitro* models. For example, LPS-induced TNFα and IL-1β production is blunted in blood monocytes obtained from healthy human volunteers after acute alcohol exposure (2 mL vodka/kg body weight; Crews, Bechara, et al., 2006; Szabo, Mandrekar, & Catalano, 1995; Szabo, Mandrekar, Dolganiuc, Catalano, & Kodys, 2001; Szabo, Verma, & Catalano, 1993). In animal models, acute ethanol exposure attenuated the TNFα, IL-1β, and IL-6 immune response to LPS (Pruett, Zheng, Fan, Matthews, & Schwab, 2004). Similarly, Szabo's group reported that ethanol (25 mM) *in vitro* added just before LPS blunts induction of TNFα (Szabo et al., 1993, 1995, 2001).

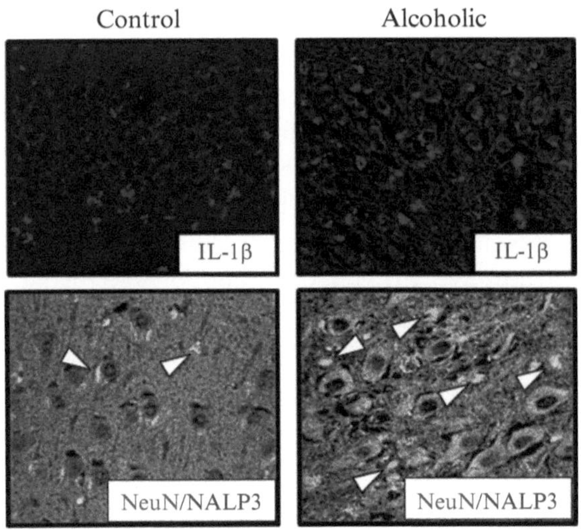

Figure 10.5 Neuronal expression of IL-1β and inflammasome proteins in the human alcoholic hippocampus. IL-1β can be induced by HMGB1-TLR signaling, including formation of the inflammasome, a unique intracellular multiprotein organelle containing specific proteases and NLRP proteins. We found increased expression of IL-1β and NLRP-inflammasome proteins within alcoholic human hippocampus. Shown is immunohistochemistry for IL-1β in control (left) and alcoholic hippocampus (right) above double immunohistochemistry sections stained for NeuN, a neuronal marker, and NLRP3, an inflammasome protein, that form yellow (white in the print version) when coexpressed consistent with ethanol induction of neuronal inflammasome formation and IL-1β release contributing to ethanol inhibition of neurogenesis. *Adapted from (Zou & Crews, 2012).*

In contrast, chronic *in vitro* ethanol exposure of astrocytes, microglia, and brain slices induces NF-κB transduction of proinflammatory genes through activation of TLR4 signaling (Blanco, Pascual, Valles, & Guerri, 2004; Blanco et al., 2005; Crews et al., 2013; Fernandez-Lizarbe et al., 2009; Zou & Crews, 2014). While it is not clear if the presence of ethanol antagonizes TLR4 receptors on all cell types, other TLR receptors are not acutely blocked by ethanol (Crews, Bechara, et al., 2006). Upregulation of TLR receptors by alcohol can lead to sensitization. In mice, binge treatment with ethanol for 10 days (5 g/kg/day) followed by LPS 24 h later when alcohol had cleared resulted in a marked increase in proinflammatory gene induction (Qin & Crews, 2012b). Ethanol treatment increased the response to LPS-induced proinflammatory cytokines in liver, blood, and brain. The responses in blood and liver were transient, but were long lasting in brain. Similarly, chronic 10-day alcohol treatment sensitized mice to the proinflammatory response to Poly:IC, a viral mimetic TLR3 agonist (Qin & Crews, 2012a). Thus, the effects of ethanol on brain neuroimmune signaling are in part related to increases in TLR receptors (see Fig. 10.4) that increase neuroimmune signaling and cytokines, such as IL-1β (see Fig. 10.5), during chronic ethanol treatment, although the presence of alcohol can blunt TLR4 responses during intoxication.

Although ethanol-induced release of HMGB1 from neurons activates TLR4 receptors and NF-κB transcription increasing synthesis of proinflammatory cytokines in brain (see Figs. 10.2, 10.4, and 10.5), a second mechanism involves ethanol-induced increases in blood proinflammatory cytokines that spread and amplify gene induction in brain (see Fig. 10.6). Recent studies indicate that ethanol in the gut releases HMGB1, which

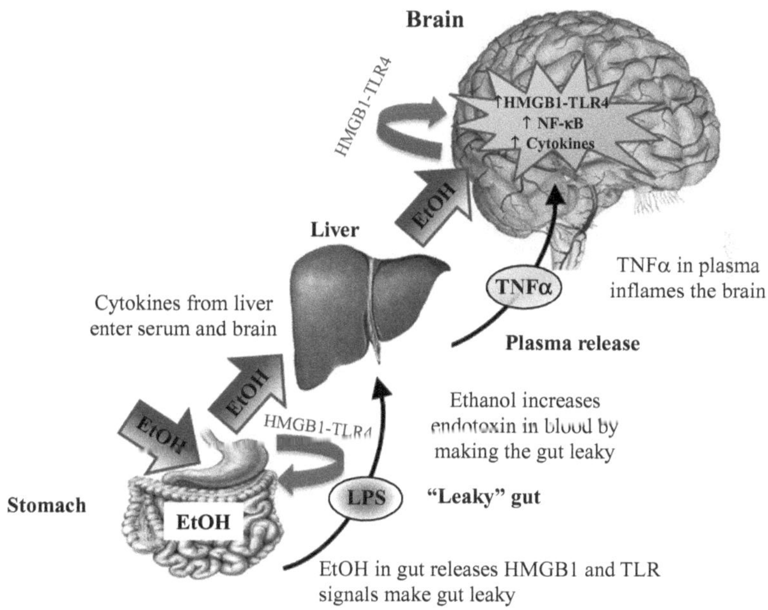

Figure 10.6 Ethanol in the gut activates systemic cytokine signaling inducing neuroimmune gene expression in the brain. Consumed EtOH enters the stomach and makes it "leaky," through the release of HMGB1 which activates gut TLR4 receptors that allows bacterial products such as lipopolysaccharide (LPS) to enter the blood. EtOH and LPS lead to induction of liver tumor necrosis factor-alpha (TNFα) and other proinflammatory cytokines. These proinflammatory cytokines in blood enter the brain and increase neuroimmune expression. Chronic ethanol increases expression of HMGB1-TLR4 receptors in brain leading to persistent and progressive increases in neuroimmune gene expression in brain.

activates TLR4 receptors causing the gut to leak endotoxin LPS-like bacterial products stimulating proinflammatory cytokine induction in the liver. Induction of liver cytokines results in increased blood TNFα and other cytokines. Qin and Crews (2012b) discovered that LPS-induced increases in serum TNFα as well as proinflammatory gene induction in brain. Proinflammatory cytokines in the blood can be transported by their receptors across the blood–brain barrier (BBB) (e.g., TNFR; Banks & Erickson, 2010; Qin et al., 2007) as well as activating endothelial cells to release cytokines (Watkins, Maier, & Goehler, 1995). Using LPS–TLR4 stimulation of proinflammatory responses in liver and other tissues induction of proinflammatory cytokines, we discovered that blood and brain levels show parallel increases in TNFα following intraperitoneal injections of LPS (Qin et al., 2007). Using transgenic mice-lacking TNF receptors, we found that LPS increased TNFα in blood, but did not increase TNFα in the brain, suggesting that LPS–TLR4 induction of TNFα in blood leads to TNF transport through its receptors across the BBB-activating proinflammatory responses in brain. Transgenic mice without the TNF receptor do not transport TNF to brain so the LPS–TLR4 proinflammatory response amplifies across peripheral tissues, but does not spread to brain. Ethanol can increase blood proinflammatory cytokines through activation of proinflammatory responses in the liver and other tissues. One mechanism appears to involve ethanol causing the gut to become permeable or "leaky" (Ferrier et al., 2006). Only high doses of ethanol, at least 2–3 g/kg ETOH intragastric doses (Ferrier et al., 2006), potentiate gut innate immune signaling, disrupting gut tight junctions, and opening sites that allow the gut biome bacteria and their endotoxins to enter portal circulation leading to the liver where they can initiate a proinflammatory response (Sims, Rowe, Rietdijk, Herbst, & Coyle, 2010). Thus, high doses of ethanol increase systemic proinflammatory responses that can spread the proinflammatory response to brain through TNFα and likely other cytokines (see Fig. 10.6). Although some *in vitro* studies have suggested ethanol can insult the BBB, most *in vivo* studies do not show BBB damage following chronic ethanol treatment. Marshall et al. (2013) assessed BBB integrity using immunohistochemistry for albumin, a 140 kDa protein that does not cross an intact BBB, and found no evidence of brain albumin indicating an intact BBB following treatment with the severe intragastric 4-day binge rat alcohol dependence model of alcoholic brain damage. Using this model, Crews and Braun (2003) found that inhibition of NF-κB protected against 4-day binge alcohol-induced brain damage and inhibition of neurogenesis, consistent with proinflammatory responses in brain mediating brain damage without BBB damage, but being induced through direct activation of proinflammatory responses in brain and/or systemic proinflammatory signals being transported across the BBB and contributing to brain proinflammatory responses.

Although proinflammatory gene expression in blood and brain parallel each other at early time points, we found the brain response to LPS is much smaller than the liver and blood responses during the first few hours. Surprisingly, the blood and liver responses to LPS return to baseline over about 8–12 h, whereas the increase in proinflammatory gene expression in brain persists for months leading to degeneration of dopamine neurons in substantia nigra (Qin et al., 2007). Binge alcohol-induced liver and blood responses have not been extensively investigated and appear to be small and transient. However, brain expression of the proinflammatory cytokine, MCP-1, persists for at least 1 week (Qin et al., 2008). As mentioned earlier, exposure of C57Bl/6 mice to 10 daily doses of ethanol followed by LPS results in increased LPS induction of proinflammatory cytokines in liver, blood, and brain, compared to control LPS-treated animals (Qin et al., 2008). In these studies, ethanol-sensitized mice to the LPS response that resulted in sustained increases in multiple proinflammatory cytokines, including TNFα, IL-1β, and MCP-1 in brain, but not in the liver. The mechanism of the sustained brain response and transient liver response is not clear, although we found IL-10, an anti-inflammatory factor that inhibits NF-κB, increased in liver 1 week after alcohol treatment, but decreased in brain (Qin et al., 2008) consistent with anti-inflammatory mechanisms contributing to the loss of the liver response. Mice pretreated with ethanol are sensitized to both the TLR4 receptor agonist LPS as well as the TLR3 agonist Poly:IC (Qin & Crews, 2012a). Similar to LPS, Poly:IC induces proinflammatory genes in brain and 24 h after 10 days of daily alcohol administration (5 g/kg/day). These findings suggest that chronic ethanol sensitizes proinflammatory TLR responses that are easily observed after the clearance of alcohol. Thus, chronic ethanol sensitizes systemic and brain responses to neuroimmune gene activation through induction of HMGB1 and TLR proteins. Ethanol-induced leaky gut occurs with high binge drinking doses, with gut ethanol exposure often being equivalent to the beverage content (i.e., 80 proof is 40% ethanol). Bacterial products enter portal circulation activating Kupffer cells, liver monocytes, which produce cytokines including TNFα. TNFα can be transported into brain activating brain neuroimmune signaling that persists for long periods (Qin et al., 2007). Thus, there are at least two mechanisms of ethanol activation of neuroimmune signaling, a direct activation within brain as well as the spread of a systemic innate immune activation to brain.

6. ETHANOL INDUCTION OF HMGB1-TLR SIGNALING IN BRAIN

Studies investigating the mechanisms of ethanol induction of proinflammatory genes in brain have led to the discovery that chronic ethanol increases expression of TLR receptors as well as the TLR4 receptor agonist HMGB1. Studies of chronic 10-day ethanol treatment of mice (Crews et al., 2013), chronic *in vitro* treatment of rat brain slice cultures (Zou & Crews, 2014), and postmortem human alcoholic brain (Crews et al., 2013) find increased expression of HMGB1, TLR4, TLR3, and TLR2 (see Fig. 10.4). There are 13 TLRs that have been identified (i.e., TLRs 1–13) in mammals (Medzhitov, 2001; Takeda, Kaisho, & Akira, 2003) with all signaling through NF-κB transcription of additional proinflammatory cytokines through paracrine and autocrine amplification. Increases in receptors and agonists are common in innate immune signaling and suggest induction of HMGB1 and its receptor, the TLR4 receptor, as well as a less well-characterized receptor, RAGE, contribute to increases in neuroimmune gene expression. Brain slice culture experiments find ethanol releases HMGB1 that increases proinflammatory gene expression that is blocked by pharmacological antagonists or knock down of TLR4 receptors (Crews et al., 2013; Zou & Crews, 2014). Studies in adolescent rats (Vetreno & Crews, 2012), adolescent mice (Coleman et al., 2014), and adult mice (Qin et al., 2008; Qin, Liu, Hong, & Crews, 2013; Qin et al., 2007) find long-lasting increases in neuroimmune gene induction. Interestingly, levels of HMGB1 and TLR receptor expression in OFC correlate with lifetime alcohol consumption (see Fig. 10.7). Moderate drinking humans consume much less alcohol than alcoholics and set the *y*-axis intercept, whereas alcoholics who are known to vary greatly in the duration and amounts of active drinking bouts show a large variation in lifetime alcohol consumption that correlates with expression of HMGB1-TLR receptor expression in their brains. This interesting correlation could only occur if ethanol induction of HMGB1-TLR receptors was persistent and cumulative with binge drinking episodes (see Fig. 10.7). Together, these studies suggest that HMGB1-TLR4 signaling is increased by chronic binge drinking, contributing to the persistent and sustained induction of HMGB1-TLR4 proinflammatory signaling in brain.

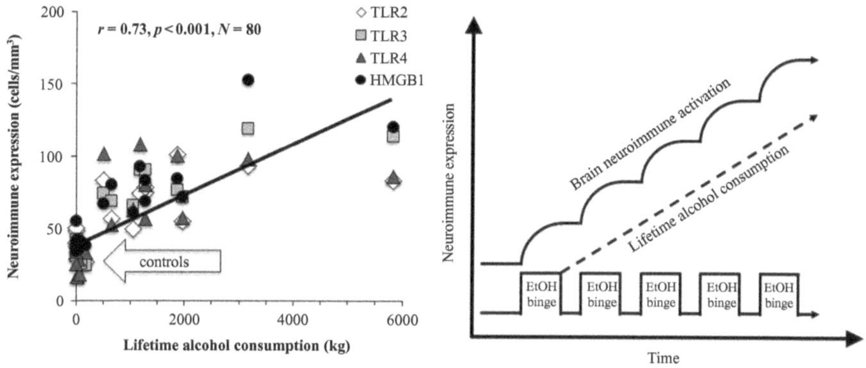

Figure 10.7 Cycles of chronic alcohol consumption lead to persistently increased neuroimmune gene expression. *Left*: Depicted are correlations of individual human TLR2 ($r=0.66$, $p<0.01$), TLR3 ($r=0.83$, $p<0.001$), TLR4 ($r=0.62$, $p<0.01$), and HMGB1 ($r=0.83$, $p<0.001$) immunoreactivity versus lifetime alcohol consumption (kg) in alcoholics and moderate drinking controls. Across human subjects, lifetime alcohol consumption positively correlated with neuroimmune signal immunoreactivity. Note that moderate drinking control subjects are clustered along the *y*-axis due to low lifetime alcohol consumption values and similar neuroimmune expression. Alcoholic subjects show considerable variation in neuroimmune expression, and over a 10-fold variation in lifetime alcohol consumption. *Right*: Repeated ethanol (EtOH) binges result in increased microglial and astrocytic activation as well as upregulated neuroimmune gene expression. In previous studies, we have shown that neuroimmune activation persists for long periods following upregulation in brain (Qin et al., 2008; Vetreno & Crews, 2012; Vetreno et al., 2013).

7. NADPH OXIDASE AND NEURODEGENERATION

One innate immune gene induced by ethanol and endotoxin is NADPH oxidase, a multisubunit enzyme that catalytically makes superoxide. NADPH oxidase was first characterized as a phagocytic oxidase in monocytes where it was hypothesized to contribute to oxidizing infectious agents. NADPH oxidase produces superoxide that can increase NF-κB transcription creating another amplifying loop of proinflammatory signaling (see Fig. 10.2). More recent studies have found that there are multiple genes and forms of NADPH oxidase all innate immune genes. Qin and Crews (2012b) discovered that LPS and ethanol can increase brain expression of NADPH oxidase subunits, particularly gp91^{phox}, the superoxide forming subunit in brain, and that ethanol treatment of mice increased superoxide formation in brain in association with neuronal death. Inhibition of oxidases reduced superoxide formation and protected against alcohol-induced neuronal death. These findings are consistent with oxidative stress through innate immune gene induction making a significant contribution to alcoholic brain damage. In mice, LPS treatment induces neuroimmune gene expression, NADPH oxidase, and oxidative stress that persists for at least 20 months and leads to neurodegeneration (Qin et al., 2013). The prolonged and persistent induction of NADPH oxidase and oxidative stress in brain could contribute to the persistent increase in NF-κB transcription, since oxidative free radicals can activate NF-κB. These studies suggest that oxidative stress contributes to alcoholic neurodegeneration.

8. NEUROIMMUNE SIGNALING, HYPEREXCITABILITY, AND NEURONAL DEATH

Excitotoxicity is associated with alcoholic neurodegeneration and HMGB1-TLR4 signaling. Chronic ethanol treatment of neurons leads to increased sensitivity to excitotoxicity (Chandler, Guzman, Sumners, & Crews, 1994). Ethanol potentiates glutamate excitotoxicity in brain slice cultures due to blockade of glial transporters (Zou & Crews, 2010). However, in neuronal primary cultures, ethanol blocks NMDA excitotoxicity consistent with many studies finding ethanol inhibition of NMDA receptors (Chandler, Harris, & Crews, 1998). Similar to ethanol blocking TLR4 receptor signaling when present, ethanol blocks NMDA receptors when present (Chandler et al., 1998). Although ethanol can block NMDA responses, glutamate excitotoxicity is increased by ethanol and TNFα in brain slice cultures due in part to glial loss of glutamate uptake (Zou & Crews, 2006) and perhaps release by ethanol. Further, HMGB1-TLR4 signaling has been shown to activate kinase cascades that lead to phosphorylation of the NR2B subunit of NMDA receptors causing the migration of more NMDA receptors to the synapse that increase synaptic NMDA receptors, neuronal excitability, and excitotoxicity (see Fig. 10.8; Balosso, Liu, Bianchi, & Vezzani, 2014; Maroso et al., 2010). Both HMGB1-TLR4 signaling (Balosso et al., 2014) and IL1β-IL1R signaling (Viviani et al., 2003) have been shown to increase NMDA receptor mediated calcium flux, neuronal excitability, and excitotoxicity through activation of kinase cascades. IL1β-IL1R activation of Src kinase has been found to increase NMDA calcium flux, excitability, and excitotoxicity. Many studies have found tyrosine-kinase activation can increase excitability through increases in NR2B-NMDA receptor phosphorylation (see Fig. 10.8). Ron's group

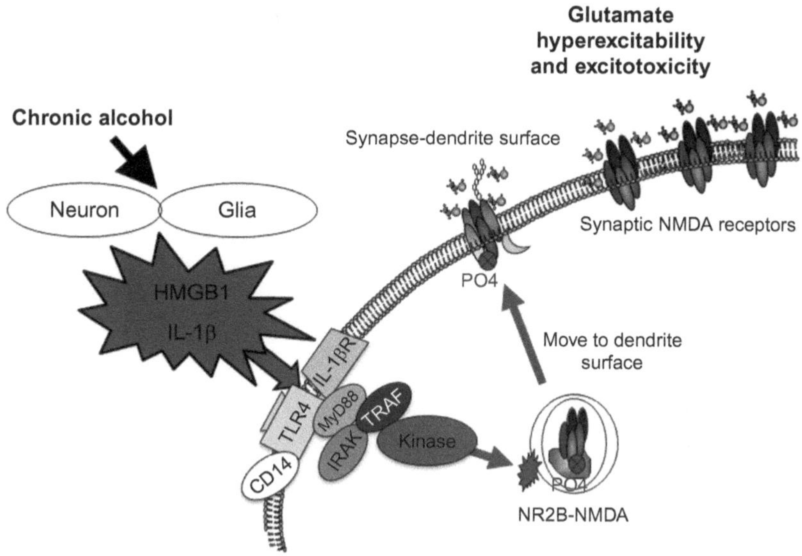

Figure 10.8 HMGB1-TLR signaling-induced hyperexcitability contributes to the neurobiology of addiction. A simplified schematic depicting how neuroimmune signaling leads to hyperexcitability and the neurobiology of addiction. Alcohol and stress activation neurons and glia in the CNS, resulting in the release of various neuroimmune signals (e.g., high-mobility group box 1 (HMGB1) and interleukin-1β (IL-1β)) that activate neuroimmune receptors (i.e., Toll-like receptors (TLRs) and receptor for advanced glycation end products (RAGE)). Neuroimmune receptor stimulation leads to activation of glutamatergic *N*-methyl-D-aspartate (NMDA) receptors (Iori et al., 2013; Maroso et al., 2010), which increases Ca^{2+} flux triggering induction of neuroimmune genes. In addition, TLR/RAGE activation leads to downstream transcription of nuclear factor kappa-light-chain-enhancer of activated B cells (NF-κB) signaling that might be accompanied by diminished cyclic AMP-responsive element binding protein (CREB) expression, which contributes to neuroimmune gene induction. These two pathways converge, leading to cycles of neuroimmune gene induction that lead to hyperexcitability, neuronal cell death, and network reorganization that culminates in addiction.

has found that ethanol increases NMDA excitability in hippocampus through kinase activation that alters receptor trafficking, leading to increased NR2B-NMDA receptors and increased excitability (Suvarna et al., 2005). Another mechanism of chronic ethanol-induced hyperexcitability is neuroimmune inhibition of glial glutamate transporters (Zou & Crews, 2005). Ethanol releases HMGB1-creating hyperexcitability that disrupts synaptic plasticity and sensitizes to excitotoxicity. HMGB1 is massively released during brain damage activating persistent neuroimmune gene induction (Kim et al., 2006). Maroso et al. (2010) found increased release of HMGB1 with hippocampal excitability that caused seizures leading to persistent increases in HMGB1 and excitability. Ethanol has modest cumulative effects with repeated chronic exposure increasing excitability and excitotoxicity due to increased neuroimmune signaling (see Fig. 10.8). Thus, the global neurodegeneration with the most severe losses in frontal cortex found in alcoholism is secondary to the persistent and progressive neuroimmune activation that occurs during alcoholism, a chronic relapsing disorder.

9. ADOLESCENCE: A MAJOR PERIOD OF RISK FOR ALCOHOL DEPENDENCE

Adolescence is a developmental period involving increased play behavior, thrill seeking, risk taking, puberty, and transitions to independence. During adolescence, the brain continues to develop, with the frontal cortex continuing to show structural changes coincident with maturation of adult behaviors and executive functions (Ernst, Romeo, & Andersen, 2009). Adolescence is also a period of experimentation as exemplified by the finding that the highest level of binge drinking among humans occurs during late adolescence. The prevalence of lifetime alcohol use disorders, as well as violence, fights, and injuries associated with alcohol use are all associated with age of drinking onset (Brown et al., 2008; Dawson, Stinson, Chou, & Grant, 2008; Sher & Gotham, 1999). The younger the age of drinking onset, the more likely alcohol use disorder will develop. Although binge drinking peaks in adolescence, the association between age of drinking onset and alcohol use disorder could represent maturation of an "alcoholic destiny" due to the emergence of conduct disorder and/or antisocial personalities, or genetic factors that identify themselves with an early onset of alcohol drinking. Alternatively, the association could be related to binge drinking-induced changes in brain that alter brain maturation thereby increasing the risk of alcohol use disorder, or both (Crews & Boettiger, 2009; Crews et al., 2007). The high prevalence of binge drinking among adolescence increases the importance of understanding how binge drinking might impact the adolescent brain. The correlation of younger age of drinking onset with increased risks of lifetime alcohol use disorder also correlates with a smaller brain and greater expression of HMGB1 and TLR4 receptors as well as other neuroimmune signaling receptors (Vetreno et al., 2013; see Fig. 10.9). The cause and effect of these associations is likely to be due to both preexisting conditions maturing into dysfunctional behavior as well as alcohol-induced factors that change the life course, and increase

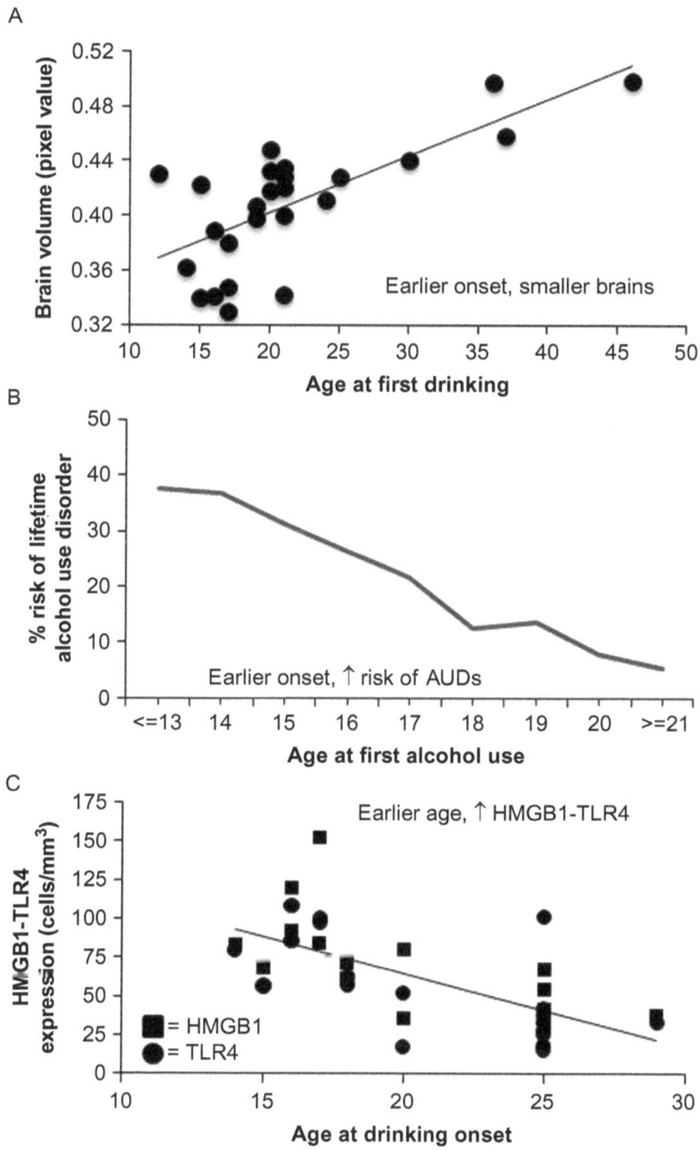

Figure 10.9 Brain regional volume, risk of alcoholism, and induction of innate immune genes correlates with age of drinking onset in humans. (A) Age of drinking onset correlates positively with gray matter reductions in the middle frontal gyrus, brainstem, and cerebellum. Depicted is correlation of pixel value of cerebellum with age at first drink. (B) An earlier age of drinking onset is predictive of an increased likelihood of developing an alcohol use disorder during an individual's lifetime. (C) TLR4 and HMGB1 expression in the postmortem human brain are negatively correlated with age of drinking onset. *Panel A: Adapted from Chanraud et al. (2007). Panel B: Adapted from Grant (1998). Panel C: Adapted from Vetreno et al. (2013).*

dysfunctional behavior, perhaps by changing brain maturation. The contributions of these two factors can only be determined by controlled experiments that do not have different genetic or other factors other than adolescent alcohol exposure. These studies cannot be done in humans, but are being done in rats where genetics and environment can be controlled. The essential need to understand the neurobiology of adolescent drinking on adulthood (NADIA) resulted in the formation of the NADIA consortium funded by NIAAA to address the contribution of adolescent alcohol abuse to adult psychopathology.

Adolescents have an immature response to alcohol, with evidence supporting unique factors that differ from the adult response to alcohol. Adolescent rats show greater ethanol-induced memory impairment on the Morris water maze and in discrimination tasks than do adults (Land & Spear, 2004; Markwiese, Acheson, Levin, Wilson, & Swartzwelder, 1998). Similarly, in humans that initiate drinking in their early 1920s are more sensitive to the effects of ethanol on multiple memory tasks compared to those that start in their late 1920s (Acheson, Stein, & Swartzwelder, 1998). The adolescent hippocampal response to ethanol compared to adults shows more potent inhibition of NMDA receptor-mediated synaptic activity (Swartzwelder, Wilson, & Tayyeb, 1995) and the induction of long-term potentiation (Martin, Tayyeb, & Swartzwelder, 1995). Adolescents, already social, are also uniquely sensitive to the social facilitative effects of ethanol (Varlinskaya & Spear, 2002). Adolescent rats are also more sensitive to binge drinking models of brain damage, particularly in the frontal cortex (Crews, Bechara, et al., 2006; Crews, Mdzinarishvili, et al., 2006). Interestingly, adolescent rats are less sensitive than adults to the sedative (Little, Kuhn, Wilson, & Swartzwelder, 1996; Silveri & Spear, 1998), motor impairing (Little et al., 1996; White, Bae, et al., 2002; White, Truesdale, et al., 2002), social inhibitory (Varlinskaya & Spear, 2002), and aversive (Anderson, Varlinskaya, & Spear, 2010) effects of ethanol. Adolescent rats also show electrophysiological differences from adults in the hippocampus, with reduced sensitivity to γ-aminobutyric acid (GABA) type A (GABAA) receptor-mediated inhibition (Carr, Spence, Peter Eriksson, Lumeng, & Li, 2003; Sullivan et al., 2006; Yan et al., 2009; Yan, Li, Madison, Wilson, & Swartzwelder, 2010). Reduced adolescent sedative sensitivity to alcohol and increased cognitive disruption is consistent with the human findings of high binge drinking coupled with high risk of horrific traffic accidents that have prompted zero tolerance for underage drinking and driving laws across the United States. The continuous increase in high binge drinking levels in human adolescents over the past decade justifies the need to know the long-term consequences of adolescent alcohol abuse.

The adolescent brain is structurally different from adult brain with both brain regional volume (Sowell et al., 2004) and prefrontal cortical synaptic changes continuing into the early 1930s in humans. Rat studies have found that the adolescent frontal cortex is more sensitive to ethanol toxicity (Crews, Bechara, et al., 2006). Multiple studies on adults following adolescent behavior done by NADIA investigators Linda Spear and Scott Swartzwelder find an altered adult that has persistent adolescent-like responses in adulthood. Studies of spatial learning in the radial arm maze find adolescence, but not adult, ethanol treatment models result in high adolescent-like sensitivity to ethanol memory impairment persisting into adulthood (Swartzwelder et al., 2014; White, Ghia, Levin, & Swartzwelder, 2000). Similarly, adolescent ethanol binge models produce a long-lasting decrease in sensitivity to the sedative/motor-impairing effects of acute ethanol, i.e., an adolescent-like low response persisting into adulthood (White, Truesdale, et al., 2002). At the cellular level, adolescent alcohol exposure, but not adult, produces an enduring decrease in the magnitude of GABA receptor-mediated tonic current in dentate granule cells (Fleming, Acheson, Moore, Wilson, & Swartzwelder, 2012; Fleming et al., 2013), which is critical for maintaining the balance of excitation and inhibition within hippocampal circuits. In addition, acute ethanol induced decreases in A-type potassium current (IA) in GABAergic hippocampal interneurons, persists after adolescent ethanol treatment resulting in adolescent-like responses in adulthood (Li et al., 2013). Levels of neurotransmitters in adolescent brain are maturing to adult levels and are altered in binge models of adolescent drinking. Expression of multiple cholinergic and dopaminergic genes peaks during adolescence (Spear, 2000). We found adolescence binge drinking models decrease the expression of multiple genes involved in cholinergic and dopaminergic neurotransmission (see Fig. 10.10; Coleman et al., 2011). Interestingly, many of these neurotransmitter genes undergo maturational decreases in expression that in animals exposed to alcohol continued the maturational decrease resulting in persistent impairment in adulthood in both mice (Coleman et al., 2011) and rats (Ehlers, Oguz, et al., 2013; Vetreno & Crews, 2012; see Fig. 10.11). Adolescent alcohol exposure, like adult alcohol exposure, induces neuroimmune genes in brain, and in humans, correlates with age of drinking onset (see Fig. 10.9). Indeed, Vetreno and Crews (2012) found that adolescent intermittent binge ethanol treatment of rats increases expression of multiple innate immune genes in adult frontal cortex. Interestingly, the critical neuroimmune signaling

A

System	Gene	PND38 EtOH (% P38 CON)	PND88 CON (% P38 CON)	PND88 EtOH (% P38 CON)
Cholinergic	**ChAT**	**45**	**36***	**15****
	Chrm1	38	29*	16**
	Chrm4	40	30	17**
	Chrna3	76	75	45*
	Chrna4	70	68	36*
	Chrnb2	37	31*	16**
	Chrnb4	39	36	19**
Dopamine	Drd1a	82	69	35*
	Drd2	75	71	35*
	Drd4	29*	45	30*
	Drd5	43	30*	15**
	Comt	98	71	36**

B

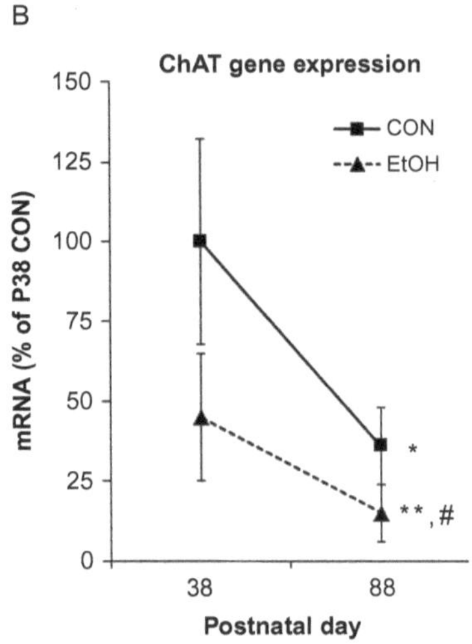

Figure 10.10 Adolescent ethanol treatment reduces choline acetyltransferase (ChAT) gene expression across development. Mice received either water or ethanol (5 g/kg) once a day for 10 days during adolescence (P28–P37). (A) The expression of cholinergic and dopaminergic neurotransmitter genes was assessed in whole brain either 24 h after treatment (P39; CON=6, EtOH=6) or in adulthood (P88; CON=5; EtOH=5) using an RT^2 Profiler Neurotransmitter Receptor and Regulator Superassay^{TM}. Gene expression levels are given as the percent of P38 CON. Genes that changed in at least on treatment group are presented. Data analysis was performed using the ΔCt method using the Data Analysis Template provided by SABioscience Corporation as published previously (Lee et al., 2008). A two-way ANOVA with Bonferroni posttests to account for multiple comparisons was used to determine statistical significance. (B) ChAT mRNA levels changed significantly across development. ChAT mRNA in adolescent CON mice (P38=100%) decreased during maturation to P80 (64% reduction). ChAT mRNA expression 24 h following ethanol treatment revealed a 55% reduction on P38, and a 85% reduction in adulthood on P88. *$p<0.05$, **$p<0.01$, relative to P38 CON (Coleman et al., 2011).

receptor TLR4 showed decreased maturational expression in controls, and adolescent binge ethanol increased expression and remained elevated into adulthood (Vetreno & Crews, 2012). In contrast, HMGB1 showed a developmental increase in expression in frontal cortex of control subjects, and adolescent binge ethanol led to greater increases in HMGB1 resulting in a persistent increase in adult HMGB1 and TLR4. High TLR4 levels in adults may represent adolescent-like HMGB1-TLR4 signaling in adults.

As mentioned earlier, HMGB1/TLR4-induced NMDA sensitivity would counteract the direct ethanol inhibitory effects on NMDA receptors resulting in adolescent-like tolerance to sedation and perhaps increased adolescent-like cognitive disruption. These mechanisms are oversimplified,

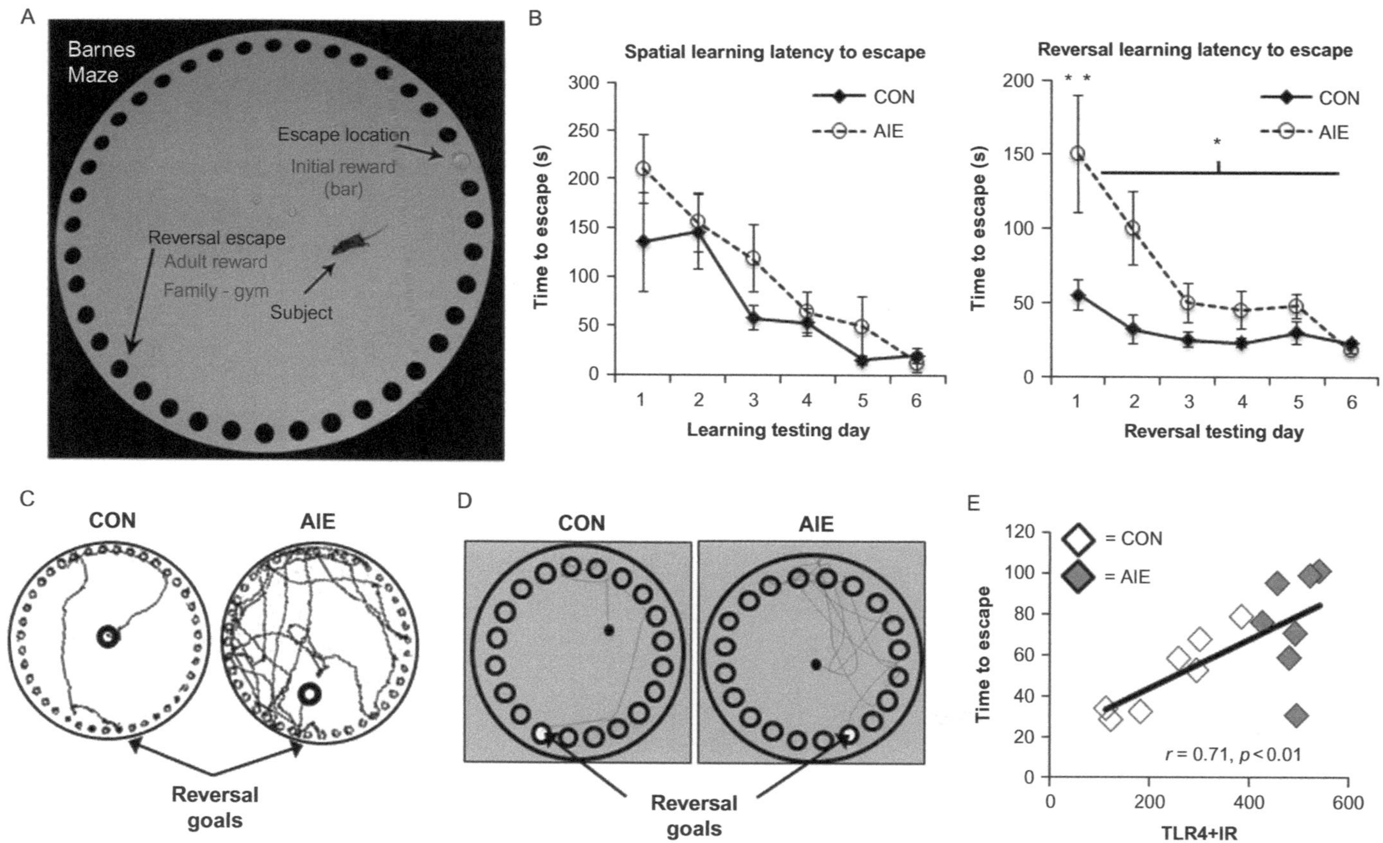

Figure 10.11 See legend on next page.

and likely more complex and in need of further study. Regardless of the molecular mechanism, NADIA studies strongly suggest that adolescent alcohol exposure increases risk for alcohol problems in adulthood. One repeated finding in the Crews laboratory is that models of binge drinking lead to long-term disruption of behavioral flexibility or the ability to adjust to changes in environment, which differs from simple learning (see Fig. 10.11). We have used reversal learning tasks and related them to changes in orbital frontal cortex, a brain region involved in predicting the outcome of initiated behaviors. Although we do not find deficits in adult learning after adolescent binge intermittent binge exposure, we have found deficits in reversal learning using in both the rats (Vetreno & Crews, 2012) and mice (Coleman et al., 2011), using both the Barnes maze and Morris water maze, respectively. These studies are consistent with the adolescent brain being vulnerable to long-lasting changes that persist through maturation into adulthood. The persistence of neuroimmune gene induction likely contributes to continuous slow degeneration as well as more specific insults on key neurotransmitters maturing in adolescence (Crews et al., 2007; Vetreno & Crews, 2012). The persistent loss of behavioral flexibility may be related to neuroimmune gene induction (see Fig. 10.11). Together, these findings of persistent loss of ability to adapt to changes and low sedative response to alcohol with increased sensitivity to cognitive disruption are all likely to promote and sustain high alcohol drinking levels that will further promote additional alcohol consumption and the chances that alcohol use disorder will develop in addition to alcoholic neurodegeneration.

Figure 10.11 Adolescent intermittent ethanol exposure causes reversal, but not spatial, learning deficits in adult rodents. (A) Photomicrograph of a mouse on the Barnes maze. The subjects and its assigned escape locations are identified. (B). Mice received either water or ethanol (5 g/kg, i.g.) once a day for 10 days during adolescence (P28–P37), and spatial learning was assessed for 6 days from P68 to P73. Adolescent ethanol exposure did not impair spatial learning, as measured by latency to escape, relative to CONs. However, adolescent ethanol exposure did impair reversal learning in adulthood as evidenced by increased time to locate and enter the reversal goal, relative to CONs. (C) Digital reconstruction of the search paths during the first trial of reversal learning in adulthood. (D) Rats received either water or ethanol (5 g/kg, i.g.) on a 2 days on/2 days off intermittent schedule from P25 to P55. Spatial and reversal learning was assessed from P64 to P75. As evidenced in the digital reconstruction, AIE treatment did not affect spatial learning, but did impair reversal learning as evidenced by the increased time to locate and enter the reversal goal. (E) TLR4 receptor expression in the adult prefrontal cortex is positively correlated with time to escape during the first trial of reversal learning on the Barnes maze (Vetreno & Crews, 2012). $*p < 0.05$, $**p < 0.01$, relative to CONs.

10. NEUROIMMUNE GENE EXPRESSION IN POSTMORTEM HUMAN ALCOHOLIC BRAIN

Although this review highlights HMGB1-TLR4 signaling, there are multiple other proinflammatory genes increased and we have found many in postmortem human brain (see Table 10.2). Our first human brain studies looked a microglia and the proinflammatory cytokine MCP-1 (CCL2), which was the cytokine most robustly induced by ethanol in brain slice cultures among those measured (Zou & Crews, 2012). We found that postmortem alcoholic human brain has increased levels of MCP-1 protein in VTA, amygdala, nucleus accumbens, and hippocampus (He & Crews, 2006; see Table 10.2). We also found increased expression of the microglial marker Iba-1. These studies indicate that the human alcoholic brain has increased neuroimmune gene expression. In later studies, we focused on the PFC, specifically the orbital frontal cortex, and found that postmortem alcoholic brain has increased levels of HMGB1, and well as TLR receptors, specifically TLR2, TLR3, and TLR4 receptors (Crews et al., 2013). In other studies, we have found increased IL-1β inflammasome markers in hippocampus of postmortem alcoholic brain that could contribute to loss of neurogenesis. In addition, NADPH oxidase is increased in alcoholic OFC consistent with increased oxidative stress as found in mice. Further, RAGE, another HMGB1 receptor, is increased in postmortem human alcoholic brain (Vetreno et al., 2013). These studies indicate that multiple neuroimmune genes are increased in alcoholic brain and likely contribute to neurodegeneration and the neurobiology of alcoholism. We investigated the relationship between alcohol drinking and neuroimmune gene expression across controls and alcoholics. Interestingly, two forms of correlations were found linking neuroimmune gene expression to alcohol consumption and alcoholism. Adolescent drinking is known to increase risk of developing alcohol dependence with risks of alcoholism decreasing with every year of delaying alcohol use across adolescence. In OFC, we found HMGB1-TLR4 expression was lower in individuals who initiated alcohol use having a negative correlation with age of drinking onset. In addition, total lifetime alcohol consumption across groups has a positive correlation with OFC expression of HMGB1, TLR4, TLR3, TLR2, and RAGE. These findings further support the role of neuroimmune signaling in alcoholic brain and alcoholic neurodegeneration.

11. SUMMARY

Alcoholic neurodegeneration is global and defuse with the most severe loss in frontal cortex. Neuroimmune gene induction by binge drinking increases neurodegeneration through increased oxidative stress, particularly NADPH oxidase-induced oxidative stress. In addition, HMGB1-TLR4 and innate immune NF-κB target genes are increased leading to persistent and sensitized neuroimmune responses to ethanol and other agents that release HMGB1 or directly stimulate TLR receptors and/or NMDA receptors. Neuroimmune signaling and glutamate excitotoxicity are linked to alcoholic neurodegeneration. Models of adolescent alcohol abuse lead to significant frontal cortical degeneration and show the most severe loss of hippocampal neurogenesis. Adolescence is a period of high risk for ethanol-induced neurodegeneration and alterations in brain structure, gene expression, and maturation of adult phenotypes. Together, these findings support the hypothesis that adolescence is a period of risk for persistent and long-lasting increases in brain neuroimmune gene expression that promote persistent and long-term increases in alcohol consumption, neuroimmune gene induction, and neurodegeneration that we find associated with alcohol use disorders.

REFERENCES

Acheson, S. K., Stein, R. M., & Swartzwelder, H. S. (1998). Impairment of semantic and figural memory by acute ethanol: Age-dependent effects. *Alcoholism, Clinical and Experimental Research*, *22*, 1437–1442.

Alfonso-Loeches, S., & Guerri, C. (2011). Molecular and behavioral aspects of the actions of alcohol on the adult and developing brain. *Critical Reviews in Clinical Laboratory Sciences*, *48*, 19–47.

Alfonso-Loeches, S., Pascual, M., Gomez-Pinedo, U., Pascual-Lucas, M., Renau-Piqueras,-J., & Guerri, C. (2012). Toll-like receptor 4 participates in the myelin disruptions associated with chronic alcohol abuse. *Glia*, *60*, 948–964.

Alfonso-Loeches, S., Pascual-Lucas, M., Blanco, A. M., Sanchez-Vera, I., & Guerri, C. (2010). Pivotal role of TLR4 receptors in alcohol-induced neuroinflammation and brain damage. *Journal of Neuroscience: The Official Journal of the Society for Neuroscience*, *30*, 8285–8295.

Altman, J., & Das, G. D. (1965). Autoradiographic and histological evidence of postnatal hippocampal neurogenesis in rats. *Journal of Comparative Neurology*, *124*, 319–335.

Alvarez-Buylla, A., & Garcia-Verdugo, J. M. (2002). Neurogenesis in adult subventricular zone. *Journal of Neuroscience*, *22*, 629–634.

Anderson, R. I., Varlinskaya, E. I., & Spear, L. P. (2010). Ethanol-induced conditioned taste aversion in male Sprague-Dawley rats: Impact of age and stress. *Alcoholism, Clinical and Experimental Research*, *34*, 2106–2115.

Balosso, S., Liu, J., Bianchi, M. E., & Vezzani, A. (2014). Disulfide-containing high mobility group box-1 promotes N-methyl-d-aspartate receptor function and excitotoxicity by activating Toll-like receptor 4-dependent signaling in hippocampal neurons. *Antioxidants & Redox Signaling*, [epub ahead of print]. http://dx.doi.org/10.1089/ars.2013.5349.

Banks, W. A., & Erickson, M. A. (2010). The blood-brain barrier and immune function and dysfunction. *Neurobiology of Disease*, *37*, 26–32.

Bechara, A., Damasio, A. R., Damasio, H., & Anderson, S. W. (1994). Insensitivity to future consequences following damage to human prefrontal cortex. *Cognition*, *50*, 7–15.

Bechara, R. I., Pelaez, A., Palacio, A., Joshi, P. C., Hart, C. M., Brown, L. A., et al. (2005). Angiotensin II mediates glutathione depletion, transforming growth factor-beta1 expression, and epithelial barrier dysfunction in the alcoholic rat lung. *American Journal of Physiology. Lung Cellular and Molecular Physiology*, *289*, L363–L370.

Bison, S., & Crews, F. (2003). Alcohol withdrawal increases neuropeptide Y immunoreactivity in rat brain. *Alcoholism, Clinical and Experimental Research*, *27*, 1173–1183.

Blanco, A. M., Pascual, M., Valles, S. L., & Guerri, C. (2004). Ethanol-induced iNOS and COX-2 expression in cultured astrocytes via NF-kappa B. *Neuroreport*, *15*, 681–685.

Blanco, A. M., Valles, S. L., Pascual, M., & Guerri, C. (2005). Involvement of TLR4/type I IL-1 receptor signaling in the induction of inflammatory mediators and cell death induced by ethanol in cultured astrocytes. *Journal of Immunology*, *175*, 6893–6899.

Blednov, Y. A., Benavidez, J. M., Geil, C., Perra, S., Morikawa, H., & Harris, R. A. (2011). Activation of inflammatory signaling by lipopolysaccharide produces a prolonged increase of voluntary alcohol intake in mice. *Brain Behavior, and Immunity*, (Suppl. 1), S92–S105.

Blednov, Y. A., Ponomarev, I., Geil, C., Bergeson, S., Koob, G. F., & Harris, R. A. (2012). Neuroimmune regulation of alcohol consumption: behavioral validation of genes obtained from genomic studies. *Addiction Biology*, *17*(1), 108–120.

Broadwater, M., Liu, W., Crews, F., & Spear, L. (2013). Persistent loss of hippocampal neurogenesis and increased cell death following adolescent, but not adult, chronic ethanol exposure. *Developmental Neuroscience*, *36*(3–4), 1–9.

Brown, S. A., Mcgue, M., Maggs, J., Schulenberg, J., Hingson, R., Swartzwelder, S., et al. (2008). A developmental perspective on alcohol and youths 16 to 20 years of age. *Pediatrics*, *121*(Suppl. 4), S290–S310.

Carr, L. G., Spence, J. P., Peter Eriksson, C. J., Lumeng, L., & Li, T. K. (2003). AA and ANA rats exhibit the R100Q mutation in the GABAA receptor alpha 6 subunit. *Alcohol*, *31*, 93–97.

Chandler, L. J., Guzman, N., Sumners, C., & Crews, F. T. (1994). Induction of nitric oxide synthase in brain glial cells: Possible interaction with ethanol in reactive neuronal injury. In F. E. Lancaster (Ed.), *Alcohol and glial cells*. Rockville: NIAAA.

Chandler, L. J., Harris, R. A., & Crews, F. T. (1998). Ethanol tolerance and synaptic plasticity. *Trends in Pharmacological Sciences*, *19*, 491–495.

Chanraud, S., Martelli, C., Delain, F., Kostogianni, N., Douaud, G., Aubin, H. J., et al. (2007). Brain morphometry and cognitive performance in detoxified alcohol-dependents with preserved psychosocial functioning. *Neuropsychopharmacology*, *32*(2), 429–438.

Chavan, S. S., Huerta, P. T., Robbiati, S., Valdes-Ferrer, S. I., Ochani, M., Dancho, M., et al. (2012). HMGB1 mediates cognitive impairment in sepsis survivors. *Molecular Medicine*, *18*, 930–937.

Coleman, L. G., Jr., He, J., Lee, J., Styner, M., & Crews, F. T. (2011). Adolescent binge drinking alters adult brain neurotransmitter gene expression, behavior, brain regional volumes, and neurochemistry in mice. *Alcoholism, Clinical and Experimental Research*, *35*, 671–688.

Coleman, L. G., Jr., Liu, W., Oguz, I., Styner, M., & Crews, F. T. (2014). Adolescent binge ethanol treatment alters adult brain regional volumes, cortical extracellular matrix protein and behavioral flexibility. *Pharmacology, Biochemistry, and Behavior*, *116*, 142–151.

Collins, M. A., Zou, J. Y., & Neafsey, E. J. (1998). Brain damage due to episodic alcohol exposure in vivo and in vitro: Furosemide neuroprotection implicates edema-based mechanism. *FASEB Journal: Official Publication of the Federation of American Societies for Experimental Biology*, *12*, 221–230.

Colton, C. A. (2009). Heterogeneity of microglial activation in the innate immune response in the brain. *Journal of Neuroimmune Pharmacology: The Official Journal of the Society on NeuroImmune Pharmacology*, *4*, 399–418.

Corso, T. D., Mostafa, H. M., Collins, M. A., & Neafsey, E. J. (1998). Brain neuronal degeneration caused by episodic alcohol intoxication in rats: Effects of nimodipine, 6,7-dinitro-quinoxaline-2,3-dione, and MK-801. *Alcoholism, Clinical and Experimental Research*, *22*, 217–224.

Crews, F. T. (2012). Immune function genes, genetics, and the neurobiology of addiction. *Alcohol Research: Current Reviews*, *34*, 355–361.

Crews, F. T., Bechara, R., Brown, L. A., Guidot, D. M., Mandrekar, P., Oak, S., et al. (2006). Cytokines and alcohol. *Alcoholism, Clinical and Experimental Research*, *30*, 720–730.

Crews, F. T., & Boettiger, C. A. (2009). Impulsivity, frontal lobes and risk for addiction. *Pharmacology, Biochemistry and Behavior*, *93*, 237–247.

Crews, F. T., & Braun, C. J. (2003). Binge ethanol treatment causes greater brain damage in alcohol-preferring P rats than in alcohol-nonpreferring NP rats. *Alcoholism, Clinical and Experimental Research*, *27*, 1075–1082.

Crews, F. T., Collins, M. A., Dlugos, C., Littleton, J., Wilkins, L., Neafsey, E. J., et al. (2004). Alcohol-induced neurodegeneration: When, where and why? *Alcoholism: Clinical and Experimental Research*, *28*, 350–364.

Crews, F., He, J., & Hodge, C. (2007). Adolescent cortical development: A critical period of vulnerability for addiction. *Pharmacology, Biochemistry and Behavior*, *86*, 189–199.

Crews, F. T., Mdzinarishvili, A., Kim, D., He, J., & Nixon, K. (2006). Neurogenesis in adolescent brain is potently inhibited by ethanol. *Neuroscience*, *137*, 437–445.

Crews, F. T., & Nixon, K. (2009). Mechanisms of neurodegeneration and regeneration in alcoholism. *Alcohol and Alcoholism*, *44*, 115–127.

Crews, F. T., Qin, L., Sheedy, D., Vetreno, R. P., & Zou, J. (2013). High mobility group box 1/Toll-like receptor danger signaling increases brain neuroimmune activation in alcohol dependence. *Biological Psychiatry*, *73*, 602–612.

Crews, F. T., Zou, J., & Qin, L. (2011). Induction of innate immune genes in brain create the neurobiology of addiction. *Brain, Behavior, and Immunity*, *25*(Suppl. 1), S4–S12.

Dawson, D. A., Stinson, F. S., Chou, S. P., & Grant, B. F. (2008). Three-year changes in adult risk drinking behavior in relation to the course of alcohol-use disorders. *Journal of Studies on Alcohol and Drugs*, *69*, 866–877.

Deshmukh, A., Rosenbloom, M. J., Pfefferbaum, A., & Sullivan, E. V. (2002). Clinical signs of cerebellar dysfunction in schizophrenia, alcoholism, and their comorbidity. *Schizophrenia Research*, *57*, 281–291.

Drake, A. I., Butters, N., Shear, P. K., Smith, T. L., Bondi, M., Irwin, M., et al. (1995). Cognitive recovery with abstinence and its relationship to family history for alcoholism. *Journal of Studies on Alcohol*, *56*, 104–109.

Ehlers, C. L., Liu, W., Wills, D. N., & Crews, F. T. (2013). Periadolescent ethanol vapor exposure persistently reduces measures of hippocampal neurogenesis that are associated with behavioral outcomes in adulthood. *Neuroscience*, *244*, 1–15.

Ehlers, C. L., Oguz, I., Budin, F., Wills, D. N., & Crews, F. T. (2013). Peri-adolescent ethanol vapor exposure produces reductions in hippocampal volume that are correlated with deficits in prepulse inhibition of the startle. *Alcoholism, Clinical and Experimental Research*, *37*, 1466–1475.

Ende, G., Walter, S., Welzel, H., Demirakca, T., Wokrina, T., Ruf, M., et al. (2006). Alcohol consumption significantly influences the MR signal of frontal choline-containing compounds. *NeuroImage*, *32*, 740–746.

Ernst, M., Romeo, R. D., & Andersen, S. L. (2009). Neurobiology of the development of motivated behaviors in adolescence: A window into a neural systems model. *Pharmacology, Biochemistry, and Behavior, 93*, 199–211.

Farina, C., Aloisi, F., & Meinl, E. (2007). Astrocytes are active players in cerebral innate immunity. *Trends in Immunology, 28*, 138–145.

Fein, G., Di Sclafani, V., Cardenas, V. A., Goldmann, H., Tolou-Shams, M., & Meyerhoff, D. J. (2002). Cortical gray matter loss in treatment-naive alcohol dependent individuals. *Alcoholism: Clinical and Experimental Research, 26*, 558–564.

Fein, G., Di Sclafani, V., & Finn, P. (2010). Sensation seeking in long-term abstinent alcoholics, treatment-naive active alcoholics, and nonalcoholic controls. *Alcoholism: Clinical and Experimental Research, 34*, 1045–1051.

Fernandez-Lizarbe, S., Pascual, M., & Guerri, C. (2009). Critical role of TLR4 response in the activation of microglia induced by ethanol. *Journal of Immunology, 183*, 4733–4744.

Ferrier, L., Berard, F., Debrauwer, L., Chabo, C., Langella, P., Bueno, L., et al. (2006). Impairment of the intestinal barrier by ethanol involves enteric microflora and mast cell activation in rodents. *American Journal of Pathology, 168*, 1148–1154.

Fleming, R. L., Acheson, S. K., Moore, S. D., Wilson, W. A., & Swartzwelder, H. S. (2012). In the rat, chronic intermittent ethanol exposure during adolescence alters the ethanol sensitivity of tonic inhibition in adulthood. *Alcoholism: Clinical and Experimental Research, 36*, 279–285.

Fleming, R. L., Li, Q., Risher, M. L., Sexton, H. G., Moore, S. D., Wilson, W. A., et al. (2013). Binge-pattern ethanol exposure during adolescence, but not adulthood, causes persistent changes in GABAA receptor-mediated tonic inhibition in dentate granule cells. *Alcoholism, Clinical and Experimental Research, 37*, 1154–1160.

Fortier, C. B., Leritz, E. C., Salat, D. H., Venne, J. R., Maksimovskiy, A. L., Williams, V., et al. (2011). Reduced cortical thickness in abstinent alcoholics and association with alcoholic behavior. *Alcoholism, Clinical and Experimental Research, 35*, 2193–2201.

Gage, F. H. (2002). Neurogenesis in the adult brain. *Journal of Neuroscience, 22*, 612–613.

Gansler, D. A., Harris, G. J., Oscar-Berman, M., Streeter, C., Lewis, R. F., Ahmed, I., et al. (2000). Hypoperfusion of inferior frontal brain regions in abstinent alcoholics: A pilot SPECT study. *Journal of Studies on Alcohol, 61*, 32–37.

Ginhoux, F., Greter, M., Leboeuf, M., Nandi, S., See, P., Gokhan, S., et al. (2010). Fate mapping analysis reveals that adult microglia derive from primitive macrophages. *Science, 330*, 841–845.

Glass, C. K., Saijo, K., Winner, B., Marchetto, M. C., & Gage, F. H. (2010). Mechanisms underlying inflammation in neurodegeneration. *Cell, 140*, 918–934.

Graeber, M. B. (2010). Changing face of microglia. *Science, 330*, 783–788.

Grant, B. F. (1998). The impact of a family history of alcoholism on the relationship between age at onset of alcohol use and DSM-IV alcohol dependence: Results from the National Longitudinal Alcohol Epidemiologic Survey. *Alcohol Health and Research World, 22*, 144–147.

Grant, B. F., & Dawson, D. A. (1997). Age of onset of alcohol use and its association with DSM-IV alcohol abuse and dependence: Results from the national longitudinal alcohol epidemiologic survey. *Journal of Substance Abuse, 9*, 103–110.

Hamelink, C., Hampson, A., Wink, D. A., Eiden, L. E., & Eskay, R. L. (2005). Comparison of cannabidiol, antioxidants, and diuretics in reversing binge ethanol-induced neurotoxicity. *Journal of Pharmacology and Experimental Therapeutics, 314*, 780–788.

Hansson, A. C., Nixon, K., Rimondini, R., Damadzic, R., Sommer, W. H., Eskay, R., et al. (2010). Long-term suppression of forebrain neurogenesis and loss of neuronal progenitor cells following prolonged alcohol dependence in rats. *International Journal of Neuropsychopharmacology/Official Scientific Journal of the Collegium Internationale Neuropsychopharmacologicum (CINP), 13*, 583–593.

Harding, A. J., Halliday, G. M., Ng, J. L., Harper, C. G., & Kril, J. J. (1996). Loss of vasopressin-immunoreactive neurons in alcoholics is dose-related and time-dependent. *Neuroscience, 72*, 699–708.

Hardingham, G. E., & Bading, H. (2010). Synaptic versus extrasynaptic NMDA receptor signalling: Implications for neurodegenerative disorders. *Nature reviews. Neuroscience, 11*, 682–696.

Harper, C., Kril, J., & Daly, J. (1987). Are we drinking our neurones away? *British Medical Journal (Clinical Research Ed), 294*, 534–536.

He, J., & Crews, F. (2007a). Neurogenesis decreases during brain maturation from adolescence to adulthood. *Pharmacology, Biochemistry, and Behavior, 86*(2), 327–333.

He, J., & Crews, F. T. (2007b). Neurogenesis decreases during brain maturation from adolescence to adulthood. *Pharmacology, Biochemistry and Behavior, 86*, 327–333.

He, J., & Crews, F. T. (2008). Increased MCP-1 and microglia in various regions of the human alcoholic brain. *Experimental Neurology, 210*, 349–358.

He, J., Nixon, K., Shetty, A. K., & Crews, F. T. (2005). Chronic alcohol exposure reduces hippocampal neurogenesis and dendritic growth of newborn neurons. *European Journal of Neuroscience, 21*, 2711–2720.

Herrera, D. G., Yague, A. G., Johnsen-Soriano, S., Bosch-Morell, F., Collado-Morente, L., Muriach, M., et al. (2003). Selective impairment of hippocampal neurogenesis by chronic alcoholism: Protective effects of an antioxidant. *Proceedings of the National Academy of Sciences of the United States of America, 100*, 7919–7924.

Huh, G. S., Boulanger, L. M., Du, H., Riquelme, P. A., Brotz, T. M., & Shatz, C. J. (2000). Functional requirement for class I MHC in CNS development and plasticity. *Science, 290*, 2155–2159.

Huttunen, H. J., & Rauvala, H. (2004). Amphoterin as an extracellular regulator of cell motility: From discovery to disease. *Journal of Internal Medicine, 255*, 351–366.

Iori, V., Maroso, M., Rizzi, M., Iyer, A. M., Vertemara, R., Carli, M., et al. (2013). Receptor for advanced glycation end products is upregulated in temporal lobe epilepsy and contributes to experimental seizures. *Neurobiology of Disease, 58*, 102–114.http://dx.doi.org/10.1016/j.nbd.2013.03.006.

Jang, M. H., Shin, M. C., Jung, S. B., Lee, T. H., Bahn, G. H., Kwon, Y. K., et al. (2002). Alcohol and nicotine reduce cell proliferation and enhance apoptosis in dentate gyrus. *Neuroreport, 13*, 1509–1513.

Jang, M. H., Shin, M. C., Kim, E. H., & Kim, C. J. (2002). Acute alcohol intoxication decreases cell proliferation and nitric oxide synthase expression in dentate gyrus of rats. *Toxicology Letters, 133*, 255–262.

Kim, J. B., Sig Choi, J., Yu, Y. M., Nam, K., Piao, C. S., Kim, S. W., et al. (2006). HMGB1, a novel cytokine-like mediator linking acute neuronal death and delayed neuroinflammation in the postischemic brain. *Journal of Neuroscience: The Official Journal of the Society for Neuroscience, 26*, 6413–6421.

Knapp, D. J., & Crews, F. T. (1999). Induction of cyclooxygenase-2 in brain during acute and chronic ethanol treatment and ethanol withdrawal. *Alcoholism: Clinical and Experimental Research, 23*, 633–643.

Kreutzberg, G. W. (1996). Microglia: A sensor for pathological events in the CNS. *Trends in Neurosciences, 19*, 312–318.

Kubota, M., Nakazaki, S., Hirai, S., Saeki, N., Yamaura, A., & Kusaka, T. (2001). Alcohol consumption and frontal lobe shrinkage: Study of 1432 non-alcoholic subjects. *Journal of Neurology, Neurosurgery and Psychiatry, 71*, 104–106.

Land, C., & Spear, N. E. (2004). Fear conditioning is impaired in adult rats by ethanol doses that do not affect periadolescents. *International Journal of Developmental Neuroscience: The Official Journal of the International Society for Developmental Neuroscience, 22*, 355–362.

Lee, J. K., McCoy, M. K., Harms, A. S., Ruhn, K. A., Gold, S. J., & Tansey, M. G. (2008). Regulator of G-protein signaling 10 promotes dopaminergic neuron survival via regulation of the microglial inflammatory response. *The Journal of Neuroscience: The Official Journal of the Society for Neuroscience, 28*(34), 8517–8528.

Li, T. K. (2008). Quantifying the risk for alcohol-use and alcohol-attributable health disorders: Present findings and future research needs. *Journal of Gastroenterology and Hepatology, 23*(Suppl. 1), S2–S8.

Li, Q., Fleming, R. L., Acheson, S. K., Madison, R. D., Moore, S. D., Risher, M. L., et al. (2013). Long-term modulation of A-type K(+) conductances in hippocampal CA1 interneurons in rats after chronic intermittent ethanol exposure during adolescence or adulthood. *Alcoholism: Clinical and Experimental Research, 37*, 2074–2085.

Little, P. J., Kuhn, C. M., Wilson, W. A., & Swartzwelder, H. S. (1996). Differential effects of ethanol in adolescent and adult rats. *Alcoholism: Clinical and Experimental Research, 20*, 1346–1351.

Liu, J., Yang, A. R., Kelly, T., Puche, A., Esoqa, C., June, H. L., Jr., et al. (2011). Binge alcohol drinking is associated with GABAA alpha2-regulated Toll-like receptor 4 (TLR4) expression in the central amygdala. *Proceedings of the National Academy of Sciences of the United States of America, 108*(11), 4465–4470.

Lonze, B. E., & Ginty, D. D. (2002). Function and regulation of CREB family transcription factors in the nervous system. *Neuron, 35*, 605–623.

Mantamadiotis, T., Lemberger, T., Bleckmann, S. C., Kern, H., Kretz, O., Martin Villalba, A., et al. (2002). Disruption of CREB function in brain leads to neurodegeneration. *Nature Genetics, 31*, 47–54.

Markwiese, B. J., Acheson, S. K., Levin, E. D., Wilson, W. A., & Swartzwelder, H. S. (1998). Differential effects of ethanol on memory in adolescent and adult rats. *Alcoholism: Clinical and Experimental Research, 22*, 416–421.

Maroso, M., Balosso, S., Ravizza, T., Liu, J., Aronica, E., Iyer, A. M., et al. (2010). Toll-like receptor 4 and high-mobility group box-1 are involved in ictogenesis and can be targeted to reduce seizures. *Nature Medicine, 16*, 413–419.

Marshall, S. A., McClain, J. A., Kelso, M. L., Hopkins, D. M., Pauly, J. R., & Nixon, K. (2013). Microglial activation is not equivalent to neuroinflammation in alcohol-induced neurodegeneration: The importance of microglia phenotype. *Neurobiology of Disease, 54*, 239–251.

Martin, D., Tayyeb, M. I., & Swartzwelder, H. S. (1995). Ethanol inhibition of AMPA and kainate receptor-mediated depolarizations of hippocampal area CA1. *Alcoholism: Clinical and Experimental Research, 19*, 1312–1316.

Medzhitov, R. (2001). Toll-like receptors and innate immunity. *Nature Reviews. Immunology, 1*, 135–145.

Michelucci, A., Heurtaux, T., Grandbarbe, L., Morga, E., & Heuschling, P. (2009). Characterization of the microglial phenotype under specific pro-inflammatory and anti-inflammatory conditions: Effects of oligomeric and fibrillar amyloid-beta. *Journal of Neuroimmunology, 210*, 3–12.

Nixon, K., & Crews, F. T. (2002). Binge ethanol exposure decreases neurogenesis in adult rat hippocampus. *Journal of Neurochemistry, 83*, 1087–1093.

Nixon, K., & Crews, F. T. (2004). Temporally specific burst in cell proliferation increases hippocampal neurogenesis in protracted abstinence from alcohol. *Journal of Neuroscience: The Official Journal of the Society for Neuroscience, 24*, 9714–9722.

Obernier, J. A., Bouldin, T. W., & Crews, F. T. (2002). Binge ethanol exposure in adult rats causes necrotic cell death. *Alcoholism: Clinical and Experimental Research, 26*, 547–557.

Obernier, J. A., White, A. M., Swartzwelder, H. S., & Crews, F. T. (2002). Cognitive deficits and CNS damage after a 4-day binge ethanol exposure in rats. *Pharmacology, Biochemistry and Behavior, 72*, 521–532.

O'Brien, C. P. (2008). Evidence-based treatments of addiction. *Philosophical Transactions of the Royal Society of London, Series B: Biological Sciences, 363*, 3277–3286, Review.

Okvist, A., Johansson, S., Kuzmin, A., Bazov, I., Merino-Martinez, R., Ponomarev, I., et al. (2007). Neuroadaptations in human chronic alcoholics: dysregulation of the NF-kappB system. *PLoS One, 2*(9), e930.

O'Neill, J., Cardenas, V. A., & Meyerhoff, D. J. (2001). Effects of abstinence on the brain: Quantitative magnetic resonance imaging and magnetic resonance spectroscopic imaging in chronic alcohol abuse. *Alcoholism: Clinical and Experimental Research, 25*, 1673–1682.

O'Neill, L. A., & Kaltschmidt, C. (1997). NF-kappa B: A crucial transcription factor for glial and neuronal cell function. *Trends in Neurosciences, 20*, 252–258.

Oscar-Berman, M., & Hutner, N. (1993). Frontal lobe changes after chronic alcohol ingestion. In W. A. Hunt, & S. J. Nixon (Eds.), *Alcohol-induced brain damage*. Rockville: NIAAA/NIH.

Oscar-Berman, M., & Marinkovic, K. (2007). Alcohol: Effects on neurobehavioral functions and the brain. *Neuropsychology Review, 17*, 239–257.

Osterholm, M. T. (2005). Preparing for the next pandemic. *New England Journal of Medicine, 352*, 1839–1842.

Pandey, S. C., Roy, A., & Mittal, N. (2001). Effects of chronic ethanol intake and its withdrawal on the expression and phosphorylation of the creb gene transcription factor in rat cortex. *Journal of Pharmacology and Experimental Therapeutics, 296*, 857–868.

Pandey, S. C., Zhang, D., Mittal, N., & Nayyar, D. (1999). Potential role of the gene transcription factor cyclic AMP-responsive element binding protein in ethanol withdrawal-related anxiety. *Journal of Pharmacology and Experimental Therapeutics, 288*, 866–878.

Parsons, O. A. (1993). Impaired neuropsychological cognitive functioning in sober alcoholics. In W. A. Hunt, & S. J. Nixon (Eds.), *Alcohol-induced brain damage*. Rockville: NIAAA/NIH.

Pascual, M., Balino, P., Alfonso-Loeches, S., Aragon, C. M., & Guerri, C. (2011). Impact of TLR4 on behavioral and cognitive dysfunctions associated with alcohol-induced neuroinflammatory damage. *Brain, Behavior, and Immunity, 25*(Suppl. 1), S80–S91.

Pfefferbaum, A., Sullivan, E. V., Mathalon, D. H., Shear, P. K., Rosenbloom, M. J., & Lim, K. O. (1995). Longitudinal changes in magnetic resonance imaging brain volumes in abstinent and relapsed alcoholics. *Alcohol, Clinical and Experimental Research, 19*, 1177–1191.

Potenza, M. N., & De Wit, H. (2010). Control yourself: Alcohol and impulsivity. *Alcoholism, Clinical and Experimental Research, 34*, 1303–1305.

Pruett, S. B., Zheng, Q., Fan, R., Matthews, K., & Schwab, C. (2004). Ethanol suppresses cytokine responses induced through Toll-like receptors as well as innate resistance to Escherichia coli in a mouse model for binge drinking. *Alcohol, 33*, 147–155.

Qin, L., & Crews, F. T. (2012a). Chronic ethanol increases systemic TLR3 agonist-induced neuroinflammation and neurodegeneration. *Journal of Neuroinflammation, 9*, 130.

Qin, L., & Crews, F. T. (2012b). NADPH oxidase and reactive oxygen species contribute to alcohol-induced microglial activation and neurodegeneration. *Journal of Neuroinflammation, 9*, 5.

Qin, L., He, J., Hanes, R. N., Pluzarev, O., Hong, J. S., & Crews, F. T. (2008). Increased systemic and brain cytokine production and neuroinflammation by endotoxin following ethanol treatment. *Journal of Neuroinflammation, 5*, 10.

Qin, L., Liu, Y., Hong, J. S., & Crews, F. T. (2013). NADPH oxidase and aging drive microglial activation, oxidative stress, and dopaminergic neurodegeneration following systemic LPS administration. *Glia, 61*, 855–868.

Qin, L., Wu, X., Block, M. L., Liu, Y., Breese, G. R., Hong, J. S., et al. (2007). Systemic LPS causes chronic neuroinflammation and progressive neurodegeneration. *Glia*, *55*, 453–462.

Quartin, A. A., Schein, R. M., Kett, D. H., & Peduzzi, P. N. (1997). Magnitude and duration of the effect of sepsis on survival Department of Veterans Affairs Systemic Sepsis Cooperative Studies Group. *JAMA: The Journal of the American Medical Association*, *277*, 1058–1063.

Raivich, G. (2005). Like cops on the beat: The active role of resting microglia. *Trends in Neurosciences*, *28*, 571–573.

Raivich, G., Bohatschek, M., Kloss, C. U., Werner, A., Jones, L. L., & Kreutzberg, G. W. (1999). Neuroglial activation repertoire in the injured brain: Graded response, molecular mechanisms and cues to physiological function. *Brain Research. Brain Research Reviews*, *30*, 77–105.

Schoenbaum, G., Roesch, M. R., Stalnaker, T. A., & Takahashi, Y. K. (2009). A new perspective on the role of the orbitofrontal cortex in adaptive behaviour. *Nature Reviews. Neuroscience*, *10*, 885–892.

Schwarz, J. M., & Bilbo, S. D. (2013). Adolescent morphine exposure affects long-term microglial function and later-life relapse liability in a model of addiction. *Journal of Neuroscience: The Official Journal of the Society for Neuroscience*, *33*, 961–971.

Shelton, M. D., & Parsons, O. A. (1987). Alcoholics' self-assessment of their neuropsychological functioning in everyday life. *Journal of Clinical Psychology*, *43*, 395–403.

Sher, K. J., & Gotham, H. J. (1999). Pathological alcohol involvement: A developmental disorder of young adulthood. *Development and Psychopathology*, *11*, 933–956.

Silveri, M. M., & Spear, L. P. (1998). Decreased sensitivity to the hypnotic effects of ethanol early in ontogeny. *Alcoholism: Clinical and Experimental Research*, *22*, 670–676.

Sims, G. P., Rowe, D. C., Rietdijk, S. T., Herbst, R., & Coyle, A. J. (2010). HMGB1 and RAGE in inflammation and cancer. *Annual Review of Immunology*, *28*, 367–388.

Sowell, E. R., Thompson, P. M., Leonard, C. M., Welcome, S. E., Kan, E., & Toga, A. W. (2004). Longitudinal mapping of cortical thickness and brain growth in normal children. *Journal of Neuroscience*, *24*, 8223–8231.

Spear, L. P. (2000). The adolescent brain and age-related behavioral manifestations. *Neuroscience and Biobehavioral Reviews*, *24*, 417–463.

Streit, W. J., Mrak, R. E., & Griffin, W. S. (2004). Microglia and neuroinflammation: A pathological perspective. *Journal of Neuroinflammation*, *1*, 14.

Sullivan, E. V., Adalsteinsson, E., Sood, R., Mayer, D., Bell, R., Mcbride, W., et al. (2006). Longitudinal brain magnetic resonance imaging study of the alcohol-preferring rat. Part I: Adult brain growth. *Alcoholism: Clinical and Experimental Research*, *30*, 1234–1247.

Sullivan, E. V., & Pfefferbaum, A. (2005). Neurocircuitry in alcoholism: A substrate of disruption and repair. *Psychopharmacology*, *180*, 583–594.

Sullivan, E. V., Rosenbloom, M. J., & Pfefferbaum, A. (2000). Pattern of motor and cognitive deficits in detoxified alcoholic men. *Alcoholism: Clinical and Experimental Research*, *24*, 611–621.

Suvarna, N., Borgland, S. L., Wang, J., Phamluong, K., Auberson, Y. P., Bonci, A., et al. (2005). Ethanol alters trafficking and functional N-methyl-D-aspartate receptor NR2 subunit ratio via H-Ras. *Journal of Biological Chemistry*, *280*, 31450–31459.

Swartzwelder, H. S., Hogan, A., Risher, M. L., Swartzwelder, R. A., Wilson, W. A., & Acheson, S. K. (2014). Effect of sub-chronic intermittent ethanol exposure on spatial learning and ethanol sensitivity in adolescent and adult rats. *Alcohol*, *48*(4), 353–360.

Swartzwelder, H. S., Wilson, W. A., & Tayyeb, M. I. (1995). Differential sensitivity of NMDA receptor-mediated synaptic potentials to ethanol in immature versus mature hippocampus. *Alcoholism: Clinical and Experimental Research*, *19*, 320–323.

Szabo, G., & Mandrekar, P. (2009). A recent perspective on alcohol, immunity, and host defense. *Alcoholism: Clinical and Experimental Research, 33*, 220–232.

Szabo, G., Mandrekar, P., & Catalano, D. (1995). Inhibition of superantigen-induced T cell proliferation and monocyte IL-1 beta, TNF-alpha, and IL-6 production by acute ethanol treatment. *Journal of Leukocyte Biology, 58*, 342–350.

Szabo, G., Mandrekar, P., Dolganiuc, A., Catalano, D., & Kodys, K. (2001). Reduced alloreactive T-cell activation after alcohol intake is due to impaired monocyte accessory cell function and correlates with elevated IL-10, IL-13, and decreased IFNgamma levels. *Alcoholism: Clinical and Experimental Research, 25*, 1766–1772.

Szabo, G., Verma, B., & Catalano, D. (1993). Selective inhibition of antigen-specific T lymphocyte proliferation by acute ethanol exposure: The role of impaired monocyte antigen presentation capacity and mediator production. *Journal of Leukocyte Biology, 54*, 534–544.

Takeda, K., Kaisho, T., & Akira, S. (2003). Toll-like receptors. *Annual Review of Immunology, 21*, 335–376.

Valles, S. L., Blanco, A. M., Pascual, M., & Guerri, C. (2004). Chronic ethanol treatment enhances inflammatory mediators and cell death in the brain and in astrocytes. *Brain Pathology, 14*, 365–371.

Varlinskaya, E. I., & Spear, L. P. (2002). Acute effects of ethanol on social behavior of adolescent and adult rats: Role of familiarity of the test situation. *Alcoholism: Clinical and Experimental Research, 26*, 1502–1511.

Vetreno, R. P., & Crews, F. T. (2012). Adolescent binge drinking increases expression of the danger signal receptor agonist HMGB1 and Toll-like receptors in the adult prefrontal cortex. *Neuroscience, 226*, 475–488.

Vetreno, R. P., Qin, L., & Crews, F. T. (2013). Increased receptor for advanced glycation end product expression in the human alcoholic prefrontal cortex is linked to adolescent drinking. *Neurobiology of Disease, 59*, 52–62.

Vezzani, A., Maroso, M., Balosso, S., Sanchez, M. A., & Bartfai, T. (2011). IL-1 receptor/Toll-like receptor signaling in infection, inflammation, stress and neurodegeneration couples hyperexcitability and seizures. *Brain, Behavior, and Immunity, 25*, 1281–1289.

Viviani, B., Bartesaghi, S., Gardoni, F., Vezzani, A., Behrens, M. M., Bartfai, T., et al. (2003). Interleukin-1beta enhances NMDA receptor-mediated intracellular calcium increase through activation of the Src family of kinases. *Journal of Neuroscience: The Official Journal of the Society for Neuroscience, 23*, 8692–8700.

Wang, H., Yang, H., Czura, C. J., Sama, A. E., & Tracey, K. J. (2001). HMGB1 as a late mediator of lethal systemic inflammation. *American Journal of Respiratory and Critical Care Medicine, 164*, 1768–1773.

Watkins, L. R., Maier, S. F., & Goehler, L. E. (1995). Cytokine-to-brain communication: A review & analysis of alternative mechanisms. *Life Sciences, 57*, 1011–1026.

White, A. M., Bae, J. G., Truesdale, M. C., Ahmad, S., Wilson, W. A., & Swartzwelder, H. S. (2002). Chronic-intermittent ethanol exposure during adolescence prevents normal developmental changes in sensitivity to ethanol-induced motor impairments. *Alcoholism: Clinical and Experimental Research, 26*, 960–968.

White, A. M., Ghia, A. J., Levin, E. D., & Swartzwelder, H. S. (2000). Binge pattern ethanol exposure in adolescent and adult rats: Differential impact on subsequent responsiveness to ethanol. *Alcoholism: Clinical and Experimental Research, 24*, 1251–1256.

White, A. M., Truesdale, M. C., Bae, J. G., Ahmad, S., Wilson, W. A., Best, P. J., et al. (2002). Differential effects of ethanol on motor coordination in adolescent and adult rats. *Pharmacology, Biochemistry, and Behavior, 73*, 673–677.

Yan, H., Li, Q., Fleming, R., Madison, R. D., Wilson, W. A., & Swartzwelder, H. S. (2009). Developmental sensitivity of hippocampal interneurons to ethanol: Involvement of the hyperpolarization-activated current, Ih. *Journal of Neurophysiology, 101*, 67–83.

Yan, H., Li, Q., Madison, R., Wilson, W. A., & Swartzwelder, H. S. (2010). Differential sensitivity of hippocampal interneurons to ethanol in adolescent and adult rats. *Journal of Pharmacology and Experimental Therapeutics*, *335*, 51–60.

Zou, J. Y., & Crews, F. T. (2005). TNF alpha potentiates glutamate neurotoxicity by inhibiting glutamate uptake in organotypic brain slice cultures: Neuroprotection by NF kappa B inhibition. *Brain Research*, *1034*, 11–24.

Zou, J., & Crews, F. (2006). CREB and NF-kappaB transcription factors regulate sensitivity to excitotoxic and oxidative stress induced neuronal cell death. *Cellular and Molecular Neurobiology*, *26*, 385–405.

Zou, J., & Crews, F. (2010). Induction of innate immune gene expression cascades in brain slice cultures by ethanol: Key role of NF-kappaB and proinflammatory cytokines. *Alcoholism: Clinical and Experimental Research*, *34*, 777–789.

Zou, J., & Crews, F. T. (2012). Inflammasome-IL-1beta signaling mediates ethanol inhibition of hippocampal neurogenesis. *Frontiers in Neuroscience*, *6*, 77.

Zou, J. Y., & Crews, F. T. (2014). Release of neuronal HMGB1 by ethanol through decreased HDAC activity activates brain neuroimmune signaling. *PLoS One*, *9*, e87915.

CHAPTER ELEVEN

Converging Actions of Alcohol on Liver and Brain Immune Signaling

Gyongyi Szabo*[,1], Dora Lippai[†]
*Department of Medicine, University of Massachusetts Medical School, Worcester, Massachusetts, USA
[†]2nd Department of Internal Medicine, Semmelweis University, Budapest, Hungary
[1]Corresponding author: e-mail address: gyongyi.szabo@umassmed.edu

Contents

Abstract

Chronic excessive alcohol consumption results in inflammation in multiple organs, including the brain. While the contribution of neuroinflammation to alcohol-related cognitive dysfunction and behavioral alterations is established, the mechanisms by which alcohol triggers inflammation in the brain are only partially understood. There are acute and long-term alterations in brain function due to intercellular and intracellular changes of different cell types as a result of alcohol consumption. This review

International Review of Neurobiology, Volume 118
ISSN 0074-7742
http://dx.doi.org/10.1016/B978-0-12-801284-0.00011-7

focuses on the alcohol-induced proinflammatory cellular and molecular changes in the central nervous system. Alcohol passes through the blood–brain barrier and alters neurotransmission. Alcohol use activates microglia and astrocyte, contributing to neurodegeneration and impaired regeneration. Alcohol-induced cell injury in the brain results in release of damage-associated molecular patterns, such as high mobility group box 1, that trigger inflammatory changes through activation of pattern recognition receptors. In addition, alcohol consumption increases intestinal permeability and results in increased levels of pathogen-associated molecular pattern such as endotoxin in the systemic circulation that triggers PRRs and inflammation. The Toll-like receptor-4 pathway that activates nuclear factor-κB and secretion of proinflammatory cytokines, tumor necrosis factor-α, interleukin-1-beta, and chemokines, including monocyte chemotactic protein-1, has been suggested to contribute to alcohol-induced neuroinflammation. Alcohol-induced IL-1β secretion also requires Nod-like receptor-mediated inflammasome and caspase-1 activation, and, consistent with this, disruption of IL-1/IL-1-receptor signaling prevents alcohol-induced neuroinflammation. Delicate regulators of inflammatory gene expressions are micro-RNAs (miRs) that have recently been identified in alcohol-related neuroinflammation. Alcohol induces miR155, a regulator of inflammation in the brain, and deficiency in miR-155 in mice was protective from neuroinflammatory changes. These observations suggest that manipulation of miR pathways and cytokine induction may reduce alcohol-induced proinflammatory processes.

1. INTRODUCTION

The ancient observation that alcohol consumption leads to altered behavior combined with the fact that alcoholism still poses a major public health concern according to WHO triggered broad range of investigations to reveal the underlying mechanisms of alcohol addiction and alcohol-induced neuropathology. The behavioral effects of alcohol are exerted via functional changes in the brain and excessive use of alcohol results in dementia and ataxia via neurodegeneration.

The pathogenesis of alcohol-induced changes in the brain involves neuroinflammatory and neurodegenerative changes that are partially mediated by innate immune responses (Alfonso-Loeches, Pascual-Lucas, Blanco, Sanchez-Vera, & Guerri, 2010; Qin & Crews, 2012). Alcohol ingestion results in neuroinflammation (Valles, Blanco, Pascual, & Guerri, 2004) and neurodegeneration in humans and animals (Crews & Nixon, 2009; He & Crews, 2008; Qin et al., 2008). Cellular changes, inflammasome activation, proinflammatory cytokine secretion, reactive oxygen species (ROS) production, apoptosis, and changes in miRNA expression are all part of the underlying mechanism (Lippai, Bala, Csak, Kurt-Jones, & Szabo, 2013;

Lippai, Bala, Petrasek, et al., 2013). Inflammatory cytokines affect neurons and are associated with disease symptoms (Tracey, 2010). Toll-like receptor-4 (TLR4) signaling pathway plays an important role in the alcohol-induced production of inflammatory cytokines in different brain regions (Alfonso-Loeches et al., 2010; Lippai, Bala, Petrasek, et al., 2013). Most studies suggest that the extent of inflammation, neurodegeneration has site specificity, as the frontal, prefrontal cortical, the limbic and hippocampal as well as the cerebellar areas seem to be more affected by alcohol use than other brain regions (Alfonso-Loeches et al., 2010; Lippai, Bala, Csak, et al., 2013; Qin & Crews, 2012).

2. PATHOLOGY AND CELLULAR CHARACTERISTICS OF NEUROINFLAMMATION AFTER ALCOHOL EXPOSURE

2.1. Microglia and astrocytes

Microglia and astrocytes are the target of pathogen/danger-associated molecular patterns (P/DAMPs), and upon challenge, they become morphologically and functionally activated (Guillemin & Brew, 2004; Ridet, Malhotra, Privat, & Gage, 1997; Streit, Graeber, & Kreutzberg, 1988). It has been shown that chronic alcohol administration results in increase in the number of microglia and astrocytes as well as in their functional activation in the brain, creating an environment for proinflammatory cellular and cytokine activation (Guillemin & Brew, 2004; Ridet et al., 1997; Streit et al., 1988).

As important microhomeostasis regulators, astrocytes are responsible to protect neurons from ROS. When activated, they contribute to neuroinflammation by producing cytokines (Pignataro, Varodayan, Tannenholz, Protiva, & Harrison, 2013). The activation marker of astrocytes, GFAP, is induced by alcohol in alcohol-fed rodents as well as in alcoholic human brains (Qin & Crews, 2012). Alcohol induces NADPH oxidase activity and ROS production in the brain of alcoholics (Qin & Crews, 2012). Furthermore, alcohol induces oxidoreductase gene expression in astrocytes that might be a salvage mechanism to protect neurons from the harmful effects of excessive ROS production induced by alcohol (Pignataro et al., 2013).

As resident monocytes of the brain, microglia are responsible for removing the debris and protecting neurons from harmful and toxic effects; however, when activated they produce inflammatory cytokines, chemokines, and DAMPs inducing inflammation and cytotoxicity (Nakajima & Kohsaka, 2001). Microglia when become activated and phagocytic express immunomolecules and secrete proinflammatory molecules and cytotoxic factors (Guillemin & Brew, 2004; Ridet et al., 1997; Streit et al., 1988). When harmful P/DAMPs are continuously present, as are in chronic alcoholism, the fine balance of neuroprotection versus neurotoxicity is in favor of proinflammatory activation of microglia.

2.2. Neurons

Alcohol consumption enhances neural death and decrease of cerebellar granule cells by alcohol has been observed in culture (Pantazis, Dohrman, Goodlett, Cook, & West, 1993) Oxidative stress plays an important role in inducing neural apoptosis (Zou & Crews, 2014). The survival and functions of neurons are affected by proinflammatory mediators resulting in further damage of neurons as well as disease symptoms in animal models (Tracey, 2010).

Ethanol activates TLR4 within the lipid rafts of glial cells, inducing the production of inflammatory mediators, which trigger ROS generation, neuronal apoptosis, and causing neuroinflammation. It has been proposed that ethanol induces a physical association between TLR2 and TLR4 receptors, as a result TLR2 potentiates the effects of ethanol on the TLR4 response reflected by inducible NO synthase (Fernandez-Lizarbe, Montesinos, & Guerri, 2013). Consistent with this is the observation that TLR4 deficiency protects against ethanol-induced glial activation, induction of inflammatory mediators, and apoptosis (Alfonso-Loeches et al., 2010).

Alcohol-induced neuroinflammation is mediated by interleukin-1-beta (IL-1β). IL-1β production requires caspase-1 activation by inflammasomes. Alcohol upregulates and activates the NLRP3/ASC inflammasome, leading to caspase-1 activation and IL-1β increment. Alcohol upregulates iNOS, COX-2, and IL-1 RI with the stimulation of IRAK and MAP kinases, including ERK1/2, p-38, and JNK, which trigger the downstream activation of oxidant-sensitive transcription factors NF-κB and AP-1, resulting in enhancement of both caspase-3 and apoptosis (Lippai, Bala, Petrasek, et al., 2013; Valles et al., 2004).

Additionally, nonapoptotic neurodegenerative processes may also be involved in repetitive ethanol intoxication as PARP-1 elevations in the absence of caspase-3 activation may be potentiating mechanisms such as parthanatos (Tajuddin, Przybycien-Szymanska, Pak, Neafsey, & Collins, 2013). Neural death induces degeneration. Degeneration is associated with increased nuclear factor-κB (NF-κB) proinflammatory transcription and decreased CREB transcription (Crews & Nixon, 2009). Microglia activation during neuroinflammation is one of the key components in neurodegeneration. Though in the binge model, even in the absence of classical neuroinflammation, neurodegeneration occurs with partially activated microglia (Marshall et al., 2013).

2.3. Cytokines and chemokines

The innate immune system and inflammation are major contributors of alcohol-induced organ damage affecting different organs, including the brain (Kawaguchi et al., 2011; Qin et al., 2008; Szabo, Mandrekar, Petrasek, & Catalano, 2011). The pathogenesis of neuroinfectious (Holley & Kielian, 2012) and neurodegenerative diseases (Glass, Saijo, Winner, Marchetto, & Gage, 2010) is highly affected by innate immune signaling pathways. Alcoholism causes neuroinflammation (Valles et al., 2004) and neurodegeneration in both animals and humans (Crews & Nixon, 2009; He & Crews, 2008; Qin et al., 2008). Alcohol increases the levels of the proinflammatory cytokines, monocyte chemotactic protein-1 (MCP-1), tumor necrosis factor-α (TNFα), and IL-1β in the brain (Crews & Nixon, 2009; He & Crews, 2008; Lippai, Bala, Petrasek, et al., 2013). MCP-1 is an alcohol-induced chemokine that promotes recruitment of monocytes into sites of tissue damage and inflammation, and increased levels of MCP-1 are associated with increases in the numbers and activation of microglia cells in alcohol-fed mice (Guillemin & Brew, 2004; Leonard & Yoshimura, 1990; Ridet et al., 1997; Streit et al., 1988). Tumor necrosis factor-alpha, a potent proinflammatory cytokine, is produced at increased levels in the brain and cerebellum after chronic alcohol intake (Alfonso-Loeches et al., 2010; Sriram et al., 2006; Valles et al., 2004). IL-1β, a secreted cytokine, is potently increased in the brain after alcohol exposure in animal models and in humans (Blanco, Valles, Pascual, & Guerri, 2005; Qin et al., 2008; Valles et al., 2004). It has been suggested that even a single alcohol binge induces IL-1β in the brain resulting in sustained neuroinflammation. Lippai et al. demonstrated that alcohol-induced IL-1β production in the brain is mediated by the NLRP3 inflammasome activation that is required for the proinflammatory IL-1β secretion (Lippai, Bala, Petrasek, et al., 2013).

Acute inflammation is homeostatically regulated by anti-inflammatory cytokines, IL-10 and TGFβ, produced by monocytes, macrophages, and microglia in the late phase of inflammation (Letterio & Roberts, 1998;

Szabo, Verma, Fogarasi, & Catalano, 1992). Chronic alcohol, however, is associated with prolonged proinflammatory changes that are thought to be linked to an attenuated induction of anti-inflammatory cytokines in the brain. Consistent with this notion, in a mouse model of chronic alcohol-induced neuroinflammation, the levels of IL-10 were decreased in alcohol-fed mice compared to control diet-fed controls (Qin et al., 2008).

Inflammatory cytokines have features of autocrine amplification and contribute to the production of other cytokines. For example, intracranial injection of IL-1β in mice resulted in increased proinflammatory cytokine, MCP-1, and TNFα protein levels; furthermore, it augmented IL-1β mRNA production in the cerebellum (Lippai, Bala, Petrasek, et al., 2013). In contrast, IL-1β receptor blockade prevented alcohol-induced neuroinflammation suggesting that disruption of the proinflammatory amplification loop can control alcohol-induced neuroinflammation (Lippai, Bala, Petrasek, et al., 2013). The TLR4 signaling pathway has an important role in triggering alcohol-induced neuroinflammation in rodents via proinflammatory cytokine production (Alfonso-Loeches et al., 2010; Lippai, Bala, Petrasek, et al., 2013).

3. MOLECULAR MECHANISMS OF ALCOHOL-INDUCED NEUROINFLAMMATION

3.1. Pattern recognition receptors sense danger signals in the brain (TLRs and NLRs)

Pattern recognition receptors (PRRs), including the TLR and Nod-like receptor (NLR) families, sense evolutionary conserved molecular patterns as "danger" signals and in response, activate innate immunity and proinflammatory pathways. TLR and NLR receptors, including TLR2, TLR4, TLR6, TLR9, NLRC4, NLRP1, and NLRP3, are upregulated in the alcoholic brain (Lippai, Bala, Petrasek, et al., 2013).

TLRs are expressed on the cell surface (TLR1–TLR6) or in endosomes (TLR3, TLR7–TLR9) and upon activation induce production of proinflammatory cytokines and/or type-1 interferons. Recruitment of the MyD88 adapter molecule to TLRs results in downstream activation of IRAK1/4 and NF-κB activation leading to proinflammatory cytokine production. The TLRs MyD88 (TLR7/8 and 9) or TRIF (TLR3 and TLR4) and signal through IRF3 to induce type-1 interferons (Szabo et al., 2006). TLR4 detects bacterial lipopolysaccharide (LPS) and it can be activated by

DAMPs that are induced by alcohol. TLR4 activation contributes to microglia accumulation, astrogliosis, and inflammatory cytokine TNFα, MCP-1, and pro-IL-1β production (Alfonso-Loeches et al., 2010; Fernandez-Lizarbe, Pascual, & Guerri, 2009) and is involved in the pathogenesis of alcohol-induced neuroinflammation and neurodegeneration (Alfonso-Loeches et al., 2010; Lippai, Bala, Petrasek, et al., 2013).

The multiprotein complex, inflammasome, consists of an NLR (NLRs: NLRP1, NLRP3, NLRC4); an adapter molecule, ASC; and procaspase-1 (Mankan, Kubarenko, & Hornung, 2012). Inflammatory stimuli, including DAMPs and PAMPs, upregulate the expression of the inflammasome components via TLRs (Mankan et al., 2012; Fig. 11.1). Alcohol upregulates and activates NLRP3/ASC inflammasome evidenced by caspase-1 activation and increased mature IL-1β levels in mouse brains (Lippai, Bala, Petrasek, et al., 2013). NLRP3 inflammasome activation requires two signals, a priming signal to increase expression of inflammasome components and target proteins, and a second signal that leads to inflammasome activation and subsequent IL-1β secretion (Gross, Thomas, Guarda, & Tschopp, 2011). During this process, procaspase-1 is converted to a proteolytic effector enzyme,

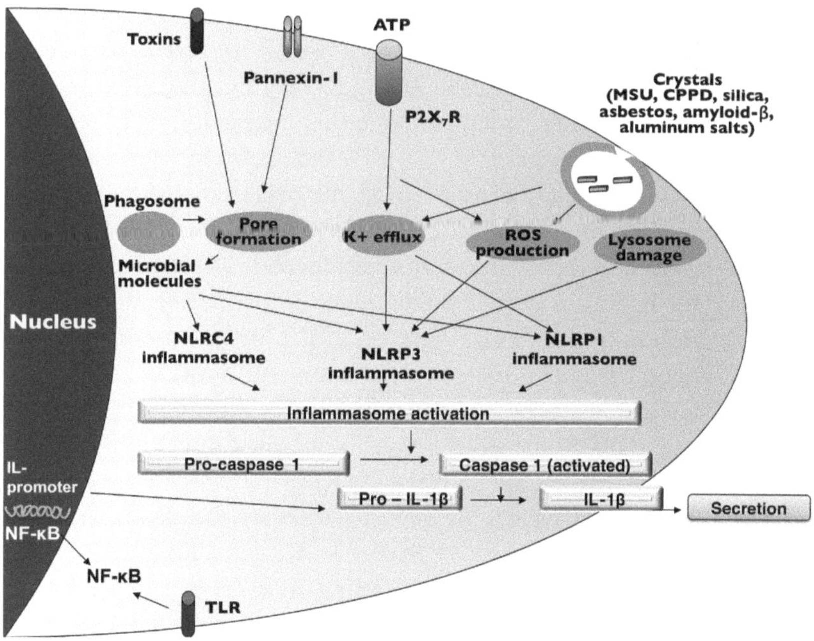

Figure 11.1 Inflammasome activation and IL-1β production.

caspase-1 which then cleaves pro-IL-1β to mature, secretable IL-1β (Mankan et al., 2012). Suggestive of a feed-forward loop, alcohol-induced neuroinflammation is prevented by recombinant IL-1ra by inhibiting inflammasome activation, IL-1β, and TNFα production in the brain (Lippai, Bala, Petrasek, et al., 2013). TLR activation via DAMPs and PAMPs leads to NF-κB activation (Zhang & Ghosh, 2001). NF-κB induces the transcription of TNFα, MCP1, and IL-1β (Thompson & Van Eldik, 2009; Zhang & Ghosh, 2001), all of which are increased in alcohol-induced neuroinflammation (Alfonso-Loeches et al., 2010; Lippai, Bala, Petrasek, et al., 2013; Qin & Crews, 2012).

3.2. Exogenous and endogenous danger signals are induced by alcohol use

PAMPs are recognized by the innate immune system via PRRs (Aderem & Ulevitch, 2000; Akira & Takeda, 2004; Szabo et al., 2006; Trinchieri & Sher, 2007). PAMPs, including the bacterial endotoxin LPS, activate the immune system and are responsible for the initiation and maintenance of inflammatory responses. As alcohol disrupts the barrier function of the intestine that hosts the microbiome, bacterial components can enter the portal and/or systemic circulation. For example, the levels of LPS, a component of Gram-negative bacteria, are increased in the serum of alcoholics and LPS can induce activation of the PRR, TLR4 (Aderem & Ulevitch, 2000; Akira & Takeda, 2004; Szabo et al., 2006; Trinchieri & Sher, 2007).

LPS is increased in the portal and systemic circulation of patients with alcoholic cirrhosis (Keshavarzian et al., 1999) and in chronic alcohol-fed mice (Hritz et al., 2008). High concentration of endotoxin can disrupt the blood–brain barrier (BBB) (Jin, Nation, Li, & Nicolazzo, 2013) although a recent study demonstrated no detectable endotoxin levels in the brain of mice after chronic alcohol feeding in spite of the presence of increased endotoxin in the systemic circulation (Lippai, Bala, Petrasek, et al., 2013). Previous studies are nonconclusive on the effect of alcohol on the integrity of the BBB and it is likely that the extent of alcohol use and/or cofactors affect the ultimate effects of alcohol on the BBB. Some studies have found impairment in permeability (Haorah, Heilman, et al., 2005; Haorah, Knipe, Leibhart, Ghorpade, & Persidsky, 2005) while others suggest no increase (Banks & Robinson, 2010; Elmas, Kucuk, Kalayci, Cevik, & Kaya, 2001; Singh, Jiang, Gupta, & Benlhabib, 2007). We detected no increase of LPS in the brain (Lippai, Bala, Petrasek, et al., 2013); however, endotoxin might influence the brain secondarily via organ interactions or could induce signal transduction in the endothelium of brain capillaries that are in close contact with other cells, including astrocytes.

DAMPs are molecules that are actively excreted or passively released by stressed or dying cells and further enhance inflammatory or cell-death signaling (Rubartelli & Lotze, 2007). One of the most recently studied DAMP in neuroinflammation is high mobility group box 1 (HMGB1), a nuclear protein, that after its cytoplasmic translocation, phosphorylation, and/or acetylation is released and results in activation of TLRs and the inflammasome (Bianchi, 2009; Ghiringhelli et al., 2009).

As a metabolite of alcohol degradation, acetate is increased in the blood after alcohol consumption that is a major source of energy for neurons (Israel, Orrego, & Carmichael, 1994). Acetate can reduce the activity of histone deacetylases (HDACs) and result in the acetylation of different endogenous molecules (Soliman & Rosenberger, 2011), including HMGB1. Alcohol reduces HDAC activity in rat brains; consequently, acetyl-HMGB1 is increased in the brain after alcohol consumption (Lippai, Bala, Petrasek, et al., 2013; Zou & Crews, 2014). Acetyl-HMGB1 is secretable and is increased in alcoholic rodent brains, therefore, could act as a danger molecule inducing neuroinflammation (Lippai, Bala, Petrasek, et al., 2013). According to recent studies, HMGB1 is actively secreted from neurons upon alcohol challenge (Zou & Crews, 2014). The alcohol-exposed brain is primed to react to the increased HMGB1 levels because of the increased expression of all of the HMGB1-sensing receptors including TLR4, TLR2, TLR9, and RAGE receptor in the brain (Bianchi, 2009; Klune, Dhupar, Cardinal, Billiar, & Tsung, 2008; Lippai, Bala, Petrasek, et al., 2013; Oh et al., 2009; Youn & Shin, 2006).

Recent studies suggest a cross regulation between HMGB1 and the inflammasome. Inflammasome components are necessary for HMGB1 release (Willingham et al., 2009) and IL-1β contributes to HMGB1 release from monocytes and macrophages (Gardella et al., 2002; Wang et al., 1999). Reversely, HMGB1 can induce the multiprotein inflammasome complex leading to IL-1β production (Ghiringhelli et al., 2009; Xiang et al., 2011). HMGB1 signals through TLR2, TLR4, TLR9, RAGE, or IL-1R (Bianchi, 2009; Klune et al., 2008; Oh et al., 2009; Youn & Shin, 2006; Fig. 11.2). The mRNA expression of these receptors is induced by alcohol in the cerebellum of mice (Lippai, Bala, Petrasek, et al., 2013). Caspase-3-mediated neuronal apoptosis that is increased in alcohol consumption (Alfonso-Loeches et al., 2010) is mediated by HMGB1 (Kim et al., 2011). Furthermore, HMGB1 causes iNOS, TNFα, COX2, and IFNγ

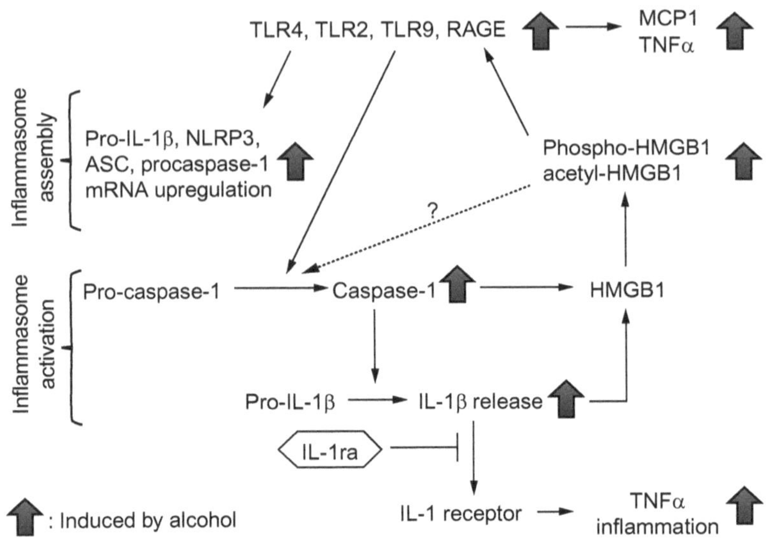

Figure 11.2 Working model for alcohol-induced inflammasome activation in the brain.

production in primary microglia (Kim et al., 2011), which has been observed in alcoholic brains (Alfonso-Loeches et al., 2010).

Cold-inducible RNA-binding protein (CIRP), a possible DAMP molecule, has been shown to be produced by the microglia cell line BV2 upon alcohol challenge (Rajayer et al., 2013). CIRP has also recently been discovered to play as a possible role in alcohol-induced neuroinflammation as DAMP (Rajayer et al., 2013). Alcohol induced the level of CIRP in alcohol-fed mice in a binge drinking model; furthermore, TNFα and IL-1β levels were attenuated in alcohol-fed CIRP-KO mice (Rajayer et al., 2013). CIRP along with other DAMPs is acting upon the TLR4-mediated pathway to induce inflammatory response (Zhou, Yang, Ji, Qiang, & Wang, 2014).

3.3. Micro-RNAs

Micro-RNAs (miRNAs) are evolutionarily conserved, small noncoding RNAs that are involved in various biological processes such as development, differentiation, innate, and adaptive immune responses (Cardoso, Guedes, Pereira de Almeida, & Pedroso de Lima, 2012). miRNAs have been implicated in the pathogenesis of predominantly neurodegenerative or neuroinflammatory diseases (Schonrock & Gotz, 2012; Thounaojam, Kaushik, & Basu, 2013). Mature miRNAs regulate posttranscriptional gene expression via repressing translation, inducing mRNA degradation (Valencia-Sanchez, Liu, Hannon, & Parker, 2006) or posttranslational stabilization of mRNA (Bala et al., 2011).

miR-155 expression leads to cell survival and modification of inflammation (Tili, Croce, & Michaille, 2009). miR-155 targets SHIP1 (a negative regulator of TNFα) and SOCS1 (a negative regulator of cytokines), which leads to increased inflammation (Schulte, Westermann, & Vogel, 2013). Furthermore, TLR stimulation induces miR-155 in macrophages, dendritic cells, B- and T-cells (Ceppi et al., 2009; Tili et al., 2009), and in microglial cell lines (Cardoso et al., 2012). In dendritic cells, miR-155 silencing resulted in increased IL-1β production (Ceppi et al., 2009). Inhibitory effect of mir-155 on its proinflammatory targets MyD88 and TAB2 (O'Neill, Sheedy, & McCoy, 2011) can lead to decreased NF-κB activation (Verstrepen et al., 2008), which is involved in the pathogenesis of alcohol-induced neuroinflammation (Qin & Crews, 2012). miR-155 is induced by NF-κB activation in a TLR4-dependent manner (Lippai, Bala, Petrasek, et al., 2013). Alcohol consumption upregulated miR-155 and miR-132 expression in the cerebellum in mice (Lippai, Bala, Petrasek, et al., 2013). Furthermore, miR-155 deficiency protected mice from alcohol-induced increase in inflammatory cytokines TNFα and MCP1, but had no effect on IL-1β production and caspase-1 activation (Lippai, Bala, Petrasek, et al., 2013). As pro-IL-1β mRNA induction is NF-κB mediated (He, Franchi, & Nunez, 2013), its levels stayed unchanged in miR-155-KO mice secondary to lack of NF-κB activation, suggesting that the inflammasome-mediated pathway has a distinctive regulatory pattern from that of TNFα and MCP-1 regulation.

Additional importance of miRNAs is their possible action in a process that leads to alcohol dependence through their role in myelination and cell adhesion in synapse formation, miR-203, miR-586, miR-146a, and miR-519b-3p, miR665 targeting components in myelin sheath, and miR-194 targeting myelin gene regulatory factor. Neurogenesis-related genes such as TIMP metallopeptidase inhibitor 2 (TIMP2) and TGFβ2 are also regulated by multiple miRNAs (Pignataro, Varodayan, Tannenholz, & Harrison, 2009; Wilke, Sganga, Barhite, & Miles, 1994). miRNAs were suggested to have a role in adaptive mechanism in the brain of chronic alcohol abusers (Lewohl et al., 2011). This effect is mediated through Dicer molecule. Ribonuclease Dicer processes precursor miRNAs and siRNA into their short mature forms. Dicer knockdown results in reduced expression of mature miRNA (Kumar, Lu, Mercer, Golub, & Jacks, 2007). Targeting and downregulation of Dicer by miRNAs in a negative feedback manner avoids excess deregulation of miRNAs (Lewohl et al., 2011).

miRNAs possibly contribute to alcohol tolerance via BK channel (Pietrzykowski et al., 2008). BK channel influences neuronal excitability and transmitter release and is potentiated by alcohol (Pietrzykowski et al., 2008). Alcohol increases miR-9 in the neurons with BK channel and leads to change its isoform consistent with the development of tolerance (Pietrzykowski et al., 2008).

3.4. Oxidative stress

Alcohol activation of NADPH oxidase leads to excessive production of ROS. ROS can be toxic to neurons. Alcohol induces gp91^{phox}, the catalytic subunit of NADPH oxidase which then produces superoxide (Qin & Crews, 2012). Furthermore, p47^{phox} KO mice are partially protected from alcohol-induced neuroinflammatory changes (Lippai, Bala, Petrasek, et al., 2013).

4. CROSSTALK BETWEEN ORGANS

4.1. Gut–liver–brain axis

While consumed alcohol has some direct effects on the gastrointestinal tract upon contact with the mucosa, most of the biological effects of alcohol are related to its systemic distribution and delivery through the blood. It has been shown that alcohol affects the microbiome in the GI tract and alcoholics have altered and increased bacterial load in their gut (Hauge, Persson, & Danielsson, 1997). Not only the composition of the microbiome changes with alcohol use, but also the integrity of the gut mucosa is impaired by alcohol. Alcohol results in decreased expression of tight junction proteins between gut epithelial cells resulting in increased gut permeability (Keshavarzian et al., 1999). For example, alcohol inhibits the expression of zona occludin-1 protein (Tang et al., 2008). Bacterial endotoxin, LPS, enters the portal circulation through alcohol-disrupted barrier of gut and results in increased circulating levels in alcoholics (Beck & Dinda, 1981; Bhonchal et al., 2008; Bode & Bode, 2003; Bode, Kugler, & Bode, 1987; Hritz et al., 2008; Keshavarzian et al., 1999; Kirpich et al., 2013; Tamai et al., 2000). LPS and alcohol contribute to alcoholic liver disease and DAMP secretion in hepatocytes in the liver (Hritz et al., 2008). The proinflammatory cytokines and DAMPs might enhance the inflammatory changes in the brain via direct and indirect effects. For example, radiolabeled TNFα is being transported through the BBB via endothelial receptors in a saturable manner (Pan & Kastin, 2002; Fig. 11.3), therefore TNFα produced by other organs, i.e., from alcoholic liver, could penetrate the BBB.

LPS most likely has an indirect role in neuroinflammation, as neuroimmune reflexes can sense and respond to peripheral insults (Tracey, 2010). Moreover, vagotomy diminishes IL-1β response in the brain, but not on the periphery after intraperitoneal injection with LPS (Laye et al., 1995).

The liver is a source of excessive inflammatory cytokine production released to the serum in alcoholics (Szabo et al., 2011). The signaling and regulatory pathways might be similar to regulate gene expression in different organs, as there are similarities of gene expression profiles induced by alcohol in astrocytes and in hepatocytes (Pignataro et al., 2013) or other cell types (Lippai, Bala, Petrasek, et al., 2013; Petrasek et al., 2012). Alcoholics with cirrhosis and liver failure have more brain damage than patients without liver disease. High ammonia level is widely known for its neurotoxic effect and can activate peripheral BZ receptors, contributing to astrocyte swelling and subsequent coma (Palomero-Gallagher & Zilles, 2013).

Finally, inflammation induced in the CNS has remote effects on other organs. For example, administration of IL-1β to the CNS resulted in

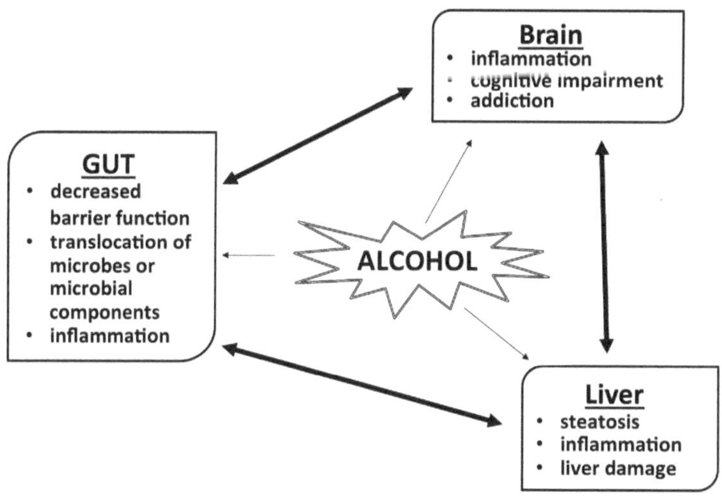

Figure 11.3 Chronic alcohol use affects multiple organs.

increased recruitment of neutrophils in the liver (Campbell et al., 2010). In a recent study, we demonstrated that IL-1β administration to the brain resulted in increased proinflammatory cytokine levels not only in the brain but also in the liver, supporting the notion of interorgan communications (Lippai, Bala, Petrasek, et al., 2013).

5. THERAPEUTIC TARGETS

5.1. General consideration

Prevention and cessation of alcohol along with proper nutrition is the first line of treatment. However, this might not be enough due to the availability, the addictive effect, and the long-term effects of alcohol that also pose a major economical problem. Therefore, efforts should focus on finding a safe, effective, and economically beneficial treatment for the people suffering from the harmful consequences of alcohol use on the body, including the central nervous system.

5.2. TNFα system

Targeting the inflammatory cytokine TNFα with recombinant anti-TNFα treatment seemed a reasonable option as TNFα is induced by alcohol in the organs, including brain. However, in clinical trials of anti-TNFα administration in alcoholic liver disease did not bring the expected beneficial results as there was a major drawback in the already immune compromised alcoholic patients, getting life-threatening opportunistic infections (Naveau et al., 2004; Tilg et al., 2003).

5.3. IL-1β system

Another proinflammatory cytokine, IL-1β, is also induced in alcoholics in the brain as well as in mice with short- or long-term alcohol administration (Nguyen et al., 1998; Qin et al., 2008). A disease-modifying anti-inflammatory medication, the recombinant IL-1ra, anakinra, is used safely and effectively in active rheumatoid arthritis. rIL-1ra has also been suggested for diseases with induced IL-1β production, including diabetes mellitus, gout, or recurring-fever syndromes (Goldbach-Mansky, 2012). Reduced sedation and motor impairment recovery time was observed in short-term ethanol-fed mice with a pretreatment of rIL-1ra (Wu et al., 2011). rIL-ra prevented and protected from advancement of alcohol-induced liver disease as well as alcohol-induced neuroinflammation in chronic alcohol feeding in mice (Lippai, Bala, Petrasek, et al., 2013; Petrasek et al., 2012). The circuit of neuroinflammation was interrupted at the level of IL-1 receptor. Prevention of excessive activation of the inflammasome complex, release of proinflammatory cytokines, HMGB1 activation, as well as neuroinflammation in the brain was achieved in the murine model (Lippai, Bala, Petrasek, et al., 2013). Further animal and human studies are warranted with rIL-ra in alcoholism to assess its protective and potential therapeutic role.

5.4. miR-155

Deficiency of miR-155 in mice results in a phenotype that is protected from alcohol-induced neuroinflammation (Lippai, Bala, Petrasek, et al., 2013); therefore, it is plausible that miR-155 silencing might have a therapeutic role in the improvement of alcohol-induced neuroinflammation. Further work in this field is warranted.

5.5. HMGB1

HMGB1 has been shown to be involved in alcohol-induced neuroinflammation (Lippai, Bala, Petrasek, et al., 2013). In a postischemic brain injury model, significant neuroprotection has been achieved with nasal introduction of HMGB1 siRNA (Kim et al., 2012). Targeting HMGB1 blocks alcohol-mediated neuroimmune activation (Zou & Crews, 2014). It is plausible that targeting DAMPs, including HMGB1, might be beneficial in alcoholism.

5.6. TLR4 pathway

TLR4 inhibition by siRNA in central amygdala results in inhibition of binge drinking (Liu et al., 2011); furthermore, alcohol preference is decreased in mice lacking CD14, which is important for TLR4 signaling (Zou & Crews, 2014). LPS injection increases voluntary intake of alcohol in mice (Blednov et al., 2011). The opiate receptor antagonist, naltrexone, which is used in alcoholism has been shown to be a TLR4 antagonist (Hutchinson et al., 2008; O'Mara & Wesley, 1994).

6. CONCLUSIONS

Alcohol, as the most commonly used substance of abuse worldwide, is associated with multiple health effects. Excessive and prolonged alcohol use is clearly associated with harmful effects on brain both on cognitive functions and behavior. Overall, much progress was achieved in recent years in understanding mechanisms by which alcohol consumption results in neuroinflammation. Alcohol directly and indirectly via induction of danger molecules and sterile inflammation promotes sustained inflammation in the brain. Better understanding of the triggers and mechanisms of alcohol-induced inflammation and tissue damage will guide future therapeutic interventions to ameliorate the devastating and permanent damage of alcohol in the central nervous system.

REFERENCES

Aderem, A., & Ulevitch, R. J. (2000). Toll-like receptors in the induction of the innate immune response. *Nature*, *406*(6797), 782–787. http://dx.doi.org/10.1038/35021228.

Akira, S., & Takeda, K. (2004). Toll-like receptor signalling. *Nature Reviews. Immunology*, *4*(7), 499–511. http://dx.doi.org/10.1038/nri1391.

Alfonso-Loeches, S., Pascual-Lucas, M., Blanco, A. M., Sanchez-Vera, I., & Guerri, C. (2010). Pivotal role of TLR4 receptors in alcohol-induced neuroinflammation and brain damage. *Journal of Neuroscience: The Official Journal of the Society for Neuroscience*, *30*(24), 8285–8295. http://dx.doi.org/10.1523/JNEUROSCI.0976-10.2010.

Bala, S., Marcos, M., Kodys, K., Csak, T., Catalano, D., Mandrekar, P., et al. (2011). Up-regulation of microRNA-155 in macrophages contributes to increased tumor necrosis factor {alpha} (TNF{alpha}) production via increased mRNA half-life in alcoholic liver disease. *Journal of Biological Chemistry*, *286*(2), 1436–1444. http://dx.doi.org/10.1074/jbc.M110.145870.

Banks, W. A., & Robinson, S. M. (2010). Minimal penetration of lipopolysaccharide across the murine blood-brain barrier. *Brain, Behavior, and Immunity*, *24*(1), 102–109. http://dx.doi.org/10.1016/j.bbi.2009.09.001.

Beck, I. T., & Dinda, P. K. (1981). Acute exposure of small intestine to ethanol: Effects on morphology and function. *Digestive Diseases and Sciences*, *26*(9), 817–838.

Bhonchal, S., Nain, C. K., Prasad, K. K., Nada, R., Sharma, A. K., Sinha, S. K., et al. (2008). Functional and morphological alterations in small intestine mucosa of chronic alcoholics. *Journal of Gastroenterology and Hepatology*, *23*(7 Pt. 2), e43–e48. http://dx.doi.org/JGH5080 [pii].

Bianchi, M. E. (2009). HMGB1 loves company. *Journal of Leukocyte Biology*, *86*(3), 573–576. http://dx.doi.org/10.1189/jlb.1008585.

Blanco, A. M., Valles, S. L., Pascual, M., & Guerri, C. (2005). Involvement of TLR4/type I IL-1 receptor signaling in the induction of inflammatory mediators and cell death induced by ethanol in cultured astrocytes. *Journal of Immunology (Baltimore, Md.: 1950)*, *175*(10), 6893–6899. http://dx.doi.org/175/10/6893 [pii].

Blednov, Y. A., Benavidez, J. M., Geil, C., Perra, S., Morikawa, H., & Harris, R. A. (2011). Activation of inflammatory signaling by lipopolysaccharide produces a prolonged increase of voluntary alcohol intake in mice. *Brain, Behavior, and Immunity*, *25*(Suppl. 1), S92–S105. http://dx.doi.org/10.1016/j.bbi.2011.01.008.

Bode, C., & Bode, J. C. (2003). Effect of alcohol consumption on the gut. *Best Practice & Research Clinical Gastroenterology*, *17*(4), 575–592. http://dx.doi.org/S1521691803000349 [pii].

Bode, C., Kugler, V., & Bode, J. C. (1987). Endotoxemia in patients with alcoholic and non-alcoholic cirrhosis and in subjects with no evidence of chronic liver disease following acute alcohol excess. *Journal of Hepatology*, *4*(1), 8–14. http://dx.doi.org/S0168-8278(87)80003-X [pii].

Campbell, S. J., Meier, U., Mardiguian, S., Jiang, Y., Littleton, E. T., Bristow, A., et al. (2010). Sickness behaviour is induced by a peripheral CXC-chemokine also expressed in multiple sclerosis and EAE. *Brain Behavior and Immunity*, *24*(5), 738–746. http://dx.doi.org/10.1016/j.bbi.2010.01.011.

Cardoso, A. L., Guedes, J. R., Pereira de Almeida, L., & Pedroso de Lima, M. C. (2012). miR-155 modulates microglia-mediated immune response by down-regulating SOCS-1 and promoting cytokine and nitric oxide production. *Immunology*, *135*(1), 73–88. http://dx.doi.org/10.1111/j.1365-2567.2011.03514.x.

Ceppi, M., Pereira, P. M., Dunand-Sauthier, I., Barras, E., Reith, W., Santos, M. A., et al. (2009). MicroRNA-155 modulates the interleukin-1 signaling pathway in activated human monocyte-derived dendritic cells. *Proceedings of the National Academy of Sciences of the United States of America*, *106*(8), 2735–2740. http://dx.doi.org/10.1073/pnas.0811073106.

Crews, F. T., & Nixon, K. (2009). Mechanisms of neurodegeneration and regeneration in alcoholism. *Alcohol and Alcoholism (Oxford, Oxfordshire)*, *44*(2), 115–127. http://dx.doi.org/10.1093/alcalc/agn079.

Elmas, I., Kucuk, M., Kalayci, R. B., Cevik, A., & Kaya, M. (2001). Effects of profound hypothermia on the blood-brain barrier permeability in acute and chronically ethanol treated rats. *Forensic Science International*, *119*(2), 212–216. http://dx.doi.org/S0379073800004291 [pii].

Fernandez-Lizarbe, S., Montesinos, J., & Guerri, C. (2013). Ethanol induces TLR4/TLR2 association, triggering an inflammatory response in microglial cells. *Journal of Neurochemistry*, *126*(2), 261–273. http://dx.doi.org/10.1111/jnc.12276.

Fernandez-Lizarbe, S., Pascual, M., & Guerri, C. (2009). Critical role of TLR4 response in the activation of microglia induced by ethanol. *Journal of Immunology (Baltimore, Md.: 1950)*, *183*(7), 4733–4744. http://dx.doi.org/10.4049/jimmunol.0803590.

Gardella, S., Andrei, C., Ferrera, D., Lotti, L. V., Torrisi, M. R., Bianchi, M. E., et al. (2002). The nuclear protein HMGB1 is secreted by monocytes via a non-classical, vesicle-mediated secretory pathway. *EMBO Reports*, *3*(10), 995–1001. http://dx.doi.org/10.1093/embo-reports/kvf198.

Ghiringhelli, F., Apetoh, L., Tesniere, A., Aymeric, L., Ma, Y., Ortiz, C., et al. (2009). Activation of the NLRP3 inflammasome in dendritic cells induces IL-1beta-dependent adaptive immunity against tumors. *Nature Medicine*, *15*(10), 1170–1178. http://dx.doi.org/10.1038/nm.2028.

Glass, C. K., Saijo, K., Winner, B., Marchetto, M. C., & Gage, F. H. (2010). Mechanisms underlying inflammation in neurodegeneration. *Cell*, *140*(6), 918–934. http://dx.doi.org/10.1016/j.cell.2010.02.016.

Goldbach-Mansky, R. (2012). Immunology in clinic review series; focus on autoinflammatory diseases: Update on monogenic autoinflammatory diseases: The role of interleukin (IL)-1 and an emerging role for cytokines beyond IL-1. *Clinical and Experimental Immunology*, *167*(3), 391–404. http://dx.doi.org/10.1111/j.1365-2249.2011.04533.x.

Gross, O., Thomas, C. J., Guarda, G., & Tschopp, J. (2011). The inflammasome: An integrated view. *Immunological Reviews*, *243*(1), 136–151. http://dx.doi.org/10.1111/j.1600-065X.2011.01046.x.

Guillemin, G. J., & Brew, B. J. (2004). Microglia, macrophages, perivascular macrophages, and pericytes: A review of function and identification. *Journal of Leukocyte Biology*, *75*(3), 388–397. http://dx.doi.org/10.1189/jlb.0303114.

Haorah, J., Heilman, D., Knipe, B., Chrastil, J., Leibhart, J., Ghorpade, A., et al. (2005). Ethanol-induced activation of myosin light chain kinase leads to dysfunction of tight junctions and blood-brain barrier compromise. *Alcoholism, Clinical and Experimental Research*, *29*(6), 999–1009. http://dx.doi.org/00000374-200506000-00010 [pii].

Haorah, J., Knipe, B., Leibhart, J., Ghorpade, A., & Persidsky, Y. (2005). Alcohol-induced oxidative stress in brain endothelial cells causes blood-brain barrier dysfunction. *Journal of Leukocyte Biology, 78*(6), 1223–1232. http://dx.doi.org/jlb.0605340 [pii].

Hauge, T., Persson, J., & Danielsson, D. (1997). Mucosal bacterial growth in the upper gastrointestinal tract in alcoholics (heavy drinkers). *Digestion, 58*(6), 591–595.

He, J., & Crews, F. T. (2008). Increased MCP-1 and microglia in various regions of the human alcoholic brain. *Experimental Neurology, 210*(2), 349–358. http://dx.doi.org/10.1016/j.expneurol.2007.11.017.

He, Y., Franchi, L., & Nunez, G. (2013). TLR agonists stimulate Nlrp3-dependent IL-1beta production independently of the purinergic P2X7 receptor in dendritic cells and in vivo. *Journal of Immunology (Baltimore, Md.: 1950), 190*(1), 334–339. http://dx.doi.org/10.4049/jimmunol.1202737.

Holley, M. M., & Kielian, T. (2012). Th1 and Th17 cells regulate innate immune responses and bacterial clearance during central nervous system infection. *Journal of Immunology (Baltimore, Md.: 1950), 188*(3), 1360–1370. http://dx.doi.org/10.4049/jimmunol.1101660.

Hritz, I., Mandrekar, P., Velayudham, A., Catalano, D., Dolganiuc, A., Kodys, K., et al. (2008). The critical role of toll-like receptor (TLR) 4 in alcoholic liver disease is independent of the common TLR adapter MyD88. *Hepatology (Baltimore, Md.), 48*(4), 1224–1231. http://dx.doi.org/10.1002/hep.22470.

Hutchinson, M. R., Zhang, Y., Brown, K., Coats, B. D., Shridhar, M., Sholar, P. W., et al. (2008). Non-stereoselective reversal of neuropathic pain by naloxone and naltrexone: Involvement of toll-like receptor 4 (TLR4). *European Journal of Neuroscience, 28*(1), 20–29. http://dx.doi.org/10.1111/j.1460-9568.2008.06321.x.

Israel, Y., Orrego, H., & Carmichael, F. J. (1994). Acetate-mediated effects of ethanol. *Alcoholism, Clinical and Experimental Research, 18*(1), 144–148.

Jin, L., Nation, R. L., Li, J., & Nicolazzo, J. A. (2013). Species-dependent blood-brain barrier disruption of lipopolysaccharide: Amelioration by colistin in vitro and in vivo. *Antimicrobial Agents and Chemotherapy,* [Epub ahead of print].

Kawaguchi, M., Takahashi, M., Hata, T., Kashima, Y., Usui, F., Morimoto, H., et al. (2011). Inflammasome activation of cardiac fibroblasts is essential for myocardial ischemia/reperfusion injury. *Circulation, 123*(6), 594–604. http://dx.doi.org/10.1161/CIRCULATIONAHA.110.982777.

Keshavarzian, A., Holmes, E. W., Patel, M., Iber, F., Fields, J. Z., & Pethkar, S. (1999). Leaky gut in alcoholic cirrhosis: A possible mechanism for alcohol-induced liver damage. *American Journal of Gastroenterology, 94*(1), 200–207. http://dx.doi.org/S0002927098006789 [pii].

Kim, S. W., Lim, C. M., Kim, J. B., Shin, J. H., Lee, S., Lee, M., et al. (2011). Extracellular HMGB1 released by NMDA treatment confers neuronal apoptosis via RAGE-p38 MAPK/ERK signaling pathway. *Neurotoxicity Research, 20*(2), 159–169. http://dx.doi.org/10.1007/s12640-010-9231-x.

Kim, I. D., Shin, J. H., Kim, S. W., Choi, S., Ahn, J., Han, P. L., et al. (2012). Intranasal delivery of HMGB1 siRNA confers target gene knockdown and robust neuroprotection in the postischemic brain. *Molecular Therapy: The Journal of the American Society of Gene Therapy, 20*(4), 829–839. http://dx.doi.org/10.1038/mt.2011.291.

Kirpich, I. A., Feng, W., Wang, Y., Liu, Y., Beier, J. I., Arteel, G. E., et al. (2013). Ethanol and dietary unsaturated fat (corn oil/linoleic acid enriched) cause intestinal inflammation and impaired intestinal barrier defense in mice chronically fed alcohol. *Alcohol (Fayetteville, NY), 47*(3), 257–264. http://dx.doi.org/10.1016/j.alcohol.2013.01.005.

Klune, J. R., Dhupar, R., Cardinal, J., Billiar, T. R., & Tsung, A. (2008). HMGB1: Endogenous danger signaling. *Molecular Medicine (Cambridge, Mass.), 14*(7–8), 476–484. http://dx.doi.org/10.2119/2008-00034.Klune.

Kumar, M. S., Lu, J., Mercer, K. L., Golub, T. R., & Jacks, T. (2007). Impaired microRNA processing enhances cellular transformation and tumorigenesis. *Nature Genetics*, *39*(5), 673–677. http://dx.doi.org/ng2003 [pii].

Laye, S., Bluthe, R. M., Kent, S., Combe, C., Medina, C., Parnet, P., et al. (1995). Subdiaphragmatic vagotomy blocks induction of IL-1 beta mRNA in mice brain in response to peripheral LPS. *American Journal of Physiology*, *268*(5 Pt. 2), R1327–R1331.

Leonard, E. J., & Yoshimura, T. (1990). Human monocyte chemoattractant protein-1 (MCP-1). *Immunology Today*, *11*(3), 97–101.

Letterio, J. J., & Roberts, A. B. (1998). Regulation of immune responses by TGF-beta. *Annual Review of Immunology*, *16*, 137–161. http://dx.doi.org/10.1146/annurev.immunol.16.1.137.

Lewohl, J. M., Nunez, Y. O., Dodd, P. R., Tiwari, G. R., Harris, R. A., & Mayfield, R. D. (2011). Up-regulation of microRNAs in brain of human alcoholics. *Alcoholism, Clinical and Experimental Research*, *35*(11), 1928–1937. http://dx.doi.org/10.1111/j.1530-0277.2011.01544.x.

Lippai, D., Bala, S., Csak, T., Kurt-Jones, E. A., & Szabo, G. (2013). Chronic alcohol-induced microRNA-155 contributes to neuroinflammation in a TLR4-dependent manner in mice. *PLoS One*, *8*(8), e70945. http://dx.doi.org/10.1371/journal.pone.0070945.

Lippai, D., Bala, S., Petrasek, J., Csak, T., Levin, I., Kurt-Jones, E. A., et al. (2013). Alcohol-induced IL-1beta in the brain is mediated by NLRP3/ASC inflammasome activation that amplifies neuroinflammation. *Journal of Leukocyte Biology*, *94*(1), 171–182. http://dx.doi.org/10.1189/jlb.1212659.

Liu, J., Yang, A. R., Kelly, T., Puche, A., Esoga, C., June, H. L., Jr., et al. (2011). Binge alcohol drinking is associated with GABAA alpha2-regulated toll-like receptor 4 (TLR4) expression in the central amygdala. *Proceedings of the National Academy of Sciences of the United States of America*, *108*(11), 4465–4470. http://dx.doi.org/10.1073/pnas.1019020108.

Mankan, A. K., Kubarenko, A., & Hornung, V. (2012). Immunology in clinic review series; focus on autoinflammatory diseases: Inflammasomes: Mechanisms of activation. *Clinical and Experimental Immunology*, *167*(3), 369–381. http://dx.doi.org/10.1111/j.1365-2249.2011.04534.x.

Marshall, S. A., McClain, J. A., Kelso, M. L., Hopkins, D. M., Pauly, J. R., & Nixon, K. (2013). Microglial activation is not equivalent to neuroinflammation in alcohol-induced neurodegeneration: The importance of microglia phenotype. *Neurobiology of Disease*, *54*, 239–251. http://dx.doi.org/10.1016/j.nbd.2012.12.016.

Nakajima, K., & Kohsaka, S. (2001). Microglia: Activation and their significance in the central nervous system. *Journal of Biochemistry*, *130*(2), 169–175.

Naveau, S., Chollet-Martin, S., Dharancy, S., Mathurin, P., Jouet, P., Piquet, M. A., et al. (2004). A double-blind randomized controlled trial of infliximab associated with prednisolone in acute alcoholic hepatitis. *Hepatology*, *39*(5), 1390–1397.

Nguyen, K. T., Deak, T., Owens, S. M., Kohno, T., Fleshner, M., Watkins, L. R., et al. (1998). Exposure to acute stress induces brain interleukin-1beta protein in the rat. *Journal of Neuroscience: The Official Journal of the Society for Neuroscience*, *18*(6), 2239–2246.

Oh, Y. J., Youn, J. H., Ji, Y., Lee, S. E., Lim, K. J., Choi, J. E., et al. (2009). HMGB1 is phosphorylated by classical protein kinase C and is secreted by a calcium-dependent mechanism. *Journal of Immunology (Baltimore, Md.: 1950)*, *182*(9), 5800–5809. http://dx.doi.org/10.4049/jimmunol.0801873.

O'Mara, N. B., & Wesley, L. C. (1994). Naltrexone in the treatment of alcohol dependence. *Annals of Pharmacotherapy*, *28*(2), 210–211.

O'Neill, L. A., Sheedy, F. J., & McCoy, C. E. (2011). MicroRNAs: The fine-tuners of toll-like receptor signalling. *Nature Reviews. Immunology*, *11*(3), 163–175. http://dx.doi.org/10.1038/nri2957.

Palomero-Gallagher, N., & Zilles, K. (2013). Neurotransmitter receptor alterations in hepatic encephalopathy: A review. *Archives of Biochemistry and Biophysics*, *536*(2), 109–121. http://dx.doi.org/10.1016/j.abb.2013.02.010.

Pan, W., & Kastin, A. J. (2002). TNFalpha transport across the blood-brain barrier is abolished in receptor knockout mice. *Experimental Neurology*, *174*(2), 193–200. http://dx.doi.org/10.1006/exnr.2002.787.

Pantazis, N. J., Dohrman, D. P., Goodlett, C. R., Cook, R. T., & West, J. R. (1993). Vulnerability of cerebellar granule cells to alcohol-induced cell death diminishes with time in culture. *Alcoholism, Clinical and Experimental Research*, *17*(5), 1014–1021.

Petrasek, J., Bala, S., Csak, T., Lippai, D., Kodys, K., Menashy, V., et al. (2012). IL-1 receptor antagonist ameliorates inflammasome-dependent alcoholic steatohepatitis in mice. *Journal of Clinical Investigation*, *122*(10), 3476–3489. http://dx.doi.org/10.1172/JCI60777.

Pietrzykowski, A. Z., Friesen, R. M., Martin, G. E., Puig, S. I., Nowak, C. L., Wynne, P. M., et al. (2008). Posttranscriptional regulation of BK channel splice variant stability by miR-9 underlies neuroadaptation to alcohol. *Neuron*, *59*(2), 274–287. http://dx.doi.org/10.1016/j.neuron.2008.05.032.

Pignataro, L., Varodayan, F. P., Tannenholz, L. E., & Harrison, N. L. (2009). The regulation of neuronal gene expression by alcohol. *Pharmacology & Therapeutics*, *124*(3), 324–335. http://dx.doi.org/10.1016/j.pharmthera.2009.09.002.

Pignataro, L., Varodayan, F. P., Tannenholz, L. E., Protiva, P., & Harrison, N. L. (2013). Brief alcohol exposure alters transcription in astrocytes via the heat shock pathway. *Brain and Behavior*, *3*(2), 114–133. http://dx.doi.org/10.1002/brb3.125.

Qin, L., & Crews, F. T. (2012). NADPH oxidase and reactive oxygen species contribute to alcohol-induced microglial activation and neurodegeneration. *Journal of Neuroinflammation*, *9*(5), 2094–2095. http://dx.doi.org/10.1186/1742-2094-9-5.

Qin, L., He, J., Hanes, R. N., Pluzarev, O., Hong, J. S., & Crews, F. T. (2008). Increased systemic and brain cytokine production and neuroinflammation by endotoxin following ethanol treatment. *Journal of Neuroinflammation*, *5*. http://dx.doi.org/10.1186/1742-2094-5-10. http://dx.doi.org/10-2094-5-10.

Rajayer, S. R., Jacob, A., Yang, W. L., Zhou, M., Chaung, W., & Wang, P. (2013). Cold-inducible RNA-binding protein is an important mediator of alcohol-induced brain inflammation. *PLoS One*, *8*(11), e79430. http://dx.doi.org/10.1371/journal.pone.0079430.

Ridet, J. L., Malhotra, S. K., Privat, A., & Gage, F. H. (1997). Reactive astrocytes: Cellular and molecular cues to biological function. *Trends in Neurosciences*, *20*(12), 570–577. http://dx.doi.org/S0166-2236(97)01139-9 [pii].

Rubartelli, A., & Lotze, M. T. (2007). Inside, outside, upside down: Damage-associated molecular-pattern molecules (DAMPs) and redox. *Trends in Immunology*, *28*(10), 429–436. http://dx.doi.org/S1471-4906(07)00207-4 [pii].

Schonrock, N., & Gotz, J. (2012). Decoding the non-coding RNAs in Alzheimer's disease. *Cellular and Molecular Life Sciences: CMLS*, *69*(21), 3543–3559. http://dx.doi.org/10.1007/s00018-012-1125-z.

Schulte, L. N., Westermann, A. J., & Vogel, J. (2013). Differential activation and functional specialization of miR-146 and miR-155 in innate immune sensing. *Nucleic Acids Research*, *41*(1), 542–553. http://dx.doi.org/10.1093/nar/gks1030.

Singh, A. K., Jiang, Y., Gupta, S., & Benlhabib, E. (2007). Effects of chronic ethanol drinking on the blood brain barrier and ensuing neuronal toxicity in alcohol-preferring rats subjected to intraperitoneal LPS injection. *Alcohol and Alcoholism (Oxford, Oxfordshire)*, *42*(5), 385–399. http://dx.doi.org/agl120 [pii].

Soliman, M. L., & Rosenberger, T. A. (2011). Acetate supplementation increases brain histone acetylation and inhibits histone deacetylase activity and expression. *Molecular and Cellular Biochemistry*, *352*(1–2), 173–180. http://dx.doi.org/10.1007/s11010-011-0751-3.

Sriram, K., Matheson, J. M., Benkovic, S. A., Miller, D. B., Luster, M. I., & O'Callaghan, J. P. (2006). Deficiency of TNF receptors suppresses microglial activation and alters the susceptibility of brain regions to MPTP-induced neurotoxicity: Role of TNF-alpha. *FASEB Journal: Official Publication of the Federation of American Societies for Experimental Biology*, *20*(6), 670–682. http://dx.doi.org/20/6/670 [pii].

Streit, W. J., Graeber, M. B., & Kreutzberg, G. W. (1988). Functional plasticity of microglia: A review. *Glia*, *1*(5), 301–307. http://dx.doi.org/10.1002/glia.440010502.

Szabo, G., Dolganiuc, A., Szabo, G., Dolganiuc, A., Szabo, G., Dolganiuc, A., et al. (2006). Pattern recognition receptors: A contemporary view on liver diseases. *Hepatology (Baltimore, Md.)*, *44*(2), 287–298. http://dx.doi.org/10.1002/hep.21308.

Szabo, G., Mandrekar, P., Petrasek, J., & Catalano, D. (2011). The unfolding web of innate immune dysregulation in alcoholic liver injury. *Alcoholism, Clinical and Experimental Research*, *35*(5), 782–786. http://dx.doi.org/10.1111/j.1530-0277.2010.01398.x.

Szabo, G., Verma, B. K., Fogarasi, M., & Catalano, D. E. (1992). Induction of transforming growth factor-beta and prostaglandin E2 production by ethanol in human monocytes. *Journal of Leukocyte Biology*, *52*(6), 602–610.

Tajuddin, N. F., Przybycien-Szymanska, M. M., Pak, T. R., Neafsey, E. J., & Collins, M. A. (2013). Effect of repetitive daily ethanol intoxication on adult rat brain: Significant changes in phospholipase A2 enzyme levels in association with increased PARP-1 indicate neuroinflammatory pathway activation. *Alcohol (Fayetteville, N.Y.)*, *47*(1), 39–45. http://dx.doi.org/10.1016/j.alcohol.2012.09.003.

Tamai, H., Kato, S., Horie, Y., Ohki, E., Yokoyama, H., & Ishii, H. (2000). Effect of acute ethanol administration on the intestinal absorption of endotoxin in rats. *Alcoholism, Clinical and Experimental Research*, *24*(3), 390–394.

Tang, Y., Banan, A., Forsyth, C. B., Fields, J. Z., Lau, C. K., Zhang, L. J., et al. (2008). Effect of alcohol on miR-212 expression in intestinal epithelial cells and its potential role in alcoholic liver disease. *Alcoholism, Clinical and Experimental Research*, *32*(2), 355–364. http://dx.doi.org/10.1111/j.1530-0277.2007.00584.x.

Thompson, W. L., & Van Eldik, L. J. (2009). Inflammatory cytokines stimulate the chemokines CCL2/MCP-1 and CCL7/MCP-3 through NFkB and MAPK dependent pathways in rat astrocytes [corrected]. *Brain Research*, *1287*, 47–57. http://dx.doi.org/10.1016/j.brainres.2009.06.081.

Thounaojam, M. C., Kaushik, D. K., & Basu, A. (2013). MicroRNAs in the brain: It's regulatory role in neuroinflammation. *Molecular Neurobiology*, *47*(3), 1034–1044. http://dx.doi.org/10.1007/s12035-013-8400-3.

Tilg, H., Jalan, R., Kaser, A., Davies, N. A., Offner, F. A., Hodges, S. J., et al. (2003). Anti-tumor necrosis factor-alpha monoclonal antibody therapy in severe alcoholic hepatitis. *Journal of Hepatology*, *38*(4), 419–425. http://dx.doi.org/S0168827802004427 [pii].

Tili, E., Croce, C. M., & Michaille, J. J. (2009). miR-155: On the crosstalk between inflammation and cancer. *International Reviews of Immunology*, *28*(5), 264–284. http://dx.doi.org/10.1080/08830180903093796.

Tracey, K. J. (2010). Understanding immunity requires more than immunology. *Nature Immunology*, *11*(7), 561–564. http://dx.doi.org/10.1038/ni0710-561; 10.1038/ni0710-561.

Trinchieri, G., & Sher, A. (2007). Cooperation of toll-like receptor signals in innate immune defence. *Nature Reviews. Immunology*, 7(3), 179–190. http://dx.doi.org/nri2038 [pii].

Valencia-Sanchez, M. A., Liu, J., Hannon, G. J., & Parker, R. (2006). Control of translation and mRNA degradation by miRNAs and siRNAs. *Genes & Development*, *20*(5), 515–524. http://dx.doi.org/20/5/515 [pii].

Valles, S. L., Blanco, A. M., Pascual, M., & Guerri, C. (2004). Chronic ethanol treatment enhances inflammatory mediators and cell death in the brain and in astrocytes. *Brain Pathology (Zurich, Switzerland)*, *14*(4), 365–371.

Verstrepen, L., Bekaert, T., Chau, T. L., Tavernier, J., Chariot, A., & Beyaert, R. (2008). TLR-4, IL-1R and TNF-R signaling to NF-kappaB: Variations on a common theme. *Cellular and Molecular Life Sciences: CMLS*, *65*(19), 2964–2978. http://dx.doi.org/10.1007/s00018-008-8064-8.

Wang, H., Bloom, O., Zhang, M., Vishnubhakat, J. M., Ombrellino, M., Che, J., et al. (1999). HMG-1 as a late mediator of endotoxin lethality in mice. *Science (New York, N.Y.)*, *285*(5425), 248–251. http://dx.doi.org/7585 [pii].

Wilke, N., Sganga, M., Barhite, S., & Miles, M. F. (1994). Effects of alcohol on gene expression in neural cells. *EXS*, *71*, 49–59.

Willingham, S. B., Allen, I. C., Bergstralh, D. T., Brickey, W. J., Huang, M. T., Taxman, D. J., et al. (2009). NLRP3 (NALP3, cryopyrin) facilitates in vivo caspase-1 activation, necrosis, and HMGB1 release via inflammasome-dependent and -independent pathways. *Journal of Immunology (Baltimore, Md: 1950)*, *183*(3), 2008–2015. http://dx.doi.org/10.4049/jimmunol.0900138.

Wu, Y., Lousberg, E. L., Moldenhauer, L. M., Hayball, J. D., Robertson, S. A., Coller, J. K., et al. (2011). Attenuation of microglial and IL-1 signaling protects mice from acute alcohol-induced sedation and/or motor impairment. *Brain, Behavior, and Immunity*, *25*(Suppl. 1), S155–S164. http://dx.doi.org/10.1016/j.bbi.2011.01.012.

Xiang, M., Shi, X., Li, Y., Xu, J., Yin, L., Xiao, G., et al. (2011). Hemorrhagic shock activation of NLRP3 inflammasome in lung endothelial cells. *Journal of Immunology (Baltimore, Md.: 1950)*, *187*(9), 4809–4817. http://dx.doi.org/10.4049/jimmunol.1102093.

Youn, J. H., & Shin, J. S. (2006). Nucleocytoplasmic shuttling of HMGB1 is regulated by phosphorylation that redirects it toward secretion. *Journal of Immunology (Baltimore, Md.: 1950)*, *177*(11), 7889–7897. http://dx.doi.org/177/11/7889 [pii].

Zhang, G., & Ghosh, S. (2001). Toll-like receptor-mediated NF-kappaB activation: A phylogenetically conserved paradigm in innate immunity. *Journal of Clinical Investigation*, *107*(1), 13–19. http://dx.doi.org/10.1172/JCI11837.

Zhou, M., Yang, W. L., Ji, Y., Qiang, X., & Wang, P. (2014). Cold-inducible RNA-binding protein mediates neuroinflammation in cerebral ischemia. *Biochimica Et Biophysica Acta*, *1840*(7), 2253–2261. http://dx.doi.org/S0304-4165(14)00088-9 [pii].

Zou, J. Y., & Crews, F. T. (2014). Release of neuronal HMGB1 by ethanol through decreased HDAC activity activates brain neuroimmune signaling. *PLoS One*, *9*(2), e87915. http://dx.doi.org/10.1371/journal.pone.0087915.

CHAPTER TWELVE

Opportunities for the Development of Neuroimmune Therapies in Addiction

Lara A. Ray*,[1], Daniel J.O. Roche*, Keith Heinzerling†, Steve Shoptaw†
*Department of Psychology, University of California, Los Angeles, Los Angeles, California, USA
†Department of Family Medicine, University of California, Los Angeles, Los Angeles, California, USA
[1]Corresponding author: e-mail address: lararay@psych.ucla.edu

Contents

Abstract

Studies have implicated neuroinflammatory processes in the pathophysiology of various psychiatric conditions, including addictive disorders. Neuroimmune signaling represents an important and relatively poorly understood biological process in drug addiction. The objective of this review is to update the field on recent developments in neuroimmune therapies for addiction. First, we review studies of neuroinflammation in relation to alcohol and methamphetamine dependence followed by a section on neuroinflammation and accompanying neurocognitive dysfunction in HIV infection and concomitant substance abuse. Second, we provide a review of pharmacotherapies with neuroimmune properties and their potential development for the treatment of addictions. Pharmacotherapies covered in this review include ibudilast, minocycline, doxycycline, topiramate, indomethacin, rolipram, anakinra (IL-1Ra), peroxisome proliferator-activated receptor agonists, naltrexone, and naloxone. Lastly, summary

International Review of Neurobiology, Volume 118
ISSN 0074-7742
http://dx.doi.org/10.1016/B978-0-12-801284-0.00012-9

and future directions are provided with recommendations for how to efficiently translate preclinical findings into clinical studies that can ultimately lead to novel and more effective pharmacotherapies for addiction.

1. INTRODUCTION

Multiple studies implicate neuroinflammatory processes in the pathophysiology of various psychiatric conditions (Hirsch & Hunot, 2009; Sidoryk-Wegrzynowicz, Wegrzynowicz, Lee, Bowman, & Aschner, 2011), including addictive disorders. As carefully reviewed in this issue of *International Review of Neurobiology*, neuroimmune signaling represents an important and relatively poorly understood biological process in drug addiction. As the field begins to more fully understand and appreciate the contribution of the innate immune system to addiction etiology and maintenance, these discoveries set up opportunities for the development of novel treatments for addiction targeting neuroimmune dysfunction. In this chapter, we briefly review findings implicating neuroinflammation in alcohol dependence, methamphetamine (MA) dependence, and HIV. We then discuss specific pharmacotherapies with neuroimmune properties and their development potential for the indication of alcohol and/or drug use disorders. Lastly, we place these findings in the context of medication development for addiction, including efforts to effectively translate preclinical findings into more efficacious treatments.

1.1. Neuroinflammation and alcohol dependence

Several studies have demonstrated that neuroinflammation plays a role in alcohol use and abuse, with chronic alcohol use being associated with microglia activation and increased innate immune cell signaling (Mayfield, Ferguson, & Harris, 2013). Glial cell line-derived neurotrophic factor (GDNF) is a protein that is essential for the maintenance and survival of dopamine (DA) neurons (Boger et al., 2006) and can inhibit microglial activation (Rocha, Cristovão, Campos, Fonseca, & Baltazar, 2012). Additionally, preclinical evidence suggests that infusion of GDNF into the ventral tegmental area (VTA) blocks the acquisition and expression of alcohol-induced conditioned place preference (Barak, Ahmadiantehrani, Kharazia, & Ron, 2011; Barak, Carnicella, Yowell, & Ron, 2011), rapidly reduces alcohol intake (Carnicella, Ahmadiantehrani, Janak, & Ron, 2009, Carnicella, Amamoto, &

Ron, 2009; Carnicella, Kharazia, Jeanblanc, Janak, & Ron, 2008), and blocks alcohol reinstatement following extinction (Carnicella et al., 2008). Furthermore, endogenous levels of GDNF have been found to negatively regulate the rewarding effect of alcohol after a period of abstinence (Carnicella, Ahmadiantehrani, et al., 2009; Carnicella, Amamoto, et al., 2009). In one human study, GDNF serum levels measured peripherally were found to be significantly reduced in alcohol-dependent patients versus healthy controls and to be negatively associated with measures of tolerance and withdrawal (Heberlein et al., 2010). It has been hypothesized that GDNF functions to reduce these alcohol-related behaviors in animal models by reversing an alcohol-induced allostatic DA deficiency in the mesolimbic system caused by prolonged excessive alcohol consumption and repeated withdrawal (Barak, Ahmadiantehrani, et al., 2011; Barak, Carnicella, et al., 2011). Furthermore, there is evidence that pharmacological inhibition of phosphodiesterase-4 (PDE4), an enzyme that hydrolyses cyclic adenosine monophosphate (cAMP), decreases alcohol intake in mice (Hu et al., 2011) and rat (Wen et al., 2012) models of alcoholism, as well as reduces neuroinflammation and neuronal death in rats (Wang et al., 2012). In addition, a recent study found that another phosphodiesterase, PDE10A, mRNA levels correlated with greater alcohol self-administration during a relapse model and with ethanol preference after acquisition (Logrip & Zorrilla, 2012), suggesting that inhibition of PDE10A may have behavioral effects on alcohol ingestion. PDE4 inhibitors are highly relevant because by increasing cAMP levels, PDE4 inhibitors show a broad spectrum of anti-inflammatory effects in almost all inflammatory cells (Page & Spina, 2011). While these studies are promising, they rely primarily on animal models. One of the primary limitations for the translation of these findings to human samples is the relative difficulty in assessing accurate central markers of inflammation coupled with the unclear nature of the relationship between peripheral (i.e., more readily accessible) and central makers of inflammation.

1.2. Neuroinflammation and MA dependence

Preclinical studies have shown that MA has multiple effects on neuroimmune activities. MA activates microglia, and blocking this glial activation subsequently attenuates MA-induced neurodegeneration (Flora et al., 2002; Ladenheim et al., 2000; Thomas, Francescutti-Verbeem, & Kuhn, 2008; Thomas & Kuhn, 2005). Importantly, MA-induced microglial activation precedes the development of pathological changes in striatal DA neurons (LaVoie, Card, & Hastings, 2004), suggesting that microglial activation is involved in the development of MA-induced neurological changes and is not merely a reaction to neurodegeneration.

In a human imaging study, a marker for activated microglia was significantly increased in abstinent MA users versus nonusing controls, and binding levels correlated inversely with the duration of MA abstinence (Sekine et al., 2008). MA-dependent women exhibited severe reductions in glial tricarboxylic acid cycle rate compared to healthy control subjects in a magnetic resonance spectroscopy study, providing further evidence of *in vivo* glial cell dysfunction in MA users (Sailasuta, Abulseoud, Harris, & Ross, 2010). Furthermore, emerging research suggests that microglial activation may mediate MA-induced synaptic plasticity (Narita et al., 2006), thereby contributing to the prolonged susceptibility to drug relapse. Among human MA users, increased plasma levels of proinflammatory cytokines (IFN-α, IL-1β, IL-2, IL-6, and TNF-α) and chemokines (MCP-1, MIP-1α, and MIP-1β) were significantly associated with greater neurocognitive dysfunction (Loftis, Choi, Hoffman, & Huckans, 2011). Together, these results suggest that medications that counteract MA-induced neuroinflammation and microglial activation may reduce MA-induced neurodegeneration, thereby improving neurocognition and treatment outcomes in MA dependence and perhaps other substance use disorders as well.

In addition to the potential negative impact of glial-mediated neuroinflammation on MA-related neurodegeneration, glial cells also produce neurotrophic factors that may ameliorate DA dysfunction in MA dependence. For example, GDNF selectively protects DA neurons, but not serotonergic neurons, from MA-induced neurodegeneration (Cass, 1996), and increased GDNF expression in the putamen actually regenerates DA neurons and restores DA functioning in a nonhuman primate model of Parkinson's disease (Kells et al., 2010). GDNF is found at high levels in the striatum including the nucleus accumbens (NAcc), and GFRα1 and Ret, the receptors for GDNF, are highly expressed in DA neurons in the VTA (Carnicella & Ron, 2009). Preclinical studies suggest that increased GDNF expression and the activation of the GDNF pathway reduce the biochemical and behavioral response to a variety of drugs of abuse including cocaine, opioids, alcohol, and MA. GDNF expression is increased in the NAcc in mice following MA administration, and treatment with the peptide Leu-Ile, which is a GDNF inducer, blocked the development of MA conditioned place preference and behavioral sensitization in wild type but not heterozygous GDNF knockout (GDNF +/−) mice (Niwa et al., 2007).

GDNF +/− mice have lower levels of GDNF and exhibit greater MA conditioned place preference (Niwa et al., 2007). GDNF +/− mice acquire stable MA self-administration behavior more quickly than wild-type mice, exhibit greater motivation to self-administer MA (increased dose–response curve for MA self-administration and higher break point on progressive ratio schedule), and display greater reinstatement of prime- and cue-induced drug seeking following extinction, an effect that remained even 6 months after extinction training (Yan et al., 2007). In humans, polymorphisms in the GDNF gene have been associated with age of onset of MA dependence and addiction severity in Japanese MA users (Yoshimura et al., 2011). Together, these studies suggest that increasing GDNF is a promising approach to treating MA dependence due to its neurotrophic and neuroprotective effects that may restore DA functioning (Gramage & Herradon, 2011) and provide at least one mechanism that could not only reduce the reinforcing effect of MA but also reduce the use of MA itself (Carnicella & Ron, 2009; Ghitza et al., 2010).

1.3. Neuroinflammation and HIV

Neuroinflammation and accompanying neurocognitive dysfunction are also major clinical issues in HIV infection, and they are exacerbated by concomitant substance abuse. HIV-associated neurocognitive disorders (HANDs) are common even with antiretroviral therapy, with 52% of patients in a recent HIV clinical cohort exhibiting at least some level of neuropsychological impairment (Heaton et al., 2010). HIV does not directly infect central nervous system (CNS) neurons, and HAND is thought to result primarily from the infection and subsequent activation of CNS macrophages and microglia, which then secrete many of the same proinflammatory cytokines that are also secreted in response to MA and chronic alcohol use, including TNF-α, IL-1β, IL-6, and MCP-1 (Yadav & Collman, 2009). HIV proteins gp120 and Tat are also neurotoxic and combined with MA exhibit synergistic toxicity on striatal DA neurons and the blood–brain barrier leading to enhanced CNS penetration by HIV (Silverstein et al., 2011). Not surprisingly, MA abuse increases the risk for neurocognitive impairment among HIV-infected persons (Carey et al., 2006; Rippeth et al., 2004), especially with HIV/Hepatitis C Virus (HCV) coinfection (Cherner et al., 2005; Letendre et al., 2007). Greater cognitive dysfunction is associated with poor HIV clinical outcomes including medication nonadherence (Becker, Thames, Woo, Castellon, & Hinkin, 2011) and worsened quality of life

(Parsons, Braaten, Hall, & Robertson, 2006). Therefore, medications that reduce neuroinflammation in HIV-infected substance users may improve HIV and substance-related clinical outcomes via improvements in neurocognitive functioning.

2. NEUROIMMUNE TREATMENTS

This section provides a review of pharmacotherapies with neuroimmune properties and their potential development for the treatment of addictions. Pharmacotherapies covered in this section include ibudilast, minocycline, doxycycline, topiramate, indomethacin, rolipram, anakinra (IL-1Ra), peroxisome proliferator-activated receptor (PPAR) agonists, naltrexone, and naloxone. A summary of the medications, their potential neuroimmune targets, and preclinical and clinical findings is provided in Table 12.1.

2.1. Ibudilast

Ibudilast (IBUD; MN-166/AV411) is a nonselective phosphodiesterase inhibitor with preferential inhibition of PDE3A, PDE4, PDE10, and PDE11 (Gibson et al., 2006) that also inhibits glial cell activation (Suzumura, Ito, Yoshikawa, & Sawada, 1999) and production of macrophage migration inhibitory factor (Cho et al., 2010). IBUD has been used clinically for over 20 years in Asia for the treatment of bronchial asthma and, more recently, for poststroke dizziness and ocular allergies for which it has proven to be safe and well tolerated (Rolan, Hutchinson, & Johnson, 2009). IBUD increases expression of GDNF in *in vitro* studies (Mizuno et al., 2004) suggesting that IBUD may ameliorate DA dysfunction among MA users and alcohol-dependent patients via the induction of GDNF expression. IBUD also reduces microglial activation *in vitro* in preclinical studies (Suzumura, Ito, & Mizuno, 1999, 2003; Suzumura et al., 1999). IBUD dose-dependently protected against microglial activation and the subsequent cerebrovascular white matter lesions following bilateral ligation of the carotid arteries (an animal model of vascular dementia/cognitive impairment) in rats (Wakita et al., 2003). IBUD also suppressed activated microglia-induced neuronal cell death *in vitro* via inhibiting production of proinflammatory cytokines (IL-1β, IL-6, and TNF-α), reactive oxygen species, and nitric oxide and via increasing the secretion of anti-inflammatory mediators (IL-10, nerve growth factor, neurotrophin-4, and GDNF) by microglial cells (Mizuno et al., 2004).

Table 12.1 Neuroimmune medications with potential for the treatment of addictive disorders

Medication	Potential immune targets	Findings in animal studies	Findings in human studies
Ibudilast	Glia, PDE	↓ MA self-administration, locomotor sensitization, and reinstatement	Clinical trials underway for MA, alcohol, and opioid dependence
Tetracyclines	Glia, NMDA receptors, oxidative stress signaling, NO	↓ MA conditioned place preference, MA-induced DA release ↓ Alcohol self-administration ↑ Alcohol-induced motor impairment	↓ D-Amphetamine subjective reward
Topiramate	T cells or antigen-presenting cells	↓ Alcohol self-administration[a]	Mixed evidence supporting use for alcohol dependence[a]
Indomethacin	COX-2, iNOS	↓ Alcohol-induced apoptosis and cognitive/motor dysfunction	Did not affect acute response to alcohol or pentobarbital
Rolipram	PDE	↓ Alcohol self-administration and preference	
Anakinra	IL-1 receptor	↓ Alcohol-induced sedation and liver damage	
Thiazolidinediones and fibrates	PPAR	↓ Alcohol self-administration, reinstatement, and withdrawal ↓ MA locomotor sensitization ↓ Nicotine-induced DA release, nicotine self-administration, and reinstatement	

Continued

Table 12.1 Neuroimmune medications with potential for the treatment of addictive disorders—cont'd

Medication	Potential immune targets	Findings in animal studies	Findings in human studies
(+)(−) Naltrexone and Naloxone	TLR4	↓ Cocaine and amphetamine locomotor activity ↓ Alcohol-induced apoptosis, motor impairment, and sedation	

[a]Unclear if medication effects are related to immune function.

Preclinical studies have found that IBUD has significant effects on behavior in multiple rodent models of MA dependence including reinstatement, locomotor sensitization, and self-administration. Importantly, IBUD demonstrated a dose-dependent effect on behavior in all three models with the greatest effect at higher doses. In the MA-reinstatement model, rats were trained to lever press for MA after which MA infusions were discontinued and lever pressing extinguished. IBUD significantly reduced MA prime- and stress-induced reinstatement of active lever pressing (Beardsley, Shelton, Hendrick, & Johnson, 2010) suggesting that IBUD may be effective in reducing relapse during clinical treatment for MA dependence. While both high and low IBUD doses reduced stress-induced reinstatement, only the higher IBUD dose reduced prime-induced reinstatement. Our research group is actively engaged in studies of IBUD for the indications of MA (ClinicalTrial.Gov identifier: NCT01860807, NCT01217970) and alcohol dependence (ClinicalTrial.Gov identifier: NCT02025998). Additionally, a clinical study of IBUD for opiate dependence is underway (ClinicalTrial.Gov identifier: NCT00723177). In brief, IBUD is a potentially promising medication for addiction with supportive preclinical studies and several human trials underway. Importantly, the neuroinflammatory actions of IBUD represent novel targets in the field of psychiatry, and addiction in particular.

2.2. Minocycline and doxycycline

Minocycline is a tetracycline antibiotic typically used to treat acne. Recent studies have focused on minocycline as a therapeutic agent for psychiatric disorders in light of its antioxidant properties, which in turn are thought to target deficits in oxidative defense associated with psychiatric disorders (Dean, Data-Franco, Giorlando, & Berk, 2012). For addiction, in particular, studies have highlighted the glutamatergic and DA effects of minocycline, which in turn have been implicated in addiction etiology (Kalivas & Volkow, 2005). In addition, the effects of minocycline on neuroimmune and cytokine expression have been emphasized as potential therapeutic targets for this medication (Fan et al., 2007; Mishra & Basu, 2008).

Preclinical studies have found that minocycline attenuated NMDA receptor antagonist-induced cognitive impairment in rodents (Fujita et al., 2008; Munzar, Li, Nicholson, Wiley, & Balster, 2002). Further, minocycline was protective against the deleterious effects of MA on DA transporter levels in monkeys (Hashimoto et al., 2007). Additional preclinical studies suggested that minocycline blocked the rewarding effects of MA (Fujita, Kunitachi, Iyo, & Hashimoto, 2012) and reduced ethanol administration in mice (Agrawal, Hewetson, George, Syapin, & Bergeson, 2011). A study of doxycycline, another anti-inflammatory mediator in the tetracycline derivative family, also observed reductions in alcohol consumption in mice, along with increased sensitivity to the motor-impairing effects of alcohol (McIver, Muccigrosso, & Haydon, 2012). One study to date has examined the effects of minocycline in healthy human volunteers and found that minocycline reduced the subjective rewarding effects of dextroamphetamine, increased reaction times on a Go No-Go task, and reduced plasma levels of cortisol compared to placebo (Sofuoglu, Mooney, Kosten, Waters, & Hashimoto, 2011). In summary, while there are no studies of clinical populations to date, the preclinical literature suggests that tetracycline antibiotic drugs such as minocycline, and possibly doxycycline, may have therapeutic effects for addiction and that these effects may be, at least in part, mediated by their neuroinflammatory properties.

2.3. Topiramate

Topiramate is an anticonvulsant medication with demonstrated clinical effects on drinking outcomes among alcohol-dependent individuals (Johnson & Ait-Daoud, 2010). In addition, studies have found that topiramate has anti-inflammatory properties and that it decreases alcohol consumption in animal models (Breslin, Johnson, & Lynch, 2010; Zalewska-Kaszubska et al., 2013). A preclinical study found that topiramate inhibited the production of several proinflammatory cytokines, including IL-17, IFN-γ, TNF, IL-6, and IL-10, which are generally produced by either T cells or antigen-presenting cells (Bhat et al., 2010). This study demonstrates that GABAergic drugs such as topiramate can act on T cells or antigen-presenting cells to suppress inflammatory signalling. Notably, no studies to date have implicated the anti-inflammatory properties of topiramate in its clinical or preclinical efficacy for alcohol or other substance use disorders. Nonetheless, recognizing that these anti-inflammatory effects are present for medications with known clinical efficacy, such as topiramate and opioid antagonists (discussed below), provides intriguing evidence to suggest that neuroinflammatory process may be common mechanisms across efficacious pharmacotherapies for addiction.

2.4. Indomethacin

Indomethacin is a nonsteroidal anti-inflammatory and a cyclooxygenase (COX-2) enzyme inhibitor. A preclinical study found that administration of indomethacin prevented ethanol-induced behavioral deficits caused by the ethanol-induced COX-2 and nitric oxide synthase (iNOS) expression and subsequent neuronal death. These findings indicate that indomethacin had protective effects against ethanol-induced brain damage by reducing inflammatory signalling and, in turn, prevented ethanol-related cognitive and motor decrements (Pascual, Blanco, Cauli, Minarro, & Guerri, 2007). However, a human pharmacology study found that indomethacin pretreatment did not alter the effects of alcohol or pentobarbital on subjective drug ratings, heart rate, and cognitive/psychomotor performance (Pickworth, Fant, & Henningfield, 1997). Perhaps the neuroprotective effects of indomethacin may be better observed after chronic alcohol consumption in humans, yet extensive controlled studies are needed before establishing whether indomethacin may have therapeutic value for substance use disorders.

2.5. Rolipram

The cAMP signaling cascade is thought to subserve the behavioral responses to alcohol. As previously described, PDE4 catalyzes the hydrolysis of cAMP and regulates intracellular cAMP levels. Rolipram is a selective PDE4 inhibitor thought to represent a novel treatment option for alcoholism due to its effects on the cAMP cascade. A recent preclinical study found that rolipram acutely reduced ethanol self-administration in a dose-dependent fashion and also, after chronic dosing, reduced ethanol preference and consumption (Wen et al., 2012). This is consistent with a previous study showing that acute rolipram administration substantially reduced ethanol consumption and preference in mice (Hu et al., 2011). Together, these findings suggest that rolipram as well as other PDE4 inhibitors, may have potential for the treatment of alcoholism and perhaps other substance use disorders.

2.6. Anakinra (IL-1Ra)

Anakinra is an interleukin (IL-1) receptor antagonist frequently used in the treatment of rheumatoid arthritis. Studies have found that anakinra (IL-1Ra) crosses the blood–brain barrier in rodents (Shavit, Wolf, Goshen, Livshits, & Yirmiya, 2005) and also reduces CNS inflammation in humans (Goldbach-Mansky et al., 2006). Perhaps most intriguing, an animal study found that anakinra reduced alcohol-induced sedation in mice (Wu et al., 2011). More recently, a rodent study found that IL-1Ra's inhibition of IL-1 signaling was associated with a significant reduction in alcohol-induced liver inflammation, fat accumulation, and damage (Petrasek et al., 2012). Although preliminary, these intriguing studies advance IL-1 receptor antagonists as possible therapeutics for alcoholism. One ongoing trial of Anakinra for patients with severe and acute hepatitis and alcoholism was identified through a search of ClinicalTrial.Gov database (ClinicalTrial.Gov identifier: NCT01809132).

2.7. PPAR agonists

PPARs are ligand-activated nuclear receptors that function as transcription factors. There are three known PPAR isoforms, PPARα, PPARβ/δ, and PPARγ, which are located throughout most peripheral tissues, as well as in neurons and glia in the brain (Gofflot et al., 2007; Moreno, Farioli-Vecchioli, & Ceru, 2004; Sarruf et al., 2009; Woods et al., 2003). Activation of PPAR attenuates innate immune signaling, thereby mediating anti-inflammatory and neuroprotective processes (Berger & Moller, 2002; Kapadia, Yi, & Vemuganti, 2008; Landreth & Heneka, 2001; Pistis & Melis, 2010). Importantly, PPARα and PPARγ receptors are densely expressed in the lateral hypothalamus, are located in VTA DA neurons, and can modulate DA release from the VTA into the NAcc (Melis et al., 2010, 2008; Moreno et al., 2004; Sarruf et al., 2009), all of which suggest a potential role in addiction-related processes. In support, recent studies that have pharmacologically manipulated PPAR in rodents and nonhuman primates have provided promising results advocating PPAR agonists as potential addiction-related treatments. Two selective PPARγ agonists, the thiazolidinediones pioglitazone and rosiglitazone, reduced alcohol consumption, abolished reinstatement of alcohol-seeking behavior, and reduced alcohol withdrawal symptoms in rats (Stopponi et al., 2011). Pioglitazone and ciglitazone, which is another thiazolidinedione selective for PPARγ, blocked the expression of locomotor sensitization to MA in mice (Maeda et al., 2007). Additionally, in rodents and nonhuman primates, clofibrate, a fibrate medication and selective PPARα agonist, blocked nicotine-induced VTA firing and DA release in the NAcc at the molecular level, and at the behavioral level it blocked the acquisition of nicotine-seeking behavior in nondependent animals, decreased nicotine self-administration in dependent animals, and prevented relapse to nicotine seeking in abstinent animals (Panlilio et al., 2012). Thus, in rodents and nonhuman primates, PPAR agonists may be effective in reducing the motivational and salient properties of multiple drugs (i.e., MA, alcohol, and nicotine) by modulating neurotransmission within the common reward pathway by which drugs of abuse are thought to exert their positively reinforcing effects. One ongoing clinical study of pioglitazone was identified as an adjunct treatment to opioid and nicotine dependence (ClinicalTrial.Gov identifier: NCT01395797). While these medications have not yet been tested in humans for this purpose, fibrates and thiazolidinediones are already approved for use in humans as treatments for elevated cholesterol and diabetes, respectively, and thus, repositioning these medications for use as addiction pharmacotherapies may provide a fast and economically feasible alternative in treatment development.

2.8. Naltrexone/naloxone

Naltrexone is an opioid antagonist approved for the treatment of alcoholism and heroin dependence and is currently under investigation for treatment of nicotine dependence. Naloxone is also an opioid receptor antagonist with similar affinity to mu opioid receptors as naltrexone, but relatively lower affinity to kappa and delta opioid receptors. Each of these medications is available in two isomers: the (−) isomer is the common opioid receptor antagonist form of each drug, whereas the (+) isomer does not bind (or has significantly reduced binding affinity) to opioid receptors (Hutchinson et al., 2008, 2011). However, both the (+) and (−) forms of each drug are antagonists at the Toll-like receptor 4 (TLR4; Hutchinson et al.,

2008). Interestingly, despite its inability to antagonize opioid receptors, (+) naloxone was found to reduce stimulant-induced locomotor activity (Chatterjie, Alexander, Sechzer, & Lieberman, 1996; Chatterjie, Sechzer, Lieberman, & Alexander, 1998), which is congruent with findings suggesting that TLR4 contributes to the acute effects of drugs of abuse (Hutchinson et al., 2012) and the ability of opioid receptor antagonists to affect such responses (Wu et al., 2012). The activation of TLR4 predominantly contributes to glial activation and the subsequent release of numerous proinflammatory cytokines (Mayfield et al., 2013). Importantly, these TLR4-related processes are involved in the behavioral and neuroinflammatory effects of drugs of abuse (Mayfield et al., 2013), as TLR4 activation has been shown to be integral to alcohol-induced glial activation and proinflammatory signaling (Alfonso-Loeches, Pascual-Lucas, Blanco, Sanchez-Vera, & Guerri, 2010; Blanco, Pascual, Valles, & Guerri, 2004; Blanco, Valles, Pascual, & Guerri, 2005; Fernandez-Lizarbe, Pascual, & Guerri, 2009), as well as alcohol's behavioral effects in rodents (Wu et al., 2012). Furthermore, in rodents, naltrexone attenuates proinflammatory TLR4-related signaling (Hutchinson et al., 2011) and blocks ethanol-induced glial activation and neuronal death (Qin & Crews, 2012), while (+) naloxone reduces acute alcohol-induced sedation and motor impairment (Wu et al., 2012). In sum, these findings may indicate that TLR4 signaling is involved in both the acute behavioral and chronic inflammatory effects of alcohol and other drugs of abuse and also that such TLR4-mediated processes may be ameliorated by the opioid receptor antagonists naltrexone and naloxone.

2.9. Summary and conclusions

The literature on the role of neuroinflammatory processes in psychiatric disorders broadly, and addiction, in particularly, is in its infancy. As the field quickly develops a more refined understanding of the effects of the innate immune system in addiction etiology, the opportunities for intervention become clearer and hopefully more targeted. One of the important recognitions from this review is that while largely predicated on preclinical studies, there is compelling evidence to suggest that medications modulating neuroinflammatory processes represent promising alternatives for addiction treatment and do so by targeting novel pathways. In addition, one quickly recognizes that medications with established efficacy for addiction, such as naltrexone and topiramate, also have neuroinflammatory properties. As such, a plausible question is the degree to which such inflammatory effects contribute to their clinical efficacy.

Importantly, the promise of newer and more effective treatments for substance use disorders ought to be considered in light of an efficient pathway from preclinical to clinical science. As outlined recently by Litten and colleagues, the effective translation of treatments from bench to bedside involves carefully addressing translational questions at all levels of analyses (Litten et al., 2012). To that end, proof-of-concept laboratory studies offer an important bridge between preclinical findings and clinical application to treatment-seeking samples. Human laboratory models can be used to guide identification of medications with promise of efficacy by collecting both safety and alcohol/drug interaction data along with initial demonstration of subjective responses to alcohol/drug, cue-reactivity, and self-administration models (Plebani et al., 2012). Human laboratory models can also aid in the effective translation of preclinical findings by elucidating the biobehavioral mechanisms by which pharmacotherapies may be efficacious for addiction (Ray, Hutchison, & Tartter, 2010; Ray, Mackillop, & Monti, 2010). Such findings on safety and mechanisms are vital to deciding whether to invest resources for efficacy testing for a putative addiction medication. Our team has used human laboratory paradigms to test several medications for addiction, including naltrexone (Ray, Bujarski, Chin, & Miotto, 2012; Ray & Hutchison, 2007), topiramate (Miranda et al., 2008; Ray et al., 2009), quetiapine (Moallem & Ray, 2012; Ray, Chin, Heydari, & Miotto, 2011), and varenicline (Ray et al., 2014, 2013). Given the new opportunities presented by recent discoveries on the role of neuroinflammation in addiction as well as new advancements in the technology of medication development, including the refinement of powerful human laboratory models, the stage is set for the discovery of novel treatments for substance use disorders. The ultimate goal is to develop treatments for addiction that are more efficacious than the available ones and to further boost the efficacy of these novel compounds through personalized approaches, including genomics and behavioral science. Together, these approaches have the potential to mitigate the many costs of addiction to the individual and to society at large.

REFERENCES

Agrawal, R. G., Hewetson, A., George, C. M., Syapin, P. J., & Bergeson, S. E. (2011). Minocycline reduces ethanol drinking. *Brain, Behavior, and Immunity*, *25*(Suppl. 1), S165–S169. http://dx.doi.org/10.1016/j.bbi.2011.03.002.

Alfonso-Loeches, S., Pascual-Lucas, M., Blanco, A. M., Sanchez-Vera, I., & Guerri, C. (2010). Pivotal role of TLR4 receptors in alcohol-induced neuroinflammation and brain damage. *Journal of Neuroscience*, *30*(24), 8285–8295. http://dx.doi.org/10.1523/jneurosci.0976-10.2010.

Barak, S., Ahmadiantehrani, S., Kharazia, V., & Ron, D. (2011). Positive autoregulation of GDNF levels in the ventral tegmental area mediates long-lasting inhibition of excessive alcohol consumption. *Translational Psychiatry*, *419*, 456–459. http://dx.doi.org/10.1038/tp.2011.57.

Barak, S., Carnicella, S., Yowell, Q. V., & Ron, D. (2011). Glial cell line-derived neurotrophic factor reverses alcohol-induced allostasis of the mesolimbic dopaminergic system: Implications for alcohol reward and seeking. *Journal of Neuroscience*, *31*(27), 9885–9894. http://dx.doi.org/10.1523/JNEUROSCI.1750-11.2011.

Beardsley, P. M., Shelton, K. L., Hendrick, E., & Johnson, K. W. (2010). The glial cell modulator and phosphodiesterase inhibitor, AV411 (ibudilast), attenuates prime- and stress-induced methamphetamine relapse. *European Journal of Pharmacology*, *637*(1–3), 102–108. http://dx.doi.org/10.1016/j.ejphar.2010.04.010.

Becker, B. W., Thames, A. D., Woo, E., Castellon, S. A., & Hinkin, C. H. (2011). Longitudinal change in cognitive function and medication adherence in HIV-infected adults. *AIDS and Behavior*, *15*(8), 1888–1894. http://dx.doi.org/10.1007/s10461-011-9924-z.

Berger, J., & Moller, D. E. (2002). The mechanisms of action of PPARs. *Annual Review of Medicine*, *53*(1), 409–435.

Bhat, R., Axtell, R., Mitra, A., Miranda, M., Lock, C., Tsien, R. W., et al. (2010). Inhibitory role for GABA in autoimmune inflammation. *Proceedings of the National Academy of Sciences of the United States of America*, *107*(6), 2580–2585. http://dx.doi.org/10.1073/pnas.0915139107.

Blanco, A. M., Pascual, M., Valles, S. L., & Guerri, C. (2004). Ethanol-induced iNOS and COX-2 expression in cultured astrocytes via NF-kappa B. *Neuroreport*, *15*(4), 681–685.

Blanco, A. M., Valles, S. L., Pascual, M., & Guerri, C. (2005). Involvement of TLR4/type I IL-1 receptor signaling in the induction of inflammatory mediators and cell death induced by ethanol in cultured astrocytes. *Journal of Immunology*, *175*(10), 6893–6899.

Boger, H. A., Middaugh, L. D., Huang, P., Zaman, V., Smith, A. C., Hoffer, B. J., et al. (2006). A partial GDNF depletion leads to earlier age-related deterioration of motor function and tyrosine hydroxylase expression in the substantia nigra. *Experimental Neurology*, *202*, 336–347. http://dx.doi.org/10.1016/j.expneurol.2006.06.006.

Breslin, F. J., Johnson, B. A., & Lynch, W. J. (2010). Effect of topiramate treatment on ethanol consumption in rats. *Psychopharmacology*, *207*(4), 529–534. http://dx.doi.org/10.1007/s00213-009-1683-4.

Carey, C. L., Woods, S. P., Rippeth, J. D., Gonzalez, R., Heaton, R. K., & Grant, I. (2006). Additive deleterious effects of methamphetamine dependence and immunosuppression on neuropsychological functioning in HIV infection. *AIDS and Behavior*, *10*(2), 185–190. http://dx.doi.org/10.1007/s10461-005-9056-4.

Carnicella, S., Ahmadiantehrani, S., Janak, P. H., & Ron, D. (2009). GDNF is an endogenous negative regulator of ethanol-mediated reward and of ethanol consumption after a period of abstinence. *Alcoholism, Clinical and Experimental Research*, *33*(6), 1012–1024. http://dx.doi.org/10.1111/j.1530-0277.2009.00922.x.

Carnicella, S., Amamoto, R., & Ron, D. (2009). Excessive alcohol consumption is blocked by glial cell line-derived neurotrophic factor. *Alcohol*, *43*(1), 35–43. http://dx.doi.org/10.1016/j.alcohol.2008.12.001.

Carnicella, S., Kharazia, V., Jeanblanc, J., Janak, P. H., & Ron, D. (2008). GDNF is a fast-acting potent inhibitor of alcohol consumption and relapse. *Proceedings of the National Academy of Sciences of the United States of America*, *105*(23), 8114–8119. http://dx.doi.org/10.1073/pnas.0711755105.

Carnicella, S., & Ron, D. (2009). GDNF—A potential target to treat addiction. *Pharmacology & Therapeutics*, *122*(1), 9–18.

Cass, W. A. (1996). GDNF selectively protects dopamine neurons over serotonin neurons against the neurotoxic effects of methamphetamine. *Journal of Neuroscience*, *16*(24), 8132–8139.

Chatterjie, N., Alexander, G. J., Sechzer, J. A., & Lieberman, K. W. (1996). Prevention of cocaine-induced hyperactivity by a naloxone isomer with no opiate antagonist activity. *Neurochemical Research*, *21*(6), 691–693.

Chatterjie, N., Sechzer, A., Lieberman, K., & Alexander, G. (1998). Dextro-naloxone counteracts amphetamine-induced hyperactivity. *Pharmacology, Biochemistry, and Behavior*, *59*(2), 271–274.

Cherner, M., Letendre, S., Heaton, R. K., Durelle, J., Marquie-Beck, J., Gragg, B., et al. (2005). Hepatitis C augments cognitive deficits associated with HIV infection and methamphetamine. *Neurology*, *64*(8), 1343–1347. http://dx.doi.org/10.1212/01.WNL.0000158328.26897.0D.

Cho, Y., Crichlow, G. V., Vermeire, J. J., Leng, L., Du, X., Hodsdon, M. E., et al. (2010). Allosteric inhibition of macrophage migration inhibitory factor revealed by ibudilast. *Proceedings of the National Academy of Sciences of the United States of America*, *107*(25), 11313–11318. http://dx.doi.org/10.1073/pnas.1002716107.

Dean, O. M., Data-Franco, J., Giorlando, F., & Berk, M. (2012). Minocycline: Therapeutic potential in psychiatry. *CNS Drugs*, *26*(5), 391–401. http://dx.doi.org/10.2165/11632000-000000000-00000.

Fan, R., Xu, F., Previti, M. L., Davis, J., Grande, A. M., Robinson, J. K., et al. (2007). Minocycline reduces microglial activation and improves behavioral deficits in a transgenic model of cerebral microvascular amyloid. *Journal of Neuroscience*, *27*(12), 3057–3063. http://dx.doi.org/10.1523/JNEUROSCI.4371-06.2007.

Fernandez-Lizarbe, S., Pascual, M., & Guerri, C. (2009). Critical role of TLR4 response in the activation of microglia induced by ethanol. *Journal of Immunology*, *183*(7), 4733–4744. http://dx.doi.org/10.4049/jimmunol.0803590.

Flora, G., Lee, Y. W., Nath, A., Maragos, W., Hennig, B., & Toborek, M. (2002). Methamphetamine-induced TNF-alpha gene expression and activation of AP-1 in discrete regions of mouse brain: Potential role of reactive oxygen intermediates and lipid peroxidation. *Neuromolecular Medicine*, *2*(1), 71–85. http://dx.doi.org/10.1385/NMM:2:1:71.

Fujita, Y., Ishima, T., Kunitachi, S., Hagiwara, H., Zhang, L., Iyo, M., et al. (2008). Phencyclidine-induced cognitive deficits in mice are improved by subsequent subchronic administration of the antibiotic drug minocycline. *Progress in Neuro-Psychopharmacology & Biological Psychiatry*, *32*(2), 336–339. http://dx.doi.org/10.1016/j.pnpbp.2007.08.031.

Fujita, Y., Kunitachi, S., Iyo, M., & Hashimoto, K. (2012). The antibiotic minocycline prevents methamphetamine-induced rewarding effects in mice. *Pharmacology, Biochemistry, and Behavior*, *101*(2), 303–306. http://dx.doi.org/10.1016/j.pbb.2012.01.005.

Ghitza, U. E., Zhai, H., Wu, P., Airavaara, M., Shaham, Y., & Lu, L. (2010). Role of BDNF and GDNF in drug reward and relapse: A review. *Neuroscience & Biobehavioral Reviews*, *35*(2), 157–171.

Gibson, L. C., Hastings, S. F., McPhee, I., Clayton, R. A., Darroch, C. E., Mackenzie, A., et al. (2006). The inhibitory profile of Ibudilast against the human phosphodiesterase enzyme family. *European Journal of Pharmacology*, *538*(1–3), 39–42. http://dx.doi.org/10.1016/j.ejphar.2006.02.053.

Gofflot, F., Chartoire, N., Vasseur, L., Heikkinen, S., Dembele, D., Le Merrer, J., et al. (2007). Systematic gene expression mapping clusters nuclear receptors according to their function in the brain. *Cell*, *131*(2), 405–418.

Goldbach-Mansky, R., Dailey, N. J., Canna, S. W., Gelabert, A., Jones, J., Rubin, B. I., et al. (2006). Neonatal-onset multisystem inflammatory disease responsive to interleukin-1beta inhibition. *New England Journal of Medicine*, *355*(6), 581–592. http://dx.doi.org/10.1056/NEJMoa055137.

Gramage, E., & Herradon, G. (2011). Connecting Parkinson's disease and drug addiction: Common players reveal unexpected disease connections and novel therapeutic approaches. *Current Pharmaceutical Design*, *17*(5), 449–461. http://dx.doi.org/10.2174/138161211795164103.

Hashimoto, K., Tsukada, H., Nishiyama, S., Fukumoto, D., Kakiuchi, T., & Iyo, M. (2007). Protective effects of minocycline on the reduction of dopamine transporters in the striatum after administration of methamphetamine: A positron emission tomography study in conscious monkeys. *Biological Psychiatry*, *61*(5), 577–581. http://dx.doi.org/10.1016/j.biopsych.2006.03.019.

Heaton, R. K., Clifford, D. B., Franklin, D. R., Jr., Woods, S. P., Ake, C., Vaida, F., et al. (2010). HIV-associated neurocognitive disorders persist in the era of potent antiretroviral therapy: CHARTER Study. *Neurology*, *75*(23), 2087–2096. http://dx.doi.org/10.1212/WNL.0b013e318200d727.

Heberlein, A., Muschler, M., Wilhelm, J., Frieling, H., Lenz, B., Groschl, M., et al. (2010). BDNF and GDNF serum levels in alcohol-dependent patients during withdrawal. *Progress in Neuro-Psychopharmacology & Biological Psychiatry*, *34*(6), 1060–1064. http://dx.doi.org/10.1016/j.pnpbp.2010.05.025.

Hirsch, E. C., & Hunot, S. (2009). Neuroinflammation in Parkinson's disease: A target for neuroprotection? *Lancet Neurology*, *8*(4), 382–397. http://dx.doi.org/10.1016/S1474-4422(09)70062-6.

Hu, W., Lu, T., Chen, A., Huang, Y., Hansen, R., Chandler, L. J., et al. (2011). Inhibition of phosphodiesterase-4 decreases ethanol intake in mice. *Psychopharmacology*, *218*(2), 331–339. http://dx.doi.org/10.1007/s00213-011-2290-8.

Hutchinson, M. R., Northcutt, A., Hiranita, T., Wang, X., Lewis, S., Thomas, J., et al. (2012). Opioid activation of toll-like receptor 4 contributes to drug reinforcement. *The Journal of Neuroscience*, *32*(33), 11187–11200.

Hutchinson, M. R., Shavit, Y., Grace, P. M., Rice, K. C., Maier, S. F., & Watkins, L. R. (2011). Exploring the neuroimmunopharmacology of opioids: An integrative review of mechanisms of central immune signaling and their implications for opioid analgesia. *Pharmacological Reviews*, *63*(3), 772–810.

Hutchinson, M. R., Zhang, Y., Brown, K., Coats, B. D., Shridhar, M., Sholar, P. W., et al. (2008). Non-stereoselective reversal of neuropathic pain by naloxone and naltrexone: Involvement of toll-like receptor 4 (TLR4). *European Journal of Neuroscience*, *28*(1), 20–29.

Johnson, B. A., & Ait-Daoud, N. (2010). Topiramate in the new generation of drugs: Efficacy in the treatment of alcoholic patients. *Current Pharmaceutical Design*, *16*(19), 2103–2112. http://dx.doi.org/10.2174/138161210791516404.

Kalivas, P. W., & Volkow, N. D. (2005). The neural basis of addiction: A pathology of motivation and choice. *The American Journal of Psychiatry*, *162*(8), 1403–1413. http://dx.doi.org/10.1176/appi.ajp.162.8.1403.

Kapadia, R., Yi, J.-H., & Vemuganti, R. (2008). Mechanisms of anti-inflammatory and neuroprotective actions of PPAR-gamma agonists. *Frontiers in Bioscience: A Journal and Virtual Library*, *13*, 1813–1826.

Kells, A. P., Eberling, J., Su, X., Pivirotto, P., Bringas, J., Hadaczek, P., et al. (2010). Regeneration of the MPTP-lesioned dopaminergic system after convection-enhanced delivery of AAV2-GDNF. *Journal of Neuroscience*, *30*(28), 9567–9577. http://dx.doi.org/10.1523/JNEUROSCI.0942-10.2010.

Ladenheim, B., Krasnova, I. N., Deng, X., Oyler, J. M., Polettini, A., Moran, T. H., et al. (2000). Methamphetamine-induced neurotoxicity is attenuated in transgenic mice with a null mutation for interleukin-6. *Molecular Pharmacology*, *58*(6), 1247–1256.

Landreth, G. E., & Heneka, M. T. (2001). Anti-inflammatory actions of peroxisome proliferator-activated receptor gamma agonists in Alzheimer's disease. *Neurobiology of Aging*, *22*(6), 937–944.

LaVoie, M. J., Card, J. P., & Hastings, T. G. (2004). Microglial activation precedes dopamine terminal pathology in methamphetamine-induced neurotoxicity. *Experimental Neurology*, *187*(1), 47–57. http://dx.doi.org/10.1016/j.expneurol.2004.01.010.

Letendre, S., Paulino, A. D., Rockenstein, E., Adame, A., Crews, L., Cherner, M., et al. (2007). Pathogenesis of hepatitis C virus coinfection in the brains of patients infected with HIV. *The Journal of Infectious Diseases*, *196*(3), 361–370. http://dx.doi.org/10.1086/519285.

Litten, R. Z., Egli, M., Heilig, M., Cui, C., Fertig, J. B., Ryan, M. L., et al. (2012). Medications development to treat alcohol dependence: A vision for the next decade. *Addiction Biology*, *17*(3), 513–527. http://dx.doi.org/10.1111/j.1369-1600.2012.00454.x.

Loftis, J. M., Choi, D., Hoffman, W., & Huckans, M. S. (2011). Methamphetamine causes persistent immune dysregulation: A cross-species, translational report. *Neurotoxicity Research*, *20*(1), 59–68. http://dx.doi.org/10.1007/s12640-010-9223-x.

Logrip, M. L., & Zorrilla, E. P. (2012). Stress history increases alcohol intake in relapse: Relation to phosphodiesterase 10A. *Addiction Biology*, *17*(5), 920–933. http://dx.doi.org/10.1111/j.1369-1600.2012.00460.x.

Maeda, T., Kiguchi, N., Fukazawa, Y., Yamamoto, A., Ozaki, M., & Kishioka, S. (2007). Peroxisome proliferator-activated receptor gamma activation relieves expression of behavioral sensitization to methamphetamine in mice. *Neuropsychopharmacology*, *32*(5), 1133–1140.

Mayfield, J., Ferguson, L., & Harris, R. A. (2013). Neuroimmune signaling: A key component of alcohol abuse. *Current Opinion in Neurobiology*, *23*(4), 513–520.

McIver, S. R., Muccigrosso, M. M., & Haydon, P. G. (2012). The effect of doxycycline on alcohol consumption and sensitivity: Consideration for inducible transgenic mouse models. *Experimental Biology and Medicine (Maywood, N.J.)*, *237*(10), 1129–1133. http://dx.doi.org/10.1258/ebm.2012.012029.

Melis, M., Carta, S., Fattore, L., Tolu, S., Yasar, S., Goldberg, S. R., et al. (2010). Peroxisome proliferator-activated receptors-alpha modulate dopamine cell activity through nicotinic receptors. *Biological Psychiatry*, *68*(3), 256–264.

Melis, M., Pillolla, G., Luchicchi, A., Muntoni, A. L., Yasar, S., Goldberg, S. R., et al. (2008). Endogenous fatty acid ethanolamides suppress nicotine-induced activation of mesolimbic dopamine neurons through nuclear receptors. *The Journal of Neuroscience*, *28*(51), 13985–13994.

Miranda, R., Jr., MacKillop, J., Monti, P. M., Rohsenow, D. J., Tidey, J., Gwaltney, C., et al. (2008). Effects of topiramate on urge to drink and the subjective effects of alcohol: A preliminary laboratory study. *Alcoholism, Clinical and Experimental Research*, *32*(3), 489–497. http://dx.doi.org/10.1111/j.1530-0277.2007.00592.x.

Mishra, M. K., & Basu, A. (2008). Minocycline neuroprotects, reduces microglial activation, inhibits caspase 3 induction, and viral replication following Japanese encephalitis. *Journal of Neurochemistry*, *105*(5), 1582–1595. http://dx.doi.org/10.1111/j.1471-4159.2008.05238.x.

Mizuno, T., Kurotani, T., Komatsu, Y., Kawanokuchi, J., Kato, H., Mitsuma, N., et al. (2004). Neuroprotective role of phosphodiesterase inhibitor ibudilast on neuronal cell death induced by activated microglia. *Neuropharmacology*, *46*(3), 404–411. http://dx.doi.org/10.1016/j.neuropharm.2003.09.009.

Moallem, N., & Ray, L. A. (2012). Quetiapine improves response inhibition in alcohol dependent patients: A placebo-controlled pilot study. *Pharmacology, Biochemistry, and Behavior*, *100*(3), 490–493. http://dx.doi.org/10.1016/j.pbb.2011.10.012.

Moreno, S., Farioli-Vecchioli, S., & Ceru, M. (2004). Immunolocalization of peroxisome proliferator-activated receptors and retinoid X receptors in the adult rat CNS. *Neuroscience*, *123*(1), 131–145.

Munzar, P., Li, H., Nicholson, K. L., Wiley, J. L., & Balster, R. L. (2002). Enhancement of the discriminative stimulus effects of phencyclidine by the tetracycline antibiotics doxycycline and minocycline in rats. *Psychopharmacology*, *160*(3), 331–336. http://dx.doi.org/10.1007/s00213-001-0989-7.

Narita, M., Miyatake, M., Shibasaki, M., Shindo, K., Nakamura, A., Kuzumaki, N., et al. (2006). Direct evidence of astrocytic modulation in the development of rewarding effects induced by drugs of abuse. *Neuropsychopharmacology*, *31*(11), 2476–2488. http://dx.doi.org/10.1038/sj.npp.1301007.

Niwa, M., Nitta, A., Yamada, Y., Nakajima, A., Saito, K., Seishima, M., et al. (2007). An inducer for glial cell line-derived neurotrophic factor and tumor necrosis factor-alpha protects against methamphetamine-induced rewarding effects and sensitization. *Biological Psychiatry*, *61*(7), 890–901. http://dx.doi.org/10.1016/j.biopsych.2006.06.016.

Page, C. P., & Spina, D. (2011). Phosphodiesterase inhibitors in the treatment of inflammatory diseases. *Handbook of Experimental Pharmacology*, *204*, 391–414. http://dx.doi.org/10.1007/978-3-642-17969-3_17.

Panlilio, L. V., Justinova, Z., Mascia, P., Pistis, M., Luchicchi, A., Lecca, S., et al. (2012). Novel use of a lipid-lowering fibrate medication to prevent nicotine reward and relapse: Preclinical findings. *Neuropsychopharmacology*, *37*(8), 1838–1847.

Parsons, T. D., Braaten, A. J., Hall, C. D., & Robertson, K. R. (2006). Better quality of life with neuropsychological improvement on HAART. *Health and Quality of Life Outcomes*, *4*, 11. http://dx.doi.org/10.1186/1477-7525-4-11.

Pascual, M., Blanco, A. M., Cauli, O., Minarro, J., & Guerri, C. (2007). Intermittent ethanol exposure induces inflammatory brain damage and causes long-term behavioural alterations in adolescent rats. *European Journal of Neuroscience*, *25*(2), 541–550. http://dx.doi.org/10.1111/j.1460-9568.2006.05298.x.

Petrasek, J., Bala, S., Csak, T., Lippai, D., Kodys, K., Menashy, V., et al. (2012). IL-1 receptor antagonist ameliorates inflammasome-dependent alcoholic steatohepatitis in mice. *Journal of Clinical Investigation*, *122*(10), 3476–3489. http://dx.doi.org/10.1172/JCI60777.

Pickworth, W. B., Fant, R. V., & Henningfield, J. E. (1997). Inhibition of prostaglandin synthesis and effects of ethanol and pentobarbital in humans. *Journal of Pharmacology and Experimental Therapeutics*, *283*(1), 274–280.

Pistis, M., & Melis, M. (2010). From surface to nuclear receptors: The endocannabinoid family extends its assets. *Current Medicinal Chemistry*, *17*(14), 1450–1467.

Plebani, J. G., Ray, L. A., Morean, M. E., Corbin, W. R., MacKillop, J., Amlung, M. T., et al. (2012). Human laboratory paradigms in alcohol research. *Alcoholism, Clinical and Experimental Research*, *36*(6), 972–983. http://dx.doi.org/10.1111/j.1530-0277.2011.01704.x.

Qin, L., & Crews, F. T. (2012). Chronic ethanol increases systemic TLR3 agonist-induced neuroinflammation and neurodegeneration. *Journal of Neuroinflammation*, *9*, 130. http://dx.doi.org/10.1186/1742-2094-9-130.

Ray, L. A., Bujarski, S., Chin, P. F., & Miotto, K. (2012). Pharmacogenetics of naltrexone in Asian Americans: A randomized placebo-controlled laboratory study. *Neuropsychopharmacology*, *37*(2), 445–455. http://dx.doi.org/10.1038/npp.2011.192.

Ray, L. A., Chin, P. F., Heydari, A., & Miotto, K. (2011). A human laboratory study of the effects of quetiapine on subjective intoxication and alcohol craving. *Psychopharmacology*, *217*(3), 341–351. http://dx.doi.org/10.1007/s00213-011-2287-3.

Ray, L. A., Courtney, K. E., Ghahremani, D. G., Miotto, K., Brody, A., & London, E. D. (2014). Varenicline, low dose naltrexone, and their combination for heavy-drinking smokers: Human laboratory findings. *Psychopharmacology*. http://dx.doi.org/10.1007/s00213-014-3519-0.

Ray, L. A., & Hutchison, K. E. (2007). Effects of naltrexone on alcohol sensitivity and genetic moderators of medication response: A double-blind placebo-controlled study.

Archives of General Psychiatry, 64(9), 1069–1077. http://dx.doi.org/10.1001/archpsyc.64.9.1069.
Ray, L. A., Hutchison, K. E., & Tartter, M. (2010). Application of human laboratory models to pharmacotherapy development for alcohol dependence. *Current Pharmaceutical Design, 16*(19), 2149–2158. http://dx.doi.org/10.2174/138161210791516422.
Ray, L. A., Lunny, K., Bujarski, S., Moallem, N., Krull, J. L., & Miotto, K. (2013). The effects of varenicline on stress-induced and cue-induced craving for cigarettes. *Drug and Alcohol Dependence, 131*(1–2), 136–142. http://dx.doi.org/10.1016/j.drugalcdep.2012.12.015.
Ray, L. A., Mackillop, J., & Monti, P. M. (2010). Subjective responses to alcohol consumption as endophenotypes: Advancing behavioral genetics in etiological and treatment models of alcoholism. *Substance Use and Misuse, 45*(11), 1742–1765. http://dx.doi.org/10.3109/10826084.2010.482427.
Ray, L. A., Miranda, R., Jr., MacKillop, J., McGeary, J., Tidey, J. W., Rohsenow, D. J., et al. (2009). A preliminary pharmacogenetic investigation of adverse events from topiramate in heavy drinkers. *Experimental and Clinical Psychopharmacology, 17*(2), 122–129. http://dx.doi.org/10.1037/a0015700.
Rippeth, J. D., Heaton, R. K., Carey, C. L., Marcotte, T. D., Moore, D. J., Gonzalez, R., et al. (2004). Methamphetamine dependence increases risk of neuropsychological impairment in HIV infected persons. *Journal of the International Neuropsychological Society: JINS, 10*(1), 1–14. http://dx.doi.org/10.1017/S1355617704101021.
Rolan, P., Hutchinson, M., & Johnson, K. (2009). Ibudilast: A review of its pharmacology, efficacy and safety in respiratory and neurological disease. *Expert Opinion on Pharmacotherapy, 10*(17), 2897–2904. http://dx.doi.org/10.1517/14656560903426189 (Review).
Rocha, S. M., Cristovão, A. C., Campos, F. L., Fonseca, C. P., & Baltazar, G. (2012). Astrocyte-derived GDNF is a potent inhibitor of microglial activation. *Neurobiology of Disease, 47*(3), 407–415. http://dx.doi.org/10.1016/j.nbd.2012.04.014.
Sailasuta, N., Abulseoud, O., Harris, K. C., & Ross, B. D. (2010). Glial dysfunction in abstinent methamphetamine abusers. *Journal of Cerebral Blood Flow and Metabolism, 30*(5), 950–960. http://dx.doi.org/10.1038/jcbfm.2009.261.
Sarruf, D. A., Yu, F., Nguyen, H. T., Williams, D. L., Printz, R. L., Niswender, K. D., et al. (2009). Expression of peroxisome proliferator-activated receptor-gamma in key neuronal subsets regulating glucose metabolism and energy homeostasis. *Endocrinology, 150*(2), 707–712. http://dx.doi.org/10.1210/en.2008-0899.
Sekine, Y., Ouchi, Y., Sugihara, G., Takei, N., Yoshikawa, E., Nakamura, K., et al. (2008). Methamphetamine causes microglial activation in the brains of human abusers. *Journal of Neuroscience, 28*(22), 5756–5761. http://dx.doi.org/10.1523/JNEUROSCI.1179-08.2008.
Shavit, Y., Wolf, G., Goshen, I., Livshits, D., & Yirmiya, R. (2005). Interleukin-1 antagonizes morphine analgesia and underlies morphine tolerance. *Pain, 115*(1–2), 50–59. http://dx.doi.org/10.1016/j.pain.2005.02.003.
Sidoryk-Wegrzynowicz, M., Wegrzynowicz, M., Lee, E., Bowman, A. B., & Aschner, M. (2011). Role of astrocytes in brain function and disease. *Toxicologic Pathology, 39*(1), 115–123. http://dx.doi.org/10.1177/0192623310385254.
Silverstein, P. S., Shah, A., Gupte, R., Liu, X., Piepho, R. W., Kumar, S., et al. (2011). Methamphetamine toxicity and its implications during HIV-1 infection. *Journal of Neurovirology, 17*(5), 401–415. http://dx.doi.org/10.1007/s13365-011-0043-4.
Sofuoglu, M., Mooney, M., Kosten, T., Waters, A., & Hashimoto, K. (2011). Minocycline attenuates subjective rewarding effects of dextroamphetamine in humans. *Psychopharmacology, 213*(1), 61–68. http://dx.doi.org/10.1007/s00213-010-2014-5.
Stopponi, S., Somaini, L., Cippitelli, A., Cannella, N., Braconi, S., Kallupi, M., et al. (2011). Activation of nuclear PPARγ receptors by the antidiabetic agent pioglitazone suppresses alcohol drinking and relapse to alcohol seeking. *Biological Psychiatry, 69*(7), 642–649.
Suzumura, A., Ito, A., & Mizuno, T. (2003). Phosphodiesterase inhibitors suppress IL-12 production with microglia and T helper 1 development. *Multiple Sclerosis, 9*(6), 574–578.

Suzumura, A., Ito, A., Yoshikawa, M., & Sawada, M. (1999). Ibudilast suppresses TNFalpha production by glial cells functioning mainly as type III phosphodiesterase inhibitor in the CNS. *Brain Research*, *837*(1–2), 203–212. http://dx.doi.org/10.1016/S0006-8993(99)01666-2.

Thomas, D. M., Francescutti-Verbeem, D. M., & Kuhn, D. M. (2008). Methamphetamine-induced neurotoxicity and microglial activation are not mediated by fractalkine receptor signaling. *Journal of Neurochemistry*, *106*(2), 696–705. http://dx.doi.org/10.1111/j.1471-4159.2008.05421.x.

Thomas, D. M., & Kuhn, D. M. (2005). MK-801 and dextromethorphan block microglial activation and protect against methamphetamine-induced neurotoxicity. *Brain Research*, *1050*(1–2), 190–198. http://dx.doi.org/10.1016/j.brainres.2005.05.049.

Wakita, H., Tomimoto, H., Akiguchi, I., Lin, J. X., Ihara, M., Ohtani, R., et al. (2003). Ibudilast, a phosphodiesterase inhibitor, protects against white matter damage under chronic cerebral hypoperfusion in the rat. *Brain Research*, *992*(1), 53–59. http://dx.doi.org/10.1016/j.brainres.2003.08.028.

Wang, C., Yang, X. M., Zhuo, Y. Y., Zhou, H., Lin, H. B., Cheng, Y. F., et al. (2012). The phosphodiesterase-4 inhibitor rolipram reverses Aβ-induced cognitive impairment and neuroinflammatory and apoptotic responses in rats. *The International Journal of Neuropsychopharmacology*, *15*(6), 749–766. http://dx.doi.org/10.1017/S1461145711000836.

Wen, R. T., Zhang, M., Qin, W. J., Liu, Q., Wang, W. P., Lawrence, A. J., et al. (2012). The phosphodiesterase-4 (PDE4) inhibitor rolipram decreases ethanol seeking and consumption in alcohol-preferring Fawn-Hooded rats. *Alcoholism, Clinical and Experimental Research*, *36*(12), 2157–2167. http://dx.doi.org/10.1111/j.1530-0277.2012.01845.x.

Woods, J. W., Tanen, M., Figueroa, D. J., Biswas, C., Zycband, E., Moller, D. E., et al. (2003). Localization of PPARδ in murine central nervous system: Expression in oligodendrocytes and neurons. *Brain Research*, *975*(1), 10–21.

Wu, Y., Lousberg, E. L., Moldenhauer, L. M., Hayball, J. D., Coller, J. K., Rice, K. C., et al. (2012). Inhibiting the TLR4-MyD88 signalling cascade by genetic or pharmacological strategies reduces acute alcohol-induced sedation and motor impairment in mice. *British Journal of Pharmacology*, *165*(5), 1319–1329. http://dx.doi.org/10.1111/j.1476-5381.2011.01572.x.

Wu, Y., Lousberg, E. L., Moldenhauer, L. M., Hayball, J. D., Robertson, S. A., Coller, J. K., et al. (2011). Attenuation of microglial and IL-1 signaling protects mice from acute alcohol-induced sedation and/or motor impairment. *Brain, Behavior, and Immunity*, *25*(Suppl. 1), S155–S164. http://dx.doi.org/10.1016/j.bbi.2011.01.012.

Yadav, A., & Collman, R. G. (2009). CNS inflammation and macrophage/microglial biology associated with HIV-1 infection. *Journal of Neuroimmune Pharmacology: The Official Journal of the Society on NeuroImmune Pharmacology*, *4*(4), 430–447. http://dx.doi.org/10.1007/s11481-009-9174-2.

Yan, Y., Yamada, K., Niwa, M., Nagai, T., Nitta, A., & Nabeshima, T. (2007). Enduring vulnerability to reinstatement of methamphetamine-seeking behavior in glial cell line-derived neurotrophic factor mutant mice. *The FASEB Journal*, *21*(9), 1994–2004. http://dx.doi.org/10.1096/fj.06-7772com.

Yoshimura, T., Usui, H., Takahashi, N., Yoshimi, A., Saito, S., Aleksic, B., et al. (2011). Association analysis of the GDNF gene with methamphetamine use disorder in a Japanese population. *Progress in Neuro-Psychopharmacology and Biological Psychiatry*, *35*(5), 1268–1272. http://dx.doi.org/10.1016/j.pnpbp.2011.04.003.

Zalewska-Kaszubska, J., Bajer, B., Gorska, D., Andrzejczak, D., Dyr, W., & Bienkowski, P. (2013). Effect of repeated treatment with topiramate on voluntary alcohol intake and beta-endorphin plasma level in Warsaw alcohol high-preferring rats. *Psychopharmacology*, *225*(2), 275–281. http://dx.doi.org/10.1007/s00213-012-2812-z.

CHAPTER THIRTEEN

NeuroHIV and Use of Addictive Substances

Sulie L. Chang*,†,1, Kaitlyn P. Connaghan*, Yufeng Wei*, Ming D. Li‡
*Institute of NeuroImmune Pharmacology, Seton Hall University, South Orange, New Jersey, USA
†Department of Biological Sciences, Seton Hall University, South Orange, New Jersey, USA
‡Department of Psychiatry and Neurobehavioral Sciences, University of Virginia, Charlottesville, Virginia, USA
[1]Corresponding author: e-mail address: sulie.chang@shu.edu

Contents

Abstract

In the past three decades, substance abuse has been identified as a key comorbidity of human immunodeficiency virus-1 (HIV-1) infection. Many studies have found that the use and abuse of addictive substances hastens the progression of HIV-1 infection and HIV-associated neurocognitive disorders. Advances in highly active antiretroviral therapy (HAART) in the mid-1990s have been successful in limiting the HIV-1 viral load and maintaining a relatively healthy immune response, allowing the life expectancy of patients infected with HIV to approach that of the general population. However, even with HAART, HIV-1 viral proteins are still expressed and eradication of the virus, particularly in the brain, the key reservoir organ, does not occur. In the post-HAART era, the

International Review of Neurobiology, Volume 118
ISSN 0074-7742
http://dx.doi.org/10.1016/B978-0-12-801284-0.00013-0

clinical challenge in the treatment of HIV infection is inflammation of the central nervous system (CNS) and its subsequent neurological disorders. To date, various explicit and implicit connections have been identified between the neuronal circuitry involved in immune responses and brain regions affected by and implicated in substance abuse. This chapter discusses past and current medical uses of prototypical substances of abuse, including morphine, alcohol, cocaine, methamphetamine, marijuana, and nicotine, and the evidence that systemic infections, particularly HIV-1 infection, cause neurological dysfunction as a result of inflammation in the CNS, which can increase the risk of substance abuse.

1. BRAIN–IMMUNE INTERACTIONS: INDUCTION OF NEUROINFLAMMATION BY SYSTEMIC INFECTIONS

In the past three to four decades, the bidirectional communication between the brain and the immune system has been a focus of considerable research. Both immune and neuronal products coexist in lymphoid and neuronal tissues (Aller et al., 2001; Tomaszewska & Przekop, 1997). Various immune cells have been shown to produce hormones and neuropeptides, including ACTH (Ottaviani, Franchini, & Genedani, 1999), endorphins (Morch & Pedersen, 1995), enkephalin (Jankovic & Radulovic, 1992; Wybran, 1985), prolactin (Gala, 1991; Yu-Lee, 1997, 2002), growth hormone (Gala, 1991; Meazza, Pagani, Travaglino, & Bozzola, 2004), catecholamines (Elenkov, 2007), and acetylcholine (Hosoi & Nomura, 2004), and the release of various neurotransmitters in immune system organs is triggered by nerve impulses from the central nervous system (CNS) to peripheral tissue (Pacheco, Riquelme, & Kalergis, 2010; Qiu, Peng, & Wang, 1996). In contrast, cytokines, such as interleukin-6, normally produced by immune cells, have been shown to be synthesized by and secreted from anterior pituitary cells when stimulated by the bacterial endotoxin, lipopolysaccharide (LPS), a model compound which mimics bacterial infection (Fuchs, Damm, Gerstberger, Roth, & Rummel, 2013).

Since 1927, there have been many reports suggesting that microglial cells are immune effectors in the CNS that respond to pathological conditions and neurological disorders, such as Alzheimer's disease (AD), Parkinson's disease, multiple sclerosis, and acquired immune deficiency syndrome (AIDS) dementia complex, by releasing cytotoxic molecules, including proinflammatory cytokines, reactive oxygen intermediates, proteinases, and complement proteins. For example, cytokines produced by glia have been implicated in synaptic formation, long-term potentiation, and neurogenesis (Pivneva, 2008). Evidence also suggests that microglia are able to secrete neurotrophic or neuron survival factors during inflammation or injury (Pivneva, 2008).

Hormones, neurotransmitters, and neuropeptides can affect the immune system and modulate immune responses (Sternberg, 2006). For example, glucocorticoids, androgens, progesterone, ACTH, and opioids have been shown to decrease immune responses, whereas growth hormone, prolactin, and insulin increase immune responses. Possible molecular mechanisms that could determine a molecule's effects on the immune system include intracellular signal transduction, selection of lymphocytes during ontogeny, contribution to immunospecificity, immune cell trafficking, recirculation and homing, control of gene expression, production of cytokines and their receptors, and immune cell interactions (Sternberg, 2006).

In the early 1990s, data indicated that neuropeptides released in the brain in response to immune challenges could trigger behavioral responses and possibly cause behavioral disorders (de Wied, Diamant, & Fodor, 1993). Conversely, cytokines and chemokines secreted by peripheral immune cells can have profound effects on behavior (Espinosa & Bermudez-Rattoni, 2001). Proinflammatory cytokines can activate the hypothalamus–pituitary–adrenal (HPA) axis and other CNS areas to induce "sickness behavior" during the acute phase response (Mulla & Buckingham, 1999). In addition, there is mounting evidence that the immune system plays a vital role in normal "homeostatic" brain processes, including sleep, metabolism, memory, and development. Thus, it appears that, in response to an immune challenge, both the immune and nervous systems coordinate to first activate and amplify local inflammatory responses to contain or eliminate invading pathogens, and subsequently to terminate inflammation and restore host homeostasis.

Deregulation of immune mediators in the brain, including, but not limited to, increased cytokine exposure during brain development, acts as a "vulnerability" factor for later brain pathology leading to behavior disorders. For example, there is strong evidence correlating early CNS infection with the development of schizophrenia (Suvisaari, Mautemps, Haukka, Hovi, & Lonnqvist, 2003). Infections early in life can alter cytokine expression and glial activation in response to a subsequent immune challenge in adulthood (Bland, Beckley, Watkins, Maier, & Bilbo, 2010), and a neonatal *Escherichia coli* infection has been shown to impair memory formation in adulthood in the presence of a subsequent immune challenge with LPS (Bilbo & Schwarz, 2009).

A number of studies have shown that injection of LPS leads to deregulation of cytokine and chemokine expression and secretion in the CNS, which is enhanced in the morphine tolerant state (Staikos, Malellari, & Chang, 2008) or in the persistent presence of human immunodeficiency virus-1 (HIV-1) viral proteins (Chen, Zhou, Beltran, Malellari, & Chang, 2005; Homji, Mao, Langsdorf, & Chang, 2012). Studies have shown that viral infections, including HIV infection, lead to CNS inflammation, which subsequently causes alterations in neurotransmitter-dependent pathways associated with compulsive behavior (Lindl, Marks, Kolson, & Jordan-Sciutto, 2010).

Since the advent of highly active antiretroviral therapy (HAART), the life expectancy of HIV-1-infected individuals has increased greatly. However, HIV-infected individuals continue to exhibit HIV-1-associated dysfunctions. A number of transgenic rodent models have been developed to study HIV-associated disorders. For example, a noninfectious HIV-1 transgenic (HIV-1Tg) mouse model was created in 1991 that expressed HIV-1 in the skin, tail, and muscle, and manifests renal disease, lymphoproliferation, thymic hypoplasia, dry, scaly hyperkeratotic skin, and early death (Dickie et al., 1991). Kim et al. (2003) produced a transgenic Tat line of mice that manifest neurological disorders similar to those seen in long-term HIV-1 patients, and Toggas et al. (1994) similarly created gp120 transgenic mice which exhibit neuropathological abnormalities.

The noninfectious HIV-1Tg rat was developed as a model for AIDS-related pathology and immune dysfunction by transferring a noninfectious HIV-1 *gag–pol* virus, with a deleted 3-kb SphI–MscI fragment containing the 3′-region of *gag* and the 5′-region of *pol*, into F344 rats. Because the HIV-1Tg rat is noninfectious, it is not suitable for studies on viral progression or replication, or for studying the impact of HAART on viral replication. However, it is a valuable model of studying NeuroAIDS as it mimics the disease state of patients on HAART. It is an ideal model for investigating the efficacy of therapeutic treatments which reduce neurological dysfunction in HIV-infected individuals in the post-HAART era. The use of this rodent model for investigating neurologically related issues has been well established (Kass & Chang, 2010; Moran, Aksenov, Booze, Webb, & Mactutus, 2012; Moran, Booze, Webb, & Mactutus, 2013; Royal, Wang, Jones, Tran, & Bryant, 2007; Royal et al., 2012). For example, Moran et al. (2012, 2013) used this model to test alterations in sensorimotor gating and behavior resulting from HIV-1 infection, including changes in dopamine (DA) function. Royal et al. (2007, 2012) used the HIV-1Tg rat to test the effects of vitamin A deficiency on HIV-1-associated neuroinflammation and mu opioid receptor (MOR) expression as well as peripheral and CNS immune responses in HIV infection. Also, Rao et al. (2011) found that HIV-1Tg rats exhibit neurological markers for neuroinflammation which are associated with cognitive impairment and identified neuroinflammation as a target for improving such impairments in HIV-positive populations (Rao et al., 2011).

The HIV-1Tg rat displays characteristics similar to human HIV-1 patients on HAART. Both HAART patients and the HIV-1Tg rat experience age-related immune response alterations (Reid et al., 2001), T-cell abnormalities (Reid et al., 2004), kidney failure (Ray et al., 2003), changes in behavior, and neuropathology (Reid et al., 2001). HIV-1Tg rats exhibit learning deficits even before they develop symptomatic signs of HIV infection (LaShomb, Vigorito, & Chang, 2008; Vigorito, LaShomb, & Chang, 2007), which is consistent with the cognitive impairment reported in HIV-1 patients receiving HAART and suggests that, despite the lack of viral replication, the persistent presence of HIV-1 viral proteins continues to affect brain and immune cells, causing neurological and immunological damage (Rao et al., 2011).

Using RNA deep sequencing, Li et al. (2013) sequenced 72 RNA samples from the prefrontal cortex (PFC), hippocampus (HIP), and striatum (STR) of HIV-1Tg and F344 control rats. Following deep-sequencing analysis of 50-bp paired-end reads of RNA-Seq, Bowtie/Tophat/Cufflinks suites were used to align these reads against the Rn4 rat reference genome and to quantify the relative abundance of each transcript. Statistical and bioinformatics analyses of each brain region between the two strains revealed that immune response-related pathways are altered in the HIV-1Tg rat, with brain region differences, suggesting that the persistent presence of HIV viral proteins causes inflammation in the brain of the HIV-1Tg rat (Li et al., 2013). In addition, analysis of serum cytokine levels revealed that, while LPS induces a significant increase in tumor necrosis factor-alpha (TNF-α) and interleukin-1 beta in both F344 and HIV-1Tg rats, the increase in these cytokines was significantly greater in the HIV-1Tg rats (Chang & Connaghan, 2012; Rao et al., 2011).

Expression of proinflammatory cytokines in the brain also leads to an increase in microglial cells in the HIV-1Tg rats (S. L. Chang and K. Hauser, unpublished data). Chang and Connaghan (2012) proposed the possibility that a positive feedback interaction exists between opioid receptor-dependent pathways and HIV progression and that this interaction is, at least partly, moderated by HIV-induced neuroinflammation (Chang & Connaghan, 2012).

Receptors for cytokines, hormones, neurotransmitters, and neuropeptides are found in both immune and brain cells and include opionergic, dopaminergic, cholinergic, and cannabinoid receptors. Those that mediate the actions of addictive substances will be discussed later in this chapter.

2. MEDICAL USES AND ADVERSE EFFECTS OF ADDICTIVE SUBSTANCES

While the detrimental effects of addictive substances, such as morphine, alcohol, cocaine, methamphetamine (METH), marijuana (MJ), and nicotine, are well known (National Institute on Drug Abuse, 2007), these substances have either been used for therapeutic purposes or have the potential for medicinal use because they also possess beneficial effects. All of these substances exert their biological or physiological effects by acting on one of the endogenous neurotransmitter-dependent pathways, including the opioid receptor-, DA receptor-, DA transporter-, acetylcholine receptor-, and cannabis (CB)-dependent pathways (Filip et al., 2006; Hyman, Malenka, & Nestler, 2006; Landa, Sulcova, & Slais, 2006).

Opioids, such as morphine, have been used historically in the medical field to treat acute and chronic pain because of their analgesic effects. For example, dentists often prescribe an opioid, such as hydrocodone or codeine, to treat the acute pain that patients experience after having a tooth pulled (Deyo et al., 2011). Long-term opioid therapy is used for more persistent pain, such as chronic back pain (Deyo et al., 2011), neuropathic pain (Largent-Milnes, Guo, Wang, Burns, & Vanderah, 2008), and pain associated with HIV-1 infection (Krashin, Merrill, & Trescot, 2012; Silverberg et al., 2012). While opioids are the most effective painkillers, they also have risks and side effects. Interactions between opioids and the endocrine system cause reduced secretion of sex hormones, thereby deregulating the menstrual cycle in women and decreasing the sex drive in men (Katz, 2005). Opioids also cause histamine release, which results in itching, and decreased white blood cell counts, affecting immune functions (Feldman, Meyer, & Quenzer, 1997). Most opioid-associated deaths result from respiratory depression caused by activation of opioid receptors in the brain stem (Trigo, Martin-Garcia, Berrendero, Robledo, & Maldonado, 2010). In animal models, chronic exposure to morphine causes desensitization of the HPA axis (Chang, Moldow, House, & Zadina, 1996; Chang, Patel, Romero, Thompson, & Zadina, 1996; House et al., 2001) and inhibits the release of the final product of the HPA axis, glucocorticoid. Inhibition of this endogenous anti-inflammatory molecule in individuals addicted to opioids could be one of the factors responsible for the high susceptibility of these individuals to opportunistic infections (Ocasio, Jiang, House, & Chang, 2004).

Alcohol is another substance that has been used to mitigate pain and induce sedation. While fermentation of alcohol dates back to 10,000 BC (Granet, 1957; Hanson, 1995; Patrick, 1952), alcohol has only recently caught the interest of researchers for its possible health benefits. Light-to-moderate alcohol consumption has been reported to reduce the risk of vascular issues (O'Keefe, Bybee, & Lavie, 2007), ischemic stroke (Schwarz & Wetterling, 2002), congestive heart failure (Djousse & Gaziano, 2008), peripheral arterial disease (Athyros et al., 2007), blood flow restriction (Di Minno et al., 2012), and nonalcoholic fatty liver disease (Dunn et al., 2012; Moriya et al., 2011). Blomster et al. (2014) reported that moderate consumption of alcoholic beverages, particularly wine, is correlated with decreased cardiovascular risk and "all-cause mortality"; however, there is a dose-dependent increase in cardiovascular risk and all-cause mortality in individuals who reported heavy alcohol consumption, despite the type of alcohol consumed. In another study, Jones et al. (2013) investigated the relationship between alcohol consumption and endocrine and cognitive responses and response to stress (Jones et al., 2013). They found that whereas alcohol intake, in general, has beneficial effects on heart rate, cardiac output, vascular resistance, and average blood pressure, in the context of mental stress, heavy consumption of alcohol causes a greater increase in the production of the stress hormone, cortisol. Thus, the beneficial effects of low-to-moderate alcohol consumption may be negated by the neurotoxic effects of increased levels of cortisol induced by heavy drinking. Further, excessive consumption of alcohol can result in reversible drug-induced dementia, also known as a "black-out" (Perry et al., 2006). Long-term abuse of alcohol results in liver damage, inflammation of the pancreas and stomach, and neurodegenerative disorders such as Korsakoff's syndrome. Research also suggests that alcohol consumption may increase the risk for various forms of cancer, such as breast and esophageal cancers (Seitz & Becker, 2007).

Nicotine, alcohol, and caffeine are the three most used psychoactive substances in the world (Crocq, 2003). Typically, nicotine is only available in the form of tobacco, which also contains many types of other chemical componunds. Most diseases commonly associated with nicotine are a result of the tar found in tobacco. Smoking tobacco increases carbon monoxide and decreases the flow of oxygen to the heart and lungs. Smoking increases the risk for arterial thrombosis and atherosclerosis in the heart, abdomen, and neck (Howard et al., 1998), thus, increasing the possibility of stroke, pulmonary disease, and emphysema (Bhat et al., 2008). In recent years, tobacco use has become more and more stigmatized due, at least in part, to the implementation of federal regulations that make it illegal to smoke tobacco in public places, such as restaurants.

Biomedical research has uncovered some potential neuroprotective properties of nicotine. Numerous epidemiological studies indicate that individuals with mental illness have a substantially higher incidence of cigarette smoking than the general population (Smith, Mazure, & McKee, 2014). Based on the hypothesis that many mental illnesses are a result of neurodegeneration or deregulation, it has been suggested that cigarette smoking may be a form of self-medication (Leonard, Mexal, & Freedman, 2007). Delineating the potential medical uses of nicotine, and making the substance available in a purer and less dangerous form, may reduce the rate of cigarette smoking among those with mental illness by providing alternative methods by which they can attain the benefits of nicotine.

Nicotine is a natural alkaloid substance present in cigarette smoke and is an agonist for the nicotinic acetylcholine receptors (nAChRs). Nicotine has been shown to have positive effects on learning and memory through activation of nAChRs (Bartus, 2000; Picciotto & Zoli, 2002; Riljak & Langmeier, 2005). Individuals with neurocognitive disorders, such as AD, have decreased neuronal nAChRs, which may be one of the causes of the cognitive deficits and neurodegeneration seen in AD (Marutle, Warpman, Bogdanovic, Lannfelt, & Nordberg, 1999). When patients with AD are treated with nicotine, functional nAChRs are significantly increased, and learning and memory function, as well as overall cognitive ability, improves (Picciotto, Caldarone, King, & Zachariou, 2000; Teaktong, Graham, Johnson, Court, & Perry, 2004). Several studies have reported that nicotine also has a neuroprotective effect by guarding neurons from cytotoxic glutamate and amyloid-β (Aβ)-mediated cell death via nAChRs (Kihara et al., 1997; Marin, Maus, Desagher, Glowinski, & Premont, 1994; Owman, Fuxe, Janson, & Kahrstrom, 1989). In addition, Li, Kane, Matta, Blaner, and Sharp (2000) have shown that, in the rat hippocampus and brainstem, nicotine significantly enhances the secretion of transthyretin, a substance important in clearing Aβ monomers and reducing Aβ deposits. In another study, Gutala, Wang, Hwang, Haq, and Li (2006) showed that chronic nicotine treatment increases the expression of the amyloid precursor protein and its homologue, amyloid precursor-like protein 2, in several rat brain regions as well as in human SH-SY5Y neuroblastoma cells.

Cocaine has been used for psychoactive and medicinal purposes for a number of years. Sigmund Freud promoted the use of cocaine for the treatment of depression (Calatayud & Gonzalez, 2003). Subsequently, cocaine has been used for its local anesthetic properties in otolaryngeal, nasal, dental (Calatayud & Gonzalez, 2003), and corneal surgery (Goerig, Bacon, & van Zundert, 2012). While cocaine proved to be an effective anesthetic, its high risk for abuse initiated the creation of more nonaddictive local anesthetics, such as procaine (novocaine) and lidocaine. Cocaine use is associated with ischemic or hemorrhagic strokes, seizures, pulmonary edema and hemorrhage, renal failure, cardiac complications such as myocardial infarction, myocardosis, and ruptured aorta (Phillips et al., 2009), and cortical microischemia in the brain (Ren et al., 2012).

METH is rarely, if ever, used in its pure form for any potential medicinal benefits. Instead, a variety of slightly altered chemical alternatives have been used to treat neurologically based disorders, such as attention-deficit hyperactivity disorder (ADHD) and narcolepsy (Berman, Kuczenski, McCracken, & London, 2009). In rare cases, an amphetamine-based drug may be prescribed as a short-term aid for weight loss (Cummings, Parham, Strain, & American Dietetic Association, 2002). Like cocaine, the use of METH alternatives carries a high risk of abuse and dependence. As a result of its neurotoxicity, METH may also induce persistent mood disorders, such as depression (Sekine et al., 2006). In addition, METH damages nigrostriatal dopaminergic terminals and induces hyperthermia, which exacerbates METH neurotoxicity (Friend & Keefe, 2013; Kikuchi-Utsumi et al., 2013).

MJ, also known as CB, is the most frequently used illicit substance in the United States (National Institute on Drug Abuse, 2014). Currently, MJ is classified as a Schedule I substance by the federal government, which means that the substance has a high abuse liability and no established medical use. While the use, sale, transportation, cultivation, and possession of CB is punishable at the federal level, various states have made progress in the legalization of the substance for medical use. Washington and Colorado were the first states in the continental United States to legalize MJ for recreational use by individuals who meet state regulatory requirements (e.g., $\geq$21 years old in Washington). Cannabinoids, such as tetrahydrocannabinol (THC) and cannabidiol, are two of the active ingredients in CB.

Cannabinoids act on the endocannabinoid (EC) system to produce analgesic, neuroprotective, appetite stimulatory, and anti-inflammatory effects (Aggarwal et al., 2009; Robson, 2014). Dronabinol, a synthetic cannabinoid (SC), is also used for the treatment of nausea associated with chemotherapy (Hernandez, Sheyner, Stover, & Stewart, 2013). Due to these medicinally relevant effects, CB has been used successfully to treat symptoms such as HIV/AIDS-associated anorexia (Lutge, Gray, & Siegfried, 2013) and chronic pain associated with neuropathy, fibromyalgia, and rheumatoid arthritis (Lynch & Campbell, 2011; Martin-Sanchez, Furukawa, Taylor, & Martin, 2009). In addition, cannabinoids have shown promising potential in the treatment of multiple sclerosis, spinal cord injuries, Tourette's syndrome, epilepsy, and glaucoma (Ben Amar, 2006; Borgelt, Franson, Nussbaum, & Wang, 2013; Clark, Capuzzi, & Fick, 2011; Lynch & Campbell, 2011). A majority of the risks associated with the use of MJ, such as memory and attention impairment, are short term and reversible. Equivocal research has been presented regarding the role that MJ plays in the development of schizophrenia (D'Souza, Sewell, & Ranganathan, 2009; Fernandez-Espejo, Viveros, Nunez, Ellenbroek, & Rodriguez de Fonseca, 2009).

In summary, while addictive substances can be used to alleviate various ailments and modulate particular neurological systems when used at therapeutic doses, the use and abuse of these substances comes with a substantial list of adverse side effects. While some of these adverse effects are more uncomfortable than they are dangerous, neuroinflammation and neuronal damage, particularly those associated with systemic infections such as HIV, may have a more profound impact on daily living due to impairment in cognitive functioning and deregulation of neurotransmitter-dependent pathways, leading to neuropathology.

3. HIV-1 INFECTION AND HIV-ASSOCIATED NEUROPATHOLOGY

HIV-1 infects cells of the immune system, such as macrophages and T-lymphocytes (Cicala et al., 2002; Clapham & McKnight, 2001; Jiang & Aiken, 2006; Torensma & Figdor, 2002; Trkola, 2004; Verani, Gras, & Pancino, 2005), and causes apoptosis of $CD4^+$ T cells, monocytes, and macrophages (Lindl et al., 2010).

The CNS is highly vulnerable to HIV-1 infection, particularly the STR, PFC, and HIP (Aksenov et al., 2001; Maragos et al., 2002), although the mechanism by which the virus infiltrates the brain remains to be determined (Gonzalez-Scarano & Martin-Garcia, 2005; Gorry et al., 2003). Infection with HIV-1 results in neurodegeneration and neuronal apoptosis, even though the virus typically does not infect neurons (Gonzalez-Scarano & Martin-Garcia, 2005; Kaul, Zheng, Okamoto, Gendelman, & Lipton, 2005; McArthur et al., 2003).

Both viral- and cell-encoded factors secreted by infected immune cells contribute to the development of HIV-associated neuropathologies and subsequent cognitive deficits (Genis et al., 1992; Giulian, Vaca, & Noonan, 1990; Lindl et al., 2010; Pulliam, Herndier, Tang, & McGrath, 1991). The neuropathology caused by HIV-1 infection includes progressive cognitive decline, motor dysfunction, striatal pathology, and gliosis (Khurdayan et al., 2004; Yoshioka & Itoyama, 1997).

HIV-associated neurocognitive disorders (HANDs) are divided into three subgroups: asymptomatic neurocognitive impairment (ANI), minor neurocognitive disorder (MND), and HIV-associated dementia (HAD). ANI individuals do not develop clinical symptoms that affect their everyday lives, but they perform poorly on neuropsychological tests; MND individuals exhibit poor performance on tests similar to the ANI subgroup, but they also exhibit clinical symptoms that affect their daily activities. HAD individuals demonstrate even worse cognitive performance than those with ANI and MND, and also have significant problems functioning at home and at work (Antinori, Arendt, et al., 2007; McArthur, Steiner, Sacktor, & Nath, 2010).

Viral proteins, such as gp120 and Tat, are encoded in the HIV-1 genome. These proteins can interact with macrophages and microglia, leading to the expression and release of cell-encoded factors, such as neurotoxins, cytokines, and chemokines (Gendelman, Genis, Jett, Zhai, & Nottet, 1994; Lindl et al., 2010; Rafalowska, 1998; Yoshioka & Itoyama, 1997). Some neurotrophic factors that are modulated by macrophage-secreted cytokines (Chao, Hu, Sheng, & Peterson, 1995; Sevigny et al., 2004) may be key factors mediating the development of HAND (Sevigny et al., 2004, 2007; Wesselingh et al., 1997). Chemokine receptors, such as CXCR4 (Vergote et al., 2006) and CCR5 (Alkhatib et al., 1996; Bleul et al., 1996; Locati & Murphy, 1999), play a key role in mediating HIV-1 infection. CXCR4 appears to promote HAND (Vergote et al., 2006), whereas the CCR5 receptor appears to protect neurons (Hesselgesser et al., 1998; Kaul & Lipton, 1999; Meucci et al., 1998).

Treatment with antiretroviral drugs can, at least temporarily, reverse HIV dementia (Gendelman et al., 1998; Gisslen, Svennerholm, Fuchs, & Hagberg, 1998; Portegies, 1995), suggesting that there is a therapeutic window in which HIV encephalopathy and functional impairment can be reversed. Although the development of HAART has slowed and, in some cases, halted viral replication, as many as 40% of individuals infected with HIV-1 still experience HAND (Antinori, Trotta, et al., 2007; McArthur, 2004; Sacktor, 2002; Sacktor et al., 2001), most likely because of the continued presence of viral proteins and the related inflammation in the CNS.

4. HIV-1 INFECTION, NeuroHIV, AND THE USE OF ADDICTIVE SUBSTANCES

4.1. Addiction and neuroHIV

Addiction is a complex and dynamic disorder with neurological, psychological, and social consequences and implications. It is widely accepted that addiction is a disease of the brain involving short-term and persistent changes in the brain's reward circuitry following chronic or repeated use of substances which activate the reward circuitry (Koob, 2013; Koob & Le Moal, 2005, 2008). Research investigating the neurological mechanisms underlying substance abuse and addiction has strongly supported the belief that chronic use of addictive substances is linked to plasticity or abnormalities in neurotransmitter-dependent pathways, such as the mesolimbic dopaminergic reward pathway and the endogenous opioid system (EOS) (Herz, 1998). Recently, the EOS was shown to interact with another reward system, the endogenous cannabinoid system (ECS) (Fattore et al., 2004).

Neuroinflammation also leads to alterations in neurotransmitter-dependent pathways associated with addictive behavior (Chang & Connaghan, 2012; Homji et al., 2012; Li et al., 2013; Trigo et al., 2010). Neuroinflammation can cause neuronal damage leading to deregulation of the dopaminergic system. Inflammatory cytokines have been shown to alter the expression of the MOR, an important component of the EOS.

Many individuals contract the HIV virus via intravenous drug use or substance-related risky sexual behavior. This is one possible explanation for the increased incidence of substance use and abuse in the HIV-positive population. However, it is also possible that HIV-induced changes in brain structure and function increase the liability of drugs of abuse. Even though HAART successfully reduces the plasma viral load, it is not efficient in attenuating HIV-induced neuroinflammation (Anthony & Bell, 2008). Patients on HAART continue to experience neurological impairment, especially in the HIP and temporal cortex, regions associated with memory, spatial recognition, motivation, and movement (Brew, Crowe, Landay, Cysique, & Guillemin, 2009). Thus, HIV-induced inflammation in the CNS may be the key mechanism underlying the use of addictive substances.

4.2. Endogenous opioid system

In medicine, the EOS is primarily targeted for its analgesic effects; however, the EOS has also been linked to addiction (Trigo et al., 2010). The EOS is comprised of three naturally occurring opioid peptides—beta endorphins, dynorphins, and enkephalins—and three classes of opioid receptors—mu (MOR), delta (DOR), and kappa (KOR) (Holden, Jeong, & Forrest, 2005). The MOR and KOR have been shown to play important roles in substance abuse and the development of addiction (Wang, Sun, Tao, Chi, & Liu, 2010). Systemic infections, including HIV-1 infection, deregulate the EOS, primarily by interaction with the MOR (Hauser et al., 2005). Naltrexone, an opioid antagonist with a high affinity for the MOR, has been shown to reduce the risk of relapse in alcoholics (Pettinati et al., 2006) and to promote abstinence in individuals with a history of amphetamine abuse (Jayaram-Lindstrom, Hammarberg, Beck, and Franck, 2008; Jayaram-Lindstrom, Konstenius, et al., 2008). Evidence also suggests that activation of the KOR by substances of abuse may be linked to compulsive drug use (Wee & Koob, 2010).

4.3. HIV-1 and the dopaminergic system

DA and DA receptors in the mesolimbic DA pathway, another neuronal reward circuit, have been associated with substance abuse and HIV-1 infection. The HIV-1 proteins, Tat and gp120, are toxic to DA neurons (Gurwell et al., 2001; Nath et al., 2000), and HIV-1 impacts DA neurons in subcortical structures, particularly those in the basal ganglia (Urbina & Jones, 2004). Both HIV-1 patients (Wang et al., 2004) and HIV-1Tg rats (Liu et al., 2009) have been reported to have fewer DA receptors. Volkow et al. (2001) found that individuals with fewer DA receptors are at a higher risk for substance abuse due to their increased sensitivity to the pleasurable effects of those substances. This may be due to the fact that individuals with lower DA availability experience less pleasure from typically normal activities, such as eating and sex.

While the inverse relationship between the use of addictive substances and the reduction in DA density has not been established, individuals addicted to various DA-associated substances tend to have less DA receptors

(Volkow, Fowler, Wang, Swanson, & Telang, 2007; Volkow et al., 1990, 1996; Wang et al., 1997). The majority of substances which produce dependence increase the bioavailability of DA in the mesolimbic DA pathway. When substances which increase DA levels are introduced, the experience of pleasure is far more substantial, perhaps as a result of a rebound action.

In summary, the presence of HIV-1 viral proteins leads to a decrease in DA receptors (Liu et al., 2009) and may explain the behavioral evidence that indicates that HIV-1Tg rats are more prone to addiction (Chang & Connaghan, 2012; Chang & Vigorito, 2006).

4.4. Morphine and neuroHIV

Various studies have reported that the progression of HIV-1 infection is hastened in individuals receiving morphine (El-Hage, Podhaizer, Sturgill, & Hauser, 2011; Hahn et al., 2010; McCarthy, Wetzel, Sliker, Eisenstein, & Rogers, 2001). Morphine increases the expression of chemokine receptors (Mahajan, Schwartz, Shanahan, Chawda, & Nair, 2002; Rogers & Peterson, 2003), which are major coreceptors for the HIV-1 virus (Horuk et al., 1997), thereby increasing the susceptibility of HIV-infected individuals to opportunistic infections. Chang and Connaghan (2012) proposed the possibility that a positive feedback interaction exists between the MOR and HIV progression, and that this interaction is, at least partly, moderated by HIV-induced neuroinflammation (Chang & Connaghan, 2012). Using tail-flick latency as a measure of the antinociceptive effects of morphine, Vigorito and Chang (2006) demonstrated that HIV-1Tg rats exhibit an increased sensitivity to morphine's antinociceptive properties. Tail-flick latencies are significantly longer in the HIV-1Tg rat following treatment with morphine, and the ED50 of morphine is lower in HIV-1Tg rats than in control rats (Vigorito & Chang, 2006), suggesting that the HIV-1Tg rat is more prone to the antinociceptive and rewarding properties of opioids. One of the possible molecular mechanisms underlying the increase in tail-flick latency is the increased expression of the MOR in the HIV-1Tg rat (Chang, Beltran, & Swarup, 2007; Vigorito & Chang, 2006).

Chang and Connaghan (2012) provided further behavioral evidence of a link between morphine and neuroHIV. In this study, conditioned place preference (CPP) was used to determine the rewarding effects of stimuli based on operant conditioning. The preferred chamber was first established for drug-naïve HIV-1Tg and F344 control rats. The rats were then administered either morphine or saline (control) and placed into their nonpreferred chambers. If the rats find the drug rewarding, they will spend time in their originally nonpreferred chamber after it has been paired with the drug. All the rats treated with morphine established a preference for their originally nonpreferred chamber, indicating that morphine is rewarding for both HIV-1Tg and control rats. Over a 7-day period, the control rats showed a decreased preference for the drug-paired chamber and eventually came to prefer their originally preferred chamber. The HIV-1Tg rats, however, continued to show a preference for the morphine-paired compartment for up to 14 days, suggesting that they are more likely to partake in drug-seeking behavior despite exposure to unreinforced drug cues (Chang & Connaghan, 2012).

These findings are of great importance in terms of the abuse liability of opioid drugs and other substances which activate the EOS in HIV-infected populations, and support the hypothesis that HIV-1-induced alterations in the structure and function of CNS reward pathways may increase the likelihood that HIV-1-positive individuals will become addicted to opioid drugs. These results are also supported by clinical evidence which shows that HIV-1-infected patients are more likely to abuse substances than the general population (Avdoshina, Biggio, Palchik, Campbell, & Mocchetti, 2010; Molitor, Truax, Ruiz, & Sun, 1998; Nath et al., 2002).

4.5. Alcohol and HIV-1

Alcohol abuse is a common problem in HIV-infected patients and contributes to their noncompliance with antiretroviral medication. As many as 50% of HIV-infected patients report a history of alcohol abuse (Samet, Phillips, Horton, Traphagen, & Freedberg, 2004). This abuse may be due to changes in neurological and metabolic functioning caused by HIV infection. Blednov et al. (2011) investigated the possibility that immune signaling plays a role in alcohol consumption by injecting mice with either LPS or control vehicle and then giving them access to alcohol in a continuous two-bottle choice test, 1 month following the LPS injection. The mice injected with LPS consumed more alcohol at 1 week and 1 month post-LPS than did the control animals (Blednov et al., 2011). Further, the LPS decreased DA neuron activation in the ventral tegmental area and decreased the aversive effects of alcohol as measured by the conditioned place aversion test (Blednov et al., 2011).

In another study, using Long Evans rats as the study subjects, Vigorito et al. (2011) reported that a "liking" for ethanol (EtOH) was induced in rats with daily home cage exposure to EtOH. Treatment with LPS further increased the animal's preference for EtOH, but did not reliably affect the drinking of EtOH during the sign-tracking procedure. These results suggest that systemic activation of the immune system by a bacterial endotoxin can influence subsequent EtOH intake (Blednov et al., 2011) by enhancing the hedonic impact (liking) of EtOH, but not by affecting compulsive responding (wanting) in the presence of EtOH.

Data comparing the effects of EtOH on HIV-1Tg and F344 control rats suggest that HIV-1 viral proteins increase the sensitivity to the effects of EtOH by suppressing the production of EtOH-metabolizing enzymes. EtOH-metabolizing enzymes are activated following an initial binge EtOH exposure in healthy control rats, but not in HIV-1Tg rats. Blood ethanol concentration is deregulated and cytokine/chemokine gene expression is altered following a 3-day binge treatment with EtOH only in the HIV-1Tg rats. HIV-1Tg rats also show a decrease in locomotor activity at 2 and 24 h following the last binge treatment, possibly due to significant EtOH-induced alterations in the PFC, an area of the brain responsible for locomotor activity.

4.6. Nicotine and HIV-1

Epidemiologic studies show that smokers are at a lower risk of developing AD and other neurodegenerative disorders than are nonsmokers (Mihailescu & Drucker-Colin, 2000a, 2000b). One hypothesis is that nAChR subtypes are involved in nicotine's protective actions against neuronal damage produced by various stimuli. In AD patients, there is a decrease in neuronal nAChRs, which may be the cause of the underlying cognitive deficits and neurodegeneration associated with AD (Marutle et al., 1999). Administration of nicotine to AD patients causes a significant increase in nAChRs, with improvement in learning and memory function and in cognitive abilities (Picciotto et al., 2000; Teaktong et al., 2004). By interacting with nAChRs, nicotine can protect neurons from cytotoxic glutamate and Aβ-mediated cell death (Kihara et al., 1997; Marin et al., 1994; Owman et al., 1989).

In rats, HIV-1 gp120 binds to nAChRs in muscle cells and neurons (Bracci, Lozzi, Rustici, & Neri, 1992) and prevents HIV-1 gp120-induced alterations in cholinergic neurotransmission. Although the inhibitory mechanism is not clear, HIV-1 gp120 and nicotine may compete for nAChRs and, thus, act as antagonists (Ballester et al., 2012).

Nicotine is an important modulator of cellular interactions and immune functions (Sopori, 2002). Both nicotine and acetylcholine inhibit pro-inflammatory cytokine release, expression of cyclooxygenase isoform 2, and prostaglandin E2 expression (De Simone, Ajmone-Cat, Carnevale, &

Minghetti, 2005; Sharifzadeh et al., 2005). HIV-1 gp120 and nicotine share several signaling pathways, including those mediated by MAPKs and NF-κB, which are responsible for expression of proinflammatory cytokines. Nicotine can suppress gp120-induced TNF-α in murine microglial cells by inhibiting the activation of p44/p42 MAPKs, which can be suppressed by α-bungarotoxin, a selective α7 nAChR antagonist (Giunta et al., 2004). Very recently, a Toll-like receptor pathway that plays a key role in innate immunity and in the production of proinflammatory cytokines was shown to be modulated by nicotine through α7 nAChRs (Cui et al., 2012, 2013).

Nicotine produces unique effects in the HIV-infected population. Li et al. (2013) compared cortical, hippocampal, and striatal gene expression in HIV-1Tg rats using deep-sequencing analysis of RNA transcripts in brain regions related to learning and memory. They found that there are differences in neural pathways related to immune responses, neuronal health, and neurotransmission in HIV-1Tg rats (Li et al., 2013). In a subsequent study, RNA deep-sequencing analysis was also used to determine whether the altered gene expression observed in the HIV-1Tg rats could be corrected by nicotine. Cao et al. (2013) found that nicotine restores expression of about 20% of the altered genes in each brain region and modulates distinct pathways in different brain regions. The most significantly restored pathways are the Wnt/b catenin signaling and ephrin B signaling pathways in the PFC, cAMP-responsive element-binding protein signaling and glutathione metabolism pathways in the HIP, and the tricarboxylic acid cycle and calcium signaling pathway in the STR. These findings indicate that cholinergic modulators, such as nicotine, can have beneficial effects on HIV-1-induced neurologic deficits.

4.7. Interactive effects of alcohol and nicotine in neuroHIV

Alcohol consumption is often paired with cigarette smoking. However, the findings from studies on the combined effects of nicotine and alcohol are equivocal. Abreu-Villaca et al. (2007) found that administration of both nicotine and EtOH to C57BL/6 mice for two consecutive days produces additive memory and learning deficits compared with the effects of nicotine or EtOH alone (Abreu-Villaca et al., 2007), whereas Gould and Lommock (2003) found that administration of EtOH alone causes impairment in contextual fear conditioning and learning deficiencies that are ameliorated by nicotine (Gould & Lommock, 2003). More recently, Gulick and Gould (2011) found that mice given EtOH show a deficit in learning that is reversed by infusion of nicotine into the anterior cingulate cortex and the dorsal and ventral

HIP (Gulick & Gould, 2011). Rezvani and Levin (2002) reported that, in rats, neither EtOH nor nicotine alone affects memory, whereas simultaneous administration of nicotine and EtOH causes memory impairment in a dose-dependent manner (Rezvani & Levin, 2002). These same researchers also found that EtOH blocks nicotine's attention-enhancing effects in female Sprague–Dawley rats (Rezvani & Levin, 2003).

Despite the advent of HAART, HIV patients still develop cognitive deficits. Considering that a significant number of HIV-infected individuals use both tobacco and other addictive substances, including alcohol, delineating the interactions between alcohol and nicotine at the molecular level would help in the development of new strategies to treat HIV-1 patients.

4.8. Cocaine and HIV-1

Cocaine can interact with the EOS through the KOR (Shippenberg, Zapata, & Chefer, 2007), the MOR (Azaryan, Clock, & Cox, 1996; Unterwald, 2001), and the HPA axis (Zhou et al., 2002). In addition, the EOS has been implicated in drug-seeking behavior (Leri, Zhou, Goddard, Cummins, & Kreek, 2006; Leri et al., 2009), craving, and relapse (Ghitza et al., 2010; Gorelick et al., 2005) in cocaine addiction. Thus, alterations in the EOS, especially the expression of the KOR and MOR, can potentially exacerbate the risk of cocaine addiction.

The chemokine, CCR5, is elevated in HIV-1-infected patients (Giovannetti et al., 1999) and has been identified as a target for gene therapies aimed at reducing the deleterious effects of the HIV-1 virus (Nazari & Joshi, 2008; Nazari, Ma, & Joshi, 2008). However, CCR5 deficiency has also been linked to modifications in the dopaminergic reward pathway (Choi, Lee, & Hong, 2012).

Glutamate and glutamatergic receptors, such as the *N*-methyl-D-aspartate receptor (NMDAR), have also been implicated in the relationship between cocaine and HIV-1 (Buch et al., 2011). More specifically, intracerebroventricular injection of the HIV-1 protein, Tat, has been shown to increase extrasynaptic NMDAR (Samet et al., 2004), and NMDARs are influential in neuronal plasticity, which is paramount to learning and memory (Li & Tsien, 2009; Samet et al., 2004). Further, Agatsuma, Dang, Li, and Hiroi (2010) found that NMDARs are necessary in order to produce cocaine-induced CPP in NMDAR knockout mice.

Past research suggests that HIV infection, while potentially dangerous to the whole CNS, is particularly detrimental to "DA-rich" brain regions, which are also implicated in substance abuse and addiction (Purohit, Rapaka, & Shurtleff, 2011), despite treatment with HAART (Kumar et al., 2009). Dopamine transporter (DAT) functions can be impaired by the HIV-1 viral proteins, Tat and gp120, leading to elevated DA levels in dopaminergic synapses in the early asymptomatic stage of HIV infection (Purohit et al., 2011). Cocaine strongly binds to the DAT with submicromolar affinity (Amara & Kuhar, 1993; Han & Gu, 2006; Ritz, Lamb, Goldberg, & Kuhar, 1987) and impairs DA reuptake, resulting in overflow of extracellular levels of DA (Reith, Li, & Yan, 1997). Recently, Ferris, Frederick-Duus, Fadel, Mactutus, and Booze (2009) reported that Tat significantly reduces local DAT functions within 5 h of administration, whereas cocaine significantly elevates DA levels in both Tat- and Veh-treated animals with no noticeable difference in DA overflow (Ferris et al., 2009). After 48 h of Tat administration, cocaine facilitates degeneration of nerve terminals and reduces the update and release of DA (Ferris et al., 2009). This is particularly important in terms of cocaine-addicted HIV-1 patients as well as for individuals with HAND. DA agonists have been proposed as potential treatment options for HIV-1 patients who use cocaine since some HAND symptoms are similar to those seen in Parkinson's disease; however, therapy with these kinds of agents may exacerbate neurodegeneration.

4.9. Methamphetamine and HIV-1

METH is an addictive psychostimulant that can cause dopaminergic neurotoxicity, stereotypic behavior, and hyperthermia. In the past two decades, studies have shown that there is a high risk of transmission of HIV associated with METH use. Although METH is considered more potent than cocaine and other drugs of abuse in increasing synaptic levels of DA (Di Chiara & Imperato, 1988), only a few studies have investigated how HIV infection affects the use of METH. Liu et al. (2009) compared the physiological and behavioral parameters as well as DA receptor and DAT expression in the STR and PFC of young adult male HIV-1Tg rats with those of F344 control animals following daily injections of METH for 6 days. They found that METH induced an increase in body temperature, rearing, and stereotypical head movement over days (behavioral sensitization) in both the HIV-1Tg and F344 rats, but with greater changes occurring in HIV-1Tg rats. METH attenuated the expression of the DA receptor, D1R, in the PFC of the HIV-1Tg rats, but not in the F344 rats (Liu et al., 2009). In another study, HIV-1Tg rats were shown to be more prone to both the behavioral sensitization and acute physiological effects (hyperthermia) of METH than the control rats (Kass, Liu, Vigorito, Chang, & Chang, 2010). Like cocaine, there is an additive effect in neurodegeneration when METH is used by HIV-positive individuals. Neuronal damage caused by the combination of METH and Tat is, at least partly, moderated by disruption of D1 and NMSD receptor signaling (Aksenov, Aksenova, Mactutus, & Booze, 2012).

Kesby, Hubbard, Markou, and Semenova (2012) found that a gp120 transgenic mouse model exhibits increased sensitivity to the rewarding effects of METH, as evidenced by the development of CPP to lower doses of METH than control mice (Kesby et al., 2012). The fact that the gp120 transgenic mice, which express only one of the nine HIV-1 viral proteins, exhibit similar neurological and behavioral effects as HIV-1Tg rats, which express seven of the nine HIV-1 viral proteins, indicates that there is a common factor, possibly neuroinflammation, that links the presence of HIV-1 viral proteins to the increased use of METH.

4.10. Marijuana and HIV-1

Just as opioids act on the EOS, cannabinoids, such as MJ, act on the ECS, which is comprised of two main receptors, CB_1 and CB_2. CB_1 receptors are located in most neuronal cells, whereas CB_2 receptors are predominantly located in immune cells and a limited number of neuronal cells (Van Sickle et al., 2005), such as astrocytes, where they play a role in neuroprotection and neuroinflammation (Molina-Holgado, Molina-Holgado, Guaza, & Rothwell, 2002; Sheng et al., 2005). CB_2 receptors modulate immune responses by altering the production of cytokines and normalizing the migration of immune cells (Miller & Stella, 2008). More specifically, CB_2 receptor ligands reduce the production of cytokines in microglia (Ehrhart et al., 2005).

CB use is greater in individuals with HIV-1 infection than in the general public (Dansak, 1997); however, the data regarding the effects of cannabinoids on HIV-1-infected individuals are controversial. One study found that chronic MJ use in the presence of HIV-1 infection increases thalamic creatine and decreases glutamate in the basal ganglia (Chang, Cloak, Yakupov, & Ernst, 2006), but there are no effects seen, in terms of cognitive functioning, in HIV-positive chronic MJ smokers. Cristiani, Pukay-Martin, and Bornstein (2004) reported that, in asymptomatic HIV-positive participants, MJ did not significantly affect cognitive performance, but that, in symptomatic HIV-positive individuals who used MJ chronically, there was greater cognitive impairment than in nonusers (Cristiani et al., 2004).

The addictive nature of CB has been attributed to both the interaction of the drug with the ECS, particularly the CB_1 receptor, and the interplay between the ECS and the EOS. Various research studies have demonstrated a strong connection between these two reward-related systems. For example, CB_1 knockout mice do not self-administer morphine, suggesting that CB_1 receptors are integral in the rewarding effects of opioids (Ledent et al., 1999). This is further supported by Martin, Ledent, Parmentier, Maldonado, and Valverde (2000) who found that CB_1 knockout mice do not exhibit morphine-induced CPP (Martin et al., 2000).

Ghozland et al. (2002) reported an absence of THC-induced reward in MOR knockout mice, and Justinova, Tanda, Munzar, and Goldberg (2004) reported that naltrexone, an opioid antagonist, similarly attenuates the THC-induced experience of reward. Activation of the KOR was found to be, at least partially, responsible for the aversive effects of THC (Ghozland et al., 2002).

As a result of the interaction between cannabinoids and the EOS, cannabinoid antagonists have been suggested as possible treatments for opioid withdrawal syndrome (Fattore et al., 2005).

4.11. Summary

Studies of each of these addictive substances support the hypothesis that HIV-1-induced alterations in the structure and function of the CNS reward pathways by HIV-induced neuroinflammation increase the likelihood that HIV-1-positive individuals will become addicted to one or more of these substances. This hypothesis is supported by clinical evidence which suggests that HIV-1-infected patients are more likely to abuse substances than the general population (Avdoshina et al., 2010; Molitor et al., 1998; Nath et al., 2002). According to the United Nations Program on HIV/AIDS, approximately 16 million people inject drugs and, among them, about 3 million are living with HIV worldwide (World Health Organization, 2014). Table 13.1 shows that the percentage of HIV-infected patients who use various addictive substances is greater than that in the general population in the United States.

5. CONCLUSION

Each of the substances discussed exerts its biological effects by acting on endogenous neurotransmitter-dependent pathways. Thus, each also has potential medical benefits, if used properly. In this post-HAART era, neuroHIV has become a chronic disease as a result of neuroinflammation.

Table 13.1 Use of addictive substances among HIV-1-infected individuals and the general population in 2011 in the United States

	HIV-1-infected individuals	General population
Injecting drug use	2% (Pence et al., 2008)	278,000 or <0.1% (Substance Abuse and Mental Health Services Administration, 2010)
Morphine/heroin	2% (Turner et al., 2001)	0.3% (National Survey of Drug Use and Health, 2012)
Alcohol	Heavy drinking: 63% (Miguez, Shor-Posner, Morales, Rodriguez, & Burbano, 2003; Shi et al., 2014)	Binge drinking: 24.6% Heavy drinking: 7.1% (National Institute on Alcohol Abuse and Alcoholism, 2012)
Nicotine	54% (Wallace et al., 1993) >70% (Niaura et al., 2000) 59% (Tesoriero, Gieryic, Carrascal, & Lavigne, 2010)	26.10% (National Survey of Drug Use and Health, 2012)
Methamphetamine	7% (Turner et al., 2001)	0.40% (National Survey of Drug Use and Health, 2012)
Marijuana	12% (Pence et al., 2008) 32% (Turner et al., 2001)	12.10% (National Survey of Drug Use and Health, 2012)
Cocaine	5% (Pence et al., 2008) 11% (Turner et al., 2001)	1.80% (National Survey of Drug Use and Health, 2012)
Polysubstance	7% (Pence et al., 2008)	

The indication that HIV-1-infected individuals may also have a greater risk for substance abuse and addiction that would complicate their treatment presents an urgent clinical challenge.

Recently, the National Institute on Alcohol Abuse and Alcoholism (NIAAA R01-PA-14-139 and R21-PA-14-138) and the National Institute on Drug Abuse (NIDA R01-PA-14-084 and R21-PA-14-083) provided funding opportunities for research aimed at investigating novel connections between the immune system and substance abuse disorders mediated by CNS neuroimmune signaling. More specifically, a gap has been identified in the research between potential protective and detrimental influences of neuroinflammation on substance use disorders, and the need to further investigate the effects of neuroimmune signaling on the brain in order to alter behavior associated with addiction and dependence.

The literature presented here provides support for the theory that HIV-1 is involved in sensitizing individuals to substance abuse; however, more focused research is needed to understand the mechanisms underlying relationships among these variables. While MJ has long been categorized as a substance of abuse, recent research focusing on ECs and SCs has found that these substances exert anti-inflammatory and neuroprotective properties which may be useful in treating disorders such as AD, Parkinson's disease, and HAND. Nicotine, the key ingredient in smoking tobacco, has only begun to be characterized as neuroprotective on memory and learning.

Future research needs to investigate the molecular, cellular, systemic, and behavioral changes associated HIV infection, with and without HAART, in order to identify the possible mechanisms underlying the increase in substance use disorders in the HIV-infected population. Since the HIV-1Tg rat mimics HIV patients given HAART, this noninfectious rodent model can be used to identify genes and pathways associated with drug abuse disorders in the persistent presence of HIV-1 viral proteins, and also to study the interactive effects of different substances, such as nicotine and EtOH, or MJ and METH. Based on the immune responsive pathways identified in various brain areas of the HIV-1Tg rat by Li et al. (2013), therapeutic strategies incorporating existing anti-inflammatory agents, such as nonsteroidal anti-inflammatory drugs, with mediators specific to each immune responsive pathways could be formulated. In addition, because these drugs do not readily cross the blood–brain barrier, novel nanomedical technologies could be used to enhance their delivery, efficacy, and site specificity. This reverse-translational approach would address the clinical challenges presented in this chapter, namely, that, even though HAART patients have limited HIV viral replication, the prevalence of HAND can lead to drug abuse disorders.

REFERENCES

Abreu-Villaca, Y., Medeiros, A. H., Lima, C. S., Faria, F. P., Filgueiras, C. C., & Manhaes, A. C. (2007). Combined exposure to nicotine and ethanol in adolescent mice differentially affects memory and learning during exposure and withdrawal. *Behavioural Brain Research*, *181*(1), 136–146. http://dx.doi.org/10.1016/j.bbr.2007.03.035.

Agatsuma, S., Dang, M. T., Li, Y., & Hiroi, N. (2010). N-methyl-D-aspartic acid receptors on striatal neurons are essential for cocaine cue reactivity in mice. *Biological Psychiatry*, *67*(8), 778–780. http://dx.doi.org/10.1016/j.biopsych.2009.12.023.

Aggarwal, S. K., Carter, G. T., Sullivan, M. D., ZumBrunnen, C., Morrill, R., & Mayer, J. D. (2009). Medicinal use of cannabis in the United States: Historical perspectives, current trends, and future directions. *Journal of Opioid Management*, *5*(3), 153–168.

Aksenov, M. Y., Aksenova, M. V., Mactutus, C. F., & Booze, R. M. (2012). D1/NMDA receptors and concurrent methamphetamine+ HIV-1 Tat neurotoxicity. *Journal of Neuroimmune Pharmacology*, 7(3), 599–608. http://dx.doi.org/10.1007/s11481-012-9362-3.

Aksenov, M. Y., Hasselrot, U., Bansal, A. K., Wu, G., Nath, A., Anderson, C., et al. (2001). Oxidative damage induced by the injection of HIV-1 Tat protein in the rat striatum. *Neuroscience Letters*, *305*(1), 5–8.

Alkhatib, G., Combadiere, C., Broder, C. C., Feng, Y., Kennedy, P. E., Murphy, P. M., et al. (1996). CC CKR5: A RANTES, MIP-1alpha, MIP-1beta receptor as a fusion cofactor for macrophage-tropic HIV-1. *Science*, *272*(5270), 1955–1958.

Aller, M. A., Arias, J. L., Lorente, L., Nava, M. P., Duran, H. J., & Arias, J. (2001). Neuro-immune-endocrine functional system and vascular pathology. *Medical Hypotheses*, *57*(5), 561–569. http://dx.doi.org/10.1054/mehy.2001.1408.

Amara, S. G., & Kuhar, M. J. (1993). Neurotransmitter transporters: Recent progress. *Annual Review of Neuroscience*, *16*, 73–93. http://dx.doi.org/10.1146/annurev.ne.16.030193.000445.

Anthony, I. C., & Bell, J. E. (2008). The neuropathology of HIV/AIDS. *International Review of Psychiatry*, *20*(1), 15–24. http://dx.doi.org/10.1080/09540260701862037.

Antinori, A., Arendt, G., Becker, J. T., Brew, B. J., Byrd, D. A., Cherner, M., et al. (2007). Updated research nosology for HIV-associated neurocognitive disorders. *Neurology*, *69*(18), 1789–1799. http://dx.doi.org/10.1212/01.WNL.0000287431.88658.8b.

Antinori, A., Trotta, M. P., Lorenzini, P., Torti, C., Gianotti, N., Maggiolo, F., et al. (2007). Virological response to salvage therapy in HIV-infected persons carrying the reverse transcriptase K65R mutation. *Antiviral Therapy*, *12*(8), 1175–1183.

Athyros, V. G., Liberopoulos, E. N., Mikhailidis, D. P., Papageorgiou, A. A., Ganotakis, E. S., Tziomalos, K., et al. (2007). Association of drinking pattern and alcohol beverage type with the prevalence of metabolic syndrome, diabetes, coronary heart disease, stroke, and peripheral arterial disease in a Mediterranean cohort. *Angiology*, *58*(6), 689–697. http://dx.doi.org/10.1177/0003319707306146.

Avdoshina, V., Biggio, F., Palchik, G., Campbell, L. A., & Mocchetti, I. (2010). Morphine induces the release of CCL5 from astrocytes: Potential neuroprotective mechanism against the HIV protein gp120. *Glia*, *58*(13), 1630–1639. http://dx.doi.org/10.1002/glia.21035.

Azaryan, A. V., Clock, B. J., & Cox, B. M. (1996). Mu opioid receptor mRNA in nucleus accumbens is elevated following dopamine receptor activation. *Neurochemical Research*, *21*(11), 1411–1415.

Ballester, L. Y., Capo-Velez, C. M., Garcia-Beltran, W. F., Ramos, F. M., Vazquez-Rosa, E., Rios, R., et al. (2012). Up-regulation of the neuronal nicotinic receptor alpha7 by HIV glycoprotein 120: Potential implications for HIV-associated neurocognitive disorder. *The Journal of Biological Chemistry*, *287*(5), 3079–3086. http://dx.doi.org/10.1074/jbc.M111.262543.

Bartus, R. T. (2000). On neurodegenerative diseases, models, and treatment strategies: Lessons learned and lessons forgotten a generation following the cholinergic hypothesis. *Experimental Neurology*, *163*(2), 495–529. http://dx.doi.org/10.1006/exnr.2000.7397.

Ben Amar, M. (2006). Cannabinoids in medicine: A review of their therapeutic potential. *Journal of Ethnopharmacology*, *105*(1–2), 1–25. http://dx.doi.org/10.1016/j.jep.2006.02.001.

Berman, S. M., Kuczenski, R., McCracken, J. T., & London, E. D. (2009). Potential adverse effects of amphetamine treatment on brain and behavior: A review. *Molecular Psychiatry*, *14*(2), 123–142. http://dx.doi.org/10.1038/mp.2008.90.

Bhat, V. M., Cole, J. W., Sorkin, J. D., Wozniak, M. A., Malarcher, A. M., Giles, W. H., et al. (2008). Dose-response relationship between cigarette smoking and risk of ischemic stroke in young women. *Stroke*, *39*(9), 2439–2443. http://dx.doi.org/10.1161/STROKEAHA.107.510073.

Bilbo, S. D., & Schwarz, J. M. (2009). Early-life programming of later-life brain and behavior: A critical role for the immune system. *Frontiers in Behavioral Neuroscience*, *3*, 14. http://dx.doi.org/10.3389/neuro.08.014.2009.

Bland, S. T., Beckley, J. T., Watkins, L. R., Maier, S. F., & Bilbo, S. D. (2010). Neonatal Escherichia coli infection alters glial, cytokine, and neuronal gene expression in response to acute amphetamine in adolescent rats. *Neuroscience Letters*, *474*(1), 52–57. http://dx.doi.org/10.1016/j.neulet.2010.03.006.

Blednov, Y. A., Benavidez, J. M., Geil, C., Perra, S., Morikawa, H., & Harris, R. A. (2011). Activation of inflammatory signaling by lipopolysaccharide produces a prolonged increase of voluntary alcohol intake in mice. *Brain, Behavior, and Immunity*, *25*(Suppl. 1), S92–S105. http://dx.doi.org/10.1016/j.bbi.2011.01.008.

Bleul, C. C., Farzan, M., Choe, H., Parolin, C., Clark-Lewis, I., Sodroski, J., et al. (1996). The lymphocyte chemoattractant SDF-1 is a ligand for LESTR/fusin and blocks HIV-1 entry. *Nature*, *382*(6594), 829–833.

Blomster, J. I., Zoungas, S., Chalmers, J., Li, Q., Chow, C. K., Woodward, M., et al. (2014). The relationship between alcohol consumption and vascular complications and mortality in individuals with type 2 diabetes mellitus. *Diabetes Care*, *37*, 1353–1359. http://dx.doi.org/10.2337/dc13-2727.

Borgelt, L. M., Franson, K. L., Nussbaum, A. M., & Wang, G. S. (2013). The pharmacologic and clinical effects of medical cannabis. *Pharmacotherapy*, *33*(2), 195–209. http://dx.doi.org/10.1002/phar.1187.

Bracci, L., Lozzi, L., Rustici, M., & Neri, P. (1992). Binding of HIV-1 gp120 to the nicotinic receptor. *FEBS Letters*, *311*(2), 115–118.

Brew, B. J., Crowe, S. M., Landay, A., Cysique, L. A., & Guillemin, G. (2009). Neurodegeneration and ageing in the HAART era. *Journal of Neuroimmune Pharmacology*, *4*(2), 163–174. http://dx.doi.org/10.1007/s11481-008-9143-1.

Buch, S., Yao, H., Guo, M., Mori, T., Su, T. P., & Wang, J. (2011). Cocaine and HIV-1 interplay: Molecular mechanisms of action and addiction. *Journal of Neuroimmune Pharmacology*, *6*(4), 503–515. http://dx.doi.org/10.1007/s11481-011-9297-0.

Calatayud, J., & Gonzalez, A. (2003). History of the development and evolution of local anesthesia since the coca leaf. *Anesthesiology*, *98*(6), 1503–1508.

Cao, J., Wang, S., Wang, J., Cui, W., Nesil, T., Vigorito, M., et al. (2013). RNA deep-sequencing analysis reveals that nicotine restores impaired gene expression by viral proteins in the brains of HIV-1 transgenic rats. *PLoS One*, *8*(7), e68517. http://dx.doi.org/10.1371/journal.pone.0068517.

Chang, S. L., Beltran, J. A., & Swarup, S. (2007). Expression of the mu opioid receptor in the human immunodeficiency virus type 1 transgenic rat model. *Journal of Virology*, *81*(16), 8406–8411. http://dx.doi.org/10.1128/JVI.00155-07.

Chang, L., Cloak, C., Yakupov, R., & Ernst, T. (2006). Combined and independent effects of chronic marijuana use and HIV on brain metabolites. *Journal of Neuroimmune Pharmacology*, *1*(1), 65–76. http://dx.doi.org/10.1007/s11481-005-9005-z.

Chang, S. L., & Connaghan, K. P. (2012). Behavioral and molecular evidence for a feedback interaction between morphine and HIV-1 viral proteins. *Journal of Neuroimmune Pharmacology*, 7(2), 332–340. http://dx.doi.org/10.1007/s11481-011-9324-1.

Chang, S. L., Moldow, R. L., House, S. D., & Zadina, J. E. (1996). Morphine affects the brain-immune axis by modulating an interleukin-1 beta dependent pathway. *Advances in Experimental Medicine and Biology*, *402*, 35–42.

Chang, S. L., Patel, N. A., Romero, A. A., Thompson, J., & Zadina, J. E. (1996). FOS expression induced by interleukin-1 or acute morphine treatment in the rat hypothalamus is attenuated by chronic exposure to morphine. *Brain Research*, *736*(1–2), 227–236.

Chang, S. L., & Vigorito, M. (2006). Role of HIV-1 infection in addictive behavior: A study of an HIV-1 transgenic rat model. *American Journal of Infectious Diseases*, *2*(2), 98–106.

Chao, C. C., Hu, S., Sheng, W. S., & Peterson, P. K. (1995). Tumor necrosis factor-alpha production by human fetal microglial cells: Regulation by other cytokines. *Developmental Neuroscience*, *17*(2), 97–105.

Chen, R., Zhou, H., Beltran, J., Malellari, L., & Chang, S. L. (2005). Differential expression of cytokines in the brain and serum during endotoxin tolerance. *Journal of Neuroimmunology*, *163*(1–2), 53–72. http://dx.doi.org/10.1016/j.jneuroim.2005.02.012.

Choi, D. Y., Lee, M. K., & Hong, J. T. (2012). Lack of CCR5 modifies glial phenotypes and population of the nigral dopaminergic neurons, but not MPTP-induced dopaminergic neurodegeneration. *Neurobiology of Disease*, *49C*, 159–168. http://dx.doi.org/10.1016/j.nbd.2012.08.001.

Cicala, C., Arthos, J., Selig, S. M., Dennis, G., Jr., Hosack, D. A., Van Ryk, D., et al. (2002). HIV envelope induces a cascade of cell signals in non-proliferating target cells that favor virus replication. *Proceedings of the National Academy of Sciences of the United States of America*, *99*(14), 9380–9385.

Clapham, P. R., & McKnight, A. (2001). HIV-1 receptors and cell tropism. *British Medical Bulletin*, *58*, 43–59.

Clark, P. A., Capuzzi, K., & Fick, C. (2011). Medical marijuana: Medical necessity versus political agenda. *Medical Science Monitor*, *17*(12), RA249–RA261.

Cristiani, S. A., Pukay-Martin, N. D., & Bornstein, R. (2004). Marijuana use and cognitive function in HIV-infected people. *The Journal of Neuropsychiatry and Clinical Neurosciences*, *16*, 330–335.

Crocq, M. A. (2003). Alcohol, nicotine, caffeine, and mental disorders. *Dialogues in Clinical Neuroscience*, *5*(2), 175–185.

Cui, W. Y., Wang, J., Wei, J., Cao, J., Chang, S. L., Gu, J., et al. (2012). Modulation of innate immune-related pathways in nicotine-treated SH-SY5Y cells. *Amino Acids*, *43*(3), 1157–1169. http://dx.doi.org/10.1007/s00726-011-1171-0.

Cui, W. Y., Zhao, S., Polanowska-Grabowska, R., Wang, J., Wei, J., Dash, B., et al. (2013). Identification and characterization of poly(I:C)-induced molecular responses attenuated by nicotine in mouse macrophages. *Molecular Pharmacology*, *83*(1), 61–72. http://dx.doi.org/10.1124/mol.112.081497.

Cummings, S., Parham, E. S., Strain, G. W., & American Dietetic Association. (2002). Position of the American Dietetic Association: Weight management. *Journal of the American Dietetic Association*, *102*(8), 1145–1155.

Dansak, D. A. (1997). Medical use of recreational drugs by AIDS patients. *Journal of Addictive Diseases*, *16*(3), 25–30. http://dx.doi.org/10.1300/J069v16n03_03.

De Simone, R., Ajmone-Cat, M. A., Carnevale, D., & Minghetti, L. (2005). Activation of alpha7 nicotinic acetylcholine receptor by nicotine selectively up-regulates cyclooxygenase-2 and prostaglandin E2 in rat microglial cultures. *Journal of Neuroinflammation*, *2*(1), 4.

de Wied, D., Diamant, M., & Fodor, M. (1993). Central nervous system effects of the neurohypophyseal hormones and related peptides. *Frontiers in Neuroendocrinology*, *14*(4), 251–302. http://dx.doi.org/10.1006/frne.1993.1009.

Deyo, R. A., Smith, D. H., Johnson, E. S., Donovan, M., Tillotson, C. J., Yang, X., et al. (2011). Opioids for back pain patients: Primary care prescribing patterns and use of services. *Journal of the American Board of Family Medicine*, *24*(6), 717–727. http://dx.doi.org/10.3122/jabfm.2011.06.100232.

Di Chiara, G., & Imperato, A. (1988). Drugs abused by humans preferentially increase synaptic dopamine concentrations in the mesolimbic system of freely moving rats. *Proceedings of the National Academy of Sciences of the United States of America*, *85*(14), 5274–5278.

Dickie, P., Felser, J., Eckhaus, M., Bryant, J., Silver, J., Marinos, N., et al. (1991). HIV-associated nephropathy in transgenic mice expressing HIV-1 genes. *Virology*, *185*(1), 109–119.

Di Minno, M. N., Russolillo, A., Lupoli, R., Ambrosino, P., Di Minno, A., & Tarantino, G. (2012). Omega-3 fatty acids for the treatment of non-alcoholic fatty liver disease. *World Journal of Gastroenterology*, *18*(41), 5839–5847. http://dx.doi.org/10.3748/wjg.v18.i41.5839.

Djousse, L., & Gaziano, J. M. (2008). Alcohol consumption and heart failure: A systematic review. *Current Atherosclerosis Reports*, *10*(2), 117–120.

D'Souza, D. C., Sewell, R. A., & Ranganathan, M. (2009). Cannabis and psychosis/schizophrenia: Human studies. *European Archives of Psychiatry and Clinical Neuroscience, 259*(7), 413–431. http://dx.doi.org/10.1007/s00406-009-0024-2.

Dunn, W., Sanyal, A. J., Brunt, E. M., Unalp-Arida, A., Donohue, M., McCullough, A. J., et al. (2012). Modest alcohol consumption is associated with decreased prevalence of steatohepatitis in patients with non-alcoholic fatty liver disease (NAFLD). *Journal of Hepatology, 57*(2), 384–391. http://dx.doi.org/10.1016/j.jhep.2012.03.024.

Ehrhart, J., Obregon, D., Mori, T., Hou, H., Sun, N., Bai, Y., et al. (2005). Stimulation of cannabinoid receptor 2 (CB2) suppresses microglial activation. *Journal of Neuroinflammation, 2*, 29. http://dx.doi.org/10.1186/1742-2094-2-29.

Elenkov, I. J. (2007). Effects of catecholamines on the immune response. *NeuroImmune Biology*, 7, 189–206. http://dx.doi.org/10.1016/s1567-7443(07)00210-4.

El-Hage, N., Podhaizer, E. M., Sturgill, J., & Hauser, K. F. (2011). Toll-like receptor expression and activation in astroglia: Differential regulation by HIV-1 Tat, gp120, and morphine. *Immunological Investigations, 40*, 498–522. http://dx.doi.org/10.3109/08820139.2011.561904.

Espinosa, E., & Bermudez-Rattoni, F. (2001). Behavior-immunity relationship: The role of cytokines. *Revista de Investigación Clínica, 53*(3), 240–253.

Fattore, L., Cossu, G., Spano, M. S., Deiana, S., Fadda, P., Scherma, M., et al. (2004). Cannabinoids and reward: Interactions with the opioid system. *Critical Reviews in Neurobiology, 16*(1–2), 147–158.

Fattore, L., Spano, S., Cossu, G., Deiana, S., Fadda, P., & Fratta, W. (2005). Cannabinoid CB(1) antagonist SR 141716A attenuates reinstatement of heroin self-administration in heroin-abstinent rats. *Neuropharmacology, 48*(8), 1097–1104. http://dx.doi.org/10.1016/j.neuropharm.2005.01.022.

Feldman, R. S., Meyer, J. S., & Quenzer, L. F. (1997). *Principles of neuropsychopharmacology*. Sunderland, MA: Sinauer Associates.

Fernandez-Espejo, E., Viveros, M. P., Nunez, L., Ellenbroek, B. A., & Rodriguez de Fonseca, F. (2009). Role of cannabis and endocannabinoids in the genesis of schizophrenia. *Psychopharmacology, 206*(4), 531–549. http://dx.doi.org/10.1007/s00213-009-1612-6.

Ferris, M. J., Frederick-Duus, D., Fadel, J., Mactutus, C. F., & Booze, R. M. (2009). The human immunodeficiency virus-1-associated protein, Tat1-86, impairs dopamine transporters and interacts with cocaine to reduce nerve terminal function: A no-net-flux microdialysis study. *Neuroscience, 159*(4), 1292–1299. http://dx.doi.org/10.1016/j.neuroscience.2009.01.024.

Filip, M., Golda, A., Zaniewska, M., McCreary, A. C., Nowak, E., Kolasiewicz, W., et al. (2006). Involvement of cannabinoid CB1 receptors in drug addiction: Effects of rimonabant on behavioral responses induced by cocaine. *Pharmacological Reports, 58*(6), 806–819.

Friend, D. M., & Keefe, K. A. (2013). A role for D1 dopamine receptors in striatal methamphetamine-induced neurotoxicity. *Neuroscience Letters, 555*, 243–247. http://dx.doi.org/10.1016/j.neulet.2013.08.039.

Fuchs, F., Damm, J., Gerstberger, R., Roth, J., & Rummel, C. (2013). Activation of the inflammatory transcription factor nuclear factor interleukin-6 during inflammatory and psychological stress in the brain. *Journal of Neuroinflammation, 10*, 140. http://dx.doi.org/10.1186/1742-2094-10-140.

Gala, R. R. (1991). Prolactin and growth hormone in the regulation of the immune system. *Proceedings of the Society for Experimental Biology and Medicine, 198*(1), 513–527.

Gendelman, H. E., Genis, P., Jett, M., Zhai, Q. H., & Nottet, H. S. (1994). An experimental model system for HIV-1-induced brain injury. *Advances in Neuroimmunology, 4*(3), 189–193.

Gendelman, H. E., Zheng, J., Coulter, C. L., Ghorpade, A., Che, M., Thylin, M., et al. (1998). Suppression of inflammatory neurotoxins by highly active antiretroviral therapy in human immunodeficiency virus-associated dementia. *The Journal of Infectious Diseases, 178*(4), 1000–1007.

Genis, P., Jett, M., Bernton, E. W., Boyle, T., Gelbard, H. A., Dzenko, K., et al. (1992). Cytokines and arachidonic metabolites produced during human immunodeficiency virus (HIV)-infected macrophage-astroglia interactions: Implications for the neuropathogenesis of HIV disease. *The Journal of Experimental Medicine, 176*(6), 1703–1718.

Ghitza, U. E., Preston, K. L., Epstein, D. H., Kuwabara, H., Endres, C. J., Bencherif, B., et al. (2010). Brain mu-opioid receptor binding predicts treatment outcome in cocaine-abusing outpatients. *Biological Psychiatry, 68*(8), 697–703. http://dx.doi.org/10.1016/j.biopsych.2010.05.003.

Ghozland, S., Matthes, H. W., Simonin, F., Filliol, D., Kieffer, B. L., & Maldonado, R. (2002). Motivational effects of cannabinoids are mediated by mu-opioid and kappa-opioid receptors. *The Journal of Neuroscience, 22*(3), 1146–1154.

Giovannetti, A., Ensoli, F., Mazzetta, F., De Cristofaro, M., Pierdominici, M., Muratori, D. S., et al. (1999). CCR5 and CXCR4 chemokine receptor expression and beta-chemokine production during early T cell repopulation induced by highly active anti-retroviral therapy. *Clinical and Experimental Immunology, 118*(1), 87–94.

Gisslen, M., Svennerholm, B., Fuchs, D., & Hagberg, L. (1998). Neurological efficacy of stavudine, zidovudine, and lamivudine. *Lancet, 352*(9125), 402–403.

Giulian, D., Vaca, K., & Noonan, C. A. (1990). Secretion of neurotoxins by mononuclear phagocytes infected with HIV-1. *Science, 250*(4987), 1593–1596.

Giunta, B., Ehrhart, J., Townsend, K., Sun, N., Vendrame, M., Shytle, D., et al. (2004). Galantamine and nicotine have a synergistic effect on inhibition of microglial activation induced by HIV-1 gp120. *Brain Research Bulletin, 64*(2), 165–170.

Goerig, M., Bacon, D., & van Zundert, A. (2012). Carl Koller, cocaine, and local anesthesia: Some less known and forgotten facts. *Regional Anesthesia and Pain Medicine, 37*(3), 318–324. http://dx.doi.org/10.1097/AAP.0b013e31825051f3.

Gonzalez-Scarano, F., & Martin-Garcia, J. (2005). The neuropathogenesis of AIDS. *Nature Reviews. Immunology, 5*(1), 69–81. http://dx.doi.org/10.1038/nri1527.

Gorelick, D. A., Kim, Y. K., Bencherif, B., Boyd, S. J., Nelson, R., Copersino, M., et al. (2005). Imaging brain mu-opioid receptors in abstinent cocaine users: Time course and relation to cocaine craving. *Biological Psychiatry, 57*(12), 1573–1582. http://dx.doi.org/10.1016/j.biopsych.2005.02.026.

Gorry, P. R., Ong, C., Thorpe, J., Bannwarth, S., Thompson, K. A., Gatignol, A., et al. (2003). Astrocyte infection by HIV-1: Mechanisms of restricted virus replication, and role in the pathogenesis of HIV-1-associated dementia. *Current HIV Research, 1*(4), 463–473.

Gould, T. J., & Lommock, J. A. (2003). Nicotine enhances contextual fear conditioning and ameliorates ethanol-induced deficits in contextual fear conditioning. *Behavioral Neuroscience, 117*(6), 1276–1282. http://dx.doi.org/10.1037/0735-7044.117.6.1276.

Granet, E. (1957). Noninflammatory proctalgias and allied anorectal dyscrasias. *The American Journal of Gastroenterology, 28*(4), 394–402.

Gulick, D., & Gould, T. J. (2011). Nicotine acts in the anterior cingulate, but not dorsal or ventral hippocampus, to reverse ethanol-induced learning impairments in the plus-maze discriminative avoidance task. *Addiction Biology, 16*(1), 176–188. http://dx.doi.org/10.1111/j.1369-1600.2010.00209.x.

Gurwell, J. A., Nath, A., Sun, Q., Zhang, J., Martin, K. M., Chen, Y., et al. (2001). Synergistic neurotoxicity of opioids and human immunodeficiency virus-1 Tat protein in striatal neurons in vitro. *Neuroscience, 102*(3), 555–563.

Gutala, R., Wang, J., Hwang, Y. Y., Haq, R., & Li, M. D. (2006). Nicotine modulates expression of amyloid precursor protein and amyloid precursor-like protein 2 in mouse brain and in SH-SY5Y neuroblastoma cells. *Brain Research, 1093*(1), 12–19. http://dx.doi.org/10.1016/j.brainres.2006.03.100.

Hahn, Y. K., Vo, P., Fitting, S., Block, M. L., Hauser, K. F., & Knapp, P. E. (2010). beta-Chemokine production by neural and glial progenitor cells is enhanced by HIV-1 Tat: Effects on microglial migration. *Journal of Neurochemistry, 114*(1), 97–109. http://dx.doi.org/10.1111/j.1471-4159.2010.06744.x.

Han, D. D., & Gu, H. H. (2006). Comparison of the monoamine transporters from human and mouse in their sensitivities to psychostimulant drugs. *BMC Pharmacology, 6*, 6. http://dx.doi.org/10.1186/1471-2210-6-6.

Hanson, D. J. (1995). *History of alcohol and drinking around the world.* Retrieved from, http://www2.potsdam.edu/alcohol/Controversies/1114796842.html.

Hauser, K. F., El-Hage, N., Buch, S., Berger, J. R., Tyor, W. R., Nath, A., et al. (2005). Molecular targets of opiate drug abuse in neuroAIDS. *Neurotoxicity Research, 8*(1–2), 63–80.

Hernandez, S. L., Sheyner, I., Stover, K. T., & Stewart, J. T. (2013). Dronabinol treatment of refractory nausea and vomiting related to peritoneal carcinomatosis. *The American Journal of Hospice & Palliative Care.* http://dx.doi.org/10.1177/1049909113504240.

Herz, A. (1998). Opioid reward mechanisms: A key role in drug abuse? *Canadian Journal of Physiology and Pharmacology, 76*(3), 252–258.

Hesselgesser, J., Taub, D., Baskar, P., Greenberg, M., Hoxie, J., Kolson, D. L., et al. (1998). Neuronal apoptosis induced by HIV-1 gp120 and the chemokine SDF-1 alpha is mediated by the chemokine receptor CXCR4. *Current Biology, 8*(10), 595–598.

Holden, J. E., Jeong, Y., & Forrest, J. M. (2005). The endogenous opioid system and clinical pain management. *AACN Clinical Issues, 16*(3), 291–301.

Homji, N. F., Mao, X., Langsdorf, E. F., & Chang, S. L. (2012). Endotoxin-induced cytokine and chemokine expression in the HIV-1 transgenic rat. *Journal of Neuroinflammation, 9*, 3. http://dx.doi.org/10.1186/1742-2094-9-3.

Horuk, R., Martin, A. W., Wang, Z., Schweitzer, L., Gerassimides, A., Guo, H., et al. (1997). Expression of chemokine receptors by subsets of neurons in the central nervous system. *Journal of Immunology, 158*(6), 2882–2890.

Hosoi, T., & Nomura, Y. (2004). Functional role of acetylcholine in the immune system. *Frontiers in Bioscience, 9*, 2414–2419.

House, S. D., Mao, X., Wu, G., Espinelli, D., Li, W. X., & Chang, S. L. (2001). Chronic morphine potentiates the inflammatory response by disrupting interleukin-1beta modulation of the hypothalamic-pituitary-adrenal axis. *Journal of Neuroimmunology, 118*(2), 277–285.

Howard, G., Wagenknecht, L. E., Burke, G. L., Diez-Roux, A., Evans, G. W., McGovern, P., et al. (1998). Cigarette smoking and progression of atherosclerosis: The atherosclerosis risk in communities (ARIC) study. *JAMA, 279*(2), 119–124.

Hyman, S. E., Malenka, R. C., & Nestler, E. J. (2006). Neural mechanisms of addiction: The role of reward-related learning and memory. *Annual Review of Neuroscience, 29*, 565–598. http://dx.doi.org/10.1146/annurev.neuro.29.051605.113009.

Jankovic, B. D., & Radulovic, J. (1992). Enkephalins, brain and immunity: Modulation of immune responses by methionine-enkephalin injected into the cerebral cavity. *The International Journal of Neuroscience, 67*(1–4), 241–270.

Jayaram-Lindstrom, N., Hammarberg, A., Beck, O., & Franck, J. (2008). Naltrexone for the treatment of amphetamine dependence: A randomized, placebo-controlled trial. *The American Journal of Psychiatry, 165*(11), 1442–1448. http://dx.doi.org/10.1176/appi.ajp.2008.08020304.

Jayaram-Lindstrom, N., Konstenius, M., Eksborg, S., Beck, O., Hammarberg, A., & Franck, J. (2008). Naltrexone attenuates the subjective effects of amphetamine in patients with amphetamine dependence. *Neuropsychopharmacology*, *33*(8), 1856–1863. http://dx.doi.org/10.1038/sj.npp.1301572.

Jiang, J., & Aiken, C. (2006). Maturation of the viral core enhances the fusion of HIV-1 particles with primary human T cells and monocyte-derived macrophages. *Virology*, *346*(2), 460–468.

Jones, A., McMillan, M. R., Jones, R. W., Kowalik, G. T., Steeden, J. A., Pruessner, J. C., et al. (2013). Habitual alcohol consumption is associated with lower cardiovascular stress responses—A novel explanation for the known cardiovascular benefits of alcohol? *Stress*, *16*(4), 369–376. http://dx.doi.org/10.3109/10253890.2013.777833.

Justinova, Z., Tanda, G., Munzar, P., & Goldberg, S. R. (2004). The opioid antagonist naltrexone reduces the reinforcing effects of Delta 9 tetrahydrocannabinol (THC) in squirrel monkeys. *Psychopharmacology*, *173*(1–2), 186–194. http://dx.doi.org/10.1007/s00213-003-1693-6.

Kass, M., & Chang, S. L. (2010). Transgenic rodent models. In H. E. Gendelman, I. Grant, I. P. Everall, S. A. Lipton, & S. Swindells (Eds.), *The neurology of AIDS*. New York: Oxford University Press.

Kass, M. D., Liu, X., Vigorito, M., Chang, L., & Chang, S. L. (2010). Methamphetamine-induced behavioral and physiological effects in adolescent and adult HIV-1 transgenic rats. *Journal of Neuroimmune Pharmacology*, *5*(4), 566–573. http://dx.doi.org/10.1007/s11481-010-9221-z.

Katz, N. (2005). The impact of opioids on the endocrine system. *Pain Management Rounds*, *1*(9), 1–6.

Kaul, M., & Lipton, S. A. (1999). Chemokines and activated macrophages in HIV gp120-induced neuronal apoptosis. *Proceedings of the National Academy of Sciences of the United States of America*, *96*(14), 8212–8216.

Kaul, M., Zheng, J., Okamoto, S., Gendelman, H. E., & Lipton, S. A. (2005). HIV-1 infection and AIDS: Consequences for the central nervous system. *Cell Death and Differentiation*, *12*(Suppl. 1), 878–892. http://dx.doi.org/10.1038/sj.cdd.4401623.

Kesby, J. P., Hubbard, D. T., Markou, A., & Semenova, S. (2012). Expression of HIV gp120 protein increases sensitivity to the rewarding properties of methamphetamine in mice. *Addiction Biology*. http://dx.doi.org/10.1111/adb.12023.

Khurdayan, V. K., Buch, S., El-Hage, N., Lutz, S. E., Goebel, S. M., Singh, I. N., et al. (2004). Preferential vulnerability of astroglia and glial precursors to combined opioid and HIV-1 Tat exposure in vitro. *The European Journal of Neuroscience*, *19*(12), 3171–3182.

Kihara, T., Shimohama, S., Sawada, H., Kimura, J., Kume, T., Kochiyama, H., et al. (1997). Nicotinic receptor stimulation protects neurons against beta-amyloid toxicity. *Annals of Neurology*, *42*(2), 159–163.

Kikuchi-Utsumi, K., Ishizaka, M., Matsumura, N., Watabe, M., Aoyama, K., Sasakawa, N., et al. (2013). Involvement of the alpha(1D)-adrenergic receptor in methamphetamine-induced hyperthermia and neurotoxicity in rats. *Neurotoxicity Research*, *24*(2), 130–138. http://dx.doi.org/10.1007/s12640-012-9369-9.

Kim, B. O., Liu, Y., Ruan, Y., Xu, Z. C., Schantz, L., & He, J. J. (2003). Neuropathologies in transgenic mice expressing human immunodeficiency virus type 1 Tat protein under the regulation of the astrocyte-specific glial fibrillary acidic protein promoter and doxycycline. *The American Journal of Pathology*, *162*(5), 1693–1707. http://dx.doi.org/10.1016/S0002-9440(10)64304-0.

Koob, G. F. (2013). Negative reinforcement in drug addiction: The darkness within. *Current Opinion in Neurobiology*, *23*(4), 559–563. http://dx.doi.org/10.1016/j.conb.2013.03.011.

Koob, G. F., & Le Moal, M. (2005). Plasticity of reward neurocircuitry and the 'dark side' of drug addiction. *Nature Neuroscience*, *8*(11), 1442–1444. http://dx.doi.org/10.1038/nn1105-1442.

Koob, G. F., & Le Moal, M. (2008). Addiction and the brain antireward system. *Annual Review of Psychology*, *59*, 29–53. http://dx.doi.org/10.1146/annurev.psych.59.103006.093548.

Krashin, D. L., Merrill, J. O., & Trescot, A. M. (2012). Opioids in the management of HIV-related pain. *Pain Physician*, *15*(3 Suppl.), ES157–ES168.

Kumar, A. M., Fernandez, J. B., Singer, E. J., Commins, D., Waldrop-Valverde, D., Ownby, R. L., et al. (2009). Human immunodeficiency virus type 1 in the central nervous system leads to decreased dopamine in different regions of postmortem human brains. *Journal of Neurovirology*, *15*(3), 257–274. http://dx.doi.org/10.1080/13550280902973952.

Landa, L., Sulcova, A., & Slais, K. (2006). Involvement of cannabinoid CB1 and CB2 receptor activity in the development of behavioural sensitization to methamphetamine effects in mice. *Neuro Endocrinology Letters*, *27*(1–2), 63–69.

Largent-Milnes, T. M., Guo, W., Wang, H. Y., Burns, L. H., & Vanderah, T. W. (2008). Oxycodone plus ultra-low-dose naltrexone attenuates neuropathic pain and associated mu-opioid receptor-Gs coupling. *The Journal of Pain*, *9*(8), 700–713. http://dx.doi.org/10.1016/j.jpain.2008.03.005.

LaShomb, A. L., Vigorito, M., & Chang, S. L. (2008). Further characterization of the spatial learning deficits in the human immunodeficiency virus-1 transgenic rat. *Journal of Neurovirology*, *15*, 14–24.

Ledent, C., Valverde, O., Cossu, G., Petitet, F., Aubert, J. F., Beslot, F., et al. (1999). Unresponsiveness to cannabinoids and reduced addictive effects of opiates in CB1 receptor knockout mice. *Science*, *283*(5400), 401–404.

Leonard, S., Mexal, S., & Freedman, R. (2007). Smoking, genetics and schizophrenia: Evidence for self medication. *Journal of Dual Diagnosis*, *3*(3–4), 43–59. http://dx.doi.org/10.1300/J374v03n03_05.

Leri, F., Zhou, Y., Goddard, B., Cummins, E., & Kreek, M. J. (2006). Effects of high-dose methadone maintenance on cocaine place conditioning, cocaine self-administration, and mu-opioid receptor mRNA expression in the rat brain. *Neuropsychopharmacology*, *31*(7), 1462–1474. http://dx.doi.org/10.1038/sj.npp.1300927.

Leri, F., Zhou, Y., Goddard, B., Levy, A., Jacklin, D., & Kreek, M. J. (2009). Steady-state methadone blocks cocaine seeking and cocaine-induced gene expression alterations in the rat brain. *European Neuropsychopharmacology*, *19*(4), 238–249. http://dx.doi.org/10.1016/j.euroneuro.2008.09.004.

Li, M. D., Cao, J., Wang, S., Wang, J., Sarkar, S., Vigorito, M., et al. (2013). Transcriptome sequencing of gene expression in the brain of the HIV-1 transgenic rat. *PLoS One*, *8*(3), e59582. http://dx.doi.org/10.1371/journal.pone.0059582.

Li, M. D., Kane, J. K., Matta, S. G., Blaner, W. S., & Sharp, B. M. (2000). Nicotine enhances the biosynthesis and secretion of transthyretin from the choroid plexus in rats: Implications for beta-amyloid formation. *The Journal of Neuroscience*, *20*(4), 1318–1323.

Li, F., & Tsien, J. Z. (2009). Memory and the NMDA receptors. *The New England Journal of Medicine*, *361*(3), 302–303. http://dx.doi.org/10.1056/NEJMcibr0902052.

Lindl, K. A., Marks, D. R., Kolson, D. L., & Jordan-Sciutto, K. L. (2010). HIV-associated neurocognitive disorder: Pathogenesis and therapeutic opportunities. *Journal of Neuroimmune Pharmacology*, *5*(3), 294–309. http://dx.doi.org/10.1007/s11481-010-9205-z.

Liu, X., Chang, L., Vigorito, M., Kass, M., Li, H., & Chang, S. L. (2009). Methamphetamine-induced behavioral sensitization is enhanced in the HIV-1 transgenic rat. *Journal of Neuroimmune Pharmacology*, *4*(3), 309–316. http://dx.doi.org/10.1007/s11481-009-9160-8.

Locati, M., & Murphy, P. M. (1999). Chemokines and chemokine receptors: Biology and clinical relevance in inflammation and AIDS. *Annual Review of Medicine*, *50*, 425–440.

Lutge, E. E., Gray, A., & Siegfried, N. (2013). The medical use of cannabis for reducing morbidity and mortality in patients with HIV/AIDS. *Cochrane Database of Systematic Reviews*, *4*, 1–45. http://dx.doi.org/10.1002/14651858.CD005175.pub3, CD005175.

Lynch, M. E., & Campbell, F. (2011). Cannabinoids for treatment of chronic non-cancer pain; a systematic review of randomized trials. *British Journal of Clinical Pharmacology*, *72*(5), 735–744. http://dx.doi.org/10.1111/j.1365-2125.2011.03970.x.

Mahajan, S. D., Schwartz, S. A., Shanahan, T. C., Chawda, R. P., & Nair, M. P. (2002). Morphine regulates gene expression of alpha- and beta-chemokines and their receptors on astroglial cells via the opioid mu receptor. *Journal of Immunology*, *169*(7), 3589–3599.

Maragos, W. F., Young, K. L., Turchan, J. T., Guseva, M., Pauly, J. R., Nath, A., et al. (2002). Human immunodeficiency virus-1 Tat protein and methamphetamine interact synergistically to impair striatal dopaminergic function. *Journal of Neurochemistry*, *83*(4), 955–963.

Marin, P., Maus, M., Desagher, S., Glowinski, J., & Premont, J. (1994). Nicotine protects cultured striatal neurones against N-methyl-D-aspartate receptor-mediated neurotoxicity. *Neuroreport*, *5*(15), 1977–1980.

Martin, M., Ledent, C., Parmentier, M., Maldonado, R., & Valverde, O. (2000). Cocaine, but not morphine, induces conditioned place preference and sensitization to locomotor responses in CB1 knockout mice. *The European Journal of Neuroscience*, *12*(11), 4038–4046.

Martin-Sanchez, E., Furukawa, T. A., Taylor, J., & Martin, J. L. (2009). Systematic review and meta-analysis of cannabis treatment for chronic pain. *Pain Medicine*, *10*(8), 1353–1368. http://dx.doi.org/10.1111/j.1526-4637.2009.00703.x.

Marutle, A., Warpman, U., Bogdanovic, N., Lannfelt, L., & Nordberg, A. (1999). Neuronal nicotinic receptor deficits in Alzheimer patients with the Swedish amyloid precursor protein 670/671 mutation. *Journal of Neurochemistry*, *72*(3), 1161–1169.

McArthur, J. C. (2004). HIV dementia: An evolving disease. *Journal of Neuroimmunology*, *157*(1–2), 3–10. http://dx.doi.org/10.1016/j.jneuroim.2004.08.042.

McArthur, J. C., Haughey, N., Gartner, S., Conant, K., Pardo, C., Nath, A., et al. (2003). Human immunodeficiency virus-associated dementia: An evolving disease. *Journal of Neurovirology*, *9*(2), 205–221. http://dx.doi.org/10.1080/13550280390194109.

McArthur, J. C., Steiner, J., Sacktor, N., & Nath, A. (2010). Human immunodeficiency virus-associated neurocognitive disorders: Mind the gap. *Annals of Neurology*, *67*(6), 699–714. http://dx.doi.org/10.1002/ana.22053.

McCarthy, L., Wetzel, M., Sliker, J. K., Eisenstein, T. K., & Rogers, T. J. (2001). Opioids, opioid receptors, and the immune response. *Drug and Alcohol Dependence*, *62*(2), 111–123.

Meazza, C., Pagani, S., Travaglino, P., & Bozzola, M. (2004). Effect of growth hormone (GH) on the immune system. *Pediatric Endocrinology Reviews*, *1*(Suppl. 3), 490–495.

Meucci, O., Fatatis, A., Simen, A. A., Bushell, T. J., Gray, P. W., & Miller, R. J. (1998). Chemokines regulate hippocampal neuronal signaling and gp120 neurotoxicity. *Proceedings of the National Academy of Sciences of the United States of America*, *95*(24), 14500–14505.

Miguez, M. J., Shor-Posner, G., Morales, G., Rodriguez, A., & Burbano, X. (2003). HIV treatment in drug abusers: Impact of alcohol use. *Addiction Biology*, *8*(1), 33–37. http://dx.doi.org/10.1080/1355621031000069855.

Mihailescu, S., & Drucker-Colin, R. (2000a). Nicotine and brain disorders. *Acta Pharmacologica Sinica*, *21*(2), 97–104.

Mihailescu, S., & Drucker-Colin, R. (2000b). Nicotine, brain nicotinic receptors, and neuropsychiatric disorders. *Archives of Medical Research*, *31*(2), 131–144.

Miller, A. M., & Stella, N. (2008). CB2 receptor-mediated migration of immune cells: It can go either way. *British Journal of Pharmacology*, *153*(2), 299–308. http://dx.doi.org/10.1038/sj.bjp.0707523.

Molina-Holgado, F., Molina-Holgado, E., Guaza, C., & Rothwell, N. J. (2002). Role of CB1 and CB2 receptors in the inhibitory effects of cannabinoids on lipopolysaccharide-induced nitric oxide release in astrocyte cultures. *Journal of Neuroscience Research*, *67*(6), 829–836.

Molitor, F., Truax, S. R., Ruiz, J. D., & Sun, R. K. (1998). Association of methamphetamine use during sex with risky sexual behaviors and HIV infection among non-injection drug users. *The Western Journal of Medicine*, *168*(2), 93–97.

Moran, L. M., Aksenov, M. Y., Booze, R. M., Webb, K. M., & Mactutus, C. F. (2012). Adolescent HIV-1 transgenic rats: Evidence for dopaminergic alterations in behavior and neurochemistry revealed by methamphetamine challenge. *Current HIV Research*, *10*(5), 415–424.

Moran, L. M., Booze, R. M., Webb, K. M., & Mactutus, C. F. (2013). Neurobehavioral alterations in HIV-1 transgenic rats: Evidence for dopaminergic dysfunction. *Experimental Neurology*, *239*, 139–147. http://dx.doi.org/10.1016/j.expneurol.2012.10.008.

Morch, H., & Pedersen, B. K. (1995). Beta-endorphin and the immune system—Possible role in autoimmune diseases. *Autoimmunity*, *21*(3), 161–171.

Moriya, A., Iwasaki, Y., Ohguchi, S., Kayashima, E., Mitsumune, T., Taniguchi, H., et al. (2011). Alcohol consumption appears to protect against non-alcoholic fatty liver disease. *Alimentary Pharmacology & Therapeutics*, *33*(3), 378–388. http://dx.doi.org/10.1111/j.1365-2036.2010.04520.x.

Mulla, A., & Buckingham, J. C. (1999). Regulation of the hypothalamo-pituitary-adrenal axis by cytokines. *Baillière's Best Practice & Research. Clinical Endocrinology & Metabolism*, *13*(4), 503–521.

Nath, A., Haughey, N. J., Jones, M., Anderson, C., Bell, J. E., & Geiger, J. D. (2000). Synergistic neurotoxicity by human immunodeficiency virus proteins Tat and gp120: Protection by memantine. *Annals of Neurology*, *47*(2), 186–194.

Nath, A., Hauser, K. F., Wojna, V., Booze, R. M., Maragos, W., Prendergast, M., et al. (2002). Molecular basis for interactions of HIV and drugs of abuse. *Journal of Acquired Immune Deficiency Syndromes*, *31*(Suppl. 2), S62–S69.

National Institute on Alcohol Abuse and Alcoholism (2012). *Alcohol facts and statistics*. from, http://www.niaaa.nih.gov/alcohol-health/overview-alcohol-consumption/alcohol-facts-and-statistics.

National Institute on Drug Abuse (2007). *Drugs, brains, and behavior: The science of addiction*. Retrieved from, http://www.drugabuse.gov/publications/science-addiction.

National Institute on Drug Abuse (2014). *DrugFacts: Marijuana*. Retrieved from, http://www.drugabuse.gov/publications/drugfacts/marijuana.

National Survey of Drug Use and Health (2012). *National survey on drug use and health: Trends in prevalence of various drugs for ages 12 or older, ages 12 to 17, ages 18 to 25, and ages 26 or older; 2010–2012 (in percent)*. from, http://www.drugabuse.gov/national-survey-drug-use-health.

Nazari, R., & Joshi, S. (2008). CCR5 as target for HIV-1 gene therapy. *Current Gene Therapy*, *8*(4), 264–272.

Nazari, R., Ma, X. Z., & Joshi, S. (2008). Inhibition of human immunodeficiency virus-1 entry using vectors expressing a multimeric hammerhead ribozyme targeting the CCR5 mRNA. *The Journal of General Virology*, *89*(Pt. 9), 2252–2261. http://dx.doi.org/10.1099/vir.0.2008/001222-0.

Niaura, R., Shadel, W. G., Morrow, K., Tashima, K., Flanigan, T., & Abrams, D. B. (2000). Human immunodeficiency virus infection, AIDS, and smoking cessation: The time is now. *Clinical Infectious Diseases*, *31*(3), 808–812. http://dx.doi.org/10.1086/314048.

Ocasio, F. M., Jiang, Y., House, S. D., & Chang, S. L. (2004). Chronic morphine accelerates the progression of lipopolysaccharide-induced sepsis to septic shock. *Journal of Neuroimmunology*, *149*(1–2), 90–100. http://dx.doi.org/10.1016/j.jneuroim.2003.12.016.

O'Keefe, J. H., Bybee, K. A., & Lavie, C. J. (2007). Alcohol and cardiovascular health: The razor-sharp double-edged sword. *Journal of the American College of Cardiology*, *50*(11), 1009–1014. http://dx.doi.org/10.1016/j.jacc.2007.04.089.

Ottaviani, E., Franchini, A., & Genedani, S. (1999). ACTH and its role in immune-neuroendocrine functions. A comparative study. *Current Pharmaceutical Design*, *5*(9), 673–681.

Owman, C., Fuxe, K., Janson, A. M., & Kahrstrom, J. (1989). Studies of protective actions of nicotine on neuronal and vascular functions in the brain of rats: Comparison between sympathetic noradrenergic and mesostriatal dopaminergic fiber systems, and the effect of a dopamine agonist. *Progress in Brain Research*, *79*, 267–276.

Pacheco, R., Riquelme, E., & Kalergis, A. M. (2010). Emerging evidence for the role of neurotransmitters in the modulation of T cell responses to cognate ligands. *Central Nervous System Agents in Medicinal Chemistry*, *10*(1), 65–83.

Patrick, C. H. (1952). *Alcohol, culture and society*. Durham, NC: Duke University Press.

Pence, B. W., Thielman, N. M., Whetten, K., Ostermann, J., Kumar, V., & Mugavero, M. J. (2008). Coping strategies and patterns of alcohol and drug use among HIV-infected patients in the United States Southeast. *AIDS Patient Care and STDs*, *22*(11), 869–877. http://dx.doi.org/10.1089/apc.2008.0022.

Perry, P. J., Argo, T. R., Barnett, M. J., Liesveld, J. L., Liskow, B., Hernan, J. M., et al. (2006). The association of alcohol-induced blackouts and grayouts to blood alcohol concentrations. *Journal of Forensic Sciences*, *51*(4), 896–899. http://dx.doi.org/10.1111/j.1556-4029.2006.00161.x.

Pettinati, H. M., O'Brien, C. P., Rabinowitz, A. R., Wortman, S. P., Oslin, D. W., Kampman, K. M., et al. (2006). The status of naltrexone in the treatment of alcohol dependence: Specific effects on heavy drinking. *Journal of Clinical Psychopharmacology*, *26*(6), 610–625. http://dx.doi.org/10.1097/01.jcp.0000245566.52401.20.

Phillips, K., Luk, A., Soor, G. S., Abraham, J. R., Leong, S., & Butany, J. (2009). Cocaine cardiotoxicity: A review of the pathophysiology, pathology, and treatment options. *American Journal of Cardiovascular Drugs*, *9*(3), 177–196. http://dx.doi.org/10.2165/00129784-200909030-00005.

Picciotto, M. R., Caldarone, B. J., King, S. L., & Zachariou, V. (2000). Nicotinic receptors in the brain. Links between molecular biology and behavior. *Neuropsychopharmacology*, *22*(5), 451–465.

Picciotto, M. R., & Zoli, M. (2002). Nicotinic receptors in aging and dementia. *Journal of Neurobiology*, *53*(4), 641–655. http://dx.doi.org/10.1002/neu.10102.

Pivneva, T. A. (2008). Microglia in normal condition and pathology. *Fiziologicheskiĭ Zhurnal*, *54*(5), 81–89.

Portegies, P. (1995). Review of antiretroviral therapy in the prevention of HIV-related AIDS dementia complex (ADC). *Drugs*, *49*(Suppl. 1), 25–31, discussion 38–40.

Pulliam, L., Herndier, B. G., Tang, N. M., & McGrath, M. S. (1991). Human immunodeficiency virus-infected macrophages produce soluble factors that cause histological and neurochemical alterations in cultured human brains. *The Journal of Clinical Investigation*, *87*(2), 503–512.

Purohit, V., Rapaka, R., & Shurtleff, D. (2011). Drugs of abuse, dopamine, and HIV-associated neurocognitive disorders/HIV-associated dementia. *Molecular Neurobiology*, *44*(1), 102–110. http://dx.doi.org/10.1007/s12035-011-8195-z.

Qiu, Y., Peng, Y., & Wang, J. (1996). Immunoregulatory role of neurotransmitters. *Advances in Neuroimmunology*, *6*(3), 223–231.

Rafalowska, J. (1998). HIV-1-infection in the CNS. A pathogenesis of some neurological syndromes in the light of recent investigations. *Folia Neuropathologica*, *36*(4), 211–216.

Rao, J. S., Kim, H. W., Kellom, M., Greenstein, D., Chen, M., Kraft, A. D., et al. (2011). Increased neuroinflammatory and arachidonic acid cascade markers, and reduced synaptic proteins, in brain of HIV-1 transgenic rats. *Journal of Neuroinflammation, 8*, 101. http://dx.doi.org/10.1186/1742-2094-8-101.

Ray, P. E., Liu, X. H., Robinson, L. R., Reid, W., Xu, L., Owens, J. W., et al. (2003). A novel HIV-1 transgenic rat model of childhood HIV-1-associated nephropathy. *Kidney International, 63*(6), 2242–2253. http://dx.doi.org/10.1046/j.1523-1755.2003.00028.x.

Reid, W., Abdelwahab, S., Sadowska, M., Huso, D., Neal, A., Ahearn, A., et al. (2004). HIV-1 transgenic rats develop T cell abnormalities. *Virology, 321*(1), 111–119. http://dx.doi.org/10.1016/j.virol.2003.12.010.

Reid, W., Sadowska, M., Denaro, F., Rao, S., Foulke, J., Jr., Hayes, N., et al. (2001). An HIV-1 transgenic rat that develops HIV-related pathology and immunologic dysfunction. *Proceedings of the National Academy of Sciences of the United States of America, 98*(16), 9271–9276. http://dx.doi.org/10.1073/pnas.161290298.

Reith, M. E., Li, M. Y., & Yan, Q. S. (1997). Extracellular dopamine, norepinephrine, and serotonin in the ventral tegmental area and nucleus accumbens of freely moving rats during intracerebral dialysis following systemic administration of cocaine and other uptake blockers. *Psychopharmacology, 134*(3), 309–317.

Ren, H., Du, C., Yuan, Z., Park, K., Volkow, N. D., & Pan, Y. (2012). Cocaine-induced cortical microischemia in the rodent brain: Clinical implications. *Molecular Psychiatry, 17*(10), 1017–1025. http://dx.doi.org/10.1038/mp.2011.160.

Rezvani, A. H., & Levin, E. D. (2002). Nicotine-alcohol interactions and cognitive function in rats. *Pharmacology, Biochemistry, and Behavior, 72*(4), 865–872.

Rezvani, A. H., & Levin, E. D. (2003). Nicotine-alcohol interactions and attentional performance on an operant visual signal detection task in female rats. *Pharmacology, Biochemistry, and Behavior, 76*(1), 75–83.

Riljak, V., & Langmeier, M. (2005). Nicotine an efficient tool of the neurobiological research today, the tool of treatment tomorrow? *Prague Medical Report, 106*(4), 329–348.

Ritz, M. C., Lamb, R. J., Goldberg, S. R., & Kuhar, M. J. (1987). Cocaine receptors on dopamine transporters are related to self-administration of cocaine. *Science, 237*(4819), 1219–1223.

Robson, P. J. (2014). Therapeutic potential of cannabinoid medicines. *Drug Testing and Analysis, 6*(1–2), 24–30. http://dx.doi.org/10.1002/dta.1529.

Rogers, T. J., & Peterson, P. K. (2003). Opioid G protein-coupled receptors: Signals at the crossroads of inflammation. *Trends in Immunology, 24*(3), 116–121.

Royal, W., 3rd., Wang, H., Jones, O., Tran, H., & Bryant, J. L. (2007). A vitamin A deficient diet enhances proinflammatory cytokine, Mu opioid receptor, and HIV-1 expression in the HIV-1 transgenic rat. *Journal of Neuroimmunology, 185*(1–2), 29–36. http://dx.doi.org/10.1016/j.jneuroim.2007.01.001.

Royal, W., 3rd., Zhang, L., Guo, M., Jones, O., Davis, H., & Bryant, J. L. (2012). Immune activation, viral gene product expression and neurotoxicity in the HIV-1 transgenic rat. *Journal of Neuroimmunology, 247*(1–2), 16–24. http://dx.doi.org/10.1016/j.jneuroim.2012.03.015.

Sacktor, N. (2002). The epidemiology of human immunodeficiency virus-associated neurological disease in the era of highly active antiretroviral therapy. *Journal of Neurovirology, 8*(Suppl. 2), 115–121. http://dx.doi.org/10.1080/13550280290101094.

Sacktor, N., Tarwater, P. M., Skolasky, R. L., McArthur, J. C., Selnes, O. A., Becker, J., et al. (2001). CSF antiretroviral drug penetrance and the treatment of HIV-associated psychomotor slowing. *Neurology, 57*(3), 542–544.

Samet, J. H., Phillips, S. J., Horton, N. J., Traphagen, E. T., & Freedberg, K. A. (2004). Detecting alcohol problems in HIV-infected patients: Use of the CAGE questionnaire.

AIDS Research and Human Retroviruses, 20(2), 151–155. http://dx.doi.org/10.1089/088922204773004860.
Schwarz, S., & Wetterling, T. (2002). Alcohol and stroke. Is moderate alcohol consumption protective? *Nervenarzt, 73*(8), 719–728.
Seitz, H. K., & Becker, P. (2007). Alcohol metabolism and cancer risk. *Alcohol Research & Health, 30*(1), 38–41, 44–37.
Sekine, Y., Ouchi, Y., Takei, N., Yoshikawa, E., Nakamura, K., Futatsubashi, M., et al. (2006). Brain serotonin transporter density and aggression in abstinent methamphetamine abusers. *Archives of General Psychiatry, 63*(1), 90–100. http://dx.doi.org/10.1001/archpsyc.63.1.90.
Sevigny, J. J., Albert, S. M., McDermott, M. P., McArthur, J. C., Sacktor, N., Conant, K., et al. (2004). Evaluation of HIV RNA and markers of immune activation as predictors of HIV-associated dementia. *Neurology, 63*(11), 2084–2090.
Sevigny, J. J., Albert, S. M., McDermott, M. P., Schifitto, G., McArthur, J. C., Sacktor, N., et al. (2007). An evaluation of neurocognitive status and markers of immune activation as predictors of time to death in advanced HIV infection. *Archives of Neurology, 64*(1), 97–102. http://dx.doi.org/10.1001/archneur.64.1.97.
Sharifzadeh, M., Tavasoli, M., Naghdi, N., Ghanbari, A., Amini, M., & Roghani, A. (2005). Post-training intrahippocampal infusion of nicotine prevents spatial memory retention deficits induced by the cyclo-oxygenase-2-specific inhibitor celecoxib in rats. *Journal of Neurochemistry, 95*(4), 1078–1090.
Sheng, W. S., Hu, S., Min, X., Cabral, G. A., Lokensgard, J. R., & Peterson, P. K. (2005). Synthetic cannabinoid WIN55,212-2 inhibits generation of inflammatory mediators by IL-1beta-stimulated human astrocytes. *Glia, 49*(2), 211–219. http://dx.doi.org/10.1002/glia.20108.
Shi, X., Sims, M. D., Hanna, M. M., Xie, M., Gulick, P. G., Zheng, Y. H., et al. (2014). Neutropenia during HIV infection: Adverse consequences and remedies. *International Reviews of Immunology*. http://dx.doi.org/10.3109/08830185.2014.893301.
Shippenberg, T. S., Zapata, A., & Chefer, V. I. (2007). Dynorphin and the pathophysiology of drug addiction. *Pharmacology & Therapeutics, 116*(2), 306–321. http://dx.doi.org/10.1016/j.pharmthera.2007.06.011.
Silverberg, M. J., Ray, G. T., Saunders, K., Rutter, C. M., Campbell, C. I., Merrill, J. O., et al. (2012). Prescription long-term opioid use in HIV-infected patients. *The Clinical Journal of Pain, 28*(1), 39–46. http://dx.doi.org/10.1097/AJP.0b013e3182201a0f.
Smith, P. H., Mazure, C. M., & McKee, S. A. (2014). Smoking and mental illness in the US population. *Tobacco Control*. http://dx.doi.org/10.1136/tobaccocontrol-2013-051466.
Sopori, M. (2002). Effects of cigarette smoke on the immune system. *Nature Reviews. Immunology, 2*(5), 372–377.
Staikos, L., Malellari, L., & Chang, S. L. (2008). Lipopolysaccharide-induced pro-inflammatory cytokines in the brain of rats in the morphine-tolerant state. *Journal of Neuroimmune Pharmacology, 3*(4), 236–240. http://dx.doi.org/10.1007/s11481-008-9111-9.
Sternberg, E. M. (2006). Neural regulation of innate immunity: A coordinated nonspecific host response to pathogens. *Nature Reviews. Immunology, 6*(4), 318–328. http://dx.doi.org/10.1038/nri1810.
Substance Abuse and Mental Health Services Administration (2010). *Injection drug abuse admissions to substance abuse treatment: 1992 and 2009*. Retrieved from, http://www.samhsa.gov/data/2k11/WEB_TEDS_012/WEB_TEDS_012.htm.
Suvisaari, J., Mautemps, N., Haukka, J., Hovi, T., & Lonnqvist, J. (2003). Childhood central nervous system viral infections and adult schizophrenia. *The American Journal of Psychiatry, 160*(6), 1183–1185.

Teaktong, T., Graham, A. J., Johnson, M., Court, J. A., & Perry, E. K. (2004). Selective changes in nicotinic acetylcholine receptor subtypes related to tobacco smoking: An immunohistochemical study. *Neuropathology and Applied Neurobiology*, *30*(3), 243–254.

Tesoriero, J. M., Gieryic, S. M., Carrascal, A., & Lavigne, H. E. (2010). Smoking among HIV positive New Yorkers: Prevalence, frequency, and opportunities for cessation. *AIDS and Behavior*, *14*(4), 824–835. http://dx.doi.org/10.1007/s10461-008-9449-2.

Toggas, S. M., Masliah, E., Rockenstein, E. M., Rall, G. F., Abraham, C. R., & Mucke, L. (1994). Central nervous system damage produced by expression of the HIV-1 coat protein gp120 in transgenic mice. *Nature*, *367*(6459), 188–193. http://dx.doi.org/10.1038/367188a0.

Tomaszewska, D., & Przekop, F. (1997). The immune-neuro-endocrine interactions. *Journal of Physiology and Pharmacology*, *48*(2), 139–158.

Torensma, R., & Figdor, C. G. (2002). The Achilles' heel of HIV. *Medical Hypotheses*, *58*(5), 386–387.

Trigo, J. M., Martin-Garcia, E., Berrendero, F., Robledo, P., & Maldonado, R. (2010). The endogenous opioid system: A common substrate in drug addiction. *Drug and Alcohol Dependence*, *108*(3), 183–194. http://dx.doi.org/10.1016/j.drugalcdep.2009.10.011.

Trkola, A. (2004). HIV-host interactions: Vital to the virus and key to its inhibition. *Current Opinion in Microbiology*, *7*(5), 555–559.

Turner, B. J., Fleishman, J. A., Wenger, N., London, A. S., Burnam, M. A., Shapiro, M. F., et al. (2001). Effects of drug abuse and mental disorders on use and type of antiretroviral therapy in HIV-infected persons. *Journal of General Internal Medicine*, *16*(9), 625–633.

Unterwald, E. M. (2001). Regulation of opioid receptors by cocaine. *Annals of the New York Academy of Sciences*, *937*, 74–92.

Urbina, A., & Jones, K. (2004). Crystal methamphetamine, its analogues, and HIV infection: Medical and psychiatric aspects of a new epidemic. *Clinical Infectious Diseases*, *38*(6), 890–894. http://dx.doi.org/10.1086/381975.

Van Sickle, M. D., Duncan, M., Kingsley, P. J., Mouihate, A., Urbani, P., Mackie, K., et al. (2005). Identification and functional characterization of brainstem cannabinoid CB2 receptors. *Science*, *310*(5746), 329–332. http://dx.doi.org/10.1126/science.1115740.

Verani, A., Gras, G., & Pancino, G. (2005). Macrophages and HIV-1: Dangerous liaisons. *Molecular Immunology*, *42*(2), 195–212.

Vergote, D., Butler, G. S., Ooms, M., Cox, J. H., Silva, C., Hollenberg, M. D., et al. (2006). Proteolytic processing of SDF-1alpha reveals a change in receptor specificity mediating HIV-associated neurodegeneration. *Proceedings of the National Academy of Sciences of the United States of America*, *103*(50), 19182–19187. http://dx.doi.org/10.1073/pnas.0604678103.

Vigorito, M., Casachahua, J., Michaels, C., Rivera, A. M., Anastasides, N., & Chang, S. L. (2011). In *Sign tracking of ethanol in rats: Effects of strain, environmental enrichment, and immune system activation. Paper presented at the Society for Neuroscience, Washington, DC.*

Vigorito, M., & Chang, S. L. (2006). Role of HIV-1 infection in addictive behavior: A study of a HIV-1 transgenic rat model. *American Journal of Infectious Diseases*, *2*, 98–106.

Vigorito, M., LaShomb, A. L., & Chang, S. L. (2007). Spatial learning and memory in HIV-1 transgenic rats. *Journal of Neuroimmune Pharmacology*, *2*(4), 319–328. http://dx.doi.org/10.1007/s11481-007-9078-y.

Volkow, N. D., Chang, L., Wang, G. J., Fowler, J. S., Ding, Y. S., Sedler, M., et al. (2001). Low level of brain dopamine D2 receptors in methamphetamine abusers: Association with metabolism in the orbitofrontal cortex. *The American Journal of Psychiatry*, *158*(12), 2015–2021.

Volkow, N. D., Fowler, J. S., Wang, G. J., Swanson, J. M., & Telang, F. (2007). Dopamine in drug abuse and addiction: Results of imaging studies and treatment implications. *Archives of Neurology*, *64*(11), 1575–1579. http://dx.doi.org/10.1001/archneur.64.11.1575.

Volkow, N. D., Fowler, J. S., Wolf, A. P., Schlyer, D., Shiue, C. Y., Alpert, R., et al. (1990). Effects of chronic cocaine abuse on postsynaptic dopamine receptors. *The American Journal of Psychiatry*, *147*(6), 719–724.

Volkow, N. D., Wang, G. J., Fowler, J. S., Logan, J., Hitzemann, R., Ding, Y. S., et al. (1996). Decreases in dopamine receptors but not in dopamine transporters in alcoholics. *Alcoholism, Clinical and Experimental Research*, *20*(9), 1594–1598.

Wallace, J. M., Rao, A. V., Glassroth, J., Hansen, N. I., Rosen, M. J., Arakaki, C., et al. (1993). Respiratory illness in persons with human immunodeficiency virus infection. The pulmonary complications of HIV infection study group. *The American Review of Respiratory Disease*, *148*(6 Pt. 1), 1523–1529. http://dx.doi.org/10.1164/ajrccm/148.6_Pt_1.1523.

Wang, G. J., Chang, L., Volkow, N. D., Telang, F., Logan, J., Ernst, T., et al. (2004). Decreased brain dopaminergic transporters in HIV-associated dementia patients. *Brain*, *127*(Pt. 11), 2452–2458. http://dx.doi.org/10.1093/brain/awh269.

Wang, Y. H., Sun, J. F., Tao, Y. M., Chi, Z. Q., & Liu, J. G. (2010). The role of kappa-opioid receptor activation in mediating antinociception and addiction. *Acta Pharmacologica Sinica*, *31*(9), 1065–1070. http://dx.doi.org/10.1038/aps.2010.138.

Wang, G. J., Volkow, N. D., Fowler, J. S., Logan, J., Abumrad, N. N., Hitzemann, R. J., et al. (1997). Dopamine D2 receptor availability in opiate-dependent subjects before and after naloxone-precipitated withdrawal. *Neuropsychopharmacology*, *16*(2), 174–182. http://dx.doi.org/10.1016/S0893-133X(96)00184-4.

Wee, S., & Koob, G. F. (2010). The role of the dynorphin-kappa opioid system in the reinforcing effects of drugs of abuse. *Psychopharmacology*, *210*(2), 121–135. http://dx.doi.org/10.1007/s00213-010-1825-8.

Wesselingh, S. L., Takahashi, K., Glass, J. D., McArthur, J. C., Griffin, J. W., & Griffin, D. E. (1997). Cellular localization of tumor necrosis factor mRNA in neurological tissue from HIV-infected patients by combined reverse transcriptase/polymerase chain reaction in situ hybridization and immunohistochemistry. *Journal of Neuroimmunology*, *74*(1–2), 1–8.

World Health Organization (2014). *HIV/AIDS—Injecting drug use*. from, http://www.who.int/hiv/topics/idu/en/.

Wybran, J. (1985). Enkephalins and endorphins as modifiers of the immune system: Present and future. *Federation Proceedings*, *44*(1 Pt. 1), 92–94.

Yoshioka, M., & Itoyama, Y. (1997). Pathogenesis of HIV-1-associated-neurologic diseases. *Nippon Rinsho*, *55*(4), 897–903.

Yu-Lee, L. Y. (1997). Molecular actions of prolactin in the immune system. *Proceedings of the Society for Experimental Biology and Medicine*, *215*(1), 35–52.

Yu-Lee, L. Y. (2002). Prolactin modulation of immune and inflammatory responses. *Recent Progress in Hormone Research*, *57*, 435–455.

Zhou, Y., Spangler, R., Schlussman, S. D., Yuferov, V. P., Sora, I., Ho, A., et al. (2002). Effects of acute "binge" cocaine on preprodynorphin, preproenkephalin, proopiomelanocortin, and corticotropin-releasing hormone receptor mRNA levels in the striatum and hypothalamic-pituitary-adrenal axis of mu-opioid receptor knockout mice. *Synapse*, *45*(4), 220–229. http://dx.doi.org/10.1002/syn.10101.

INDEX

Note: Page numbers followed by "*f*" indicate figures and "*t*" indicate tables.

E

F

G

CONTENTS OF RECENT VOLUMES

Volume 39

Volume 40

Volume 41

Volume 42

Volume 43

Volume 44

Volume 45

Volume 46

Volume 47

Volume 48

Volume 49

Volume 50

Volume 51

Volume 52

Volume 53

Volume 54

Volume 55

Volume 56

Volume 57

Volume 58

Volume 59

Volume 60

Volume 61

Volume 62

Volume 63

Volume 64

Volume 65

Volume 66

Volume 67

Volume 68

Volume 69

Volume 70

Volume 71

Volume 72

Volume 73

Volume 74

Volume 75

Volume 76

Volume 77

Volume 78

Volume 79

Volume 80

Volume 81

Volume 82

Volume 83

Volume 84

Volume 85

Volume 86

Volume 87

Volume 88

Volume 89

Volume 90

Volume 91

Volume 92

Volume 93

Preparation for Awakening: Self-Awakening Vs. Forced Awakening: Preparatory Changes in the Pre-Awakening Period
Mitsuo Hayashi, Noriko Matsuura and Hiroki Ikeda

Circadian and Sleep Episode Duration Influences on Cognitive Performance Following the Process of Awakening
Robert L. Matchock

The Cortisol Awakening Response in Context
Angela Clow, Frank Hucklebridge and Lisa Thorn

Causes and Correlates of Frequent Night Awakenings in Early Childhood
Amy Jo Schwichtenberg and Beth Goodlin-Jones

Pathologies of Awakenings: The Clinical Problem of Insomnia Considered From Multiple Theory Levels
Douglas E. Moul

The Neurochemistry of Awakening: Findings from Sleep Disorder Narcolepsy
Seiji Nishino and Yohei Sagawa

Index

Volume 94

5-HT_6 Medicinal Chemistry
Kevin G. Liu and Albert J. Robichaud

Patents
Nicolas Vincent Ruiz and Gloria Oranias

5-HT_6 Receptor Charactertization
Teresa Riccioni

5-HT_6 Receptor Signal Transduction: Second Messenger Systems
Xavier Codony, Javier Burgueño, Maria Javier Ramírez and José Miguel Vela

Electrophysiology of 5-HT_6 Receptors
Annalisa Tassone, Graziella Madeo, Giuseppe Sciamanna, Antonio Pisani and Paola Bonsi

Genetic Variations and Association
Massimo Gennarelli and Annamaria Cattaneo

Pharmacokinetics of 5-HT_6 Receptor Ligands
Angelo Mancinelli

Index

Volume 95

Introductory Remarks: Catechol-O-Methyltransferase Inhibition–An Innovative Approach to Enhance L-dopa Therapy in Parkinson's Disease with Dual Enzyme Inhibition
Erkki Nissinen

The Catechol-O-Methyltransferase Gene: its Regulation and Polymorphisms
Elizabeth M. Tunbridge

Distribution and Functions of Catechol-O-Methyltransferase Proteins: Do Recent Findings Change the Picture?
Timo T. Myöhänen and Pekka T. Männistö

Catechol-O-Methyltransferase Enzyme: Cofactor S-Adenosyl-L-Methionine and Related Mechanisms
Thomas Müller

Biochemistry and Pharmacology of Catechol-O-Methyltransferase Inhibitors
Erkki nissinen and Pekka T. Männisto

The Chemistry of Catechol-O-Methyltransferase Inhibitors
David A. Learmonth, László E. Kiss, and Patrício Soares-da-Silva

Toxicology and Safety of COMT Inhibitors
Kristiina Haasio

Catechol-O-Methyltransferase Inhibitors in Preclinical Models as Adjuncts of L-dopa Treatment
Concepció Marin and J. A. Obeso

Problems with the Present Inhibitors and a Relevance of New and Improved COMT Inhibitors in Parkinson's Disease
Seppo Kaakkola

Catechol-O-Methyltransferase and Pain
Oleg Kambur and Pekka T. Männistö

Index

Volume 96

The Central Role of 5-HT_6 Receptors in Modulating Brain Neurochemistry
Lee A. Dawson

5-HT_6 Receptor Memory and Amnesia: Behavioral Pharmacology – Learning and Memory Processes
Alfredo Meneses, G. Pérez-García, R. Tellez, T. Ponce-Lopez and C. Castillo

Volume 97

Volume 98

Volume 99

Volume 100

Volume 101

Volume 102

Volume 103

Volume 104

Volume 105

Volume 106

Volume 107

Volume 108

Volume 109

Volume 110

Volume 111

Volume 112

Volume 113

Volume 114

Volume 115

Volume 116

Volume 117

FSC
www.fsc.org
MIX
Papier | Fördert
gute Waldnutzung
FSC® C083411

Kolibri 360 GmbH
Ferdinand-Jühlke-Straße 7
99095 Erfurt, Deutschland
produktsicherheit@kolibri360.de